Handbook of Environmental Engineering

Volume 14

Series Editors
Lawrence K. Wang
PhD, Rutgers University, New Brunswick, New Jersey, USA
MS, University of Rhode Island, Kingston, Rhode Island, USA
MSce, Missouri University of Science and Technology, Rolla, Missouri, USA
BSCE, National Cheng Kung University, Tainan, Tiawan

Mu-Hao S. Wang
PhD, Rutgers University, New Brunswick, New Jersey, USA
MS, University of Rhode Island, Kingston, Rhode Island, USA
BSCE, National Cheng Kung University, Tainan, Tiawan

The past 35 + years have seen the emergence of a growing desire worldwide to take positive actions to restore and protect the environment from the degrading effects of all forms of pollution: air, noise, solid waste, and water. The principal intention of the Handbook of Environmental Engineering (HEE) series is to help readers formulate answers to the fundamental questions facing pollution in the modern era, mainly, (1) how serious is pollution? and (2) is the technology needed to abate it not only available, but feasible? Cutting-edge and highly practical, HEE offers educators, students, and engineers a strong grounding in the principles of Environmental Engineering, as well as effective methods for developing optimal abatement technologies at costs that are fully justified by the degree of abatement achieved. With an emphasis on using the Best Available Technologies, the authors of these volumes present the necessary engineering protocols derived from the fundamental principles of chemistry, physics, and mathematics, making these volumes a must have for environmental resources researchers.

More information about this series at http://www.springer.com/series/7645

Chih Ted Yang • Lawrence K. Wang
Editors

Advances in Water Resources Engineering

Springer

Editors
Chih Ted Yang
Borland Professor of Water Resources
Department of Civil and Environmental Engineering
Colorado State University
Fort Collins
Colorado
USA

Lawrence K. Wang
Ex-Dean & Director
Zorex Corporation
Newtonville
New York
USA

Lenox Institute of Water Technology
Newtonville
NY
USA

Krofta Engineering Corporation
Lenox
Massachusetts
USA

ISBN 978-3-319-11022-6 ISBN 978-3-319-11023-3 (eBook)
DOI 10.1007/978-3-319-11023-3
Springer Cham Heidelberg New York Dordrecht London

Library of Congress Control Number: 2014956960

Printed on acid-free paper

Springer is part of Springer Science+Business Media (www.springer.com)

Preface

The past 35+ years have seen the emergence of a growing desire worldwide that positive actions be taken to restore and protect the environment from the degrading effects of all forms of pollution—air, water, soil, thermal, radioactive, and noise. Since pollution is a direct or indirect consequence of waste, the seemingly idealistic demand for "zero discharge" can be construed as an unrealistic demand for zero waste. However, as long as waste continues to exist, we can only attempt to abate the subsequent pollution by converting it into a less noxious form. Three major questions usually arise when a particular type of pollution has been identified: (1) How serious are the environmental pollution and water resources crisis? (2) Is the technology to abate them available? And (3) do the costs of abatement justify the degree of abatement achieved for environmental protection and water resources conservation? This book is one of the volumes of the *Handbook of Environmental Engineering* series. The principal intention of this series is to help readers formulate answers to the above three questions.

The traditional approach of applying tried-and-true solutions to specific environmental and water resources problems has been a major contributing factor to the success of environmental engineering, and has accounted in large measure for the establishment of a "methodology of pollution control." However, the realization of the ever-increasing complexity and interrelated nature of current environmental problems renders it imperative that intelligent planning of pollution abatement systems be undertaken. Prerequisite to such planning is an understanding of the performance, potential, and limitations of the various methods of environmental protection available for environmental scientists and engineers. In this series of handbooks, we will review at a tutorial level a broad spectrum of engineering systems (natural environment, processes, operations, and methods) currently being utilized, or of potential utility, for pollution abatement and environmental protection. We believe that the unified interdisciplinary approach presented in these handbooks is a logical step in the evolution of environmental engineering.

Treatment of the various engineering systems presented will show how an engineering formulation of the subject flows naturally from the fundamental principles and theories of chemistry, microbiology, physics, and mathematics. This emphasis on fundamental science recognizes that engineering practice has in recent years become more firmly based on scientific principles rather than on its earlier dependency on empirical accumulation of facts. It is not intended, though, to neglect empiricism where such data lead quickly to the most economic design; certain engineering systems are not readily amenable to fundamental scientific analysis, and in these instances we have resorted to less science in favor of more art and empiricism.

Since an environmental water resources engineer must understand science within the context of applications, we first present the development of the scientific basis of a particular subject, followed by exposition of the pertinent design concepts and operations, and detailed explanations of their applications to environmental conservation or protection. Throughout the series, methods of mathematical modeling, system analysis, practical design, and calculation are illustrated by numerical examples. These examples clearly demonstrate how organized, analytical reasoning leads to the most direct and clear solutions. Wherever possible, pertinent cost data have been provided.

Our treatment of environmental water resources engineering is offered in the belief that the trained engineer should more firmly understand fundamental principles, be more aware of the similarities and/or differences among many of the engineering systems, and exhibit greater flexibility and originality in the definition and innovative solution of environmental system problems. In short, the environmental and water resources engineers should by conviction and practice be more readily adaptable to change and progress.

Coverage of the unusually broad field of environmental water resources engineering has demanded an expertise that could only be provided through multiple authorships. Each author (or group of authors) was permitted to employ, within reasonable limits, the customary personal style in organizing and presenting a particular subject area; consequently, it has been difficult to treat all subject materials in a homogeneous manner. Moreover, owing to limitations of space, some of the authors' favored topics could not be treated in great detail, and many less important topics had to be merely mentioned or commented on briefly. All authors have provided an excellent list of references at the end of each chapter for the benefit of the interested readers. As each chapter is meant to be self-contained, some mild repetition among the various texts was unavoidable. In each case, all omissions or repetitions are the responsibility of the editors and not the individual authors. With the current trend toward metrication, the question of using a consistent system of units has been a problem. Wherever possible, the authors have used the British system (fps) along with the metric equivalent (mks, cgs, or SIU) or vice versa. The editors sincerely hope that this redundancy of units' usage will prove to be useful rather than being disruptive to the readers.

The goals of the *Handbook of Environmental Engineering* series are: (1) to cover entire environmental fields, including air and noise pollution control, solid waste processing and resource recovery, physicochemical treatment processes, biological treatment processes, biotechnology, biosolids management, flotation technology, membrane technology, desalination technology, water resources, natural control processes, radioactive waste disposal, hazardous waste management, and thermal pollution control and (2) to employ a multimedia approach to environmental conservation and protection since air, water, soil, and energy are all interrelated.

Both this book (Volume 14) and its sister book (Volume 15) of the *Handbook of Environmental Engineering* series have been designed to serve as water resources engineering reference books as well as supplemental textbooks. We hope and expect they will prove of equal high value to advanced undergraduate and graduate students, to designers of water resources systems, and to scientists and researchers. The editors welcome comments from readers in all of these categories. It is our hope that the two water resources engineering books will not only provide information on water resources engineering but also serve as a basis for advanced study or specialized investigation of the theory and analysis of various water resources systems.

This book, *Advances in Water Resources Engineering, Volume 14,* covers the topics on watershed sediment dynamics and modeling, integrated simulation of interactive surface-water and groundwater systems, river channel stabilization with submerged vanes, nonequilibrium sediment transport, reservoir sedimentation and fluvial processes, minimum energy dissipation rate theory and applications, hydraulic modeling development and application, geophysical methods for the assessment of earthen dams, soil erosion on upland areas by rainfall and overland flow, geofluvial modeling methodologies and applications, and environmental water engineering glossary.

This book's sister book, *Modern Water Resources Engineering, Volume 15,* covers the topics on principles and applications of hydrology, open channel hydraulics, river ecology, river restoration, sedimentation and sustainable use of reservoirs, sediment transport, river morphology, hydraulic engineering, geographic information system (GIS), remote sensing, decision-making process under uncertainty, upland erosion modeling, machine-learning method, climate change and its impact on water resources, land application, crop management, watershed protection, wetland for waste disposal and water conservation, living machines, bioremediation, wastewater treatment, aquaculture system management and environmental protection, and glossary and conversion factors for water resources engineers.

The editors are pleased to acknowledge the encouragement and support received from Mr. Patrick Marton, Executive Editor of the Springer Science + Business Media, and his colleagues during the conceptual stages of this endeavor. We wish to thank the contributing authors for their time and effort, and for having patiently borne our reviews and numerous queries and comments. We are very grateful to our respective families for their patience and understanding during some rather trying times.

Chih Ted Yang, Fort Collins, Colorado, USA
Lawrence K. Wang, New Brunswick, New Jersey, USA

Contents

Contributors

Xudong Fu State Key Lab of Hydroscience & Engineering, School of Civil Engineering, Tsinghua University, Beijing, China

Varut Guvanasen HydroGeoLogic, Inc., Reston, VA, USA

Qiwei Han Sediment Research Department, China Institute of Water Resources and Hydroelectric Power Research, Beijing, China

Mingmin He Sediment Research Department, China Institute of Water Resources and Hydroelectric Power Research, Beijing, China

Craig J. Hickey National Center for Physical Acoustics, University of Mississippi, University, MS, USA

Peter S. Huyakorn HydroGeoLogic, Inc., Reston, VA, USA

Yong G. Lai Technical Service Center, U.S. Bureau of Reclamation, Denver, CO, USA

Tiejian Li State Key Lab of Hydroscience & Engineering, Tsinghua University, Beijing, China

A. Jacob Odgaard IIHR-Hydroscience and Engineering, University of Iowa, Iowa City, IA, USA

Mathias J. M. Römkens USDA ARS National Sedimentation Laboratory, Oxford, MS, USA

Haiyun Shi State Key Lab of Hydroscience & Engineering, Tsinghua University, Beijing, China

Francisco J.M. Simões US Geological Survey Geomorphology and Sediment Transport Laboratory, Golden, CO, USA

Bin Wang Beijing Forestry University, Beijing, China

Guangqian Wang Department of Engineering and Material Science of the NSFC, State Key Lab of Hydroscience & Engineering, Tsinghua University, Academician of Chinese Academy of Sciences, Beijing, China

Lawrence K. Wang Rutgers University, New Brunswick, NJ, USA

Lenox Institute of Water Technology, Newtonville, NY, USA

Mu-Hao Sung Wang Rutgers University, New Brunswick, NJ, USA

Lenox Institute of Water Technology, Newtonville, NY, USA

Robert R. Wells USDAARS National Sedimentation Laboratory, Oxford, MS, USA

Leti Wodajo National Center for Physical Acoustics, University of Mississippi, University, MS, USA

Guobin B. Xu State Key Laboratory of Hydraulic Engineering Simulation and Safety, Tianjin University, Tianjin, China

Chih Ted Yang Department of Civil and Environmental Engineering, Colorado State University, Fort Collins, CO, USA

Lina N. Zhao State Key Laboratory of Hydraulic Engineering Simulation and Safety, Tianjin University, Tianjin, China

Fenli Zheng Northwest Agriculture and Forestry University, Yangling, Shaanxi Province, China

List of Figures

List of Tables

Chapter 1
Watershed Sediment Dynamics and Modeling: A Watershed Modeling System for Yellow River

Guangqian Wang, Xudong Fu, Haiyun Shi and Tiejian Li

Contents

G. Wang (✉)
Department of Engineering and Material Science of the NSFC, State Key Lab of Hydroscience & Engineering, Tsinghua University, Academician of Chinese Academy of Sciences, Beijing 100084, China
e-mail: dhhwgq@tsinghua.edu.cn

X. Fu
State Key Lab of Hydroscience & Engineering, School of Civil Engineering, Tsinghua University, Beijing 100084, China
e-mail: xdfu@tsinghua.edu.cn

H. Shi · T. Li
State Key Lab of Hydroscience & Engineering, Tsinghua University, Beijing 100084, China
e-mail: shihaiyun@tsinghua.edu.cn

T. Li
e-mail: litiejian@tsinghua.edu.cn

C. T. Yang, L. K. Wang (eds.), *Advances in Water Resources Engineering*,
Handbook of Environmental Engineering, Volume 14, DOI 10.1007/978-3-319-11023-3_1

Abstract Soil erosion is the root cause of environmental and ecological degradation in the Loess Plateau of the Yellow River. Watershed sediment dynamics was fully analyzed here, and a physically based, distributed, and continuous erosion model at the watershed scale, named the Digital Yellow River Integrated Model (DYRIM), was developed. The framework, the key supporting techniques, and the formulation for natural processes were described. The physical processes of sediment yield and transport in the Loess Plateau are divided into three subprocesses, including the water yield and soil erosion on hillslopes, gravitational erosion in gullies, and hyperconcentrated flow routing in channels. For each subprocess, a physically based simulation model was developed and embedded into the whole model system. The model system was applied to simulate the sediment yield and transport in several typical years in different watersheds of the Yellow River, and the simulation results indicated that this model system is capable of simulating the physical processes of sediment yield and transport in a large-scale watershed.

Keywords Yellow river · Loess plateau · Watershed sediment dynamics and modeling · Soil erosion · Digital yellow river integrated model

Nomenclature

a and b	Coefficients that can be obtained from experiments
a_u and b_u	Matric potential coefficients of the topsoil
B	Width of the hillslope, m
BC	Left-child code of PC
C	Wave velocity coefficient
c	Nominal total cohesive strength, Pa
c'	Cohesive strength of the saturated soil, Pa
C_h	Wave velocity coefficient of the h-form diffusive wave equation
C_1	A coefficient that is related to the physicochemical property of the soil
E	Erosion rate of a hillslope, kg/s
E_{can}	Evaporation rate of canopy water, m/s
E_u	Evaporation rate of topsoil water, m/s
e_x	Soil erosion rate, kg/(m²s)
F_D	The sliding force
F_R	The sliding resistance
GC	Right-child code of PC
h	Runoff depth, m
h_u	Thickness of the topsoil layer, m
Int()	The operation of rounding
J	Slope of the hillslope
k	Coefficient related to the erodibility of the surface soil
K_{zus}	Saturated vertical hydraulic conductivity of the topsoil, m/s
L_f	The length of the failure plane
n	Manning's coefficient
P	Rainfall intensity, m/s

P_n	Net rainfall intensity, m/s
PC	Parent node code
Q_{gd}	Subsoil drainage, m^3/s
Q_{gu}	Topsoil drainage, m^3/s
q_1	Runoff per meter width at the bottom of the hillslope, m^2/s
q_{zd}	Infiltration rate from the topsoil to subsoil, m/s
q_{zu}	Infiltration rate of land surface, m/s
S and S_0	Sediment concentrations, kg/m^3
S_* and S_{0*}	Sediment transport capacities of the outlet and inlet cross-sections, kg/m^3
S_0'	Bed slope
S_{can}	Canopy storage, m
t	Time, s
v	Velocity of the water flow, m/s
v_s	Velocity of particles, m/s
w	The water content
W_d	Water storage of subsoil, m^3
W_u	Water storage of the topsoil, m^3
α	Coefficient of saturation recovery
β	The index related to the eroding efficiency of runoff
β_k	Determined by the grain composition of the soil
γ	Delayed ratio of the sediment from the water flow, which is less than 1
η	A coefficient which is 0.7–1.0 for the rising limb and 0 for the receding limb
θ	The angle of the sliding face
θ_{us}	Saturated volumetric water content of the topsoil, m^3/m^3
Θ_x	The Shields parameter denoting the strength of flow at the position x
ρ_m	Density of sediment laden flow, kg/m^3
ρ_s	Density of sediment particles, kg/m^3
σ	Normal stress, Pa
τ	Shear stress of the water flow, Pa
τ_c	Incipient shear stress, Pa
τ'	Additional cohesive strength, Pa
φ	Internal friction angle, which is assumed to be invariant with water content
ω_s	Settling velocity of sediment particles, m/s
$(1-\theta_{us})\cdot D\rho_s$	Mass of particles per layer per square meter

1 Introduction

The Yellow River is notorious for its high sediment load from the Loess Plateau, which lies in the arid and semiarid regions in the northwest China. The area of the center of loess deposits in the Loess Plateau is about 630,000 km^2. Generally, this region has an annual precipitation of 150–700 mm, while the potential evaporation can reach up to 1400–2000 mm. Precipitation primarily occurs in the flood season, and serious soil erosion occurs frequently following the storm events with high-intensity, short-duration characteristics. The amount of soil erosion due to those storm events can contribute over 70 % of annual sediment yield. Normally, severe soil loss occurs in the upland area, while channel aggradation occurs at the lower reaches.

Soil and water conservations in the Loess Plateau are of critical importance to the integrated watershed management of the Yellow River. For this reason, an integrated soil erosion model is highly desirable in order to help develop better strategies for watershed management. Excellent examples of physically based, distributed modeling systems that integrate a wide range of interacting processes (e.g., precipitation, vegetation, surface runoff, subsurface and ground flow, soil detachment, transport, and deposition) are chemicals, runoff, and erosion from agricultural management systems (CREAMS) [15], water erosion prediction project (WEPP) [11], European soil erosion model (EUROSEM)[22, 23], areas nonpoint source watershed environmental response simulation (ANSWERS) [2, 7], and Limburg soil erosion model (LISEM) [14]. However, each of these widely used erosion models has limitations for representing interacting processes in the Loess Plateau of China, as there are mainly two aspects contributing to the complexity and uniqueness of soil erosion processes in this highly erodible region. First, sediment concentration can easily reach as high as 1000 kg/m^3, which rarely occurs in other watersheds; high sediment load in runoff may increase the detachment rate in rills rather than that of weakening assumed in most erosion models [12]. Second, the steep slope of hillslopes exceeds the assumption of gentle slope in most erosion models; gravitational erosion (e.g., collapse and landslide), which rarely occurs in other watersheds, happens frequently in gullies, but it is not considered in most erosion models.

With the development of information technologies—e.g., remote sensing (RS) and geographical information system (GIS)—interacting processes in the watersheds are expected to be delineated and simulated digitally. In recent years, great efforts have been made by many researchers in China to develop physically based erosion models applicable to the Loess Plateau [2, 3, 7, 12, 14, 18, 22, 23, 32]. In these models, each watershed unit is divided into several geomorphic units from the top to the bottom of the hillslopes; then for each geomorphic unit, a different erosion module is used according to the physical processes. Since 2000, a team of researchers in Tsinghua University have been researching a physically based erosion model that can best represent the erosion processes of the Loess Plateau in the middle Yellow River watershed; and Wang et al. [34] developed a framework of a physically based, distributed-parameter, and continuous erosion model platform at the watershed scale, namely the Digital Yellow River Integrated Model (DYRIM).

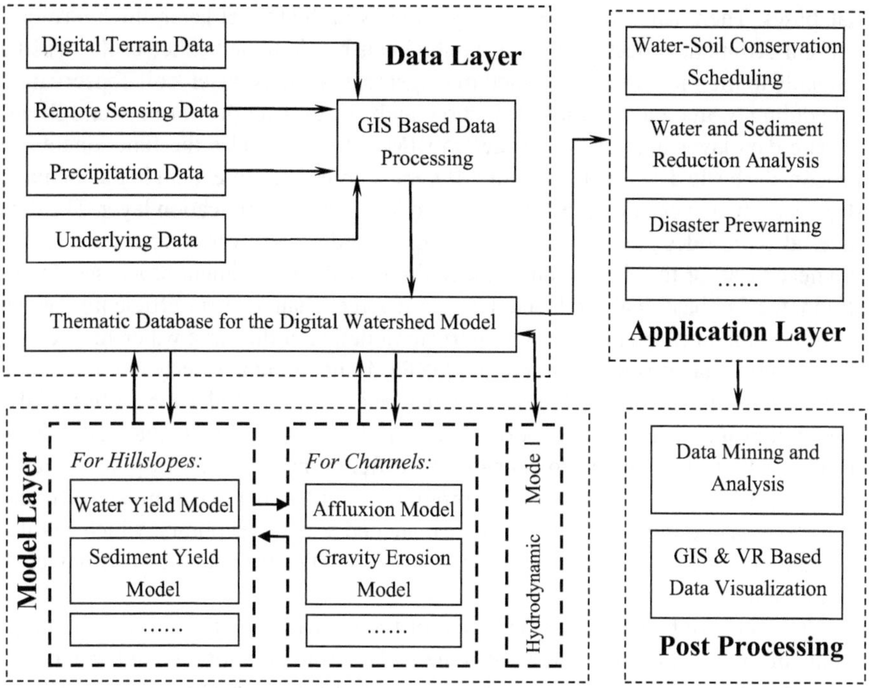

Fig. 1.1 The framework of the Digital Yellow River integrated model [34]

The DYRIM is designed to comprise a water-yield model and hydraulic soil erosion model for hillslopes, a gravitational erosion model for gullies, and a nonequilibrium sediment transport model for channels [16]. The DYRIM uses the high-resolution digital drainage network extracted from the digital elevation model (DEM) to simulate the streamflow generation and movement, and the drainage network is coded by the modified binary tree method [18]. The DYRIM also takes advantage of RS- and GIS-based parameter acquisition. Moreover, dynamic parallel computing technology is developed to speed up the simulation [19, 35, 36]. The following section provides the detailed introduction on the framework, key supporting techniques, and formulation for natural processes of the DYRIM, as well as its applications in the Yellow River watershed.

2 Framework of the DYRIM

The architecture of the DYRIM is shown in Fig. 1.1 [34]. There are four layers in the model, i.e., the data layer, model layer, application layer, and post-processing layer. The input data can be obtained from the DEM, meteorological stations, hydrological stations, or RS satellites; and they will be stored in their respective thematic databases. Then, using physically based models in the model layer, the hydrological and sediment processes in a watershed can be simulated. The results will be valuable to the integrated watershed management, such as water–soil conservation scheduling, water and sediment reduction analysis, disaster prewarning, and so on.

The data layer is the basis of the DYRIM, which provides the functionality to process the basic data obtained from different sources and store them in the thematic databases, which can be accessed by the model layer and application layer. The data layer also provides the functionality to acquire and modify various parameters that are necessary for the simulation of the hydrological and sediment processes. In addition, the database enables data to be shared and exchanged very efficiently, which facilitates the implementation of numerical modeling at the large watershed scale.

The model layer is the kernel of the DYRIM, including a water yield model and soil erosion model for hillslopes, a gravitational erosion model for gullies, and a nonequilibrium sediment transport model for channels. Hillslope channel is taken as a basic hydrological unit to consider the different hydrological response mechanisms of hillslope and channel. The program modules that can simulate different hydrological and sediment processes are managed as a model library, which enables the adoption of more modules in order to make the DYRIM become more powerful in the future.

The application layer is the objective of the DYRIM, which meets the requirements of integrated watershed management. The evaluation of soil and water conservation projects, flood early warning, and geo-disaster prevention can be realized in this layer. In addition, functionalities such as data mining and analysis, GIS and virtual reality (VR)-based data visualization, and utilization of simulation results are realized in the post-processing layer.

Overall, the physically based models are the core of the DYRIM. All of the other components are supporting techniques to ensure that these models can work properly; moreover, they also provide the functionalities to solve additional issues and challenges (e.g., watershed decomposition) in building the architecture of the DYRIM, which can elevate the efficiency and capacity of the model platform.

3 Key Supporting Techniques of the DYRIM

Database, RS, GIS, VR, and parallel computing are the major supporting techniques of the DYRIM, among which digital drainage network extraction, drainage network codification, parameter acquisition, and cluster-based parallel computing make the DYRIM differ from other fields of informatization.

3.1 Digital Drainage Network Extraction

The high-resolution digital drainage network is extracted from the DEM, which is an important format of the digital terrain data. Various geometrical parameters, such

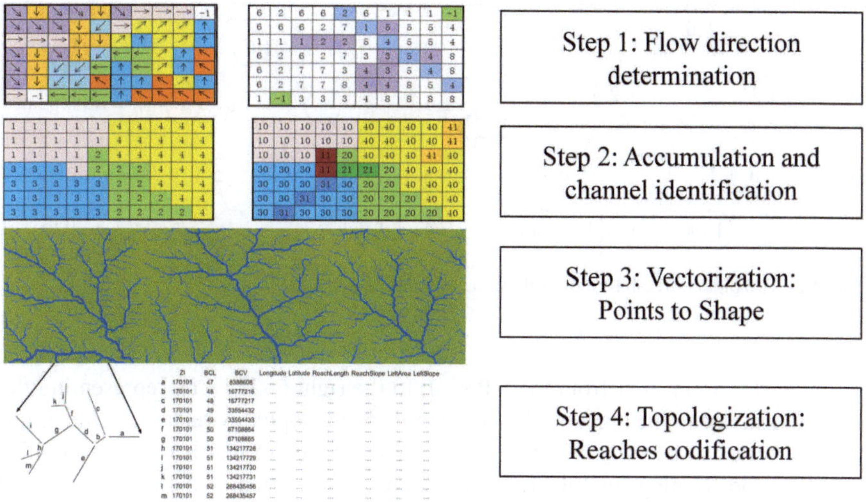

Fig. 1.2 The flowchart of digital drainage network extraction

as channel density, gradient of hillslopes, length, and gradient of channel segments, can be extracted and used to provide the basic data for physically based models. To obtain digital drainage networks from the DEM, a D8-based algorithm [25] with optimized data sorting and RAM operation was developed by Bai et al. [1], including the following four steps (see Fig. 1.2): (1) flow direction determination, (2) accumulation and channel identification, (3) vectorization, and (4) topologization. It is important to correctly identify the position of each channel head in order to obtain the true channels. In the traditional method, the density of drainage network is controlled by the critical source area (CSA), which is spatial constant and may generate false channels in the plain area; by contrast, a new algorithm for high-resolution channel head identification is proposed and integrated in the digital drainage network extraction method. A certain geomorphologic parameter is introduced to find the break point, which is regarded as the location of channel head, and a dynamic window is set for break point detection.

3.2 Drainage Network Codification

Discharge routing and sediment transport simulation should take place on the hillslope-channel units following the affluxion order from upstream to downstream in the drainage network. To make topological algorithms more effective, a structural drainage network codification method is proposed in the DYRIM [18]. A dendritic river is considered as a binary tree, and two components are proposed for a river reach codification, including the length and value components (Fig. 1.3). The length component (*L*) is the level of a node in the binary tree, representing the logical distance to the watershed outlet. The value component (*V*) is the index of a node

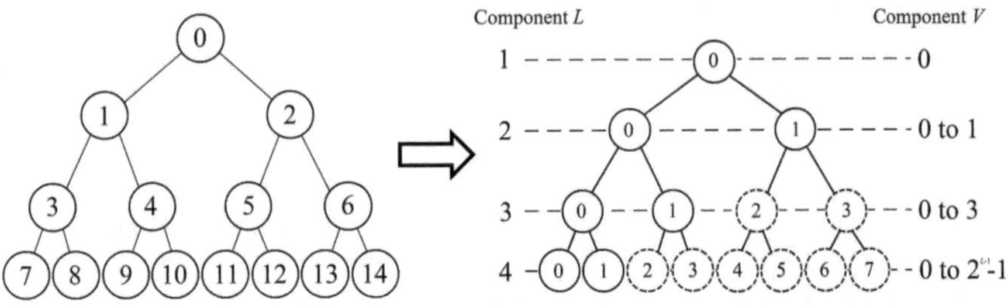

Fig. 1.3 The binary-tree-based digital drainage network [18]

in its level L and grows from the left (=0) to the right ($=2^{L-1}-1$), representing the logical distance to the main trunk. And therefore, the topology relation of the drainage network can be fully expressed by the river codes so that it is easy to realize the direct positioning for sub-watersheds. The link relation of the river reaches is defined as follows [19]:

$$\begin{aligned} BC &= 2 \times PC \\ GC &= 2 \times PC + 1. \end{aligned} \tag{1.1}$$

where

PC parent node code
BC left-child code of PC
GC right-child code of PC.

By using Eq. (1.1), all sub-watersheds in the entire watershed can be coded. Subsequently, Eq. (1.2) can be concluded as well:

$$PC = \left[\frac{BC \ or \ GC}{2} \right], \tag{1.2}$$

where

[ ] downward-rounding function.

Thus, for an arbitrary sub-watershed, its adjacent sub-watersheds upstream could be identified swiftly by Eq. (1.1), and the sub-watershed downstream could be immediately located by Eq. (1.2).

To make this structural drainage network codification method applicable to large watersheds, a policy of grading and subzoning following the pattern of a river's tributaries is adopted. This method is used in each tributary separately, and each tributary has its own grade and position number. The grade number is equal to its tributary grade and the position number increases from zero near the outlet to the upstream one by one [34].

3.3 Parameter Acquisition

The parameters of the DYRIM, including the geometrical parameters and the underlying surface parameters, are all spatially distributed. As mentioned before, the geometrical parameters are acquired from the DEM when extracting the digital drainage network, while the underlying surface parameters (e.g., vegetation cover, land use, soil type, and potential evaporation) are acquired from RS images in the format of raster data. To make the raster data match the hillslope-channel units, the central point or the polygon border of each hillslope-channel unit is used to capture the point values of the raster data [6], and the values are then counted and transformed into corresponding parameters. Moreover, the acquired parameters are all stored in their own thematic databases, which can be accessed by the model layer and application layer. Moreover, Shi et al. [30] developed an algorithm for computing spatially distributed monthly potential evaporation over the mountainous regions in order to provide the basic inputs with better accuracy for the DYRIM, and more work on improving the accuracy of relevant parameters are in progress.

3.4 Cluster-Based Parallel Computing

The physically based models in the DYRIM constitute an enormous computation mission. The time cost will be unacceptable, and the efficiency of the database will not be maximized if a serial algorithm is adopted. Moreover, the units in the DYRIM have the significant characteristic of low correlation, which meets the conditions for parallel computing. As a result, the DYRIM employs the parallel computing technology and uses message-passing interface (MPI)[21] to perform inter-processor communication [19, 35, 36]. Figure 1.4 presents the framework of the parallel computing system for watershed simulations [35]. This system can run in the Windows operating system (OS) environment on a single-core computer, a multi-core computer, or multi-computers connected by the local area network. There are four components in this system, i.e., one database, one master node, one transfer node, and any quantity of slave nodes, which can collaborate closely to accomplish a unitary simulation process. The database is the data center of the system, which stores both the original data (e.g., topography information, land use, soil type, and model parameters) needed for commencing the simulation and the final simulation results exported by all kinds of physically based models; the master node is in charge of the domain decomposition of drainage network and tasks allocation; the slave node runs physically based models for the tasks accepted from the master node; and the transfer node is responsible for the communication processes among the slave nodes.

For the dynamic parallelization of hydrological simulations, the decomposition of a watershed into a large number of sub-watersheds is necessary. Based on the binary tree codification method, Li et al. [15] developed the dynamic watershed decomposition method for dividing a drainage network into a number of

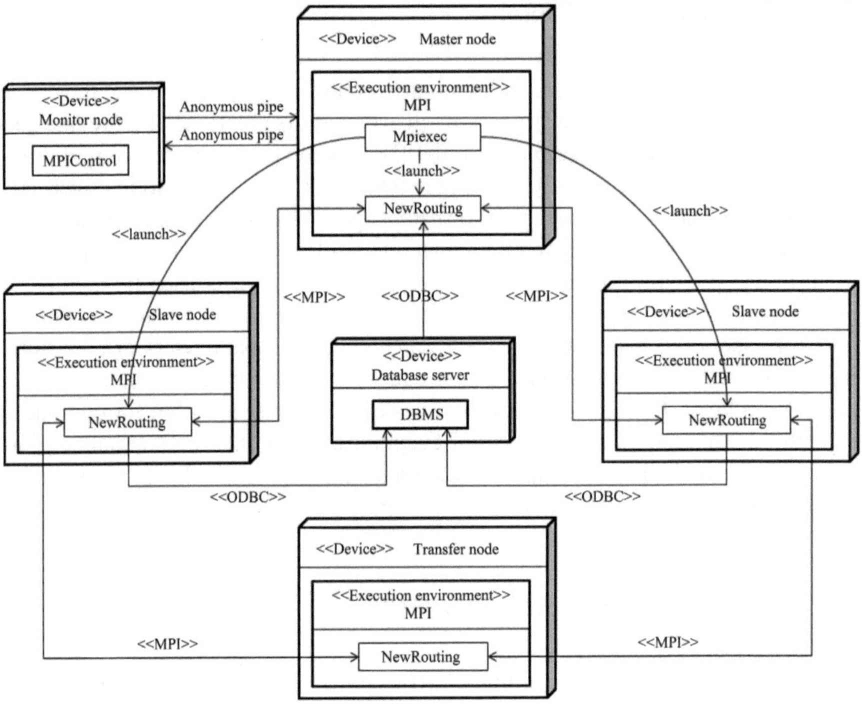

Fig. 1.4 Framework of the parallel computing system [35]

sub-watersheds, and dispatching them to each computing process. Figure 1.5 presents the diagram of a dynamic watershed decomposition, where the sub-watersheds with the boundary line colors of brown, green, and pink are dispatched to the computing processes 1, 2, and 3, respectively. From Fig. 1.5, it can be seen that there are two types of sub-watersheds. One is the headwater sub-watersheds, which do not need the input data from the upstream. The other is those sub-watersheds which need the input data from their related upstream sub-watersheds, and their simulation sequences and the data transferring paths must follow the routes from the upstream to downstream sub-watersheds. Intuitively, to minimize the simulation time, the farthermost sub-watershed (e.g., sub-watershed 1 in Fig. 1.5) from the watershed outlet should be simulated first. Figure 1.6 presents the flowchart for the dynamic watershed decomposition [19].

As mentioned above, three types of nodes are included in this parallel computing system. Figure 1.7 presents the flowchart of an execution of the master, slave, and data transfer processes [19]. The simulation procedure is driven by the dialog between the master process and slave processes through the iterative loop of request–split–new and request–new–split. Once the simulation of a certain sub-watershed is completed, the simulation results of this sub-watershed are transferred to its next downstream sub-watershed. If this downstream sub-watershed has not been decomposed and dispatched to a slave process, the simulation results will be

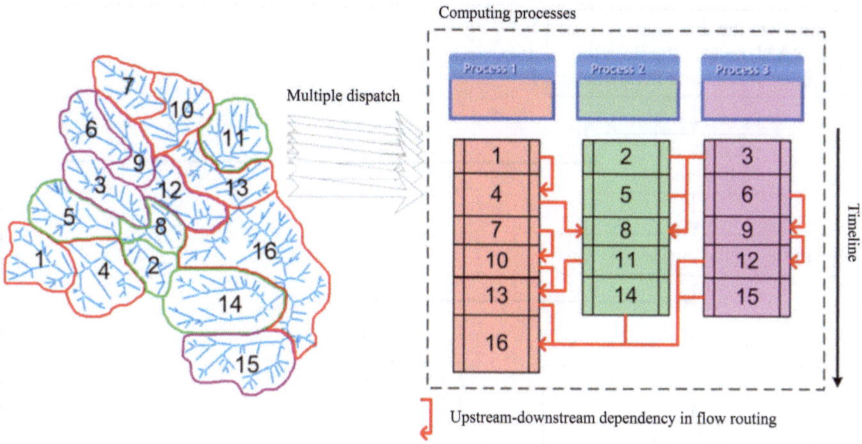

Fig. 1.5 The diagram of a dynamic watershed decomposition [19]

temporarily stored in the RAM of the data transfer process until they are requested by its downstream sub-watershed. Consequently, logical connections among split-off sub-watersheds can be achieved dynamically and efficiently [19].

4 Formulation for Natural Processes

4.1 Mechanism of Sediment Yield and Transport

Various phenomena and internal mechanisms are presented in the natural processes of sediment yield and transport in the Loess Plateau, which can be classified into several categories, such as gullied rolling loess regions, gullied loess plateau regions, dune areas, earth and rock mountains, and loess terrace regions [34]. The gullied rolling loess and gullied Loess Plateau regions are the two regions that have much in common, and represent the typical processes and mechanisms of flow and sediment transport in the coarse sediment source area of the Loess Plateau (Fig. 1.8). The terrain in this area is complicated (Fig. 1.9a); however, it can be divided into two parts, namely hillslopes and channels, which compose the hillslope-channel system. Moreover, the profile of the hillslope-channel unit is shown in Fig. 1.9b. All of the soil erosion and sediment transport processes can be categorized into three subprocesses: water yield and soil erosion on hillslopes, gravitational erosion in gullies, and hyperconcentrated flow routing in channels.

According to the experimental data based on a typical surface flow field research, the quantity of sediment erosion that forms the tiny and shallow gullies accounts for 36 % of the total, and the maximum sediment concentration of gully erosion exceeds that of a sputter erosion by 30 % [33]. However, the total quantity of the detached soil increases along the hillslope, and can be generalized as a single

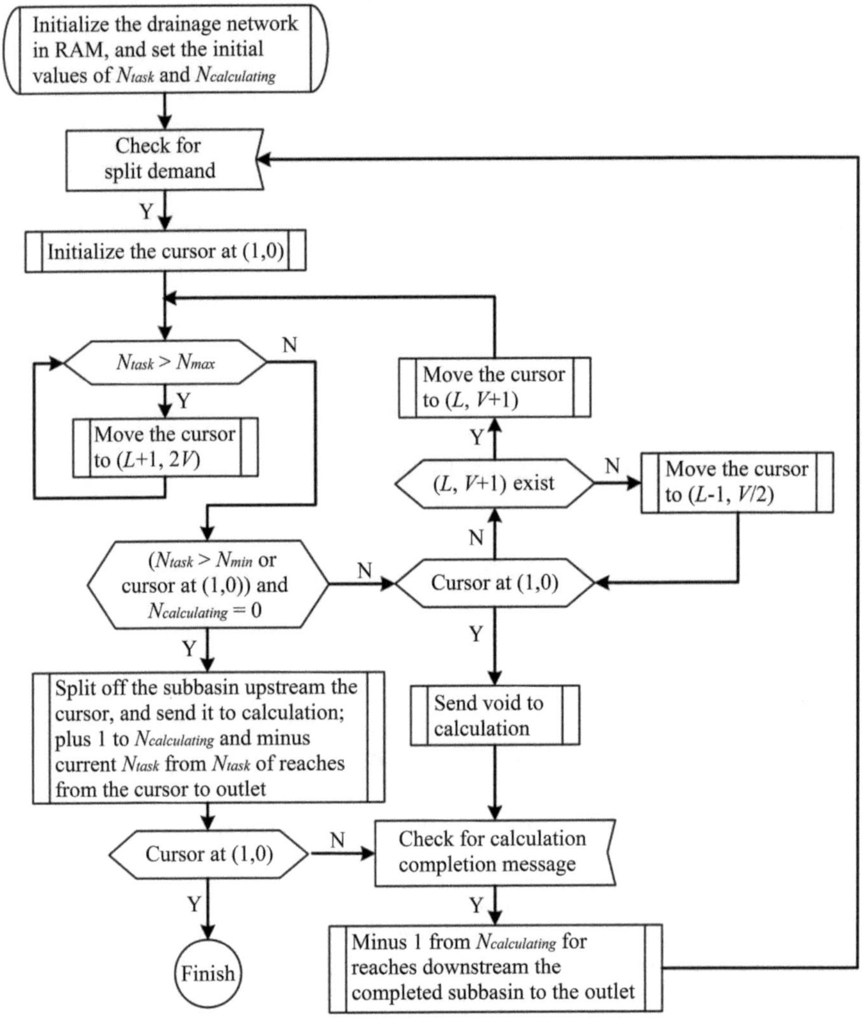

Fig. 1.6 The flowchart for a dynamic watershed decomposition [19]

erosion process affected by hydrodynamic forces. Moreover, based on the analysis of a large amount of measured data, there exists a phenomenon that the sediment discharge peak lags behind the flood peak in the Loess Plateau, and it is usually associated with the occurrence of gravitational erosion (e.g., landslides and collapses). The steep slope and the characteristics of loess soil are the main factors leading to gravitational erosion, while rainfall and runoff also play an important role in inducing the occurrence of gravitational erosion. Among all the main factors, the nature of the soil and micro-landscape are random, which ultimately makes gravitational erosion a stochastic process, which can be triggered by specific factors.

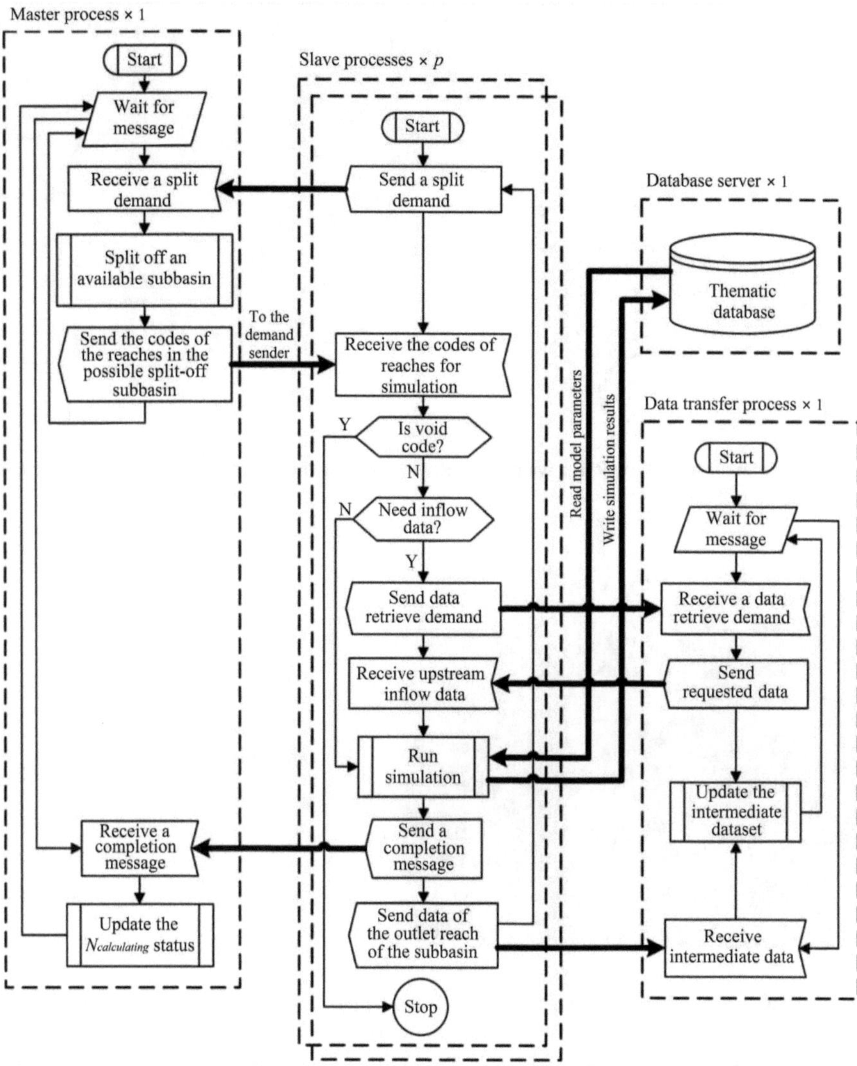

Fig. 1.7 The flowchart of execution of the master, slave, and data transfer processes [19]

In drainage networks, hillslope runoff and detachment of gravitational erosion are superposed from upstream to downstream. Thus, the flow discharge and sediment concentration increase, which finally lead to a hyperconcentrated flow. Hyperconcentrated flows in channels have some special properties: (1) due to the lag of gravitational erosion and increased sediment transport capacity of hyperconcentrated flows, the sediment discharge peak usually lags behind the flood peak, and lasts longer, (2) due to the randomness of gravitational erosion, as well as scouring and deposition in channels, the relationship between flow discharge and sediment concentration becomes unclear, and (3) for a single flood, scouring/deposition in channels and gradation adjustment make particles small at low sediment concentrations

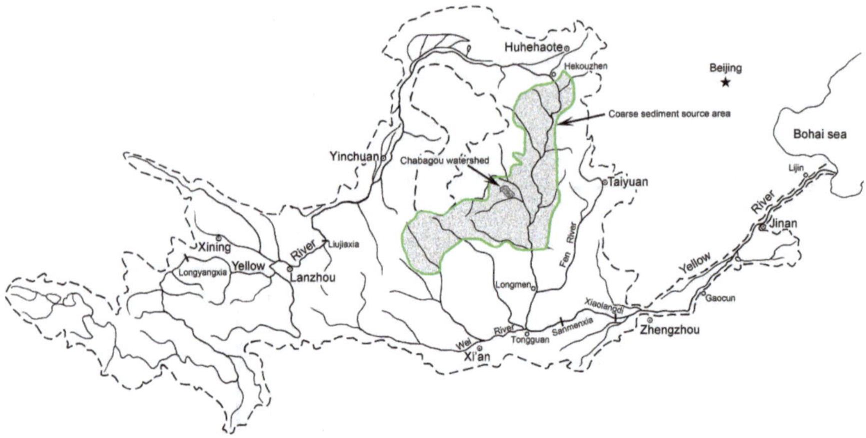

Fig. 1.8 Map of the Yellow River watershed. Region with the boundary of *green line* is the coarse sediment source area [34]

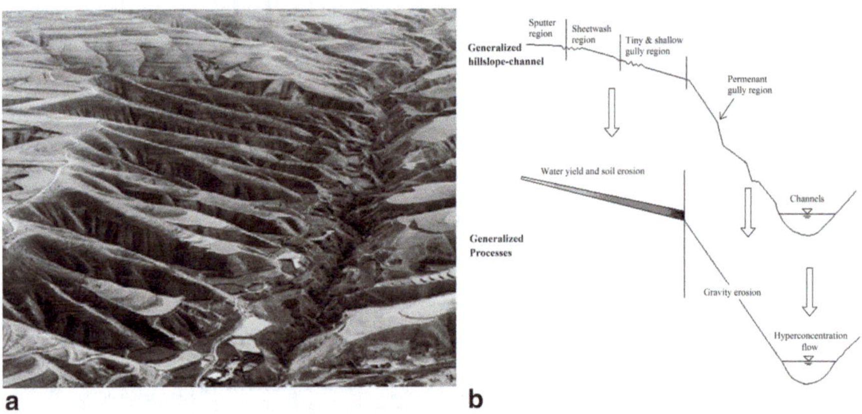

Fig. 1.9 **a** Typical hillslope-channel system [38] and **b** modeling schematic of the soil erosion and sediment transport processes [16] in the Loess Plateau of China

and coarse at high sediment concentrations. Therefore, channels should be treated as a separate part of the whole to describe the nature of hyperconcentrated flows. Major factors such as the confluence area, riverbed gradient, cross-section profile of channels, and the quantity of gravitational erosion must all be taken into account to simulate the natural processes in drainage networks.

4.2 Water Yield and Soil Erosion on Hillslopes

In the DYRIM, a conceptual water yield model is proposed to simulate continuous hillslope surface runoff. This model is established on the hillslope unit to mainly

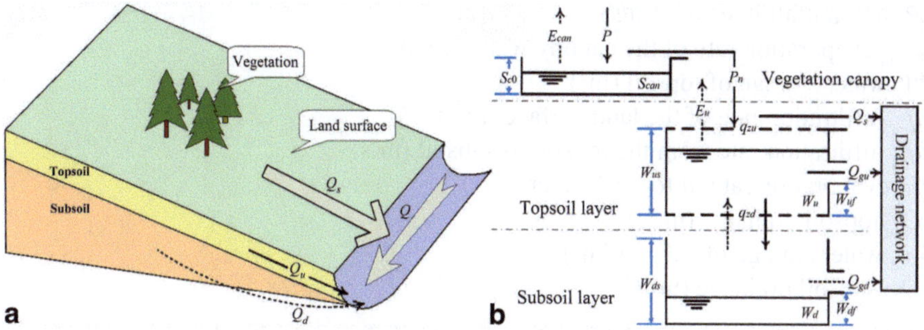

Fig. 1.10 **a** A conceptual hillslope and **b** the hydrological processes in the DYRIM [16]

represent the infiltration–excess runoff yield mechanism. The influence of ground water on a hillslope surface runoff is ignored because the unsaturated zone of the soil can be dozens of meters deep in the Loess Plateau. The hillslope soil is divided into topsoil and subsoil layers (Fig. 1.10a. A variety of hydrological processes, including vegetation interception, evapotranspiration, infiltration–excess runoff on the surface, subsurface flow in the two layers, and water exchange between the two layers, are simulated (Fig. 1.10b [16]). There are two types of parameters in the water yield model. One is the invariant parameters used for describing the properties of land use and soil type, including the field capacity of a topsoil layer, free water content of a topsoil layer, field capacity of a subsoil layer, free water content of a subsoil layer, depth of topsoil layer, and water capacity of unit leaf area index (LAI), which are influenced by the basic features of the watershed and can be determined from the literature, fieldwork, and prior studies. The other one is the adjustable parameters, including infiltration rate of ground surface, vertical infiltration rate from the topsoil layer to subsoil layer, and horizontal infiltration rates of the two soil layers, which can be calibrated and verified with the observed rainfall data and hydrologic data.

The mass conservation equations of the canopy storage, topsoil water, and subsoil water are:

$$\frac{\partial S_{\text{can}}}{\partial t} = P - P_{\text{n}} - E_{\text{can}}$$
$$\frac{\partial W_{\text{u}}}{\partial t} = A\cdot(q_{\text{zu}} - q_{\text{zd}} - E_{\text{u}}) - Q_{\text{gu}}$$
$$\frac{\partial W_{\text{d}}}{\partial t} = A\cdot q_{\text{zd}} - Q_{\text{gd}}, \tag{1.3}$$

where

S_{can} canopy storage (m)
t time (s)
P rainfall intensity (m/s)

P_n net rainfall intensity (m/s)
E_{can} evaporation rate of the canopy water (m/s)
W_u water storage of topsoil (m^3)
q_{zu} infiltration rate of the land surface (m/s)
q_{zd} infiltration rate from the topsoil to subsoil (m/s)
E_u evaporation rate of topsoil water (m/s)
Q_{gu} topsoil drainage (m^3/s)
W_d water storage of subsoil (m^3)
Q_{gd} subsoil drainage (m^3/s).

The value of q_l to simulate the soil erosion process is calculated as

$$q_l = (P_n - q_{zu}) \cdot L \text{ when } P_n > q_{zu}, \tag{1.4}$$

where

q_l runoff per meter width at the bottom of the hillslope (m^2/s).

By assuming that the moisture content at the vertical middle of the topsoil layer equals to its average value, and the surface soil is saturated during rainfall, the infiltration process from the land surface to the topsoil is generalized as one-dimensional vertical seepage, where the unsaturated Darcy's law can be used. The relative hydraulic conductivity of unsaturated to saturated soil is expressed by an exponential function of the saturation degree. The pressure drop from the land surface to the middle of the topsoil layer is equal to the differences of the gravity and matric potentials. The matric potentials are estimated by using an exponential function of the saturation degree of the soil. Therefore, the variation of infiltration rate can be calculated with the volumetric water content of the topsoil θ_u as

$$q_{zu} = K_{zus} \cdot \left(\frac{1 + \theta_u(t) / \theta_{us}}{2} \right)^{\beta_k} \left[1 + \frac{2a_u}{h_u} \left(\frac{\theta_u(t)}{\theta_{us}} \right)^{-b_u} \right], \tag{1.5}$$

where

K_{zus} saturated vertical hydraulic conductivity of the topsoil (m/s)
θ_{us} saturated volumetric water content of the topsoil (m^3/m^3)
β_k determined by the grain composition of the soil
h_u thickness of the topsoil layer (m)
a_u and b_u matric potential coefficients of the topsoil.

To calculate the variation of θ_u, all the contributing processes are simulated. The redistribution of soil water, that is, the exchange of water between the two layers, is calculated following the unsaturated soil hydrodynamics. Discharge from the topsoil layer is simulated using the Darcy's law when its water content is greater than the soil storage capacity, though it merely happens in the Loess Plateau. Soil water evaporation is simulated with soil water content and potential evaporation [13, 30].

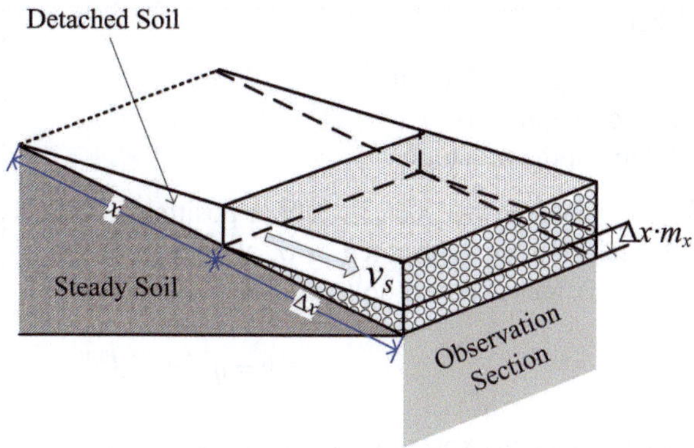

Fig. 1.11 A basic unit (*the dot-filled part*) on the surface of a conceptual hillslope for the illustration of soil erosion process [16]

Based on the simulated surface runoff, sputter erosion, sheet erosion, and tiny and shallow erosion are conceptualized to a general process of hillslope soil erosion in the DYRIM. The dot-filled part in Fig. 1.11 is a basic unit to illustrate the soil erosion process. The variable x denotes the distance from the top of the hillslope, and Δx denotes the length of the basic unit along the hillslope. It is assumed that among all the sediment particles passing the observation section, $m_x \cdot \Delta x$ layers of them are newly detached from the basic unit. That is, the eroding rate is m_x layers of particles per meter along the hillslope. To quantify the mass of newly detached sediment particles, the thickness of one layer of particles is assumed to be equal to the median diameter D of the particles. Therefore, the soil erosion rate, namely the amount of soil detached per square meter per second, can be calculated as

$$e_{\mathrm{x}} = (1-\theta_{\mathrm{us}}) D \rho_{\mathrm{s}} m_{\mathrm{x}} \mathrm{v}_{\mathrm{s}}, \tag{1.6}$$

where

e_x soil erosion rate (kg/m^{2}s)
$(1-\theta_{\mathrm{us}}) \cdot D\rho_{\mathrm{s}}$ mass of particles per layer per square meter
ρ_{s} density of the sediment particles (kg/m^3)
θ_{us} saturated volumetric water content of the topsoil (m^3/m^3)
v_{s} velocity of particles (m/s). v_{s} is estimated as:

$$\mathrm{v}_{\mathrm{s}} = \gamma \mathrm{v}, \tag{1.7}$$

where

γ delayed ratio of sediment from the water flow, which is less than 1
v velocity of the water flow (m/s)
v is calculated by using the Manning's equation

$$\mathrm{v} = h^{2/3} J^{1/2} / n, \tag{1.8}$$

where

n Manning's coefficient
h runoff depth (m)
J slope of the hillslope.

Flow discharge per meter width at position x, which can be known from the water yield model, is denoted as q_x and equal to vh. q_x is used to eliminate h from Eq. (1.8), and thus, v is solved as:

$$\mathrm{v} = q_x^{2/5} n^{-3/5} J^{3/10}. \tag{1.9}$$

Then, e_x can be expressed as

$$e_x = (1-\theta_{\mathrm{us}})\alpha D \rho_{\mathrm{s}} m_x q_x^{2/5} n^{-3/5} J^{3/10} \tag{1.10}$$

To describe the dynamics of soil erosion, m_x is proposed to be correlated to the flow strength here and the erodibility of the surface soil as

$$m_x D = k \Theta_x^{\beta}, \tag{1.11}$$

where

k coefficient related to the erodibility of the surface soil
Θ_x the Shields parameter denoting the strength of flow at the position x
β the the index related to the eroding efficiency of the runoff.

According to the study of sediment incipient motion in river channels [4], the value of the index β can be determined by the fluid and sediment characteristics. However, microtopography has a significant impact on the convergence of flow on the hillslope surface. The concentrated flow, e.g., in rills, can erode more particles than a sheet flow. Thus, microtopography is crucial to determine the amount of soil erosion, and should be taken as the main factor determining the exponent on flow strength. The Shields parameter in Eq. (1.6) is expressed as

$$\Theta_x \frac{\rho_{\mathrm{m}}}{\rho_{\mathrm{s}} - \rho_{\mathrm{m}}} \frac{hJ}{D}, \tag{1.12}$$

where

ρ_{m} density of the sediment-laden flow (kg/m^3).

Replacing h by q_x/v in Eq. (1.12) and substituting it into Eq. (1.11), we get

$$m_x D = k \left(\frac{\rho_{\mathrm{m}}}{\rho_{\mathrm{s}} - \rho_{\mathrm{m}}} \right)^{\beta} q_x^{\frac{3}{5}\beta} n^{\frac{3}{5}\beta} J^{\frac{7}{10}\beta} D^{-\beta}. \tag{1.13}$$

Substituting Eq. (1.13) into Eq. (1.10) leads to

$$e_x = (1-\theta_{us})\alpha k \rho_s D^{-\beta} \left(\frac{\rho_m}{\rho_s - \rho_m} \right)^{\beta} q_x^{\frac{3}{5}\beta+\frac{2}{5}} n^{\frac{3}{5}(\beta-1)} J^{\frac{7}{10}\beta+\frac{3}{10}} \tag{1.14}$$

Surface runoff is assumed to be generated uniformly in a hillslope so that $q_x = q_e x$, where q_e is the surface runoff rate (m/s). Then, Eq. (1.14) turns to

$$e_x = (1-\theta_{us})\alpha k \rho_s D^{-\beta} \left(\frac{\rho_m}{\rho_s - \rho_m} \right)^{\beta} q_e^{\frac{3}{5}\beta+\frac{2}{5}} x^{\frac{3}{5}\beta+\frac{2}{5}} n^{\frac{3}{5}(\beta-1)} J^{\frac{7}{10}\beta+\frac{3}{10}} \tag{1.15}$$

In the area of a hillslope, the characteristics of soil and microtopography basically do not vary largely, thus the parameters k and β can be assumed constant with x. Moreover, ρ_m increases from the top to the bottom in a hillslope, and is determined by e_x. If an expression of ρ_m is put into Eq. (1.15), e_x cannot be solved out in a simple form. Erosion rate of a whole hillslope can be obtained by integrating Eq. (1.15) along the slope and multiplying the result by the width of the hillslope. The ρ_m is assumed constant for the integration, and the integrated result is

$$E = B \cdot \int_0^L e_x \mathrm{d}x = \frac{5}{3\beta+7}(1-\theta_{us})\alpha k \rho_s D^{-\beta} \left(\frac{\rho_m}{\rho_s - \rho_m} \right)^{\beta} q_l^{\frac{3}{5}\beta+\frac{2}{5}} n^{\frac{3}{5}(\beta-1)} J^{\frac{7}{10}\beta+\frac{3}{10}} A, \tag{1.16}$$

where

E erosion rate of a hillslope (kg/s)
B width of the hillslope (m)
q_l runoff per meter width at the bottom of the hillslope (m^2/s).

The density of the sediment-laden flow, ρ_m, can be calculated by averaging the density of the clear flow at the top and that of the turbid flow at the bottom of the hillslope, because the actual range of the density of flow in a watershed is limited.

4.3 Gravitational Erosion in Gullies

Gravitational erosion is calculated according to the stability analysis-based riverbank erosion [9, 26]. However, the characteristics of unsaturated loess and the induction effect of the slope surface flow are specialized to identify gravitational erosion. The simulation is established in the gully region, and considers the collapse or sliding of the soil body as the subject, and includes the analysis of the mechanical condition of the soil body.

As shown in Fig. 1.12, the forces on the soil body include: (1) gravity (W_t), and the increment caused by water soaking is considered, (2) Anti-slide force (F_R) on

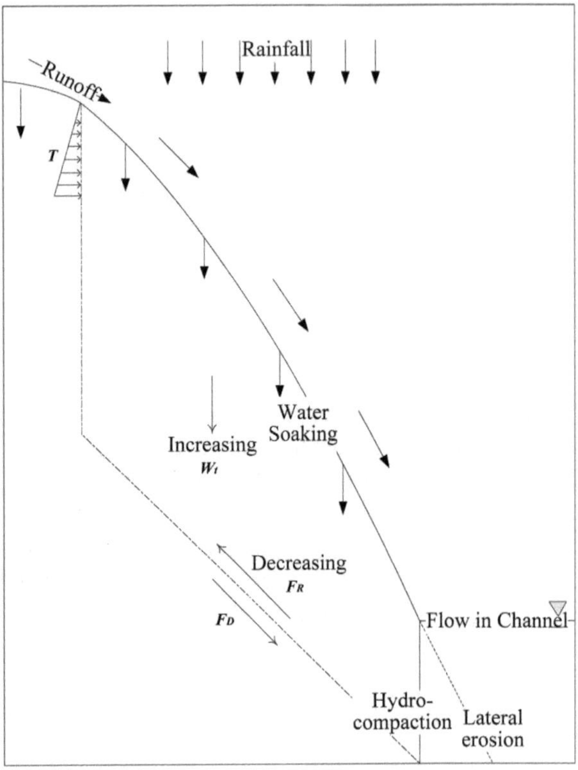

Fig. 1.12 The forces on the sliding soil body [34]

the slide-crack surface, and the reduction of cohesive strength caused by the increment of soil moisture is considered, and (3) water pressure (T) in the tension crack along the loess vertical cleavage at the top of the soil body.

To calculate the forces above, the infiltration can be obtained by the runoff yield simulation model of the system. The shear strength of the unsaturated soil (S_r) can be expressed approximately by

$$S_r = c + \sigma \tan\phi = c' + \tau' + \sigma \tan\phi, \tag{1.17}$$

where

c nominal total cohesive strength (Pa)
c' cohesive strength of the saturated soil (Pa)
τ' additional cohesive strength (Pa)
σ normal stress (Pa)
φ internal friction angle, which is assumed to be invariant with water content.

According to the results reported by Dang and Li [8], the additional cohesive strength is caused by capillary force, and has a power function with water content

$$\tau' = aw^b, \tag{1.18}$$

where

w water content
a and b coefficients that can be obtained from experiments.

Lateral erosion at the toe of the hillslope caused by the flow current in channels can be simulated by the Osman model [26]. The fallback distance in the lateral direction for the unit time interval (ΔB) can be calculated as follows:

$$\Delta B = \frac{C_1 \times (\tau - \tau_c) \times e^{-1.3\tau_c}}{\gamma_s}, \tag{1.19}$$

where

C_1 a coefficient that is related to the physicochemical property of soil
τ shear stress of the water flow (Pa)
τ_c incipient shear stress (Pa).

After that, the sliding force and sliding resistance can be expressed as follows by using the known values of soil stress and geometry:

$$\begin{aligned} F_D &= W_t \sin\theta + T\cos\theta \\ F_R &= L_f + W_t \cos\theta \tan\phi \end{aligned} \tag{1.20}$$

where

F_D the sliding force
θ the angle of the sliding face
F_R the sliding resistance
L_f the length of the failure plane.

Soil stress and geometry are time-variant with water content. Therefore, at different time steps, different assurance coefficients Fs = FR/FD will be obtained. However, in order to meet the randomicity of gravitational erosion, fuzzy analysis is adopted to gain the assurance coefficient, and finally, the membership grade of destabilization is achieved. Whether a random event happens or not is judged when the model is running.

When failure is predicted to occur, the volume of failure block per unit reach length can be calculated from geometry. Assume that the probability of failure along the channel reach is P_g; then the sediment yield caused by the gravitational erosion of each channel segment can be calculated. The detached sediment is then added into the equation for sediment transport in the channel as lateral input, which makes the sediment concentration increase to the sediment transport capacity. The lateral inputs last for several time steps until the gravitational detached sediment is used up.

4.4 Hyperconcentrated Flow Routing in Channels

From antecedent sections, the flow and sediment discharges of a hillslope as well as the amount of gravitational erosion can be obtained. They are all treated as direct inputs of subsequent models to simulate the movements of water and sediment in the river reach. Confluence and flow routing in the drainage network are simulated using a diffusive wave method, and nonequilibrium sediment transport is simulated using the method proposed by Fei and Shao [10]. And thus, water and sediment movements in a whole watershed can be simulated with the DYRIM [34].

As the channels in the coarse sediment source area have no measured cross-section profile, but only basic parameters such as the length and bed slope, the discharge routing model is based on the diffusive wave method, and the channel cross-section is assumed to be V shaped to obtain flow parameters such as stage and velocity to calculate the sediment transport.

The coefficients of the diffusive wave equation are calculated with the four-point scheme

$$
\begin{aligned}
&\langle C\rangle = \sum C_i / 4 \\
&\langle B\rangle = f(\langle Q\rangle, \langle C\rangle) \\
&\langle Q/C\rangle = \sum (Q_i / C_i)/4 \\
&i = 1,2,3,4
\end{aligned}
\tag{1.21}
$$

where

$\langle\rangle$ the value of the calculation point
$\langle C\rangle$ used to calculate the Muskingum K coefficient
$\langle B\rangle$ and $\langle Q/C\rangle$ used to calculate the Muskingum x coefficient.

It was pointed out by Cappelaere [5] that if the hydraulic slope is approximated to be the bed slope, the difference between the diffusive wave method and the kinematic wave method can be ignored. Thus, the hydraulic slope (S) is assumed to be

$$S = S_0' + \eta \frac{\partial h}{\partial x} \tag{1.22}$$

where

S_0' bed slope
η coefficient which is 0.7–1.0 for the rising limb and 0 for the receding limb.

By using the h-form diffusive wave equation, $\partial h/\partial x$ is converted to the expression of the known value Q:

$$\frac{\partial h}{\partial x} = \frac{1}{BC_{\mathrm{h}}} \frac{\partial Q}{\partial x}, \tag{1.23}$$

where

$C_h = (1/B)(\partial Q/\partial h)$ is the wave velocity coefficient of the h-form diffusive wave equation.

To prevent negative reactions, the space step should be limited as

$$\Delta x \in \left[C\Delta t - \frac{Q}{BSC}, C\Delta t + \frac{Q}{BSC} \right] \tag{1.24}$$

Therefore, with a defined time space, the channel with length L_c is divided into several segments as

$$N = Int\left(\frac{L_c}{C\Delta t} \right), N \geq 1, \tag{1.25}$$

where

C the wave velocity coefficient
Int() the operation of rounding.

Sediment transport is considered as the suspended load transport. The integrated format of the nonequilibrium sediment transport equation is

$$S = S_* + (S_0 - S_{0*})e^{\frac{-\alpha q L_c}{\omega_s}} + (S_{0*} - S_*)\frac{q}{\alpha \omega_s L_c}\left(1 - e^{\frac{-\alpha q L_c}{\omega_s}} \right), \tag{1.26}$$

where

S and S_0 sediment concentrations (kg/m^3)
S_* and S_{0*} sediment transport capacities of the outlet and inlet cross-sections (kg/m^3)
α coefficient of saturation recovery
ω_s settling velocity of the sediment particles (m/s).

4.5 Integration Based on Digital Drainage Network

Formulations for natural processes of sediment yield and transport introduced above are coupled based on the digital drainage network in the DYRIM. During simulation, water yield and sediment erosion on hillslopes, gravitational erosion in gullies, and hyperconcentrated flow routing in channels are calculated separately for different hillslope-channel units. The runoff and sediment yield of each hillslope are directly superposed on discharge in the corresponding channel segment, and the gravitational detached sediment enters the channel in the way mentioned before. In the drainage network, the simulation order of the hillslope-channel units follows the order from upper to lower reach to accord with affluxion. And therefore, calculations of different processes of sediment yield and transport are integrated in this way.

Fig. 1.13 The drainage network of the Chabagou watershed and the distribution of hydrological stations and rainfall stations [16]

5 Application of the DYRIM

5.1 Application in the Chabagou Watershed

First, the DYRIM is applied to the Chabagou watershed, which is a small watershed located in the gullied rolling loess region with a catchment area of 205 km^2 [16]. With the 50×50-m-resolution DEM, this watershed is delineated by 4763 units with an average hillslope area of 0.017 km^2. There are 6 hydrological stations and 31 rainfall stations in this watershed (Fig. 1.13).

Some model parameters can be obtained from the observed data, while some others need to be calibrated. Year 1967 was a year of high sediment yield in the Chabagou watershed. In this year, storm rainfall occurred frequently, and induced hyperconcentrated flows several times. Thus, the period from May to September, in 1967, is selected as the period for rainfall-runoff calibration; and rainfall-runoff, soil erosion, and sediment transport processes in 1967 are simulated with rainfall data from those 31 rainfall stations. According to the observed rainfall data with time steps in the vicinity of 12 min, the time step for simulation is set as 6 min. The nearest neighbor method is adopted to obtain the spatial distribution of rainfall, as shown in Fig. 1.14.

Comparison of the observed and simulated flow discharges at the Caoping station, which is used for the optimization of model parameters, is shown in Fig. 1.15.

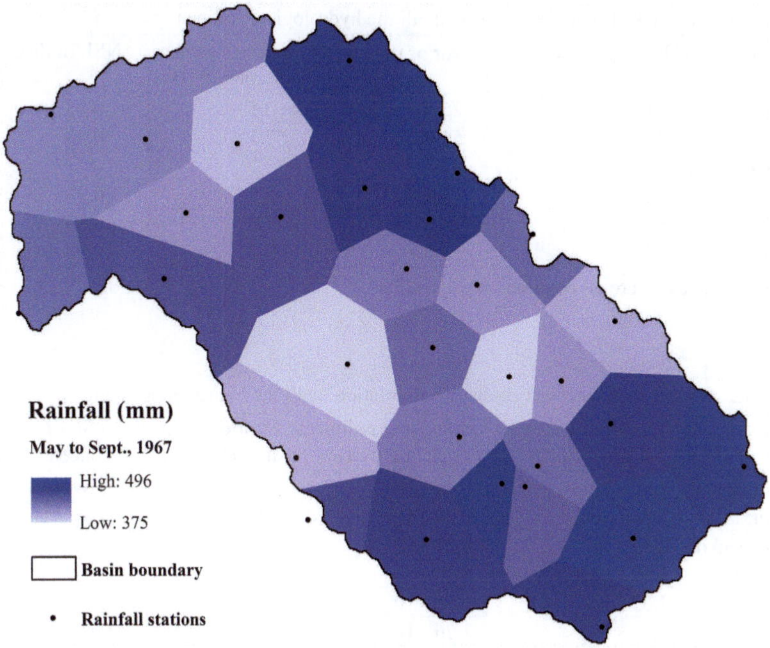

Fig. 1.14 Spatial distribution of rainfall in the simulated period [16]

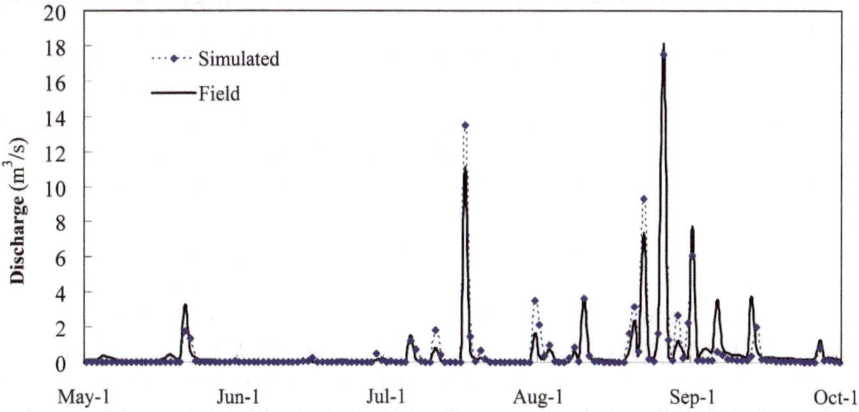

Fig. 1.15 Comparison of the observed and simulated flow discharge at the Caoping station [16]

The results of error analysis for runoff yields, flood peaks, and the Nash–Sutcliffe coefficient of efficiency (NSE) [24] of all the hydrological stations are listed in Table 1.1. There are mainly two reasons resulting in the errors of runoff simulation, especially in small tributaries. First, the observed data of all rainfall stations are not in the same time interval; thus rainfall intensities are not spatially equally expressed. Second, the spatial distribution of underlying parameters is not considered due to the lack of data.

Table 1.1 Statistics of runoff simulation in all the hydrological stations

Station	Drainage area (km^2)	Error of runoff (%)	Error of peak discharge[a] (%)	NSE of discharge
Shejiagou	4.26	25.9	−3.5	0.35
Tuoerxiang	5.74	−50.0	57.1	0.53
Sanchuankou	21.0	49.3	33.7	<0
Xizhuang	49.0	−14.2	0.2	0.73
Dujiagoucha	96.1	−28.0	1.0	0.89
Caoping	187	−3.6	−3.3	0.90

[a] The flood peaks were all counted on August 26

Table 1.2 Statistics of sediment discharge simulation in all the hydrological stations

Station	Measured sediment discharge (ton)	Simulated sediment discharge (ton)	Error of sediment discharge (%)	NSE of sediment concentration
Shejiagou	5.57×10^4	6.99×10^4	−7	0.69
Tuoerxiang	7.53×10^4	3.28×10^4	−41	0.35
Sanchuankou	2.15×10^5	4.21×10^5	96	0.43
Xizhuang	9.08×10^5	6.80×10^5	−25	0.63
Dujiagoucha	2.32×10^6	1.56×10^6	−33	0.76
Caoping	3.64×10^6	3.76×10^6	3	0.60

NSE Nash–Sutcliffe coefficient of efficiency

With the simulation results of hillslope runoff, soil erosion on hillslopes, gravitational erosion in gullies, and nonequilibrium sediment transport in channels are further simulated by using the DYRIM. The simulated sediment discharges and their NSE values at all the hydrological stations are listed in Table 1.2. Most stations have NSE greater than 0.5. Sediment concentrations at four stations (Fig. 1.16) also indicate that the simulated and observed sediment discharges are in the same order of magnitude. The distribution of hillslope erosion, gravitational erosion, and channel erosion in the whole watershed are shown in Fig. 1.17.

Overall, the simulation results reflected the soil erosion process with acceptable precision. However, distributed simulation of rain-runoff is difficult in arid and semi-arid areas due to remarkable variations of infiltration rate and soil moisture. The simulation of soil erosion process here is based on the rain-runoff simulation results that inevitably bring errors into the soil erosion simulation. This is approved by the fact that most of the simulation errors of sediment discharge and flow discharge have the same tendencies. Moreover, errors from the simulations of gravitational erosion, flow routing, and sediment transport also contribute to the errors of simulated sediment concentrations. Therefore, the simulation accuracy of sediment discharge depends on all the modules of water and sediment movement processes and their hybrids in a watershed.

5.2 Application in the Qingjian River Watershed

The DYRIM is applied to the Qingjian River watershed (109°12′–110°24′ E, 36°39′–37°19′ N), a watershed in the middle Yellow River, with a catchment area

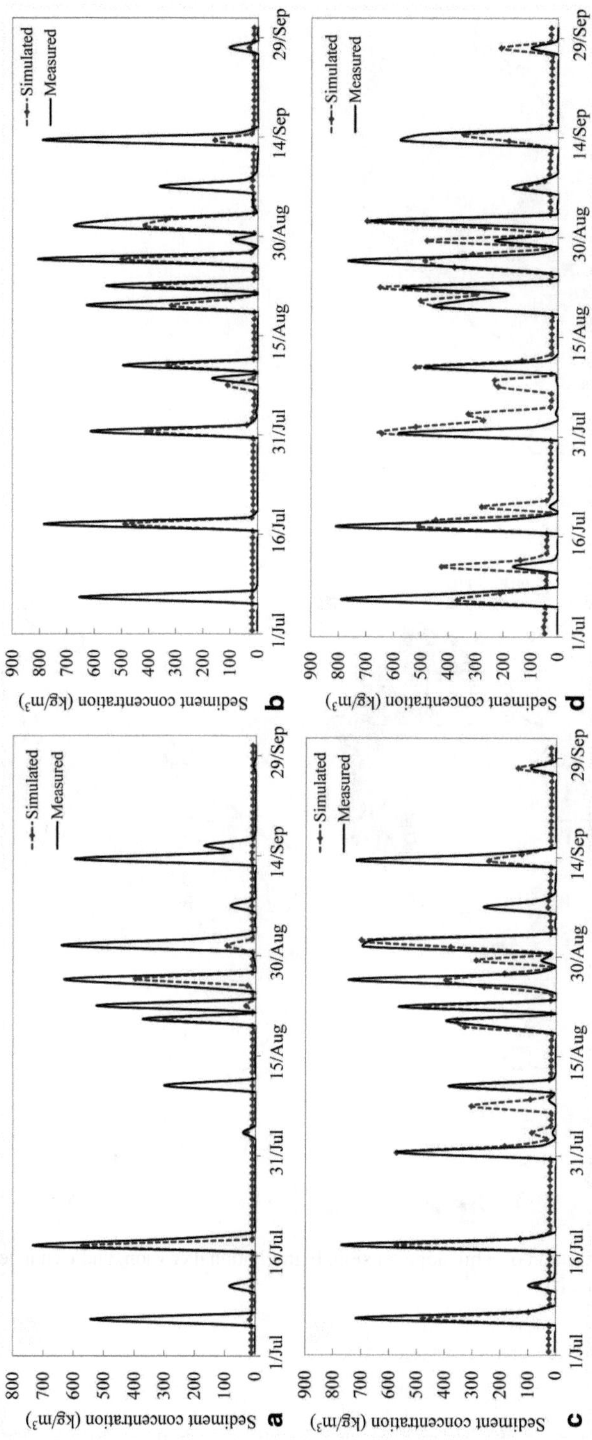

Fig. 1.16 Comparison of the observed and simulated sediment concentration: **a** Tuoerxiang, **b** Xizhuang, **c** Dujiagoucha, and **d** Caoping [16]

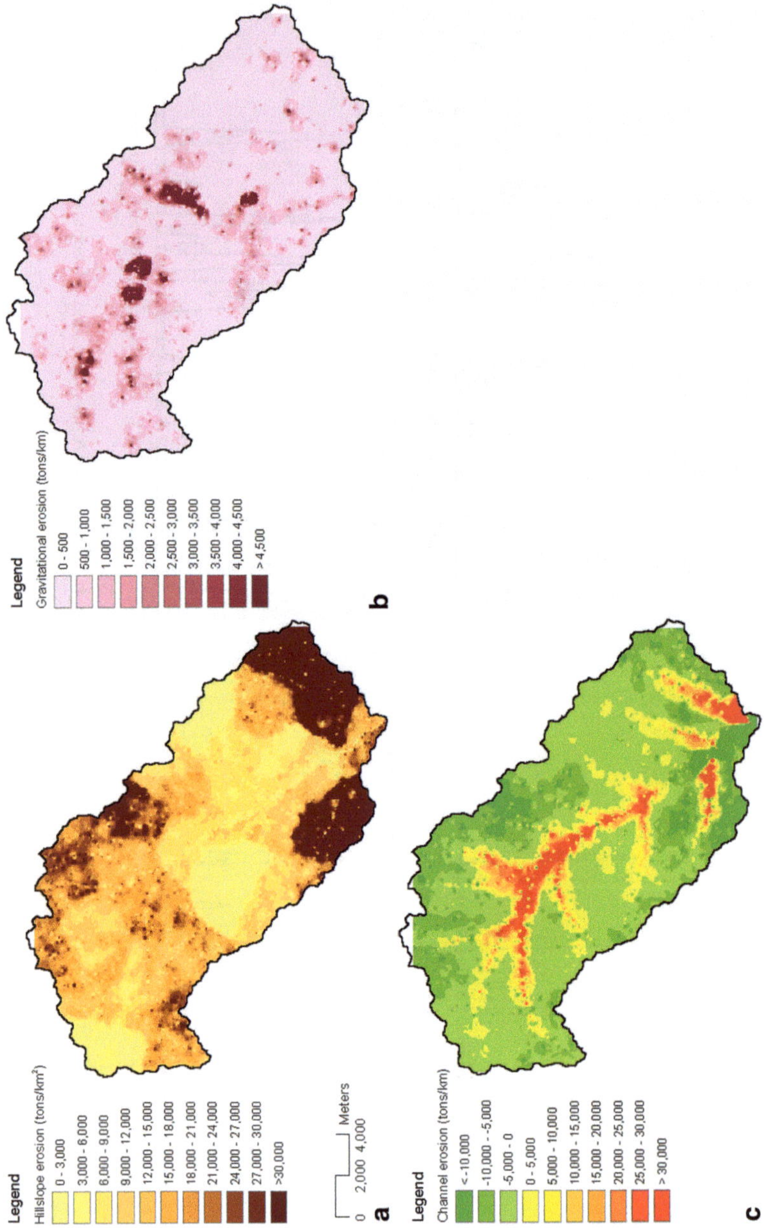

Fig. 1.17 The distribution of **a** hillslope erosion, **b** gravitational erosion, and **c** channel erosion in the Chabagou watershed

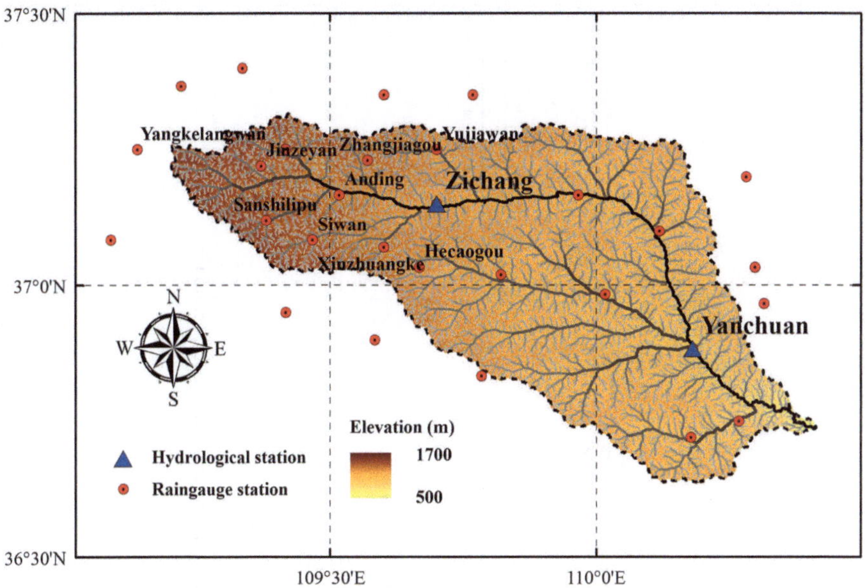

Fig. 1.18 The drainage network of the Qingjian River watershed and the distribution of hydrological stations and rainfall stations

Table 1.3 Rainfall events for model calibration and validation

Number	1	2	3	4	5
For calibration	2001.8.16–17	2002.5.10–12	2002.7.3–7	2006.8.25–27	2006.9.20–23
For validation	2001.8.17–21	2002.6.18–20	2007.9.1–2		

of 4078 km^2. With the 30×30-m-resolution DEM, this watershed is delineated by nearly 200,000 units in total. There are 29 rainfall stations inside or around this watershed; moreover, there are two hydrological stations in this watershed, named the Yanchuan station and the Zichang station, respectively (Fig. 1.18).

Divided by the Zichang station, the density of the rainfall station network in the upstream region is higher than that in the downstream region. As a result, the region in the upstream of the Zichang station is used for simulation. The observed data recorded at relevant 11 rainfall stations (their names are given in Fig. 1.18) are used as the rainfall inputs, and the observed data recorded at the Zichang station are used to evaluate the simulation results. Thus, five rainfall events occurred between 2001 and 2007 are selected for model calibration, and the other three rainfall events are selected for model validation (Table 1.3).

Comparison of the observed and simulated flow discharge (and sediment concentration) at the Zichang station during the period of model calibration is shown in Figs. 1.19 and 1.20. The results of the error analysis are listed in Tables 1.4 and 1.5. It is observed that the simulation results reflected the soil erosion process with acceptable precision as a whole.

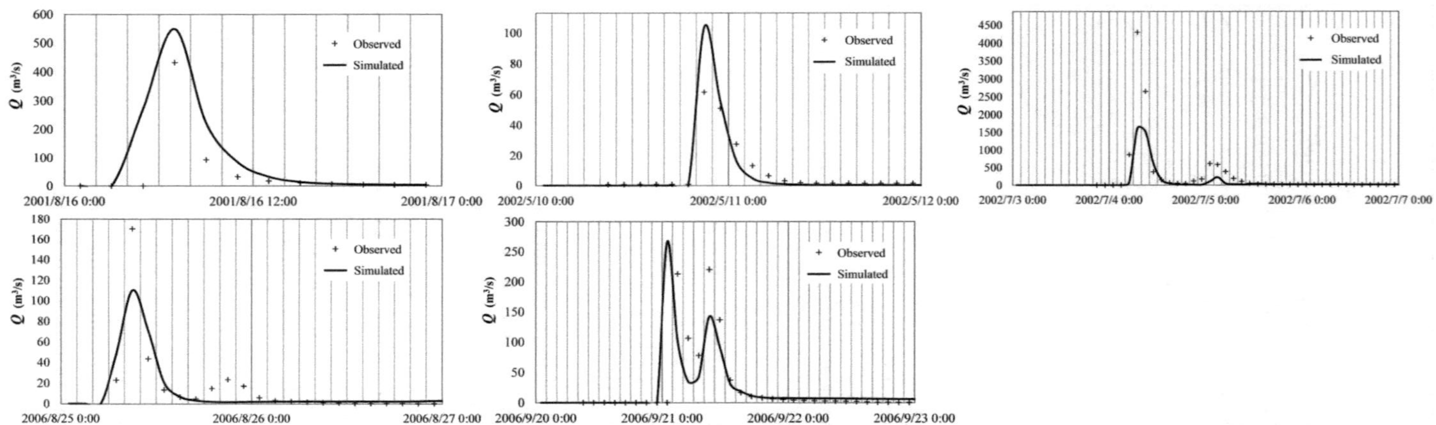

Fig. 1.19 Comparison of the observed and simulated flow discharge at the Zichang station during the period of model calibration

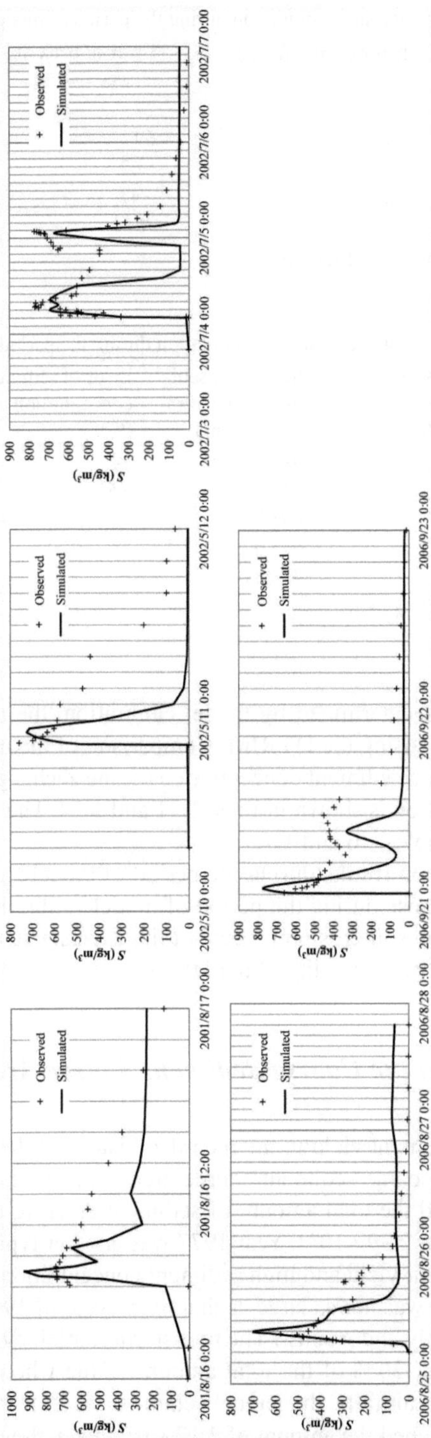

Fig. 1.20 Comparison of the observed and simulated sediment concentration at the Zichang station during the period of model calibration

Table 1.4 Statistics of flow discharge simulation during the period of model calibration

Event	Number	Error of runoff (%)	Error of peak discharge (%)	NSE of discharge
2001.8	1	97	27	036
2002.5	2	−1	69	0.67
2002.7	3	−54	−62	0.61
2006.8	4	−2	−47	0.79
2006.9	5	1	76	0.14

NSE Nash–Sutcliffe coefficient of efficiency

Table 1.5 Statistics of sediment concentration simulation during the period of model calibration

Event	Number	Measured sediment concentration (kg/m^3)	Simulated sediment concentration (kg/m^3)	Error of sediment concentration (%)	NSE of sediment concentration
2001.8	1	747	927.32	24	0.40
2002.5	2	760	725.66	−5	0.51
2002.7	3	774	695.71	−10	0.34
2006.8	4	577	696.02	21	0.66
2006.9	5	662	773.31	17	0.41

NSE Nash–Sutcliffe coefficient of efficiency

With the parameters determined by model calibration, the other three rainfall events are simulated by using the DYRIM. Comparison of the observed and simulated flow discharge (and sediment concentration) at the Zichang station during the period of model validation is shown in Figs. 1.21 and 1.22. The results of the error analysis are listed in Tables 1.6 and 1.7.

Overall, the simulation results during the period of model validation seemed to be a little worse than those during the period of model calibration. However, they still clearly reflected the soil erosion process at the Zichang station, which further proved the validity and practicability of the DYRIM.

5.3 *Application in the Coarse Sediment Source Area*

The highest annual sediment yield in the recorded history of the Loess Plateau occurred in 1967 in the coarse sediment source area. In that year, the runoff yield was 12.09×10^9 m^3, and the total amount of sediment entering into the main stem of the Yellow River was 2.39×10^9 t. Year 1977 was another typical year that had a large amount of sediment yield and high sediment concentration. However, the annual amount of rainfall was not as large as that in the year of 1967. Storm rainfalls were concentrated in July and August. The two-storm rainfall events that happened in August accounted for 30% of the total amount of the whole year. Because of the concentrated storm rainfalls, the annual sediment yield in the coarse sediment source area in 1977 reached the amount of 2.17×10^9 t even though the annual runoff yield was only in a medium level, with a value of 8.14×10^9 m^3.

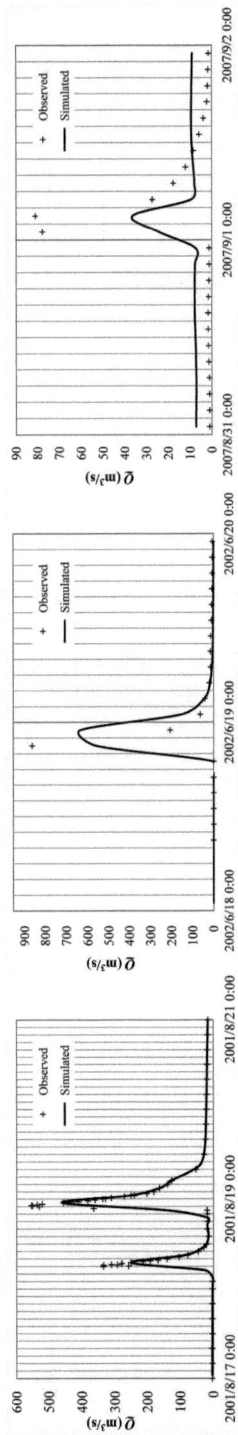

Fig. 1.21 Comparison of the observed and simulated flow discharge at the Zichang station during the period of model validation

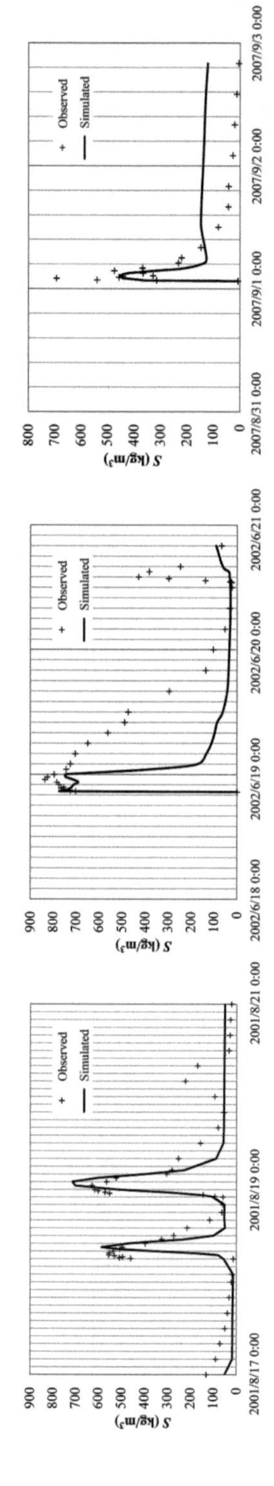

Fig. 1.22 Comparison of the observed and simulated sediment concentration at the Zichang station during the period of model validation

Table 1.6 Statistics of flow discharge simulation during the period of model validation

Event	Number	Error of runoff (%)	Error of peak discharge (%)	NSE of discharge
2001.8	1	103	47	0.33
2002.6	2	11	−27	0.60
2007.9	3	−7	−55	0.46

NSE Nash–Sutcliffe coefficient of efficiency

Table 1.7 Statistics of sediment concentration simulation during the period of model validation

Event	Number	Measured sediment concentration (kg/m^3)	Simulated sediment concentration (kg/m^3)	Error of sediment concentration (%)	NSE of sediment concentration
2001.8	1	678	771.16	14	0.40
2002.6	2	836	770.98	−8	0.28
2007.9	3	693	454.80	−34	0.69

NSE Nash–Sutcliffe coefficient of efficiency

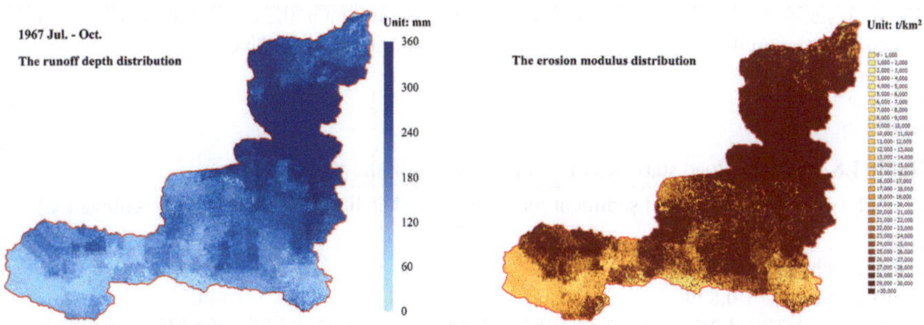

Fig. 1.23 Distributions of calculated runoff depth and erosion modulus in 1967 [34]

The distributions of runoff depth and erosion modulus calculated for the coarse sediment source area in 1967 were provided by the DYRIM (Fig. 1.23). The sediment yield of this year was calculated as 2.549×10^9 t, which is close to the measured value of 2.39×10^9 t. The sediment concentration processes for the main tributaries of 1977 were simulated, and the results for four selected tributaries are shown in Fig. 1.24; and the statistics of the sediment load of eight main tributaries in the Loess Plateau are listed in Table 1.8. Moreover, the results of the water and sediment hydrographs at the Longmen station, the outlet of the coarse sediment source area, are shown in Fig. 1.25. In general, the simulated sediment runoff had the same order of magnitude as the field data, and the simulated daily sediment load matched the trend of the field processes.

Nevertheless, there were still discrepancies between the simulated values and the observed ones based on the available data. Three possible reasons are proposed

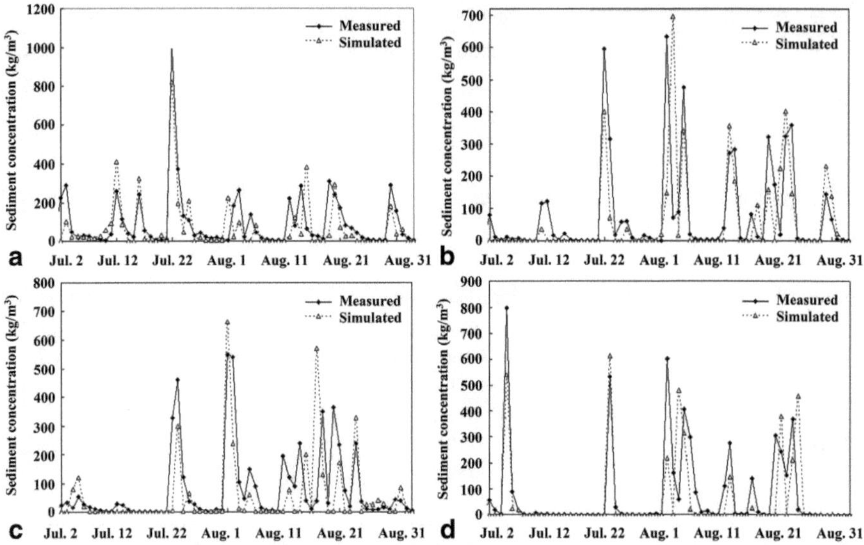

Fig. 1.24 Measured and simulated sediment concentrations in 1977 for selected tributaries: **a** Huangfu station in the Huangfuchuan River, **b** Gaoshiya station in the Gushanchuan River, **c** Wenjiachuan station in the Kuye River, **d** Shenjiawan station in the Jialu River [34]

Table 1.8 Sediment load statistics of main tributaries in the year 1977

Tributary	Measured sediment load ($\times 10^8$ t)	Simulated sediment load ($\times 10^8$ t)	Error percentage (%)
Huangfuchuan	0.26	0.33	26.92
Gushanchuan	0.839	1.21	44.22
Kuye river	1.38	1.52	10.14
Tuwei river	0.211	0.18	−14.69
Jialu river	0.121	0.17	40.5
Sanchuan river	0.465	0.69	48.38
Wuding river	2.69	3.42	27.14
Qingjian river	1.17	0.95	−18.80

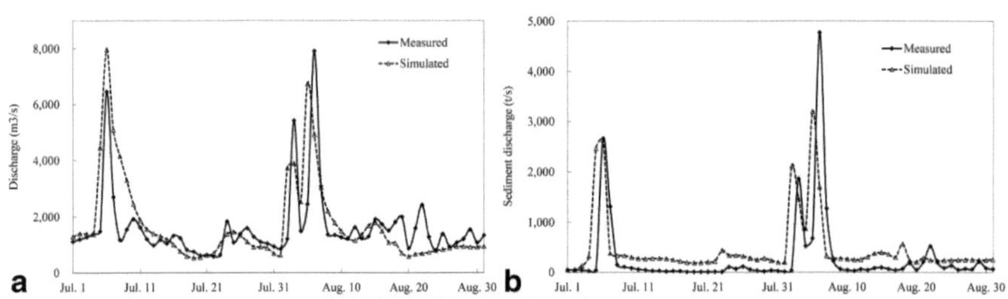

Fig. 1.25 Flow discharge and sediment load at Longmen station in 1977 [34]

here. First, it is somewhat difficult to calibrate and verify the distributed parameters for such a large watershed. Second, the temporal and spatial resolutions of the rainfall data were not high enough. Sediment yield and transport on the Loess Plateau are not only related to the quantity of rainfall but also influenced by the intensity of rainfall. However, the rainfall input adopted in the simulation was in the format of daily rainfall. Third, soil and water conservation measures (e.g., sediment trapping dams and farmland terracing) were not considered, and water consumption for mining industry and living is difficult to quantify. All of these factors mentioned above may influence the simulation accuracy.

6 Conclusions

In conclusion, the natural processes of sediment yield and transport in the Loess Plateau of the middle Yellow River, including water yield and sediment erosion on hillslopes, gravitational erosion in gullies, and hyperconcentrated flow routing in channels, were physically formulated; moreover, by taking advantage of computer and information technologies; these physically based models were well integrated in the DYRIM. Further research efforts have been made in verifying the scale effects and their mechanisms in soil erosion and sediment transport in the gullied Loess Plateau [17], developing an automatic optimization technique for model parameter calibration using the HPC job scheduling [39], and developing a method to estimate the spatially distributed rainfall by merging rain gauge measurements, satellite observations, and topographic information to providing more accurate rainfall data to the DYRIM [27].

The DYRIM can simulate all hydrological and sediment processes at the watershed scale to facilitate the investigation of sediment dynamics and modeling. Applications of the DYRIM in different watersheds in the Loess Plateau of the middle Yellow River inferred that this model was capable of simulating the processes of sediment yield and transport in a large-scale watershed. Moreover, this model can be used to evaluate the effect of the different soil and water conservation measures in this region to develop better land use and river management strategies [29]. We also tried to make the DYRIM model be extended to other watersheds, and the results indicated that DYRIM is a generic model for hydrological and sediment modeling [20, 28, 37].

Glossary

Binary tree In computer science, a binary tree is a tree data structure in which each node has at the most two children (referred to as the left child and the right child).

Channel A passage along which the water flows.

Database An organized collection of data.

Drainage network A network of channels and drains constructed on marshy or excessively wet land.

Erosion The process by which the soil and rock are removed from the Earth's surface by exogenic processes such as wind or water flow, and then transported and deposited in other locations.

Geographic information system A system designed to capture, store, manipulate, analyze, manage, and present all types of geographical data.

Gravitational erosion A type of degradation resulting from the earth's downward pull.

Gully A long, narrow valley with steep sides.

Hillslope The sloping side of a hill.

Hydrological station An agency that observes and studies the hydrologic conditions of bodies of water and territories.

Hyperconcentrated flow A two-phase flowing mixture of water and sediment in a channel, which has properties intermediate between fluvial flow and debris flow.

Leaf area index A dimensionless quantity that characterizes plant canopies.

Meteorological station A facility, either on land or sea, with instruments and equipment for measuring atmospheric conditions to provide information for weather forecasts and to study the weather and climate.

Parallel computing A form of computation in which many calculations are carried out simultaneously, operating on the principle that large problems can often be divided into smaller ones, which are then solved concurrently.

Potential evaporation The amount of evaporation that would occur if a sufficient water source was available.

Precipitation The water that falls from the clouds towards the ground, especially as rain or snow.

Remote Sensing The acquisition of information about an object or phenomenon without making physical contact with the object and thus in contrast to in situ observation.

Sediment transport The movement of solid particles (sediment), typically due to a combination of gravity acting on the sediment, and/or the movement of the fluid in which the sediment is entrained.

Watershed The area of land where all of the water that is under it or drains off of it goes into the same place; Land area that drains to a common waterway, such as a stream, lake, estuary, wetland, or ultimately the ocean.

Watershed decomposition The process by which the watershed is broken down into smaller sub-watersheds.

References

1. Bai, R., Li, T. J., Huang, Y. F., & Wang, G. Q. (2013). A hierarchical method for managing massive data of drainage networks extracted from high resolution DEM for large scale river basins. The 35th IAHR World Congress, Chengdu, China.

2. Beasley, D. B., Huggins, L. F., & Monke, E. J. (1980). ANSWERS: A model for watershed planning. Transactions of the American Society of Agricultural Engineers, 23, 938–944.
3. Cai, Q. G., Luk, S. H., & Wang, G. P. (1996). Process-based soil erosion and sediment yield model in a small basin in the hilly loess region. Acta Geographica Sinica, 51(2), 108–117.
4. Cao, Z., Pender, G., & Meng, J. (2006). Explicit formulation of the Shields diagram for incipient motion of sediment. Journal of Hydraulic Engineering, 132(10), 1097–1099.
5. Cappelaere, B. (1997). Accurate diffusive wave routing. Journal of Hydraulic Engineering, 123(3), 174–181.
6. Chen, Z. X., Wang, G. Q., Liu, J. H., & Wang, K. (2005). Parameter extract from the hydrological model based on the remote sensing image. Journal of Water Resources & Water Engineering, 16(4), 49–55.
7. Dabral, S., & Cohen, M. (2001). ANSWERS—2000: Areal nonpoint source watershed environmental response simulation. http://www.agen.ufl.edu/~klc/abe6254/answers01.pdf.
8. Dang, J. Q., & Li, J. (1996). Effect of water content on the strength of unsaturated loess. Acta University Agricultural Boreali-Occidentalis, 24(1), 57–60.
9. Darby, S. E., & Thorne, C. R. (1996). Development and testing of riverbank-stability analysis. Journal of Hydraulic Engineering, 122(8), 443–454.
10. Fei, X., & Shao, X. (2004). Sediment transport capacity of gullies in small watersheds. *Journal of Sediment Research, 1,* 1–8.
11. Flanagan, D. C., & Nearing, M. A. (1995). *USDA—water erosion prediction project: Hillslope profile and watershed model documentation.* National Soil Erosion Research Laboratory, Report No. 10, USDA-ARS, Indiana.
12. Foster, G. R., & Meyer, L. D. (1972). Transport of particles by shallow flow. *Transactions of the American Society of Agricultural Engineers, 15*(1), 99–102.
13. Fu, B. (1981). On the calculation of evaporation from soil. *Acta Meteorologica Sinica, 39*(2), 226–236.
14. Jetten, V. (2002). *Limburg soil erosion model user manual. Utrecht centre for environment and landscape dynamics*. The Netherlands: Utrecht University.
15. Knisel, W. G. (Ed.). (1980). CREAMS: A field-scale model for chemicals, runoff and erosion from agricultural management systems. USDA Conservation Research report No. 26, USDA-ARS, Washington DC.
16. Li, T. J., Wang, G. Q., Huang, Y. F., et al. (2009a). Modeling the process of hillslope soil erosion in the loess plateau. *Journal of Environmental Informatics, 14*(1), 1–10.
17. Li, T. J., Wang, G. Q., Xue, H., et al. (2009b). Soil erosion and sediment transport in the gullied Loess plateau: Scale effects and their mechanisms. *Science in China Series E-Technological Sciences, 52*(5), 1283–1292.
18. Li, T. J., Wang, G. Q., & Chen, J. (2010). A modified binary tree codification of drainage networks to support complex hydrological models. *Computers & Geosciences, 36,* 1427–1435.
19. Li, T. J., Wang, G. Q., Chen, J., et al. (2011). Dynamic parallelization of hydrological model simulations. *Environmental Modelling & Software, 26,* 1736–1746.
20. Li, T. J., Li, J. Y., Shi, H. Y., et al. (2013). A smartphone-based interactive flood warning system for ungauged mountainous regions. *Journal of Sichuan University (Engineering Science Edition), 45*(1), 23–27.
21. Message, P. I. F. (2008). MPI: A message-passing interface standard. http://www.mpi-forum.org/docs/mpi21-report.pdf.
22. Morgan, R. P. C., Quinton, J. N., Smith, R. E., Govers, G., Poesen, J. W. A., Auerswald, K., Chisci, G., Torri, D., Styczen, M. E., & Folly, A. J. V. (1998a). The European Soil Erosion Model (EUROSEM): A dynamic approach for predicting sediment transport from fields and small catchments. *Earth Surface Processes and Landforms, 23*(6), 527–544.
23. Morgan, R. P. C., Quinton, J. N., Smith, R. E., Govers, G., Poesen, J. W. A., Auerswald, K., Chisci, G., Torri, D., Styczen, M. E., & Folly, A. J. V. (1998b). *The European Soil Erosion Model (EUROSEM): Documentation and user guide*. Silsoe: Cranfield University.
24. Nash, J. E., & Sutcliffe, J. V. (1970). River flow forecasting through conceptual models, 1—A discussion of principles. *Journal of Hydrology, 10,* 282–290.

25. O'Callaghan, J. F., & Mark, D. M. (1984). The extraction of drainage networks from digital elevation data. *Computer Vision, Graphics and Image Processing, 28,* 323–344.
26. Osman, A. M., & Thorne, C. R. (1988). Riverbank stability analysis I: Theory. *Journal of Hydraulic Engineering, 114*(2), 134–150.
27. Shi, H. Y. (2013). *Computation of spatially distributed rainfall by merging raingauge measurements, satellite observations and topographic information: A case study of the 21 July 2012 rainstorm in Beijing, China.* The 35th IAHR World Congress, Chengdu, China.
28. Shi, H. Y., Fu, X. D., Wang, H., et al. (2011). Modeling and prediction of water loss and soil erosion for high-relief mountainous region: Case study of Qamdo Region. *Journal of Basic Science and Engineering, 19*(S1), 17–27.
29. Shi, H. Y., Li, T. J., Zhu, J. F., et al. (2012). *Impact of check dams on the sharp decrease of runoff in the middle Yellow River: Case study in the Huangfuchuan River basin.* The 5th International Yellow River Forum, Zhengzhou, China.
30. Shi, H. Y., Fu, X. D., Chen, J., Wang, G. Q., & Li, T. J. (2014). Spatial distribution of monthly potential evaporation over mountainous regions: A case of the Lhasa River basin in China. *Hydrological Sciences Journal.* doi:10.1080/02626667.2014.881486.
31. Tang, L. Q., Chen, G. X., & Cai, M. Y. (1990). A mathematical model of sediment yield on small watershed in the gullied-hilly Loess Plateau. *Journal of Hohai University, 18*(6), 10–16.
32. Tang, L. Q., & Chen, G. X. (1994). Soil erosion formula and its application in sediment yield calculation. *Advances in Water Science, 5*(2), 104–110.
33. Wang, X. K., Ning, Q., & Hu, W. D. (1982). The formation and process of confluence of the flow with hyperconcentration in the gullied-hilly loess area of the Yellow River basin. *Journal of Hydraulic Engineering, 7,* 26–35.
34. Wang, G. Q., Wu, B. S., & Li, T. J. (2007). Digital Yellow River model. *Journal of Hydro-Environment Research, 1*(1), 1–11.
35. Wang, H., Fu, X. D., & Wang, G. Q., et al. (2011). A common parallel computing framework for modeling hydrological processes of river basins. *Parallel Computing, 37,* 302–315.
36. Wang, H., Zhou, Y., Fu, X. D., et al. (2012). Maximum speedup ratio curve (MSC) in parallel computing of the binary-tree-based drainage network. *Computers & Geosciences, 38,* 127–135.
37. Yin, X. L. (2009). *Study on scheduling of main reservoir on Pearl River for estuarine fresh water providing based on digital basin model.* China: Tsinghua University.
38. Zhang, T. Z. (1993). *Outline of the Loess Plateau.* Beijing: China Environmental Science Press.
39. Zhang, A., Li, T. J., Li, X., et al. (2013). *A general framework for model parameter calibration using job scheduling of an HPC system.* The 35th IAHR World Congress, Chengdu, China.

Chapter 2
Integrated Simulation of Interactive Surface-Water and Groundwater Systems

Varut Guvanasen and Peter S. Huyakorn

Contents

V. Guvanasen (✉) · P. S. Huyakorn
HydroGeoLogic, Inc., 11107 Sunset Hills Road, Suite 400, Reston, VA 20190, USA
e-mail: dguvanasen@hgl.com

P. S. Huyakorn
e-mail: psh@hgl.com

C. T. Yang, L. K. Wang (eds.), *Advances in Water Resources Engineering*,
Handbook of Environmental Engineering, Volume 14, DOI 10.1007/978-3-319-11023-3_2

Abstract Effective management of watersheds and ecosystems requires a comprehensive knowledge of hydrologic processes, and the ability to predict and quantify reliably the impacts due to anthropogenic or natural changes in water availability and water quality. For integrated water resources management studies in which both surface water and groundwater are interactive, a technically rigorous and physically based approach is essential. Simulation models have been used increasingly to provide a predictive capability in support of water resources, and environmental and restoration projects. Often, simplified models are used to quantify complex hydrologic and transport processes in surface and subsurface domains. Such models incorporate restrictive assumptions relating to spatial variability, dimensionality, and interactions of components in flow and transport processes. During the past decade, with the advent of high-speed personal computers, a number of rigorous integrated surface-water/groundwater models have been developed to circumvent these limitations. In general, a typical model of an integrated hydrologic system may be divided into three interactive and interconnected domains: subsurface, overland, and channels/streams, in which water flow and transport of constituents can occur. In this chapter, the following are presented and discussed: a description of relevant processes relating to water flow and solute transport in conjunction with governing equations for all domains; procedures for model development and calibration; and two field application examples.

Keywords Integrated surface-water/groundwater modeling · Flow simulation · Transport simulation · Model calibration

Nomenclature

A_C Wetted cross-sectional area of the channel segment (L^2)
A_{GO} Area at the interface between overland and subsurface (L^2)
A_{IJ} Area through which mass influx passes from domain J to domain I (L^2)
a_L Longitudinal dispersivity (L)
a_T Transverse dispersivity (L)

a_{ijmn}	Dispersivity tensor (L)
B_C	Top width of channel (L)
b	Thickness of channel bed (L)
b	Fitting parameter (dimensionless) (Eq. 2.4c)
b_{IJ}	Distance between two centroids in domains I and J (L)
C_1, C_2	Fitting parameters (dimensionless) (Eq. 2.14b)
C_3	Fitting parameters (dimensionless) (Eq. 2.14c)
C^i	Solute concentration of species *i* (M/L^3)
C^k	Solute concentration of component *k* (M/L^3)
C_d	Weir discharge coefficient (dimensionless)
C_{int}	Canopy storage parameter (L)
$\hat{C}_k$	Concentration for species *k* vector for the transport equation
C_B^k	Specified concentration of solute *k* at the boundary (M/L^3)
C_C^{*k}	Solute concentration of component *k* of the sources (or sinks) within the channel domain (M/L^3)
C_G^{*k}	Solute concentration of component *k* of the sources (or sinks) within the subsurface domain (M/L^3)
C_I^k	Solute concentration of species *i* (M/L^3)
C_{J^+/I^-}^k	Directionally dependent concentration of component *k* in domain J, if v_{IJ} is positive, in domain I if v_{IJ} is negative (M/L^3)
C_O^{*k}	Solute concentration of component *k* of the sources (or sinks) within the overland domain (M/L^3)
C_O^i	Reference solute concentration of species i (M/L^3) corresponding to Δo and :$_o$
C_S^i	Solute concentration of species i (M/L^3) corresponding to ρ_S^i and μ_S^i
C_s^k	Concentration of component k adsorbed to the soil (M/Msoil)
D_d^k	Molecular diffusion coefficient for component *k* (L^2/T)
D_{IJ}^k	Effective dispersion coefficient of component *k* between domains I and J (L^2/T)
D_{ij}^k	Apparent hydrodynamic dispersion tensor of component *k* (L^2/T)
D_{ijB}	Dispersion coefficient tensor at the boundary (L^2/T)
d	Flow depth (L)
d_C	Depth of channel flow (L)
d_O	Depth of overland flow (L)
E_{can}	Canopy evaporation (L/T)
E_P	Reference evapotranspiration (L/T)
F_F	Forcing vector for the flow equation
F_T	Forcing vector for the transport equation
f_{Str}	Structure discharge per unit length (L^2/T)
g	Gravitation acceleration (L/T^2)
H	Specified hydraulic head at the boundary at x_{iB} (L)
h	Reference hydraulic head (or equivalent freshwater head) (L)$=\dfrac{p}{\rho_o g}+x_3$
h	Overland hydraulic head or water surface elevation (L)$=d_O+z_{LS}$
h	Hydraulic head or water surface elevation of the channel (L)$=d_C+z_C$
$\hat{h}$	Hydraulic head vector for the flow equation
h_C	Head in the channel domain (L)

h_{d} Downstream head between the two systems (L)
h_{G} Head in the subsurface domain (L)
h_{O} Head in the overland domain (L)
h_{u} Upstream head between the channel and overland domains (L)
LAI Leaf area index (dimensionless)
L_{R} Effective root length (L)
l_{UStr} Upstream reference location of the structure (L)
l_{DStr} Downstream reference location of the structure (L)
K Leakance (1/T)
K^{C} Conductance term along the length of the channel (L^3/T)
K_{ij} Hydraulic conductivity or conductance (L/T) in Eqs. (2.1), (2.5a), and (2.6a)
K_{ij}^{G} Hydraulic conductivity tensor (L/T) $= \dfrac{k_{ij}\rho_{\mathrm{o}} g}{\mu_{\mathrm{o}}}$
K_{ij}^{O} Overland conductance tensor (L/T)
K_{F} Conductance matrix for the flow equation
$K_{\mathrm{GC}}^{\mathrm{eff}}$ Effective leakance across the interface area between channel and subsurface (1/T)
K_{GO} Leakance across the interface area between overland and subsurface (1/T)
K_{T} Conductance matrix for the transport equation
k_{ij} Intrinsic permeability tensor (L^2)
k_{n} Manning's conversion factor ($\mathrm{L}^{1/3}$/T)
k_{rC} Relative channel conductance (dimensionless)
k_{rG} Relative permeability (dimensionless) which is a function of water saturation as provided by the relative permeability curve
k_{rGC} Relative leakance at the interface between channel and subsurface (dimensionless)
k_{rGO} Relative leakance at the interface between overland and subsurface (dimensionless)
k_{rO} Relative overland conductance (dimensionless)
k_{Str} Structure operation coefficient (dimensionless)
L_{C} Length of channel segment (L)
l Length along the direction of flow (L)
M_{B}^{k} Dispersive mass flux of species k per unit area (M/L^3 T)
M_{F} Mass matrix for the flow equation
M_{T} Mass matrix for the transport equation
m_{IJ}^{k} Mass influx rate per unit area from domain J to domain I of component k (M/L^2 T)
N_{P} Number of parent chemicals of solute k (dimensionless)
n_{C} Manning's roughness coefficient for channel (dimensionless)
n_{i} Unit vector (dimensionless), positive inward
n_{ij} Manning's roughness coefficient tensor for overland flow (dimensionless)
n_{R} Number of cells that contribute to the total root zone for each areal location (dimensionless)
n_{RT} Number of cells that lie within the depth interval from 0 to L_{R} at any areal location (dimensionless)

n_s Number of solutes (dimensionless)
P_C Wetted perimeter of the channel segment (L)
P_P Precipitation rate (L/T)
p Fluid pressure (M/LT^2)
p_o Reference fluid pressure (M/LT^2)
Q_B Volumetric water flux per unit area (L)
Q_{CG} Flux across the area of the interface from subsurface to channel (L^3/T)
Q_{GC} Flux across the area of the interface from channel to subsurface (L^3/T)
Q_{GO} Flux across the area of the interface from overland to subsurface (L^3/T)
Q_{OC} Flux across the total length of channel banks to/from the overland flow domain (L^3/T)
Q_i Discharge per unit width normal to the flow direction (L^2/T)
Q_{OG} Flux across the area of the interface from subsurface to overland (L^3/T)
Q_{Str} Discharge rate (L^3/T) of the structure as a function of head, h
q_C Volumetric flux per unit volume (1/T) of the overland domain and represents sources and/or sinks of water
q_{CO} Flux per unit volume of channel flow domain from the overland flow domain (1/T)
q_{CG} Flux per unit volume of channel flow domain from the subsurface (1/T)
q_G Volumetric flux per unit volume (1/T) of the subsurface domain and represents sources and/or sinks of water
q_{GC} Flux per unit volume of subsurface from the one-dimensional channel domain $= -q_{CG}$ (1/T)
q_{GO} Flux per unit volume of subsurface from the two-dimensional overland flow domain (1/T)
q_O Volumetric flux per unit volume (1/T) of the overland domain and represents sources and/or sinks of water
q_{OC} Flux per unit volume of overland flow domain from channel (1/T) $= -q_{CO}$
q_{OG} Flux per unit volume of overland flow domain from groundwater (1/T) $= -q_{GO}$
$r_F(z)$ Root extraction function (dimensionless) which typically varies logarithmically with depth
S_b Bed slope (dimensionless) at the zero-depth gradient boundary
S_e Effective water saturation (dimensionless)
S_G Degree of water saturation (dimensionless) and is determined by the moisture retention curve as a function of the pressure head
S_{Gr} Residual water saturation (dimensionless)
S_{int} Canopy storage (L)
S_{int}^{max} Canopy storage capacity (L)
S_{int}^{o} Previous time step canopy storage (L)
S_{int}^{*} Intermediate canopy storage (L)
S_O Equivalent sediment depth (L)
S_{Str} Structure unit function (dimensionless), equals unity along the length when a hydraulic structure is present, 0 otherwise
s Length along the direction maximum local slope (L)
T_{ij}^{*} Tortuosity tensor (dimensionless)

T_{pI}	Rate of transpiration for computational cell I (L/T)
t	Time (T)
V	Magnitude of the velocity vector (L/T)
V_{G}	Subsurface elementary volume (L^3)
V_{I}	Normalization volume in domain I (L3)
v_{IJ}	Water flow rate per unit area from domain J to domain I (L/T)
v_i	Darcy velocity along the ith direction (L/T)
$v_{i\text{B}}$	Specified fluid velocity at the boundary (M/L^3)
x_i	Cartesian coordinate along the ith direction (L) with x_3 being vertically upward
$x_{i\text{B}}$	Boundary coordinates (L)
Z_{Bank}	Bank elevation (L) which may be at or above the overland flow surface elevation
z	Depth coordinate from the soil surface (L) (Eq. 2.14d)
z_{C}	Channel bottom elevation (L)
z_{LS}	Land surface elevation (L)
α	Fitting parameter (1/L), (Eqs. 2.4a and 2.4b)
α_{G}	Bulk compressibility of aquifer (L^2T^2/M)
β	Fitting parameter (dimensionless) (Eqs. 2.4a and 2.4b)
β_{w}	Fluid compressibility (LT^2/M)
I^k_{CInt}	Mass transfer rate of component k between the channel and other domains (1/T)
I^k_{GInt}	Mass transfer rate of component k between subsurface and other domains (M/L^3 T)
I^k_{OInt}	Mass transfer rate of component k between overland and other domains (1/T)
γ	$1-1/\beta$ (dimensionless; Eqs. 2.4a and 2.4b)
δ	Total density factor (dimensionless) $=\dfrac{\rho_{\text{f}}-\rho_{\text{o}}}{\rho_{\text{o}}}$
$\delta(\bullet)$	Dirac delta function (1/L)
δ_{ij}	Kronecker's delta (dimensionless)
ζ	Distance along submerged channel cross section (L)
θ_{an}	Moisture content at anoxic limit (dimensionless)
θ_C	Effective porosity in the channel domain (dimensionless)
θ_{e1}	Moisture content at the end of the energy-limiting stage (above which full evaporation can occur; dimensionless)
θ_{e2}	Limiting moisture content below which evaporation is zero (dimensionless)
θ_{eG}	Effective porosity in groundwater domain (dimensionless)
θ_{fc}	Moisture content at field capacity (dimensionless)
θ_{G}	Subsurface porosity (dimensionless)
θ_{O}	Overland porosity (dimensionless)
θ_{o}	Moisture content at oxic limit (dimensionless)
θ_{wp}	Moisture content at wilting point (dimensionless)
λ^k_s	First-order decay coefficient for component k in soil (1/T)
λ^k_w	First-order decay coefficient for component k in water (1/T)

μ_f	Fluid dynamic viscosity (M/LT)
μ_o	Reference fluid dynamic viscosity (M/LT) corresponding to C_o^i
μ_S^i	Fluid dynamic viscosity of species *i* (M/LT) corresponding to C_S^i
ξ_{kj}	Fraction of parent component *j* transforming into component *k* (dimensionless)
$\rho^C{}_{f'}$	Fluid density (M/L^3) associated with q_C
$\rho^{CG}{}_{f'}$	Fluid density (M/L^3) associated with q_{CG}
$\rho^{CO}{}_{f'}$	Fluid density (M/L^3) associated with q_{CO}
$\rho^G{}_{f'}$	Fluid density (M/L^3) associated with q_G
$\rho^{GC}{}_{f'}$	Fluid density (M/L^3) associated with q_{GC}
$\rho^{GO}{}_{f'}$	Fluid density (M/L^3) associated with q_{GO}
$\rho^O{}_{f'}$	Fluid density (M/L^3) associated with q_O
$\rho^{OC}{}_{f'}$	Fluid density (M/L^3) associated with q_{OC}
$\rho^{OG}{}_{f'}$	Fluid density (M/L^3) associated with q_{OG}
ρ_B^C	Bulk density of sediment in the channel domain (M/L^3)
ρ_B^G	Bulk density of soil in the subsurface domain (M/L^3)
ρ_B^C	Bulk density of sediment in the overland domain (M/L^3)
ρ_f	Fluid density (M/L^3)
ρ_o	Reference fluid density (M/L^3) corresponding to C_o^i
ρ_S^i	Fluid density of species *i* (M/L^3) corresponding to C_S^i
ψ	Pressure head (L) $=p/(\Delta_o g)$

1 Introduction

As the global population grows, more demands are placed on one of the world's precious resources: water. With the rate of population increase of 70 million people per year, corresponding global water use is rising at an approximate rate of 30 billion m^3 per year [1]. Increased water demands give rise to global water stress. Causes that lead to global water stress include: excessive withdrawal from surface-water bodies, excessive withdrawal of groundwater from aquifers, pollution of freshwater resources, and inefficient use and management. Water resources consist of two integral systems, surface water and groundwater, both of which require rigorous management and protection. Groundwater in pristine aquifer systems usually requires little or no treatment before it is drinkable. However, if contaminated, these resources are expensive and difficult to remediate and restore. No less important is surface water in rivers, lakes, estuaries, and coastal systems, which is more visibly abundant. Surface water can have a strong impact on our everyday lives through flooding, transport, drinking water, etc. Surface-water and groundwater resources are interconnected. Baseflow in streams and rivers is derived from the contributing groundwater. Agricultural chemicals in surface water may enter into groundwater, which subsequently may emerge into streams. For these reasons, both groundwater and surface-water resources need to be protected and properly managed.

It is apparent that each water resource system is a set of interdependent water bodies and structures, each impacting on the state and performance of the others, and together contributing to the overall performance of the system [2]. There are many water resource problems that display a strong linkage between surface-water and groundwater systems. Therefore, understanding how surface-water levels are related to adjacent aquifer systems is crucial, for example, for the management of wetlands and river habitat restoration. Pollution of groundwater may influence surface-water resources and vice versa. Whether a river might flood at times of heavy rain or not will often depend on the surrounding groundwater levels. For these situations, it is desirable to consider surface-water and groundwater domains in an integrated manner and to develop appropriate tools to describe the interactions between the two domains. Therefore, water resources problems cannot be treated as isolated systems and it may be necessary to treat the entire water pathway, including overland, channel and river network, groundwater, and urban pipe and drainage systems. Those involved in the design and operation of each structure, which could be a reservoir, a diversion canal, a control structure affecting the input or output of a natural lake or wetland area, a hydropower plant, a groundwater extraction plant or an artificial recharge basin, a water or wastewater treatment plant, or a flood control levee, must examine the impacts resulting from those individual components in the system. Integrated water resource systems planning and management focus not only on the performance of individual components but also on the performance of the entire systems of components.

Computer modeling of both water resource systems has long been used as an aid to the planning and management of water resources. In 1969, Freeze and Harlan [3] proposed a blueprint for the digital modeling of the hydrologic cycle based on their assessment of the feasibility of the development of a rigorous, physically based mathematical model of the complete hydrological system. In this original blueprint, it was argued that if each of the component processes within the hydrological cycle can be described by an exact mathematical representation, then it should be possible to model the different flow and transport processes using their governing partial differential equations. Following this vision, there have been a number of models that attempt to simulate the interactions between surface water and groundwater.

Historically, groundwater and surface-water flow and transport processes were modeled separately, as their behaviors are represented by different mathematical equations and over very different time scales [4]. The interaction between them was usually taken into account as boundary conditions at respective interdomain interfaces. The simplest method, but also the least accurate, is by independently solving the surface and subsurface flow equations in succession without iteration [5, 6]. The next level of coupling is to solve the surface and groundwater flow and transport equations separately but iteratively at the same time step, interlinked by common internal boundary conditions representing the exchange between the surface and groundwater domains [7, 8]. Solution for the water flow and solute transport equations at a time step is achieved when the iteration errors fall within respective specified tolerances before the computation is advanced to the next time step. The highest level of coupling is realized by numerically solving all the flow and transport equations for surface water, groundwater, and the common internal boundary condition between the two as a set of simultaneous equations for each time step [4, 9–11]. Since this approach requires intensive computation efforts, restrictive assumptions relating to dimensionality and interaction of components of flow and transport processes were initially necessary. However, during the last decade, along with the advent of high-speed personal computers, a number of rigorous integrated surface-water/groundwater models have been developed to circumvent these limitations. This approach has been increasingly accepted and utilized by the technical community. Recent application examples reported in the literature based on this approach include: water resources evaluation and management in Florida [12], and western Australia [13]; and studies relating to the impact of water quality on water resources in Florida [14, 15], California [16], and China [17].

An integrated view of water resource systems takes into account a multitude of interactions between surface-water and groundwater components in both water quantity and quality aspects. The modeling approach described herein will attempt to address these interactions in a comprehensive and rigorous manner. In this chapter, in order to describe the flow and transport processes in a systematic manner, a typical integrated hydrologic system is divided into three interconnected domains in which flow and transport occur: the subsurface domain, the overland domain, and the channel domain. The following are presented and discussed: governing equations for the flow and transport processes in the integrated three domains; interdomain interactions; solution techniques, model development and calibration; and application examples.

2 Governing Processes and Equations

As shown Fig. 2.1, an integrated system consists of two interactive components: surface water and subsurface water. From the simulation consideration, the surface-water component is divided into two domains: stream/channel domain and overland domain. The flow and transport along streams/channels and overland are approximated by one-dimensional and two-dimensional processes, respectively. Below the ground surface is the subsurface domain. In this domain, there are two distinct but continuous zones: above the water table and below the water table. The zone above the water table is called the vadose zone (the unsaturated zone), and the zone below the water table is called the saturated zone (Fig. 2.2).

In this section, fundamental equations describing the flow and transport of water, and mass in an integrated subsurface and surface environment are described, along with boundary and initial conditions, interdomain communication, relevant process, and solution techniques. Derivation details of fundamental equations may be found in [18–21].

2.1 Flow

Flow equations for the three interactive domains are described below. The state variable that continuously spans over the three domains is hydraulic head, defined

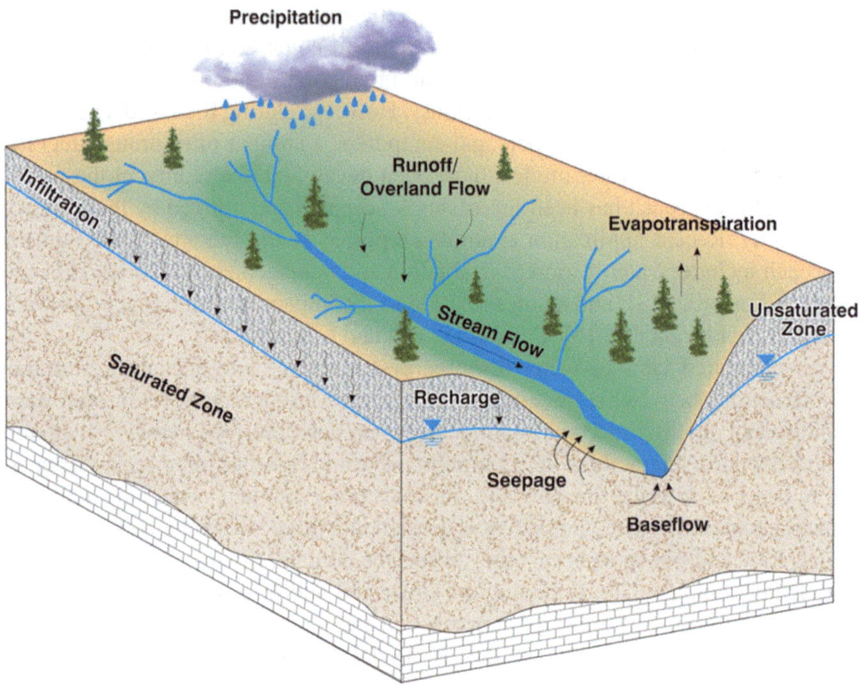

Fig. 2.1 Distribution, flow, and interaction of water on the land and in the subsurface

in the subsection below. Three capitalized prefixes and suffixes are: *G, O, and C* which denote the subsurface (groundwater), overland, and channel domains, respectively. For fluxes between any two domains, two letters are used. The domain of interest is determined by the first letter and the domain adjacent to it is denoted by the second letter. The flux is positive when the direction of the flux is from the adjacent domain to the domain of interest. As an example, *GO* indicates that the domain of interest is the subsurface domain and is communicating with the overland domain. In this case, the flux is positive from the overland domain to the subsurface domain. This convention is used throughout this chapter.

2.1.1 Subsurface Flow

In a variably saturated environment in the subsurface domain, isothermal flow may be expressed by the mixed form of the Richard's equation as (Adapted from Refs. [16, 18, 22]):

$$\frac{\partial}{\partial x_i}\left(\rho_{\mathrm{f}}\, k_{\mathrm{rG}}\, K_{ij}^{\mathrm{G}}\left(\frac{\mu_{\mathrm{o}}}{\mu_{\mathrm{f}}}\right)\frac{\partial(h+\delta x_3)}{\partial x_j}\right)=\frac{\partial}{\partial t}(\rho_{\mathrm{f}}\, S_{\mathrm{G}}\, \theta_{\mathrm{G}})-\rho_{\mathrm{f}}^{\mathrm{G}} q_{\mathrm{G}}-\rho_{\mathrm{f}}^{\mathrm{GO}} q_{\mathrm{GO}}-\rho_{\mathrm{f}}^{\mathrm{GC}} q_{\mathrm{GC}} \quad (2.1)$$

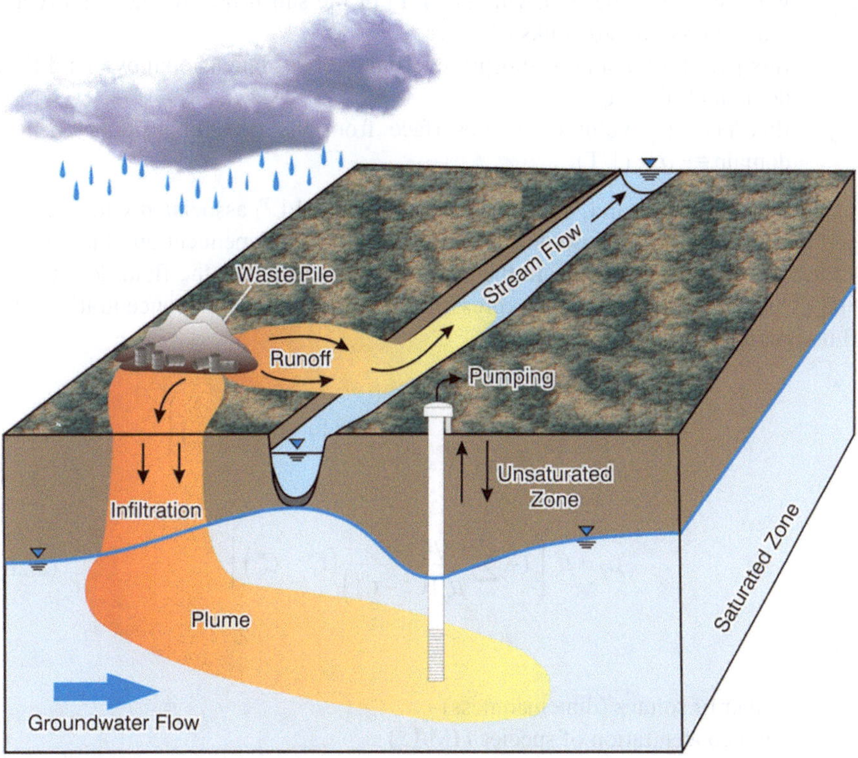

Fig. 2.2 Mass transport between different domains

where:

x_i Cartesian coordinate along the ith direction (L) with x_3 being vertically upward

ρ_f fluid density (M/L³)

k_{rG} relative permeability (dimensionless), which is a function of water saturation as provided by the relative permeability curve

K_{ij}^G hydraulic conductivity tensor (L/T) $K_{ij}^G = k_{ij}\rho_o g / \mu_o$

k_{ij} intrinsic permeability tensor (L^2)

ρ_o reference fluid density (M/L^3)

g gravitation acceleration (L/T^2)

μ_o reference fluid dynamic viscosity (M/LT)

μ_f fluid dynamic viscosity (M/LT)

h reference hydraulic head (or equivalent freshwater head) (L) $h = \frac{p}{\rho_o g} + x_3$

p fluid pressure (M/LT^2)

δ total density factor (dimensionless) $\delta = \frac{\rho_f - \rho_o}{\rho_o}$

t time (T)

S_G degree of water saturation (dimensionless) and is determined by the moisture retention curve as a function of the pressure head

θ_G porosity (dimensionless)

q_G volumetric flux per unit volume (1/T) of the subsurface domain and represents sources and/or sinks of water

q_{GO} flux per unit volume of subsurface from the two-dimensional overland flow domain (1/T) $= -q_{OG}$

q_{GC} flux per unit volume of subsurface from the one-dimensional channel domain $= -q_{CG}$ (1/T)

In Eq. (2.1), ρ_f^G, ρ_f^{GO}, and, ρ_f^{GC}, are fluid densities (M/L^3) associated with q_G, q_{GO}, and q_{GC}, respectively. These densities are directionally dependent and take on the values of densities upstream. In the case of multiple solutes, the fluid density (ρ_f) and dynamic viscosity (μ_f) are expressed as functions of solute concentration, and fluid pressure as:

$$\rho_f = \rho_o\left(1 + \sum_{i=1}^{n_S} \frac{\rho_s^i - \rho_o}{\rho_o\left(C_S^i - C_o^i\right)}\left(C^i - C_o^i\right) + \beta_w\left(p - p_o\right)\right) \quad (2.2a)$$

$$\mu_f = \mu_o\left(1 + \sum_{i=1}^{n_S} \frac{\mu_s^i - \mu_o}{\mu_o\left(C_S^i - C_o^i\right)}\left(C^i - C_o^i\right)\right) \quad (2.2b)$$

where:

n_s number of solutes (dimensionless)
C^i solute concentration of species i (M/L^3)
C_o^i reference solute concentration of species i (M/L^3) corresponding to Δ_o and $:_o$
C_s^i solute concentration of species i (M/L^3) corresponding to ρ_s^i and μ_s^i
ρ_o reference fluid density (M/L^3) corresponding to C_o^i
μ_o reference fluid dynamic viscosity (M/LT) corresponding to C_o^i
ρ_s^i fluid density of species i (M/L^3) corresponding to C_s^i
μ_s^i fluid dynamic viscosity of species i (M/LT) corresponding to C_s^i
p_o reference fluid pressure (M/LT^2)
β_w fluid compressibility (LT^2/M)

In Eq. (2.2b), it is assumed that the pressure has no effects on fluid dynamic viscosity.

The storage term in Eq. (2.1) may be expanded as:

$$\frac{\partial}{\partial t}\left(\rho_f S_G \theta_G\right) = S_G \theta_G \frac{\partial \rho_f}{\partial t} + \rho_f \theta_G \frac{\partial S_G}{\partial t} + \rho_f S_G \frac{\partial \theta_G}{\partial t} \quad (2.3a)$$

which can be recast using Eqs. (2.2a) and (2.2b), so that it becomes:

$$\frac{\partial}{\partial t}\left(\rho_f S_G \theta_G\right) = S_G \theta_G \rho_o\left(\sum_{i=1}^{n_S} \frac{\rho_f - \rho_o}{\rho_o\left(C_S^i - C_o^i\right)} \frac{\partial C^i}{\partial t} + \beta_w \rho_o g \frac{\partial h}{\partial t}\right) + \rho_f \theta_G \frac{\partial S_G}{\partial t} + \rho_f S_G\left(\alpha_G \rho_o g \frac{\partial h}{\partial t}\right) \quad (2.3b)$$

where:

α_G bulk compressibility of aquifer (L^2T^2/M)

The van Genuchten functional forms of the relative permeability curve are provided by Ref. [23] as:

$$k_{rG} = S_e^{\frac{1}{2}}\left[1-\left(1-S_e^{\frac{1}{\gamma}}\right)^{\gamma}\right]^2 \tag{2.4a}$$

$$S_e = \frac{S_G - S_{Gr}}{1-S_{Gr}} = \begin{cases} \dfrac{1}{\left[1+(\alpha\psi)^{\beta}\right]^{\gamma}} & \text{for } \psi < 0 \\ 1 & \text{for } \psi \geq 0 \end{cases} \tag{2.4b}$$

where:

S_e effective water saturation (dimensionless)
S_{Gr} residual water saturation (dimensionless)
α fitting parameter (1/L)
β fitting parameter (dimensionless)
γ $1-1/\beta$ (dimensionless)
ψ pressure head (L) $=p/(\rho_0 g)$

As an alternative, k_{rG} in Eq. (2.4a) may be expressed through the use of the Brooks–Corey function as:

$$k_{rG} = S_e^b \tag{2.4c}$$

where:

b fitting parameter (dimensionless)

Equation (2.4c) is often less nonlinear than its corresponding van Genuchten function and, therefore, might benefit from the nonlinear iterations of the Richards equation. The Brooks–Corey and van Genuchten functions for the moisture retention and relative permeability characteristics may be obtained by curve fitting to laboratory measurements, by correlation to soil type, or from soils databases [24]. It should also be noted that the solution to the Richards equation for water flow assumes that air is a passive phase.

The above nonlinear functions of a variably saturated solution are, in some cases, unnecessary, as many water resources evaluations typically begin with characterization and approximation of unconfined and confined groundwater flow. To provide a simpler level of analysis, the unsaturated zone flow equation may be based on pseudo-soil functions (in lieu of Eqs. 2.4a–2.4c) that track the unconfined water table [25]. The retention curve (Eq. 2.4b) is a step function with residual saturation (zero) above the water table and a saturation of unity below [26]. When integrated in the vertical direction in a computational cell, a pseudo-constitutive relation is developed that defines the functional relationship: (1) $S_G = S_G(\psi)$ as a straight line, with $S_G = 0$ when h = cell bottom or below and $S_G = 1$ when h = cell top or above; and (2) a relative permeability of $k_{rG} = S_G$. In addition, in order to allow transfer of water vertically between computational cells in a multilayered system above the water table, the vertical direction k_{rG} is kept at unity to allow water to flow vertically under saturated conductivity conditions.

2.1.2 Overland Flow

Assuming that the inertial effects are negligible, overland flow is characterized by the two-dimensional diffusion wave approximation to the St. Venant equations governing shallow-water flow, and averaged over the depth of the flow [21, 27]. The two dimensional flow equation is written as:

$$\frac{\partial}{\partial x_i}\left(\rho_f\, d_O\, k_{rO}\, K_{ij}^O \frac{\partial(h+\delta x_3)}{\partial x_j}\right) = \frac{\partial}{\partial t}(\rho_f\, \theta_O\, h) - \rho_f^O d_O q_O - \rho_f^{OG} d_O\, q_{OG} - \rho_f^{OC} d_O\, q_{OC} \tag{2.5a}$$

where:

h hydraulic head or water surface elevation (L) $h = d_O + z_{LS}$
d_O depth of overland flow (L)
z_{LS} land surface elevation (L)
k_{rO} relative overland conductance (dimensionless)
θ_O overland porosity (dimensionless)
q_{OG} flux per unit volume of overland flow domain from groundwater (1/T) $= -q_{GO}$
q_{OC} flux per unit volume of overland flow domain from channel (1/T) $= -q_{CO}$
K_{ij}^O overland conductance tensor (L/T)
q_O volumetric flux per unit volume (1/T) of the overland domain and represents sources and/or sinks of water.

In Eq. (2.5a), ρ_f^O, ρ_f^{OG}, and ρ_f^{OC}, are fluid densities (M/L^3) associated with q_O, q_{OG}, and q_{OC}, respectively. These densities are directionally dependent and take on the values of densities upstream. The overland conductance term, K_{ij}^O, results from manipulation of the St. Venant equations and is given for the Manning equation as [27]:

$$K_{ij}^O = \frac{k_n d_O^{\frac{2}{3}}}{n_{ij}}\left(\frac{\partial h}{\partial s}\right)^{-\frac{1}{2}} \tag{2.5b}$$

where:

k_n Manning's conversion factor ($L^{1/3}/T$)
n_{ij} Manning's roughness coefficient tensor for overland flow (dimensionless)
s length along the direction maximum local slope (L)

In Eq. (2.5b), the friction slope may alternatively be expressed by the Chezy or Darcy–Weisbach equations [28]. The overland-domain porosity and relative conductance terms are discussed in Sects. 2.5.2 and 2.5.3, respectively.

2.1.3 Channel Flow and Surface-Water Features

All features of the land surface can conceptually be represented by the overland flow surface, including surface-water features such as ponds, lakes, reservoirs, rivers, streams and canals, by the use of a sufficiently small discretization (see Sect. 2.6). However, there is a practical issue of scale in regional simulations, whereby a minimum limit must be set on the areal computational cell size of the overland flow surface as well as of the subsurface layers. Therefore, to accurately simulate surface-water features, which are smaller than the associated computational cell dimensions, and to convey water through canals or conveyance structures (whose widths are much finer than the computational cell scale), a surface-water feature/channel flow layer is added to the surficial model layer. This layer is characterized by a network of one-dimensional channels/features which communicate water within the network, as well as between it and the overland flow and subsurface domains. Flow through a network of rivers and channels is characterized by the one-dimensional diffusion wave approximation to the St. Venant equations, which is derived in a similar manner to its two-dimensional counterpart and is expressed as:

$$(1-S_{\mathrm{Str}})\frac{\partial}{\partial l}\left(\rho_{\mathrm{f}}\,k_{\mathrm{rC}}\,K^{\mathrm{C}}\frac{\partial(h+\delta x_3)}{\partial l}\right)+\rho_{\mathrm{f}}\,k_{\mathrm{Str}}S_{\mathrm{Str}}f_{\mathrm{Str}}(h)=\frac{\partial}{\partial t}(\rho_{\mathrm{f}}\,B_{\mathrm{C}}\theta_{\mathrm{C}}\,h)$$
$$-\rho_{\mathrm{f}}^{\mathrm{C}}A_{\mathrm{C}}q_{\mathrm{C}}-\rho_{\mathrm{f}}^{\mathrm{CO}}A_{\mathrm{C}}\,q_{\mathrm{CO}}-\rho_{\mathrm{f}}^{\mathrm{CG}}A_{\mathrm{C}}\,q_{\mathrm{CG}} \qquad (2.6a)$$

where:

- S_{Str} structure unit function (dimensionless), equals unity along the length when a hydraulic structure is present, 0 otherwise
- k_{Str} structure operation coefficient (dimensionless)
- f_{Str} structure discharge per unit length (L^2/T)
- B_{C} top width of channel (L)
- h hydraulic head or water surface elevation of the channel (L)$h=d_{\mathrm{C}}+z_{\mathrm{C}}$
- d_{C} depth of channel flow (L)
- z_{C} channel bottom elevation (L)
- l length along the direction of flow (L)
- k_{rC} relative channel conductance (dimensionless)
- θ_{C} channel porosity (dimensionless)
- K^{C} conductance term along the length of the channel (L^3/T)
- A_{C} channel wetted cross-sectional area (L^2)

q_{CO} flux per unit volume of channel flow domain from the overland flow domain (1/T)$=-q_{CO}$

q_{CG} flux per unit volume of channel flow domain from the subsurface (1/T)$=-q_{GC}$

q_C volumetric flux per unit volume (1/T) of the overland domain and represents sources and/or sinks of water

In Eq. (2.6a), ρ_f^C, ρ_f^{CO}, and ρ_f^{CG}, are fluid densities (M/L^3) associated with q_C, q_{CO}, and q_{CG}, respectively. These densities are directionally dependent and take on the values of densities upstream. Dimensions of this equation are volumetric mass flux per unit length of channel (M/LT), integrated over the channel's cross-sectional area of flow. The channel-domain porosity and relative conductance terms are discussed in Sects. 2.5.2 and 2.5.3, respectively.

The friction slope may be approximated by Manning's formula to provide the channel conductance terms as:

$$K^C = \frac{k_n A_C^{\frac{5}{3}}}{P_C^{\frac{2}{3}} n_C} \left(\frac{\partial h}{\partial l} \right)^{-\frac{1}{2}} \tag{2.6b}$$

where:

n_C Manning's roughness coefficient for channel (dimensionless)

P_C wetted channel perimeter (L)

Chezy's equation, or the Darcy–Weisbach relation [28] with appropriate parameters may also be used for the channel conductance term. Details for different channel geometries may be found in typical open-channel references [28, 29].

2.2 Solute Transport

2.2.1 Subsurface Transport

The partial differential equation governing three-dimensional transport of a dissolved contaminant species, k, in a variably saturated porous medium may be written in the primitive form as follows [19]:

$$\begin{aligned}\frac{\partial}{\partial x_i}\left(D_{ij}^k \frac{\partial C^k}{\partial x_j} \right) - \frac{\partial}{\partial x_i}(v_i C^k) = \frac{\partial}{\partial t}(S_w \theta_{eG} C^k) + \frac{\partial}{\partial t}(\rho_B^G C_s^k) \\ + \lambda_\alpha^k \theta_{eG} S_w C^k + \lambda_s^k \rho_B^G C_s^k - q_G C_G^{*k} - \Gamma_{GInt}^k \\ - \sum_{m=1}^{N_P} \xi_{km} \lambda_w^m \theta_{eG} S_w C^m - \sum_{m=1}^{N_P} \xi_{km} \lambda_s^m \rho_B^G C_s^m\end{aligned} \tag{2.7a}$$

where:

D_{ij}^{k} apparent hydrodynamic dispersion tensor of component k (L^2/T)
C^{k} solute concentration of component k (M/L^3)
C_{s}^{k} concentration of component k adsorbed to the soil (M/M_{soil})
v_i Darcy velocity along the ith direction (L/T)
θ_{eG} effective porosity in groundwater domain (dimensionless)
ρ_{B}^{G} bulk density of soil in the subsurface domain (M/L^3),
λ_{w}^{k} first-order decay coefficient for component k in water (1/T)
λ_{s}^{k} first-order decay coefficient for component k in soil (1/T)
C_{G}^{*k} solute concentration of component k of the sources (or sinks) within the subsurface domain (M/L^3)
I_{GInt}^{k} mass transfer rate of component k between subsurface and other domains (M/L^3 T)
N_P number of parent chemicals or solute k (dimensionless)
ξ_{kj} fraction of parent component j transforming into component k (dimensionless)

C_{G}^{*k} is specified in the case of injection. In the case of extraction, it is equal to concentration within the subsurface domain at the point of extraction. The hydrodynamic dispersion tensorial components are given by [18]:

$$D_{ij}^{k} = a_{ijmn} \frac{v_m v_n}{V} + T_{ij}^{*} D_{d}^{k} \tag{2.7b}$$

which, for isotropic media, can be written as [18, 30]:

$$D_{ij}^{k} = a_{T} V \delta_{ij} + (a_{L} - a_{T}) \frac{v_i v_j}{V} + \delta_{ij} T^{*} D_{d}^{k} \tag{2.7c}$$

where:

a_{ijmn} dispersivity tensor (L)
V magnitude of the velocity vector (L/T)
T_{ij}^{*} tortuosity tensor (dimensionless)
D_{d}^{k} molecular diffusion coefficient for component k (L^2/T)
α_L longitudinal dispersivity (L)
α_T transverse dispersivity (L)
δ_{ij} Kronecker's delta (dimensionless)

2.2.2 Overland Transport

The partial differential equation governing transport of a dissolved contaminant species, k, on the overland flow surface (vertically averaged) or along a channel segment (averaged across the channel cross section) may be written in the primitive form as follows [19]:

$$\frac{\partial}{\partial x_i}\left(d_O D_{ij}^k \frac{\partial C^k}{\partial x_j} \right) - \frac{\partial}{\partial x_i}\left(d_O v_i C^k \right) = \frac{\partial}{\partial t}\left(d_O \theta_O C^k \right) + \frac{\partial}{\partial t}\left(S_O \rho_B^O C_s^k \right)$$
$$+ \lambda_\alpha^k \theta_O d_O C^k + \lambda_s^k S_O \rho_B^O C_s^k - q_O C_O^{*k} - \Gamma_{OInt}^k$$
$$- \sum_{m=1}^{N_P} \xi_{km} \lambda_w^m \theta d_O C^m - \sum_{m=1}^{N_P} \xi_{kj} \lambda_s^m S_O \rho_B^O C_s^m \qquad (2.8)$$

where:

S_O equivalent sediment depth (L)
θ_O effective porosity in overland domain (dimensionless)
ρ_B^O bulk density of sediment in the overland domain (M/L^3)
C_O^{*k} solute concentration of component k of the sources (or sinks) within the overland domain (M/L^3)
I_{OInt}^k mass transfer rate of component k between overland and other domains (1/T)

2.2.3 Channel Transport

The partial differential equation governing transport of a contaminant species, *k*, along a channel segment (averaged across the channel cross section) may be written in the primitive form as follows [19]:

$$\frac{\partial}{\partial l}\left(A_C D_{ll}^k \frac{\partial C^k}{\partial l} \right) - \frac{\partial}{\partial l}(A_C v_l C^k) = \frac{\partial}{\partial t}(A_C \theta_C C^k) + \frac{\partial}{\partial t}(P_C S_C \rho_C^B C_s^k)$$
$$+ \lambda_\alpha^k \theta_C A_C C^k + \lambda_s^k P_C S_C \rho_C^B C_s^k - q_C C_C^{*k} - \Gamma_{CInt}^k$$
$$- \sum_{m=1}^{N_P} \xi_{km} \lambda_w^m \theta_C A_C C^m - \sum_{m=1}^{N_P} \xi_{km} \lambda_s^m P_C S_C \rho_C^B C_s^m \qquad (2.9)$$

where:

A_C wetted cross-sectional area of the channel segment (L^2)
P_C wetted perimeter of the channel segment (L)
θ_C effective porosity in the channel domain (dimensionless)
ρ_B^C bulk density of sediment in the channel domain (M/L^3)
C_C^{*k} solute concentration of component k of the sources (or sinks) within the channel domain (M/L^3)
I_{CInt}^k mass transfer rate of component k between the channel and other domains (1/T).

2.3 Boundary and Initial Conditions

In order to solve the flow and transport equations, initial and boundary conditions must be supplied. Initial conditions are prescribed as spatial distributions of hydraulic head and concentrations at time = 0. Typical boundary conditions for the flow and transport equations are given below:

$$h(x_{iB}, t) = H(x_{iB}, t) \tag{2.10a}$$

$$Q_B(x_{iB}, t) = -K_{ij} \frac{\partial H(x_{iB}, t)}{\partial x_j} n_i \tag{2.10b}$$

$$\left(v_i C^k (x_{iB}, t) - D_{ij} \frac{\partial C^k (x_{iB}, t)}{\partial x_j} \right) n_i = \left(v_{iB} C_B^k (x_{iB}, t) - D_{ijB} \frac{\partial C_B^k (x_{iB}, t)}{\partial x_j} \right) n_i \tag{2.10c}$$

$$C^k (x_{iB}, t) = C_B^k (x_{iB}, t) \tag{2.10d}$$

$$M_B^k (x_{iB}, t) = -D_{ijB} \frac{\partial C_B^k (x_{iB}, t)}{\partial x_j} n_i \tag{2.10e}$$

where:

H specified hydraulic head at the boundary at x_{iB} (L)
x_{iB} boundary coordinates (L)
Q_B volumetric water flux per unit area (L)
K_{ij} hydraulic conductivity or conductance (L/T) in Eqs. (2.1), (2.5a), and (2.6a)
n_i unit vector (dimensionless), positive inward
C_B^k specified concentration of solute k at the boundary (M/L^3)
M_B^k dispersive mass flux of species k per unit area (M/L^3 T)
D_{ijB} dispersion coefficient tensor at the boundary (L^2/T)
v_{iB} specified fluid velocity at the boundary (M/L^3)

Equations (2.10a) and (2.10b) represent prescribed hydraulic head and fluid flux, respectively, for the flow equation. Equation (2.10c) is a general boundary condition for the transport equation which states that the sum of advective and dispersive fluxes at the boundary is the same as the sum of the same fluxes immediately within the boundary. Under special conditions [18], Eq. (2.10c) becomes Eqs. (2.10d) and (2.10e) which are used for prescribing concentration and dispersive flux, respectively. These equations are applicable to all three domains.

For the overland and channel domains, surface-water boundary conditions also include zero-depth-gradient and critical-depth conditions. For the overland flow domain, the discharge rates per unit width normal to the flow direction, Q_i, at the zero-depth gradient and critical-depth boundaries, respectively, along the *i*th direction based on the Manning equation are given by:

Zero-depth gradient:

$$Q_i = \frac{1}{n_i} d_{\mathrm{O}}^{\frac{5}{3}} \sqrt{S_{\mathrm{b}}} \tag{2.10f}$$

Critical depth:

$$Q_i = \sqrt{g d_{\mathrm{O}}^{3}} \tag{2.10g}$$

where:

Q_i is the discharge per unit width normal to the flow direction (L^2/T)
S_{b} bed slope (dimensionless) at the zero-depth gradient boundary

The zero-depth-gradient condition forces the depth gradient at the boundary to be zero where flow is relatively steady. The critical-depth condition makes the depth at the boundary equal to the critical depth, as would occur at free-fall boundaries.

For the channel flow domain, the discharge rates along the flow direction, Q_C, at the zero-depth gradient and critical depth boundaries, respectively, based on the Manning equation are given by:

Zero-depth gradient:

$$Q_{\mathrm{C}} = \frac{1}{n_i} \frac{d_{\mathrm{C}}^{\frac{5}{3}}}{P_{\mathrm{C}}^{\frac{2}{3}}} \sqrt{S_{\mathrm{b}}} \tag{2.10h}$$

Critical depth:

$$Q_{\mathrm{C}} = \sqrt{\frac{g A_{\mathrm{C}}^{3}}{B_{\mathrm{C}}}} \tag{2.10i}$$

2.4 Interdomain Connections

Hydraulic connections between the subsurface, overland, and channel domains are represented by the unit interactive fluxes presented in Eqs. (2.1), (2.5a), and (2.6a). The overland/subsurface interaction term, q_{GO}, is the unit flux across the ground surface, from overland to the subsurface, which is computed as:

$$Q_{\mathrm{GO}} = -Q_{\mathrm{OG}} = q_{\mathrm{GO}} V_{\mathrm{G}} = -k_{r\mathrm{GO}} K_{\mathrm{GO}} A_{\mathrm{GO}} (h_{\mathrm{G}} - h_{\mathrm{O}}) \tag{2.11a}$$

where:

Q_{GO} flux across the area of the interface from overland to subsurface (L^3/T)
Q_{OG} flux across the area of the interface from subsurface to overland (L^3/T)
V_G subsurface elementary volume (L^3)
k_{rGO} relative leakance at the interface between overland and subsurface (dimensionless)
K_{GO} leakance across the interface area between overland and subsurface (1/T)
A_{GO} area at the interface between overland and subsurface (L^2)
h_G head in the subsurface domain (L)
h_O head in the overland domain (L)

The relative leakance term, k_{rGO}, varies from zero at the land surface elevation to unity at the top of the depression storage height above the land surface (see Sect. 2.3.2). It accounts for a fraction of the total area that is wet when water is within the depression height. V_G is typically the volume of a computational cell or element.

The channel/subsurface interaction term, q_{GC}, is the unit flux from channel to the subsurface, which is computed as:

$$Q_{GC} = -Q_{CG} = q_{GC} V_G = -k_{rGC} K_{GC}^{eff} L_C P_C (h_G - h_C) \tag{2.11b}$$

where:

Q_{GC} flux across the area of the interface from channel to subsurface (L^3/T)
Q_{CG} flux across the area of the interface from subsurface to channel (L^3/T)
k_{rGC} relative leakance at the interface between channel and subsurface (dimensionless)
K_{GC}^{eff} effective leakance across the interface area between channel and subsurface (1/T)
L_C length of channel segment (L)
P_C wetted perimeter of channel segment (L)
h_C head in the channel domain (L)

The relative leakance term, k_{rGC}, varies from zero at the land surface elevation to unity at the top of the depression storage as discussed in Sect. 2.3.3. For channels, the leakance term K_{GC}^{eff} may be a constant or may vary with flow depth within the section. For instance, a concrete lined channel may have grassy overflow storage areas, the composite being treated as a channel section. For bed conductivity varying with flow depth, the term K_{GC}^{eff} is computed as:

$$K_{GC}^{eff} = \frac{\int_{\zeta_{Start}}^{\zeta_{End}} K(\zeta) d\zeta}{b \int_{\zeta_{Start}}^{\zeta_{End}} d\zeta}, \zeta = \zeta(d) \tag{2.11c}$$

where:

$K(\cdot)$ leakance as a function of ζ (1/T)
ζ distance along submerged channel cross section (L)
d flow depth (L)
b thickness of channel bed (L)

K_{GC}^{eff} is length-weighted averaged from ζ_{Start} to ζ_{End} which is a function of channel depth. Equation (2.11c) is applicable to varying channel sediment leakances/conductivities that commonly occur in engineered systems.

The interaction term between the channel and overland domains is expressed by the equations for flow over a wide rectangular weir. Two typical situations may occur—one for a free-flowing weir, and one for a submerged weir—each represented by its own flow relationship. Note that the flow direction could be reversed, with the channel overflowing its banks. For free-flowing conditions across the channel banks, the flow is expressed as [31]:

$$\begin{aligned} Q_{OC} &= q_{OC} A_O h_O = C_d \frac{4L_C}{3}\sqrt{2g}(h_u - Z_{Bank})^{1.5}, h_d < Z_{Bank} \\ &= C_d \frac{4L_C}{3}\sqrt{2g}(h_u - h_d)(h_u - Z_{Bank})^{0.5}, h_d > Z_{Bank} \end{aligned} \quad (2.11d)$$

where:

Q_{OC} flux across the total length of channel banks to/from the overland flow domain (L^3/T)
C_d weir discharge coefficient (dimensionless)
h_u upstream head between the channel and overland domains (L)
Z_{Bank} bank elevation (L) which may be at or above the overland flow surface elevation
h_d downstream head between the two systems (L)

Equation (2.11d) is used for the interaction term, depending on overbank flow conditions, with q_{OC} equal to zero when h_u is below Z_{Bank}. Note that these flux relations are symmetric, and the same flow occurs from the overland flow surface to the channel segment for a given gradient, as would occur from the channel segment if the direction of the gradient were reversed.

A typical form of mass transfer rate of component k to a given domain from two other domains (Eqs. 2.7a, 2.8, and 2.9) is given below:

$$\Gamma_{GInt}^k = m_{GO}^k + m_{GC}^k \quad (2.12a)$$

$$\Gamma_{OInt}^k = m_{OG}^k + m_{OC}^k \quad (2.12b)$$

$$\Gamma_{CInt}^k = = m_{CO}^k + m_{CG}^k \quad (2.12c)$$

The mass influx rate per unit volume between domains I and J may be written as [18]

$$m_{IJ}^{k} = \frac{A_{IJ}}{V_I}\left(v_{IJ} C_{J^+/I^-}^{k} - D_{IJ}^{k} \frac{\partial C^k}{\partial n}\bigg|^{IJ} \right) \quad (2.12d)$$

where:

m_{IJ}^{k} mass influx rate per unit area from domain J to domain I of component k (M/L^2T)
A_{IJ} area through which mass influx passes from domain J to domain I (L^2)
V_I normalization volume in domain I (L^3)
v_{IJ} water flow rate per unit area from domain J to domain I (L/T)
C_{J^+/I^-}^{k} directionally dependent concentration of component k in domain J, if v_{IJ} is positive; in domain I, if v_{IJ} is negative (M/L^3), and
D_{IJ}^{k} effective dispersion coefficient of component k between domains I and J (L^2/T)

Note that in Eq. (2.12d), C_{J^+/I^-}^{k} depends on the flow direction and takes the concentration from the upstream domain, and that the dispersive flux represents the total dispersive flux normal to the interface area between the two domains. Equation (2.12d) may be written in an approximate finite-difference form as:

$$m_{IJ}^{k} = \frac{A_{IJ}}{V_I}\left(v_{IJ} C_{J^+/I^-}^{k} - D_{IJ}^{k} \left[\frac{C_I^k - C_J^k}{b_{IJ}} \right] \right) \quad (2.12e)$$

where:

b_{IJ} distance between two centroids in domains I and J (L)
C_I^k Solute concentration of component k at cell I (M/L^3)

2.5 *Other Relevant Processes*

2.5.1 Interception and Evapotranspiration

The simulation of interception and evapotranspiration is adapted from Refs. [10, 32]. Interception and comprehensive evapotranspiration are simulated as mechanistic processes governed by plant and climatic conditions as noted by [33, 34]. Interception is the process involving retention of a certain amount of precipitation on the leaves, branches, and stems of vegetation or on buildings and structures in urban areas. The interception process is simulated by the bucket model, in which precipitation in excess of interception storage and evaporation from interception reaches the ground surface. This is computed by keeping track of interception storage for each areal location during a given time period (a time step of Δt). The interception storage, S_{int}, varies between zero and storage capacity S_{int}^{max}:

$$0 \le S_{int} \le S_{int}^{max} \tag{2.13a}$$

The storage capacity, which depends on the vegetation type and its stage of development, is calculated from [33]:

$$S_{int}^{max} = C_{int}\, \text{LAI} \tag{2.13b}$$

where:

S_{int} canopy storage (L)
S_{int}^{max} canopy storage capacity (L)
C_{int} canopy storage parameter (L)
LAI leaf area index (dimensionless)

Note that LAI represents the cover of leaves over a unit area of the ground surface, and may be prescribed in a time-dependent manner. The interception storage is filled by rainfall and depleted by evaporation. For each time increment Δt, the actual interception storage (S_{int}) is calculated as follows:

$$S_{int}^{*} = \min(S_{int}^{max}, S_{int}^{o} + P_P\, \Delta t) \tag{2.13c}$$

$$E_{can} = \min(S_{int}^{*}, E_P\, \Delta t) \tag{2.13d}$$

$$S_{int} = S_{int}^{*} - E_{can}\, \Delta t \tag{2.13e}$$

where:

S_{int}^{o} previous time step canopy storage (L)
S_{int}^{*} intermediate canopy storage (L)
P_P precipitation rate (L/T)
E_{can} canopy evaporation (L/T)
E_P reference evapotranspiration (L/T)
Δt time increment (T)

E_P may be derived from pan measurements or computed from vegetation and climatic factors (radiation, wind, humidity, and temperature) using the Penman–Monteith equation [35] for vegetated surfaces or a bare-ground evaporation formula [36] for nonvegetated surfaces, as a function of temperature, wind, and humidity conditions. The reference evapotranspiration is computed efficiently at the start of a simulation, for further use in determining the actual evapotranspiration. The reduced rainfall rate ($\hat{P}_P$) which reaches ground surface after interception is given by

$$\hat{P}_P = P_P - \frac{S_{Int} - S_{Int}^{o}}{\Delta t} - E_{can} \ge 0 \tag{2.13f}$$

Similarly, the potential for evapotranspiration from the soil surface and below is reduced by the canopy evapotranspiration term, which is carried through in the subsequent discussions. Evapotranspiration is rigorously modeled as a combination of plant transpiration and of evaporation, and affects nodes in both surface and subsurface flow domains.

Transpiration from vegetation occurs within the root zone of the subsurface which may be above or below the water table and may involve several nodal layers. The rate of transpiration for computational cell I (T_{pI}) in the subsurface domain (see Sect. 2.6) is estimated using the following relationship that distributes the net capacity for transpiration among various factors [33]:

$$T_{pI} = f_1(\text{LAI})\, f_2(\theta_I)\, \text{RDF}_I\,(E_p - E_{can}) \tag{2.14a}$$

The vegetation-dependent function, f_1 (LAI), is expressed as:

$$f_1(\text{LAI}) = \max\left\{0,\ \min\left[1, C_2 + C_1\,\text{LAI}\right]\right\} \tag{2.14b}$$

where:

T_{pI} rate of transpiration for computational cell I (L/T)
C_1, C_2 fitting parameters (dimensionless)

The moisture-content-dependent function, $f_2(\theta)$, is expressed as:

$$f_2(\theta_I) = \begin{cases} 0 & \text{for } 0 \le \theta_I \le \theta_{wp} \\ 1 - \left[\dfrac{\theta_{fc} - \theta_I}{\theta_{fc} - \theta_{wp}}\right]^{C_3} & \text{for } \theta_{wp} < \theta_I \le \theta_{fc} \\ 1 & \text{for } \theta_{fc} < \theta_I \le \theta_o \\ \left[\dfrac{\theta_{an} - \theta_I}{\theta_{an} - \theta_o}\right]^{C_3} & \text{for } \theta_o < \theta_I \le \theta_{an} \\ 0 & \text{for } \theta_{an} \le \theta_I \end{cases} \tag{2.14c}$$

where:

C_3 fitting parameters (dimensionless)
θ_{wp} moisture content at wilting point (dimensionless)
θ_{fc} moisture content at field capacity (dimensionless)
θ_o moisture content at oxic limit (dimensionless)
θ_{an} moisture content at anoxic limit (dimensionless).

RDF_I is the value of the root distribution function (dimensionless) for computational cell I in the subsurface domain (see Sect. 2.6) which may be prescribed in a time-varying manner. The root zone term is defined by the relationship:

$$\mathrm{RDF}_{\mathrm{I}} = \frac{\int_{z_1}^{z_2} r_{\mathrm{F}}(z)\mathrm{d}z}{\int_{0}^{L_R} r_{\mathrm{F}}(z)\mathrm{d}z} \qquad (2.14\mathrm{d})$$

where:

z depth coordinate from the soil surface (L)

$r_{\mathrm{F}}(z)$ root extraction function (dimensionless) which typically varies logarithmically with depth

L_{R} effective root length (L)

The function f_1 correlates the transpiration from a cell (T_{pI}) with the leaf area index (LAI) in a linear fashion. The function f_2 correlates T_{pI} with the moisture state of the cell containing the roots and is an extension of the function of [33] to account for root processes in greater detail. Below the wilting-point moisture content, transpiration is zero; transpiration then increases to a maximum at the field-capacity moisture content. This maximum is maintained up to the oxic moisture content, beyond which the transpiration decreases to zero at the anoxic moisture content. When available, moisture is larger than the anoxic moisture content, the roots become inactive due to lack of aeration [37]. In general, $f_2(\theta_1)$is a nonlinear function of θ_1, though the ramp function is linear when $C_3 = 1$. Values of $\mathrm{RDF}_{\mathrm{I}}$ should be prescribed such that the following constraint holds among each vertical set of cells at any time:

$$\mathrm{RDF} = \sum_{I=1}^{n_{\mathrm{R}}} \mathrm{RDF}_{\mathrm{I}} = 1 \qquad (2.14\mathrm{e})$$

where:

n_{R} number of cells that contribute to the total root zone for each areal location (dimensionless)

Equation (2.14e) is obtained by expressing Eq. (2.14d) over the layers of cells in a vertical column containing the roots. As a practical matter, however, the value of *RDF* at any areal location may be less than one to account for ineffective roots. For each time step, the transpiration (T_{p}) over the effective root length is calculated as the sum of the transpiration from each of the cells at depth, as

$$T_{\mathrm{p}} = \sum_{I=1}^{n_{\mathrm{RT}}} T_{\mathrm{pI}} \qquad (2.14\mathrm{f})$$

where:

n_{RT} number of cells that lie within the depth interval from 0 to L_R at any areal location (dimensionless)

Two models are provided for evaporation. The first model assumes that evaporation occurs if the reference evapotranspiration has not been removed by the above processes of canopy evaporation and plant transpiration. Therefore, evaporation from the soil surface and subsurface soil layers is estimated as follows:

$$E_{s_I} = \alpha_I^* \, (E_P - E_{can} - T_p)\mathrm{EDF}_I \tag{2.15a}$$

The second model assumes that evaporation occurs along with transpiration, resulting from energy that penetrates the vegetation cover and is expressed as:

$$E_{s_I} = \alpha_I^* \, (E_P - E_{can})[1 - f_1(\mathrm{LAI})]\mathrm{EDF}_I \tag{2.15b}$$

in which α_I^* is the wetness factor given by

$$\alpha_I^* = \begin{cases} \left[\dfrac{\theta_I - \theta_{e2}}{\theta_{e1} - \theta_{e2}}\right] & \text{for } \theta_{e2} \le \theta_I \le \theta_{e1} \\ 1 & \text{for } \theta_I > \theta_{e1} \\ 0 & \text{for } \theta_I < \theta_{e2} \end{cases} \tag{2.15c}$$

where:

θ_{e1} moisture content at the end of the energy-limiting stage (above which full evaporation can occur) (dimensionless)

θ_{e2} limiting moisture content below which evaporation is zero (dimensionless)

Equation (2.15c) expresses the moisture availability term for the subsurface domain. For the overland domain, α_I^* is calculated as varying between unity when the elevation of flow is at or above depression storage (LS+h_{ds}) and zero for a flow elevation at the land surface (LS), thus representing the reduced evaporative area of available water on the overland flow node within the depressions.

The term EDF_I in Eq. (2.15a) or (2.15b) is the evaporation distribution function among a vertical set of nodes that includes the overland and subsurface flow domains. Two alternative conceptualizations are provided for EDF_I. For the first model, it is assumed that the capacity for evaporation ($(E_p - E_{can} - T_p)$ in Eq. (2.15a) or $(E_p - E_{can})(1 - f_1(\mathrm{LAI}))$ in Eq. (2.15b)) decreases with depth below the surface (subject to available moisture) due to the reduction of energy penetration in the soil. Therefore, an appropriate EDF_I for each cell layer may be prescribed as a function of its depth from land surface. For the second model, the capacity for evaporation is met from the land surface downward to a prescribed extinction depth (B_{soil}).

2.5.2 Depression and Storage Exclusion

The flow equations for the two-dimensional overland flow (runoff) and the one-dimensional channel (surface-water feature) flow domains (Eqs. (2.5a) and (2.6a),

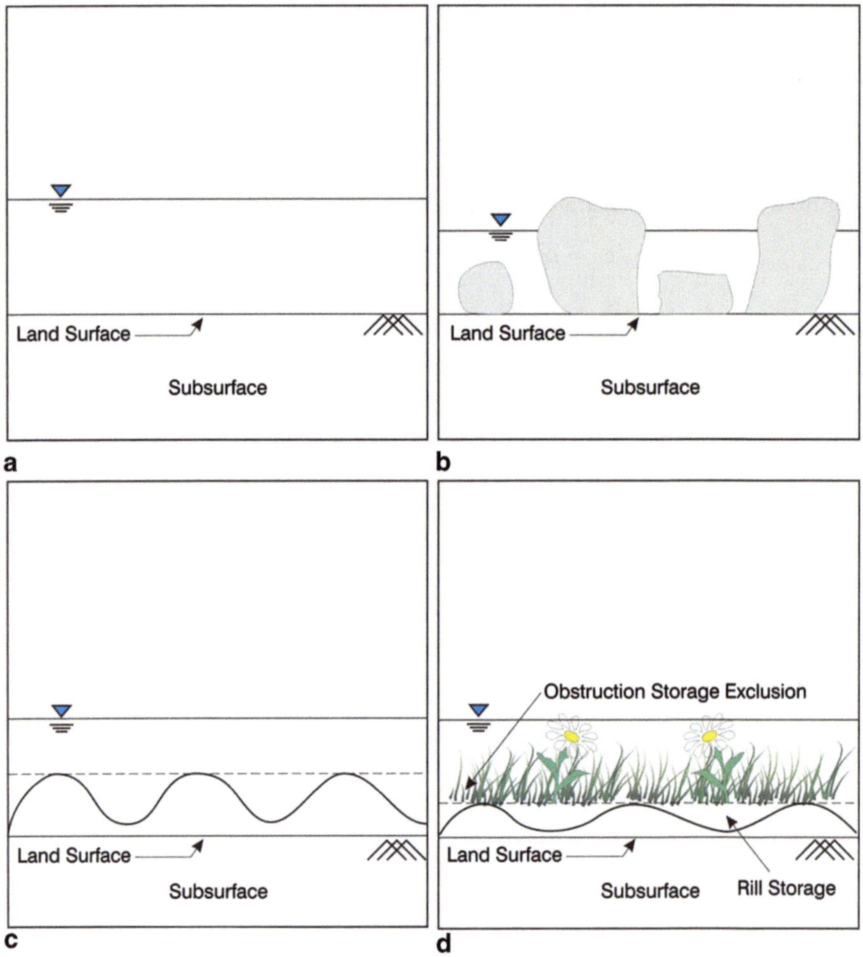

Fig. 2.3 Different types of storage in a channel, (**a**) ideal flat plane, (**b**) unlined riverbed, or natural stream, (**c**) area with depression storage, and (**d**) grassy channel

respectively) are formulated to accommodate urban or agricultural settings for analyses of large spatial scales [10, 32]. Both equations include terms for depression storage and for obstruction storage exclusion. In Fig. 2.3a, in which flow occurs over an ideal flat plane, the porosity terms (θ_O and θ_C) are unity. For unlined river beds or natural streams, the setting is much different as shown in Fig. 2.3b, with flow occurring between the obstructions in an averaged sense over the cross-sectional area. The situation shown in Fig. 2.3b is also applicable to the two-dimensional overland flow surface in urban environments with man-made obstructions such as buildings. The full area becomes available for flow and storage of water only when the water level is high enough to completely cover the obstructions. The storage capacity that is reduced by the presence of these features is termed "obstruction storage exclusion." Figure 2.3c shows the concept of "depression storage" which includes rills,

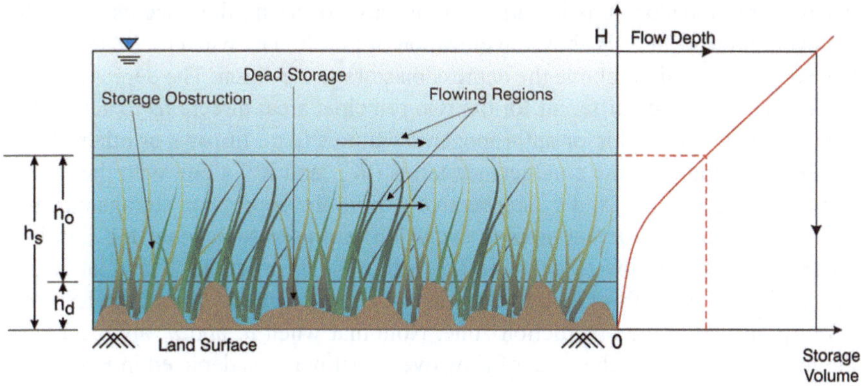

Fig. 2.4 Depression storage and obstruction storage exclusion

furrows, and other detention features. Depression storage is an important factor to account for ground unevenness as compared to the scale of the computational cell. This is the amount of storage that must be filled prior to the onset of lateral flow. Microtopographic relief is included in depression storage and can have a substantial impact on hydrograph shape [38]. For agricultural plots or grassy channels (as shown in Fig. 2.3d), the effects of depressions as well as of storage exclusion must be taken into account in the model's storage term as well as in the horizontal flow conductance term.

The storage effects of depression storage and obstruction storage exclusion may be approximated by assuming that the geometry of depressions and exclusions combined has a maximum elevation and that the horizontal area covered by surface water varies between zero and full area as the water level rises from land surface (LS, defined here as the bottom of the depressions) up to this maximum elevation (LS+h_d+h_o) as shown in Fig. 2.4. The variation of area covered by surface water with depth between LS and LS+h_d+h_o is expressed as a "volumetric height" defined as the height from LS of an equivalent volume of water without depressions or obstructions. The "volumetric height" is then used in the storage term of Eqs. (2.5a) and (2.6a) instead of the surface-water depth, to account for the reduced available storage volumes. The porosity terms are equal to the ratio of reduced volume over unobstructed volume between LS and LS+h_d+h_o.

2.5.3 Relative Conductance in Overland and Channel Domains

Horizontal flow conductance terms are also affected by the presence of depressions and storage obstruction features due to larger frictional resistance and small scale energy dissipation over the obstruction surface [32]. The depression storage height above land surface ($LS+h_d$) is used as the reference elevation for flow depth calculation in the Darcy frictional terms (left-hand-side) of Eqs. (2.5a) and (2.6a) when depression storage is included. In this case, overland flow occurs only when the overland flow head is above an elevation of LS+h_d, i.e., when the water level in the overland domain is above the depression storage elevation. The depression storage elevation may be different for the two principal areal directions to account for natural microtopographic or anthropogenic features (e.g., furrows or other plowed features). In addition, the conductance terms (K_{ij}^{O} and K^{C}) along with the relative conductance terms (k_{rO} and k_{rC}, respectively, from the depression storage elevation to the height of obstruction storage exclusion) are used to account for additional resistance losses for low-flow conditions within the obstructions. The relative conductance terms vary from zero to unity as the flow elevation varies from LS$+h_d$ to LS$+h_d+h_o$ within the obstruction zone. Note that when h_d and h_o approach zero, the formulation approaches that of flow over a flat plane as depicted in Fig. 2.3a.

2.5.4 Hydraulic Structures and Operational Rules

A hydraulic structure is a structure submerged or partially submerged in any body of water, which disrupts the natural flow of water. Hydraulic structures are used to divert, restrict, stop, or otherwise manage the natural flow of water. A dam, for instance, is a type of hydraulic structure used to hold water in a reservoir as potential energy. Another example is a weir which is a type of hydraulic structure which can be used to pool water for irrigation, or establish control of the bed (grade control). Other hydraulic structures include: levees, gates, culverts, manholes, drop structures, pumping stations, and bridges. Hydraulic structures may be present within a channel reach, and require appropriate quantification to correctly depict a river or stream system that has been influenced by development. A hydraulic structure within a channel reach can be accommodated in a general manner if its flow rate (Q) versus hydraulic head (h) type of relationship for flow over the structure length is known. When the conditions at the upstream and downstream ends of a structure are known, the term f_{Str} in Eq. (2.6a) may be expressed as:

$$\begin{aligned} f_{Str} &= -Q_{Str}(h)\,\delta(l-l_{UStr}) \\ &= +Q_{Str}(h)\,\delta(l-l_{DStr}) \end{aligned} \tag{2.16}$$

where:

Q_{Str}	discharge rate (L^3/T) of the structure as a function of head (h)
δ (C)	Dirac delta function (1/L)
l_{UStr}	upstream reference location of the structure (L)
l_{DStr}	downstream reference location of the structure (L)

Note that the hydraulic head term (h) in the expression for Q_{Str} could consist of an upstream head, a downstream head, or a combination of the two at appropriate locations.

The regulation of structures is a common practice among many waterways and may be simulated via a set of rules. Rules for operating structures include: (1) specifying the time period during which specific rules apply to a structure; (2) specifying the maximum flow rate through a structure (independent of the structure's stage–discharge relationships); (3) specifying a trigger value of head/flux at any surface or subsurface location, above/below which a structure is opened/closed (two-way rules may also be provided); and (4) specifying a range of head/flux values above/below the trigger (as the case may be) over which the structure is linearly opened/closed. A rule for flow over a structure is applied as a multiplier k_{Str}, to the flow conditions of the structure in Eq. (2.6a). Thus, when $k_{Str}=1$, the associated structure is in a fully "on" state, while when $k_{Str}=0$, the associated structure is completely off. A partially open structure may exist when water levels are within the prescribed range of the trigger value that fully opens or closes the structure.

2.6 Discretizations

The development of models for typical integrated multiple-domain hydrologic systems necessitates the use of numerical models. In a numerical model, each domain is replaced by a discretized domain consisting of an array of contiguous computational cells, nodes, and associated finite difference blocks (or cells) or finite elements. The former is the focus of this chapter.

With the finite-difference approach, the subsurface domain may be discretized using a block-centered finite-difference scheme. As shown in Fig. 2.5, the subsurface grid may be distorted vertically to reflect hydrostratigraphic boundaries. This

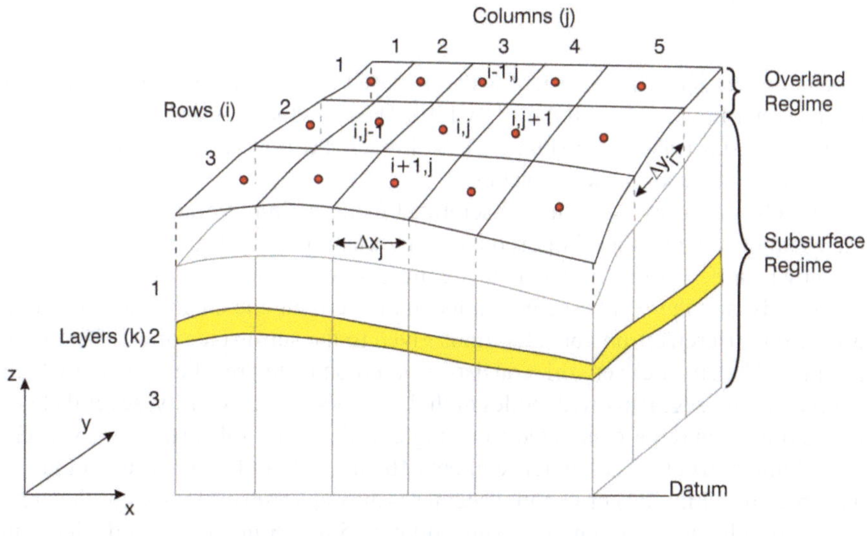

Fig. 2.5 Finite-difference discretization of the subsurface, and overland domains

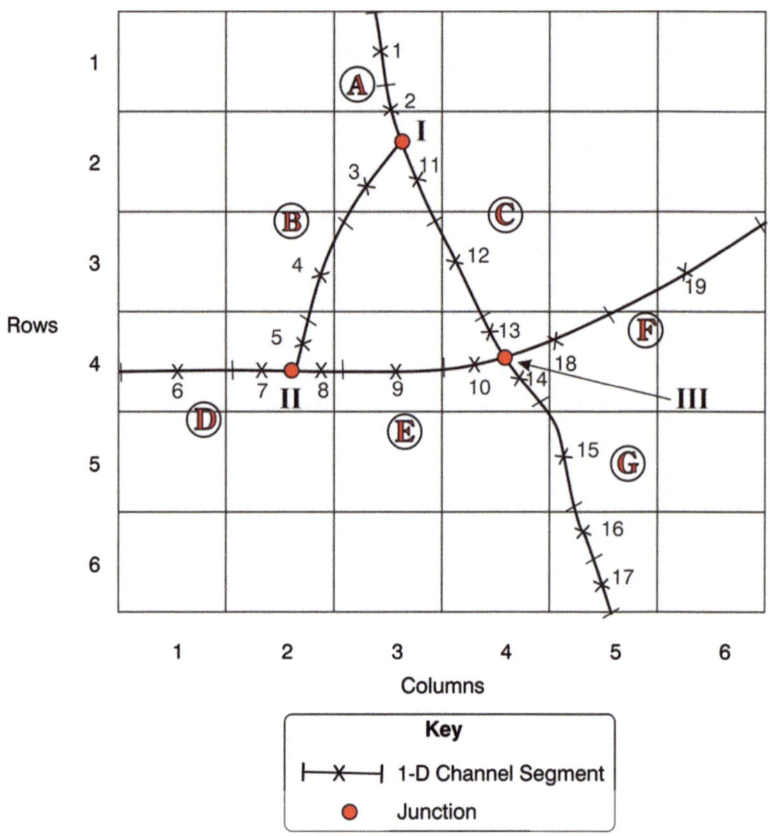

Fig. 2.6 Finite-difference discretization of the channel domain superposed on the overland or subsurface grid

type of discretization is consistent with that based on the framework of the popular groundwater flow model, MODFLOW [39]. The overland flow domain grid areally mirrors the subsurface grid (Fig. 2.5), with overland flow node elevations corresponding to the land surface elevation.

The channel flow domain is discretized using a finite volume concept with the channel network superimposed on the overland flow and subsurface domains as shown in Fig. 2.6. The channel regime discretization is independent of the areal grids, and there are no constraints on discretizing the channel network that arise from discretization considerations given to the subsurface or overland flow regimes. The channel network consists of interconnected reaches, each of which is divided into segments with nodes at their centers. For flexibility, several channel segments may be connected to a single node of the subsurface or overland flow domain grid (e.g., channel segments 16 and 17 both lie in row 6, column 5, as depicted in Fig. 2.6) or channel segments may span several nodes of the areal grid (e.g., channel segment 19 spans columns 5 and 6 in row 3, as depicted in

Fig. 2.6). Junctions are defined as channel intersections (e.g., Junction III connects channel segments 10, 13, 14, and 18 of reaches C, E, F, and G, as depicted in Fig. 2.6).

2.7 Solution Techniques

In Sects. 2.1–2.3, the governing equations for flow and transport in the three interconnected domains are presented. The governing equations are combined to form a unified system of equations for flow and transport. Equations (2.1), (2.5a), and (2.6a) govern flow of water, and Eqs. (2.7a), (2.8), and (2.9) govern solute transport, within and among the subsurface, overland, and channel domains. The governing equations can be combined to generate $1+n_s$ unified system equations of the form below:

$$[M_{\mathrm{F}}]\left\{\frac{\mathrm{d}\hat{h}}{\mathrm{d}t}\right\}+[K_{\mathrm{F}}]\{\hat{h}\} = \{F_{\mathrm{F}}\} \tag{2.17a}$$

$$[M_{\mathrm{T}}]\left\{\frac{\mathrm{d}\hat{C}_k}{\mathrm{d}t}\right\}+[K_{\mathrm{T}}]\{\hat{C}_k\} = \{F_{\mathrm{T}}\} \tag{2.17b}$$

where:

M_{F} mass matrix for the flow equation
$\hat{h}$ hydraulic head vector for the flow equation
K_{F} conductance matrix for the flow equation
F_{F} forcing vector for the flow equation
M_{T} mass matrix for the transport equation
$\hat{C}_k$ concentration for species k vector for the transport equation
K_{T} conductance matrix for the transport equation
F_{T} forcing vector for the transport equation

The matrices and vectors in the above equations take the forms below:

$$[M_{\mathrm{F/T}}]=\begin{bmatrix} M_{\mathrm{GG}} & M_{\mathrm{GO}} & M_{\mathrm{GC}} \\ M_{\mathrm{OG}} & M_{\mathrm{OO}} & M_{\mathrm{OC}} \\ M_{\mathrm{CG}} & M_{\mathrm{CO}} & M_{\mathrm{CC}} \end{bmatrix};\ [K_{\mathrm{F/T}}]=\begin{bmatrix} K_{\mathrm{GG}} & K_{\mathrm{GO}} & K_{\mathrm{GC}} \\ K_{\mathrm{OG}} & K_{\mathrm{OO}} & K_{\mathrm{OC}} \\ K_{\mathrm{CG}} & K_{\mathrm{CO}} & K_{\mathrm{CC}} \end{bmatrix} \tag{2.17c}$$

$$\{\hat{h}\} = \begin{Bmatrix} h_{\mathrm{G}} \\ h_{\mathrm{O}} \\ h_{\mathrm{C}} \end{Bmatrix};\ \{\hat{C}_k\} = \begin{Bmatrix} C_{k\mathrm{G}} \\ C_{k\mathrm{O}} \\ C_{k\mathrm{C}} \end{Bmatrix};\ \{F_{\mathrm{F/T}}\} = \begin{Bmatrix} F_{\mathrm{G}} \\ F_{\mathrm{O}} \\ F_{\mathrm{C}} \end{Bmatrix} \tag{2.17d}$$

In the matrices shown above, the off-diagonal terms represent interdomain communications. These terms are generated using Eqs. (2.11a), (2.11b), (2.11d), and (2.12a)–(2.12e). These system equations are formed based on the discretizations discussed in Sect. 2.6.

The flow matrix Eq. (2.17a) is solved first to provide velocities to the transport matrix Eq. (2.17b). In the case of density-dependent flow and transport, the flow and transport equations are solved sequentially and iteratively until convergence is achieved. Between the flow and transport equations, the former is more difficult to solve as it is highly nonlinear with pressure-dependent hydraulic properties in the variable saturated zone and flow-depth-dependent properties in the overland and channel domains. The flow equation must, therefore, be solved iteratively. In a typical transport problem with linear sorption, the transport equation is linear and can be solved without iteration.

To solve Eqs. (2.17a) and (2.17b), the subsurface components may be linked (in a time-lagged manner) with the surface components. This method is the least accurate. To improve the accuracy, the two components may be iterated within the same time step, interlinked by common internal boundary conditions, representing the exchange between the surface and subsurface domains. A third method involves solving Eqs. (2.17a) and (2.17b) by a total integration (i.e., fully coupled) procedure. In the first method, the subsystem equations are solved sequentially, first to obtain h_G and C_{kG}, followed by $\{h_O, h_C\}^T$, and $\{C_{kO}, C_{kC}\}^T$. In the second method, iteration is performed until convergence is reached. With the third method, the total integration method, Eq. (2.17a) is solved to obtain $\{h_G, h_O, h_C\}^T$ simultaneously, followed by solving Eq. (2.17b) to obtain $\{C_{kG}, C_{kO}, C_{kC}\}^T$ simultaneously. The total integration procedure is the most robust and efficient solution scheme for the coupled system of surface/subsurface equations and is theoretically most accurate and is generally recommended for most flow and transport simulations. In some cases, linked or iteratively coupled approaches allow for providing different time scales to the surface and subsurface phenomena, and several small surface-water flow time steps may be solved for each large subsurface time step solution. Linked/iteratively coupled approaches are therefore useful when surface/subsurface interactions are small. Further, when all options are available, the validity and accuracy of the linked approach may be tested before accepting/rejecting it as a feasible or more efficient alternative for solving complex systems.

Examples of models based on the time lagged without iteration approach include: subsurface flow and stream flow [5, 7], and subsurface flow and overland flow [8, 40]. Langevin et al. [6] coupled a two-dimensional overland flow and transport model with a three-dimensional saturated groundwater flow and transport model to develop a density-dependent model for a coastal area near the southern Everglades, Florida. Examples of models based on the linked with iteration approach include Refs. [7, 8].

The total integration method, even though the most accurate, may also be the most computationally demanding in terms of speed and storage. However, during the past decade, along with the advent of high-speed personal computers, a number of rigorous integrated surface-water/groundwater models have been developed to circumvent these limitations. This method has been increasingly accepted and utilized by the technical community. Recent application examples reported in the literature based on this approach include Refs. [12–14, 17, 41].

To address the issue of nonlinearity in Eqs. (2.17a) and (2.17b), iterative schemes have been proposed that are based on the modified Picard scheme [42] or the Newton–Raphson procedure with under-relaxation [43, 44]. These schemes are used to linearize the subsurface and the surface-water flow equations. The unsymmetric matrix equations generated by the Newton–Raphson method at each nonlinear iteration may be solved using the Orthomin or BiCGStab matrix solution schemes [32, 44–47]. To further reduce computational effort, time stepping may be modified during a simulation. An adaptive time-stepping scheme may be adopted such that the time-step size grows (up to a maximum) when the number of iterations required for convergence is low (solution is easy); reduces when the number of iterations is high; and is considered optimal with no change when the number of iterations required for convergence to a previous time-step is medium as compared to the prescribed maximum number of iterations [32].

2.8 Available Simulators

Currently, there are several integrated, coupled surface–subsurface codes available. These codes include: MODHMS [32], MIKE SHE [40], WASH123D [48, 49], Hydrogeosphere [50], PARFLOW-Surface Flow [51], InHM [9], and FIHM [11]. Among these, MIKE SHE, MODHMS, and PARFLOW-Surface Flow use finite difference methods [43, 52] while Hydrogeosphere, WASH123D, and InHM are based on finite element methods [43]. The widely used codes are: MODHMS, MIKE SHE, WASH123D, and Hydrogeosphere. Most codes are based on an assumption that water is slightly compressible so that the buoyancy term is retained in the flow equation. These codes are described below.

MODHMS [32] is one of the state-of-the-art simulators for addressing integrated surface/subsurface nonisothermal flow and water quality problems, within a MODFLOW-compatible framework. Governing equations for flow in surface and subsurface regimes including a mechanistic approach to evapotranspiration provide a physically based understanding of the whole hydrologic cycle. Scale effects are addressed for rill and exclusion storage adjustments as well as for surface-water bodies such as lakes, ponds, channels, rivers, and streams which may further include weirs, gates, bridges, and other hydraulic structures that affect or control flow. MIKE-SHE [39] is a fully integrated surface and subsurface flow simulator. Governing equations for flow in surface and subsurface regimes including evapotranspiration provide a physically based understanding of the whole hydrologic cycle with two-dimensional and one-dimensional surface-water representations. WASH123D [48, 49] and Hydrogeosphere [50] are fully integrated simulators for addressing conjunctive surface/subsurface flow and water quality issues in a physically based manner. Both simulators use a finite-element numerical implementation. All codes are available commercially or through special requests.

3 Model Development and Calibration

When it is determined that an integrated surface–subsurface numerical model is required, the task of model development begins. Once the purpose of the model has been established, a conceptual model [52] of the study area is developed. An appropriate computer code capable of simulating all the required processes is then selected. It is then followed by model design, parameterization, and calibration. These steps are described below.

3.1 Model Design

In an integrated surface–subsurface numerical model, the three problem domains (subsurface, overland, and channels) are replaced by arrays of discretized cells and segments. To be included in the discretized system are: hydrostratigraphic units, and hydrogeologic features such as springs, for the subsurface domain; overland flow area, and surficial features such as lakes for the overland domain; and streams, rivers, canals, and hydraulic structures for the channel domain. An example of finite-difference discretization is given in Figs. 2.5 and 2.6 and Sect. 2.6. Examples of finite-element discretization may be found in Refs. [48–50].

3.2 Model Parameterization

Following the model discretization step, all cells and segments must be assigned relevant model parameters. This process is referred to as parameterization in the literature. Model parameters are discussed in Sect. 2. Parameterization for each of the three domains based on appropriate parameters is summarized below.

3.2.1 Subsurface Domain

The subsurface domain is parameterized by hydraulic parameters for each hydrogeologic unit (hydraulic conductivity, storage coefficient, etc.), conductances (of rivers/creeks, springs, and storage-type surface-water bodies (e.g., lakes) and recharge distributions. Recharge may be estimated from local water budgets of the study area.

3.2.2 Overland Domain

The overland domain is parameterized by land surface elevations (obtained from digital elevation models) and by using land use/land cover maps with correlations between land usage and physical parameter values. Physical parameters based on land use characteristics include: surface roughness coefficient, and land surface leakance values. These parameters should be translated onto computational cells as area-weighted averages for the respective parameters.

3.2.3 Channel Domain

Channel segments and surface-water bodies are parameterized by their geometry, connectivity, bed and bank elevations, surface friction coefficient, hydraulic structures (with and without regulation) and bottom leakances.

3.2.4 Related Data/Parameters

In addition to physical parameters for the three domains, other parameters and data include: (1) stresses (extraction/injection rates for all domains); (2) precipitation (from gages or radar-derived), and (3) evapotranspiration. Parameters governing actual evapotranspiration such as field-capacity and wilting-point moisture contents are correlated to soil type, while other critical parameters such as leaf area index and root depth and density are functions of crop type or land use and stage in the growing season. These parameters should be estimated for each cell as an area-weighted average of the values for the various land use or soil types.

3.3 Model Calibration

As with other models, calibration of integrated surface–subsurface flow and transport models is an iterative process during which the model parameters are adjusted for the model to closely mimic the observed dynamics of the system under consideration. During the calibration process, model inputs such as system geometry and properties, initial and boundary conditions, and stresses are changed so that the model output matches corresponding observed or measured values. Calibration can be done by trial-and-error adjustments of model parameters or by using automated parameter estimation codes.

Integrated models tend to be intensive in terms of data requirements and each simulation run is likely to be computationally intensive. For these reasons, it is the authors' opinion that model calibration should be carried out first manually, using a systematic trial-and-error approach to take maximum advantage of prior knowledge and existing understanding of the system and data. Automated parameter estimation codes may be subsequently used to supplement the trial-and-error method and to refine the calibration.

3.3.1 Trial-and-Error Approach

For integrated models, a typical systematic trial-and-error approach involves calibration in two stages. In the first stage, the subsurface component of the model is calibrated. In the second stage, the subsurface model is expanded to include the overland flow and channel components. The main reason for segregating development and calibration into these two stages is to remove the problem of disparate time scales (between the surface and subsurface components) during the initial phase of development and calibration. In most cases, the temporal variation in the subsurface domain is not as rapid as in the overland and channel domains. In addition, in many study areas, subsurface models may have been developed and calibrated. Consequently, it is logical to develop the overall model structure based on a subsurface foundation so that one can first focus on minimizing uncertainty in features within the subsurface domain, which function as sources and sinks to the overland and channel domains. The calibration should begin with the flow component only, as it determines the velocity field required by the transport model. The calibration of the transport model should always follow that of the flow model.

3.3.1.1 Subsurface Model

Calibration of the subsurface flow model should be performed in two steps: (1) steady-state and (2) transient conditions. The steady-state conditions should include: annual average, dry and wet conditions. For the transient conditions, a multiple-year period that encompasses various system and climatic changes should be selected. Periods of calibration should be chosen such that they are representative of a wide variety of historical conditions within the study area. Given expected gaps in the historical data record, it may be necessary to have several uncontiguous calibration periods, in order to capture a complete representation of land use changes and climatic conditions within the study area.

3.3.1.2 Integrated Subsurface–Surface Model

In the integrated model development stage, the model calibration process is divided into four steps:

- *Step 1: Steady-state Calibration.* In this step, a number of steady-state calibration runs should be performed (for average, wet and dry season conditions) to identify preliminary parameters for evapotranspiration (ET), and hydraulic parameters for the channels. This calibration step is significant for capturing long-term average, high, and low baseflow conditions and is the first step towards removing large errors, which might mask parameter changes.
- *Step 2: Transient Calibration with Large Storm Events.* In this step, the model should be calibrated to a number of storm events to ensure that the integrated model is capable of generating a fast response associated with large storm events. This calibration step is significant for examining the leakance and friction characteristics of the surface in addressing peak flow and flow synchronization.

- *Step 3: Long-Term Transient Calibration.* The model calibrated in Step 2 should be further refined to represent long-term transient conditions. The calibration periods for this step should correspond to those used previously in the calibration of the subsurface model component, each of which may last for 4–10 years. These are periods that capture a complete representation of land use changes and climatic conditions within an area. The calibration runs in this calibration step will be used to further refine the evapotranspiration, hydraulic parameters, and, if necessary, subsurface model parameters.
- *Step 4: Model Verification with Independent Datasets.* The model calibrated in Step 3 should be verified by simulating an independent dataset (dataset not used as part of the calibration process) that also includes a variety of land use changes.

3.3.2 Automated Calibration

In automated calibration, the calibration is cast as an optimization problem in which parameter values are sought to minimize the objective function (typically the difference between observed and simulated variables). These optimization algorithms typically perform a local search of the parameter space, starting from an initial estimate and leading to a local optimum [53, 54]. Global algorithms have been developed that search the entire parameter space and hence result in globally optimal parameter values [55, 56]. Schoup et al. [57] utilized the Multi-Objective Shuffled Complex Evolution Metropolis (MOSCEM-UA) global optimization algorithm of Vrugt et al. [58] to calibrate an integrated surface–subsurface model for the Yaqui Valley area in Sonora, Mexico, with aggregated and competing objectives. The solution to this problem is not a single "best" parameter set, but consists of a Pareto optimal set of solutions corresponding to trade-offs among the objectives. Because of the complexity and intensive computational requirements for automated calibration of integrated surface–surface models, there are few cases reported in the literature. Other examples may be found in [59].

During the past decade, there has been a considerable progress in the development of distributed computer systems using the power of multiple processors to efficiently solve complex, high-dimensional computational problems. Parallel computing offers the possibility of solving computationally challenging optimization problems in less time than is possible using ordinary serial computing. Parallel computing implementation examples may be found in Refs. [60–62].

3.3.3 Calibration Criteria

In general, calibration criteria for the flow problem comprise two types, quantitative and qualitative criteria. The quantitative criteria are likely to consist of the following metrics: mean error (ME), mean absolute error (MAE), percent bias (PB), monthly bias, flow biases in various flow regimes, root-mean-squared error (RMSE), standard deviation of error (SD), and the largest and smallest residual errors (MaxE and MinE) [52, 63], as well as Nash-Sutcliff model efficiency coefficient [64]. These metrics are calculated by comparing simulated quantities against field observations, which are likely to include: observed potentiometric elevations at monitoring wells, observed spring discharge rates (when springs are active), observed river stages, and observed stream flows. A multi-objective approach is used to measure the performance of the model during different flow conditions (high flow, low flow, quick drainage, and slow drainage) by separate RMSE and ME evaluations. Therefore, the calibrated model should adequately express all phases of the hydrologic cycle. Key calibration targets found in watershed modeling literature [65] are on the order of $\pm 5\,\%$ of the difference between respective maximum and minimum field observations.

The qualitative criteria may include favorable and consistent similarity between simulated and observed general flow patterns of groundwater, simulated and observed potentiometric levels, and their variations across hydrostratigraphic units, simulated and observed hydrographs in creeks and rivers (e.g., runoff volume, time to peak, and peak flows), and simulated and observed general areal drainage patterns. A multi-objective approach may also be used with qualitative criteria such as specific visual characteristics of a hydrograph—slope of the rising limb, volume of runoff, and magnitude and timing of peak flow during storm periods; and rate of hydrograph recession for early quick recession and later slow recession during inter-storm periods.

Calibration criteria for the transport model are similar to those of the flow model. Quantitative metrics in this case are used to estimate the closeness between simulated and observed constituent concentrations and mass fluxes. Qualitative criteria include patterns of migration, plume configuration, and approximate speed with which mass fronts are propagated.

4 Application Examples

Two examples are presented to demonstrate the development and application of integrated subsurface and surface-water models. These two examples are from the State of Florida, USA. In both examples, MODHMS [32], discussed in Sect. 2.8, was used to perform comprehensive simulations of conjunctive surface-water/groundwater flow and transport simulations.

4.1 Example 1: Integrated Flow Model: Peace River Watershed, Florida

4.1.1 Background

The Peace River watershed (Fig. 2.7) in Southwest Florida is an important ecological, water supply, and recreation resource. Peace River flows have been in a long-term decline since the 1930s. The impact has been most pronounced in the Upper

Fig. 2.7 Location of the peace river watershed

Peace River where sections of the river have lost all flow in recent dry seasons. Factors affecting flows in the Peace River include natural phenomena as well as human impacts. Long-term natural variation in rainfall is understood to have a major influence on river flows in the Peace River and similar river systems in Florida. In addition, there are numerous human influences that impact the Peace River. These include lowering of the groundwater potentiometric surface due to groundwater pumping for industrial, agricultural, and domestic water use; structural alterations and regulation of surface water; land use and land cover changes; and reduction of wastewater discharges to the Peace River and its tributaries. Although numerous studies have been conducted to investigate and understand the phenomena that have impacted flows in the Peace River, the relative importance and quantifiable impact of these phenomena are not thoroughly understood. In 2007, the Peace River Integrated (surface-water/groundwater flow) Model (PRIM) was developed by HydroGeoLogic [66–68] for the Southwest Florida Water Management District (SWFWMD) to assess the effects of land use, water use, and climatic changes on Peace River stream flows, and also to evaluate the effectiveness of various management alternatives.

4.1.2 Site Description

The Peace River watershed covers a total area of 2350 square miles (6000 km^2) and is located in Polk, Hardee, and De Soto counties (Fig. 2.7). The Peace River is 106 miles (170 km) long. The watershed boundaries and principal sub-basins are shown in Fig. 2.7. The head waters of the Peace River originate in the northernmost group of lakes of the Saddle Creek and Peace Creek sub-basins. Surface water from the headwaters region flows to Saddle Creek and Peace Creek which form the beginning of the Peace River channel at their confluence near Bartow, south of Lake Hancock. From the confluence, the river flows south approximately 85 miles (135 km) and ultimately discharges in the Charlotte Harbor estuary. The Peace River basin is endowed with a large number of lakes, with most located in the Saddle Creek and Peace Creek sub-basins in Polk County. The majority of the lakes are the result of sinkhole activity, and are classified as either seepage lakes or drainage lakes. Seepage lakes have no surface-water outflow; the water level in these lakes is controlled by groundwater levels. Drainage lakes are lakes that lose water through surface outflows. Many of the larger lakes in the Peace River basin are drainage lakes and are hydraulically connected, often as a result of human drainage improvements.

The Peace River basin encompasses ten sub-basins. One of the major sub-basins is the Saddle Creek sub-basin, which is located in West Central Polk County, Florida, within the Peace River Watershed as seen in Fig. 2.8. The watershed covers approximately 156 square miles (400 km^2). The Saddle Creek sub-basin is characterized by a large number of sinkhole lakes. Lake Hancock, which is the largest lake in the Peace River watershed with an area of around 4500 acres (1800 hectares), is located in the southern part of Saddle Creek. Other major lakes and hydraulic connections are also shown in Fig. 2.8.

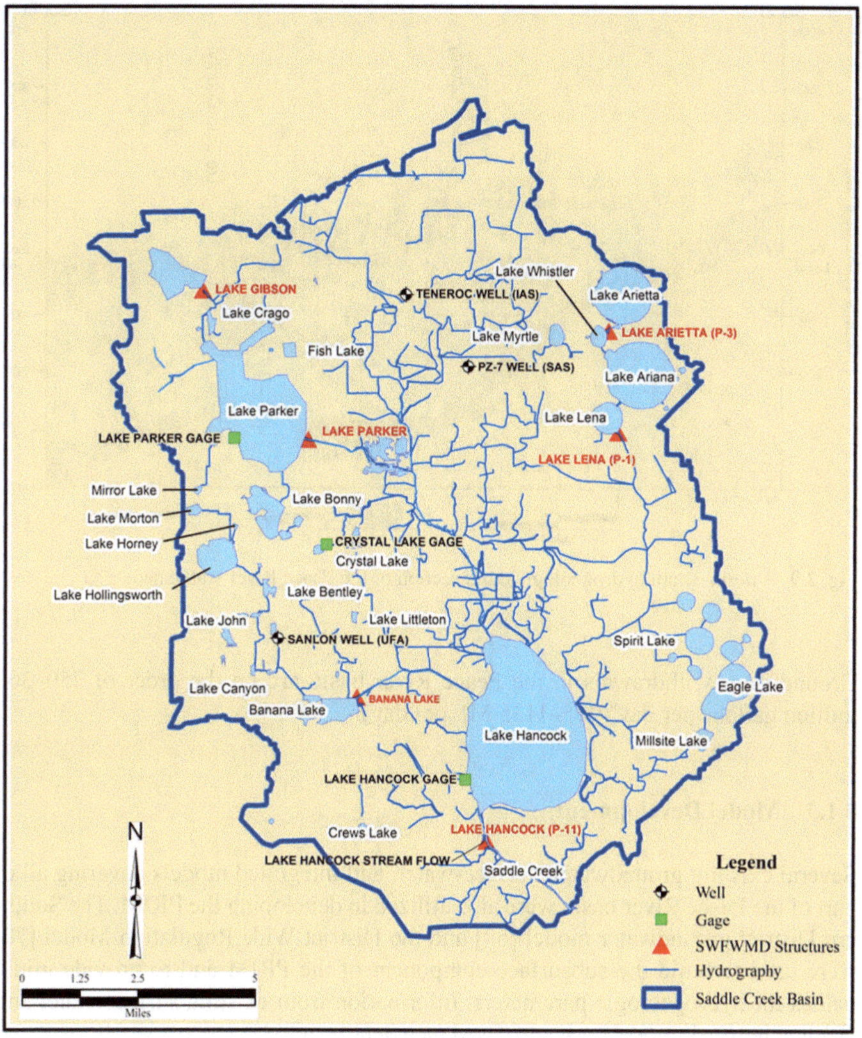

Fig. 2.8 A map of Saddle Creek showing major lakes and hydraulic structures

The watershed is underlain by three aquifer systems (Fig. 2.9). The uppermost system is the unconfined surficial aquifer system (SAS). The SAS is primarily recharged by rainfall, depleted by evapotranspiration (ET) and can provide baseflow to the surface-water features. It consists of unconsolidated sand, silt, and clayey sand. Underlying the SAS is the confined intermediate aquifer system (IAS) consisting of inter-bedded limestone, sand, and phosphatic clays of low permeability. The IAS is very thin in Saddle Creek but thickens in the southern basins of the Peace River. The IAS is underlain by the upper Floridan Aquifer (UFA) and the lower Floridan Aquifer (LFA). The UFA is the principal water supply aquifer.

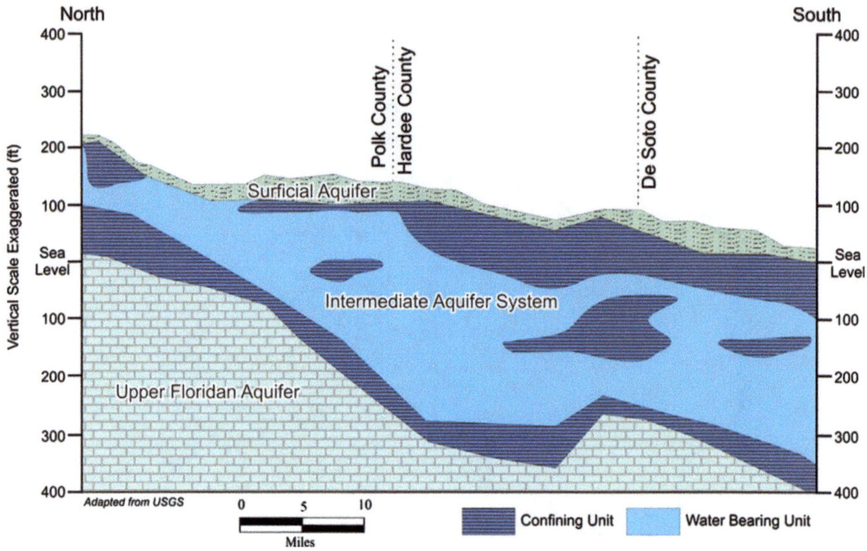

Fig. 2.9 A north–south hydrogeologic cross section of the Peace River watershed

Groundwater withdrawals in the Peace River basin are on the order of 250–300 million gallons per day (945–1135 ML per day).

4.1.3 Model Development

Several existing groundwater, surface-water, and integrated models covering all or part of the Peace River basin were also utilized in developing the PRIM. The Southern District groundwater model [69] and the District Wide Regulation Model [70] were used to build the subsurface component of the PRIM and to provide initial values for hydrogeologic parameters. Information from existing surface-water and channel network models was also used (see details in [63]).

Model development began with the development of a sub-basin pilot model and was later extended to a basin-wide scale. The Saddle Creek sub-basin (Fig. 2.8) was chosen for the development of the pilot model as it encapsulates all of the characteristics of the watershed. Because the Peace River watershed model encompasses a very large area, presentation of the details in the development of the Saddle Creek model is considered instructional to the reader.

Figure 2.10 shows the areal discretization over the Saddle Creek sub-basin. The three-dimensional subsurface grid consists of 32 columns, 40 rows, and 5 layers of grid cells. The grid conforms areally with domain boundaries by inactivating cells of the finite-difference grid which lie outside of the Saddle Creek Basin. Grid resolution is uniform at 2500 ft. (762 m) and oriented in the north and east directions. The top model layer represents the SAS, layers 2 and 3 represent the IAS where it exists, and layers 4 and 5 represent the upper and lower portions of the

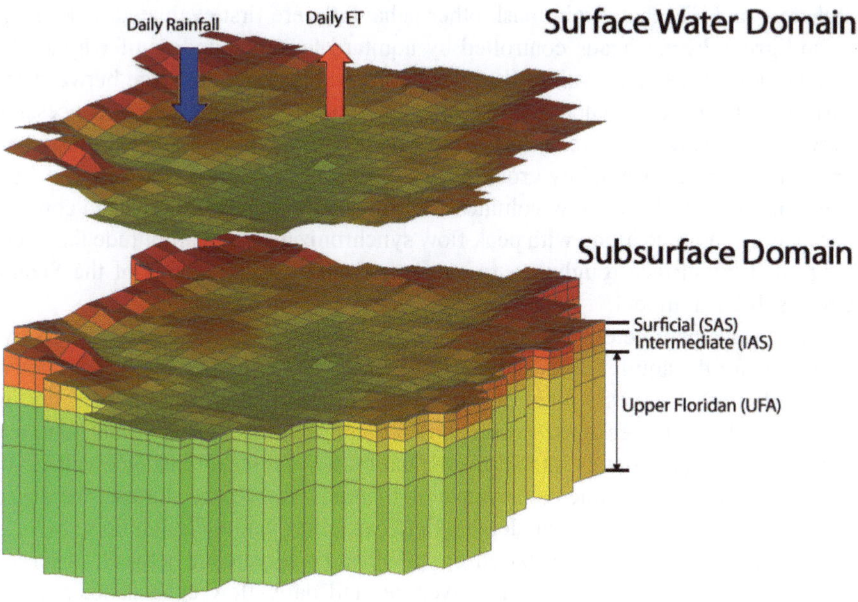

Fig. 2.10 An exploded view showing the subsurface and overland grids

UFA. The two-dimensional overland grid, which represents the land surface and coincides areally with the subsurface grid, overlies the first layer of the subsurface grid (Fig. 2.10). The grid contains 5425 active subsurface cells, 775 active overland cells, and 800 channel segments connected via 405 weirs, 285 culvert pipes, and 43 drop structures.

NEXRAD-based rainfall data was used to capture the spatial and temporal resolution of rainfall. Evapotranspiration (ET) in the Saddle Creek sub-basin was dynamically simulated as a function of potential ET [66] and the availability of water in the soil profile and shallow groundwater to sustain the ET demand. The ET demand was estimated based on crop type and land use. Rainfall was applied on a daily basis. Storages and ET losses through vegetative interception and the root zone were also taken into account.

The overland flow domain lateral boundaries are no-flow conditions as they are located along watershed boundaries. Aquifer bottom boundaries are no-flow conditions at the base of the UFA (layer 5). Outlets in the channel and overland flow domains were assigned with zero-depth-gradient boundary conditions. Lateral subsurface boundaries in the SAS were treated as no-flow conditions. Lateral subsurface boundaries in the IAS and UFA were obtained from existing groundwater models.

4.1.4 Model Calibration

The Saddle Creek sub-basin model was calibrated over the calibration period of January 1998 through December 2002. Gross water budget components (rainfall, recharge, and ET) that might mask other behavior were first evaluated. Lake stages and groundwater heads controlled by aquitard leakance and aquifer hydraulic conductivity were next evaluated to constrain the recharge/discharge between the surface and subsurface and among aquifers. Prolonged drought noflows, recession flows, or baseflows were examined next, because they affect large-scale water balances and were controlled by groundwater levels, surface-water releases, and storage. Finally, cumulative flow volumes controlled by channel leakance and conveyance were examined along with peak flow synchronization and amplitude that were controlled by surface roughness, to calibrate the flow components of the Saddle Creek sub-basin model.

Subsurface parameters altered during calibration included hydraulic conductivity values for the aquifer units and vertical leakances (vertical hydraulic conductivity divided by thickness) between the units. Overland parameters altered during calibration included leakance values and crop coefficients for the various land use types. Channel parameters altered during calibration included leakance values for channels and lakes, and the conveyance of channels. Root-zone storages were also altered by modifying root-zone depths, field capacities, and wilting points.

The model was calibrated based on satisfactory agreement between observed and simulated streamflows (weekly average, and daily flow exceedance percentiles; 10th, 50th, and 90th), and observed and simulated water levels in lakes and in groundwater observation wells. In addition, water balance parameters (cumulative streamflow, average evapotranspiration, net recharge to the subsurface, and lateral groundwater flux) must be within certain prespecified ranges. Standard quantitative calibration metrics (see Sect. 3.3.3) were used. Calibration criteria details and full calibration results are given in [64].

Examples of the calibrated model results for various stream flow targets, lake levels, and groundwater levels are presented in Figs. 2.11, 2.12, 2.13, 2.14, 2.15, 2.16, 2.17. Locations of stream gages and groundwater observation wells are given in Fig. 2.8. Good agreement is noted for fit, trend, and variation between observed and simulated conditions for all the stream, lake and groundwater observations.

The outflows from Lake Hancock into the Saddle Creek channel are regulated by the P-11 control structure located at the downstream end of the lake (Fig. 2.8). The stream gage at P-11 is the most important gage for calibration, because it captures most of the surface-water outflow from Saddle Creek. The lake stage versus time plot (Fig. 2.12) indicates that the model was successful in replicating Lake Hancock's stage favorably from January 1993 to December 2003. The streamflow plot of the P-11 gage (Fig. 2.11) demonstrates the model's ability to simulate peak flow events, characterized by a sharp increase in flows as well as periods of no flow during the drought of 2000. Also, cumulative streamflow plots at P-11 (Fig. 2.11) demonstrate a close correspondence between the observed and simulated stream flow volumes. The model overpredicted stream flow prior to 1997, but underpredicted stream flow during a strong El Niño event which lasted from mid-1997 to mid-1998.

The model generally showed good results in matching lake levels (examples in Figs. 2.12, 2.13, 2.14), and groundwater heads (examples in Figs. 2.15, 2.16, 2.17).

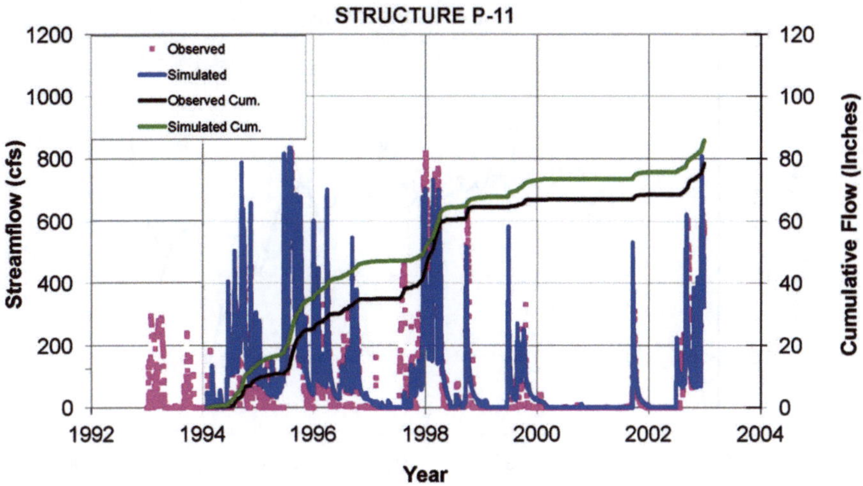

Fig. 2.11 Observed and simulated lake levels and stream flow at P-11

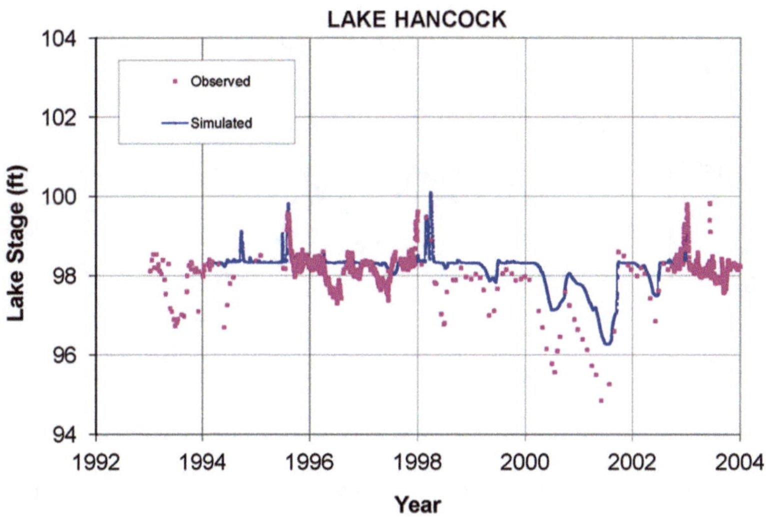

Fig. 2.12 Observed and simulated lake levels: Lake Hancock

The subsurface calibration was best for the IAS and UFA, but somewhat poorer for the SAS. The SAS has very few monitoring wells and most have an incomplete data record, making it difficult to evaluate the model performance for the shallow groundwater system. There may be a number of factors contributing to the poorer model performance in the SAS, including local heterogeneities in aquifer characteristics, as well as the presence of shallow irrigation wells that are not registered in the database and therefore not represented in the model.

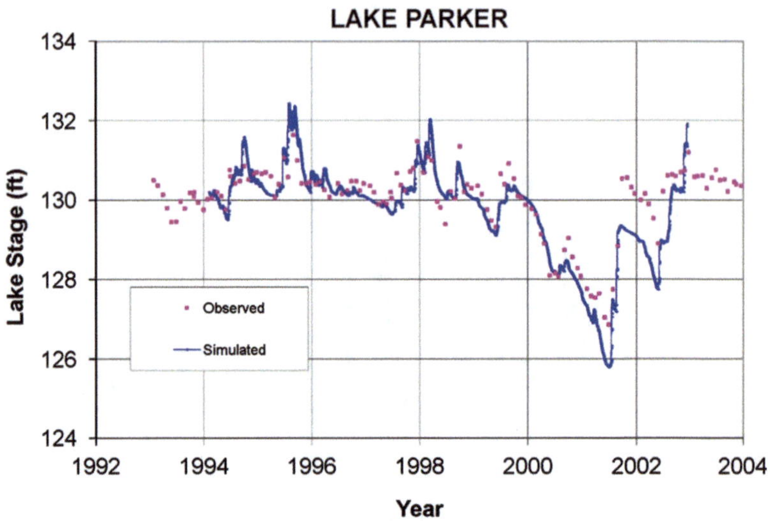

Fig. 2.13 Observed and simulated lake levels: Lake Parker

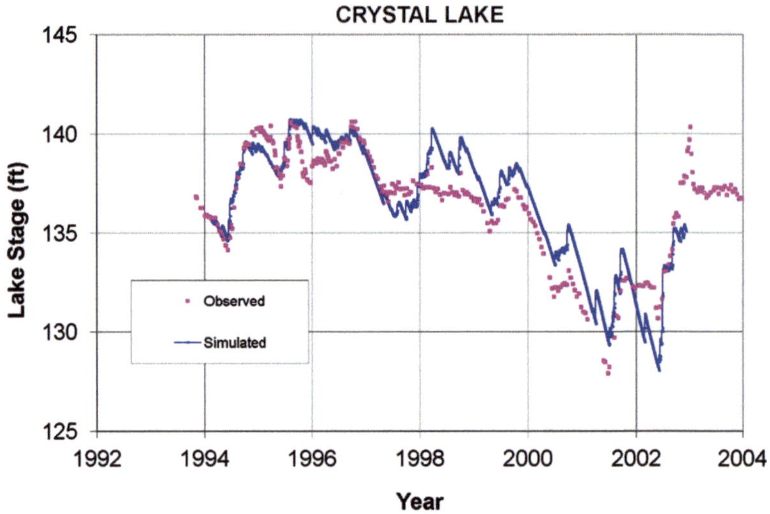

Fig. 2.14 Observed and simulated lake levels: Crystal Lake

The calibrated model performance was evaluated against data that were not utilized during calibration to validate the model's predictive behavior. The validation period was from February 1994 through December 1997. As in the calibration period, there is a good correlation between observed and simulated streamflows and water levels. Rainfall and evapotranspiration values are both higher in the validation period as it was a wetter period resulting in more water for evapotranspiration.

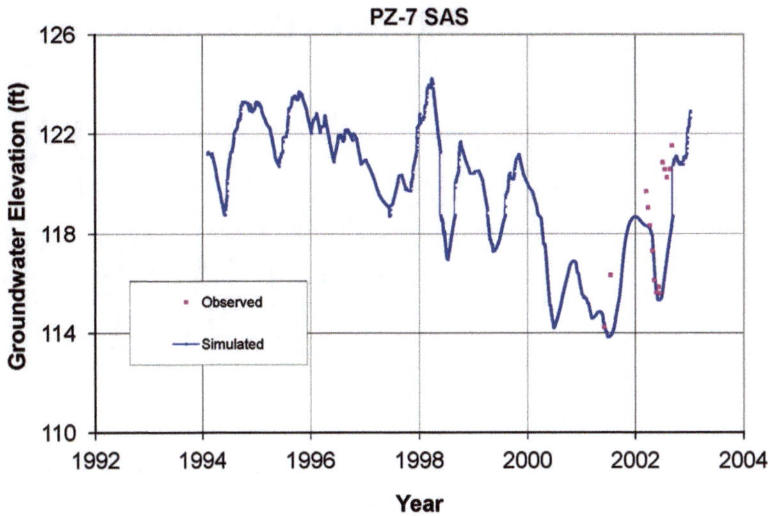

Fig. 2.15 Observed and simulated groundwater levels: PZ-7 Well (surficial aquifer system)

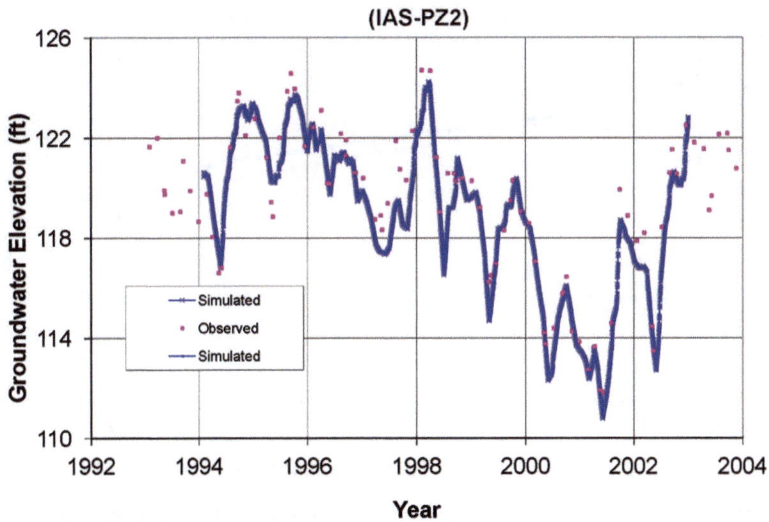

Fig. 2.16 Observed and simulated groundwater levels: Tenoroc Well (intermediate aquifer system)

Subsequent to the completion of the Saddle Creek sub-basin model, a basin-wide model was constructed by extending the Saddle sub-basin model. The basin-wide model utilized the same horizontal and vertical spatial grid discretization as the Saddle Creek sub-basin model [66]. The basin-wide model consisted of 196 rows, 102 columns, 5 layers, and 2999 reaches of channel network. The

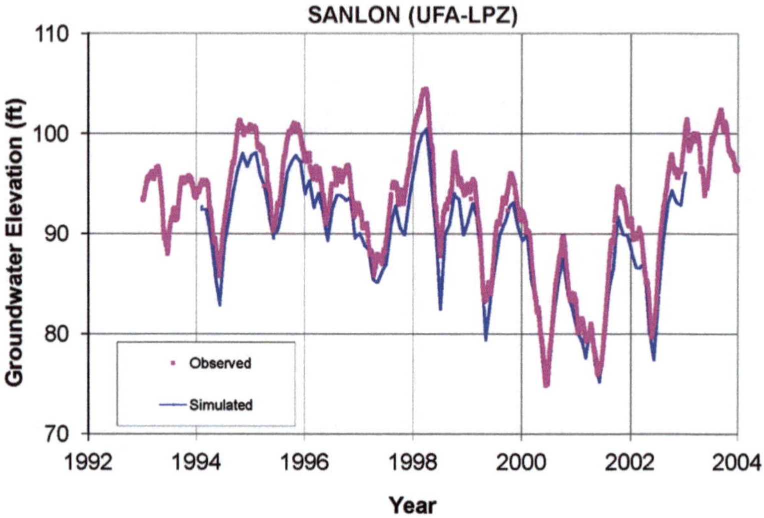

Fig. 2.17 Observed and simulated groundwater levels: Sanlon Well (upper Floridan aquifer)

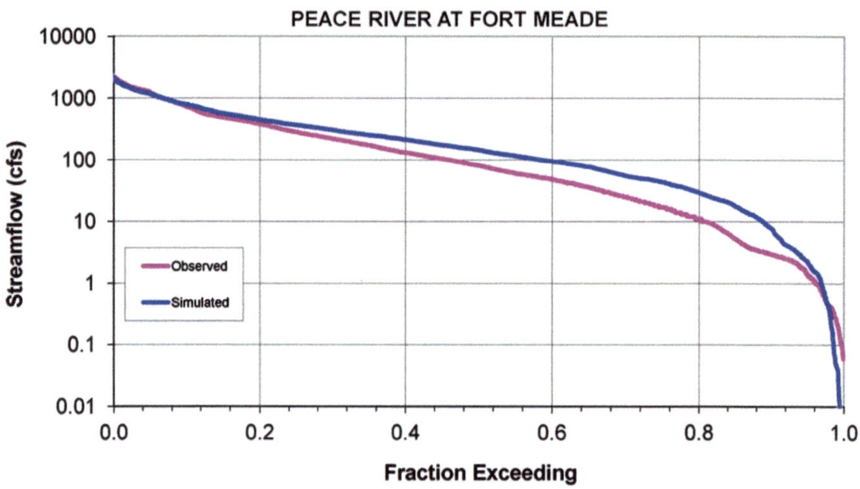

Fig. 2.18 Observed and simulated flow exceedance curves: Peace River at Fort Meade

basin-wide model was calibrated to satisfy prespecified criteria in the same manner as the Saddle Creek model. Examples are presented in Figs. 2.18, 2.19, 2.20– which show favorable comparisons between simulated and observed streamflow exceedance curves at Fort Meade, Zolfo Springs, and Arcadia (see Fig. 2.7 for gage locations along the Peace River), respectively. More details may be found in Refs. [12, 66, 67]. Predictive simulations and application examples are given in Ref. [68].

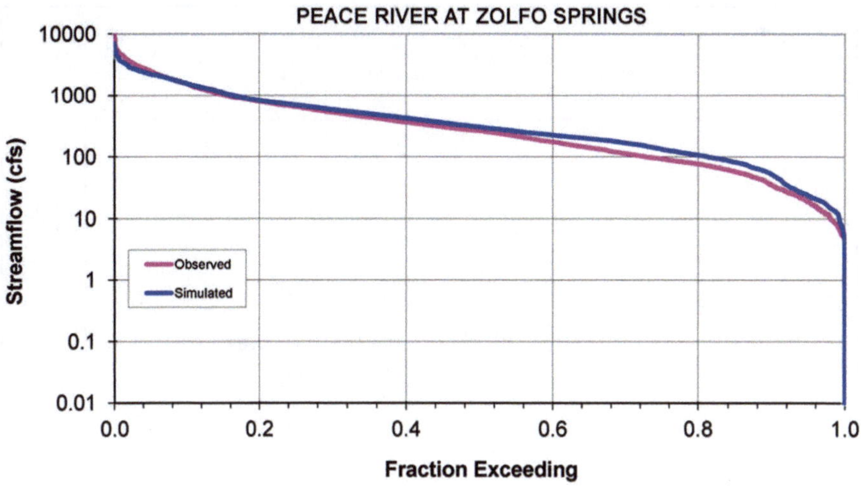

Fig. 2.19 Observed and simulated flow exceedance curves: Peace River at Zolfo Springs

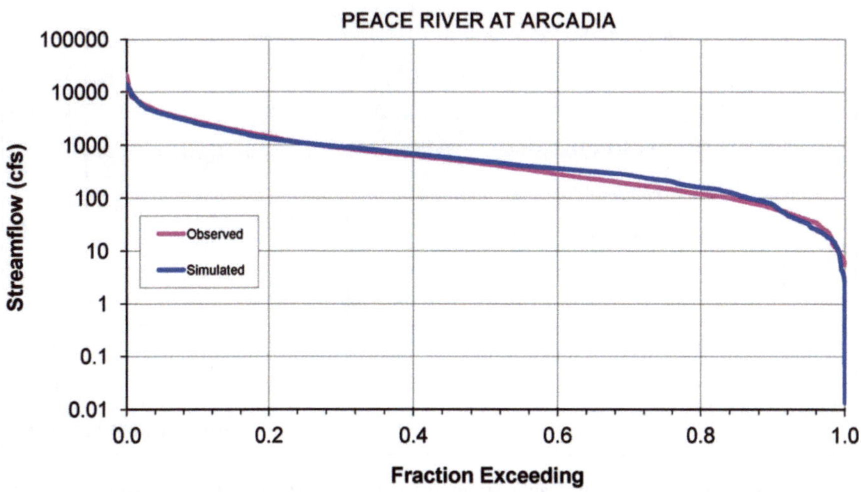

Fig. 2.20 Observed and simulated flow exceedance curves: Peace River at Arcadia

4.2 *Example 2: Integrated Flow and Transport: Everglades National Park Detention Basins, Florida*

4.2.1 Background

Everglades National Park (ENP) in south Florida, with the largest subtropical wilderness in the USA, was created to protect a fragile ecosystem instead of safeguarding

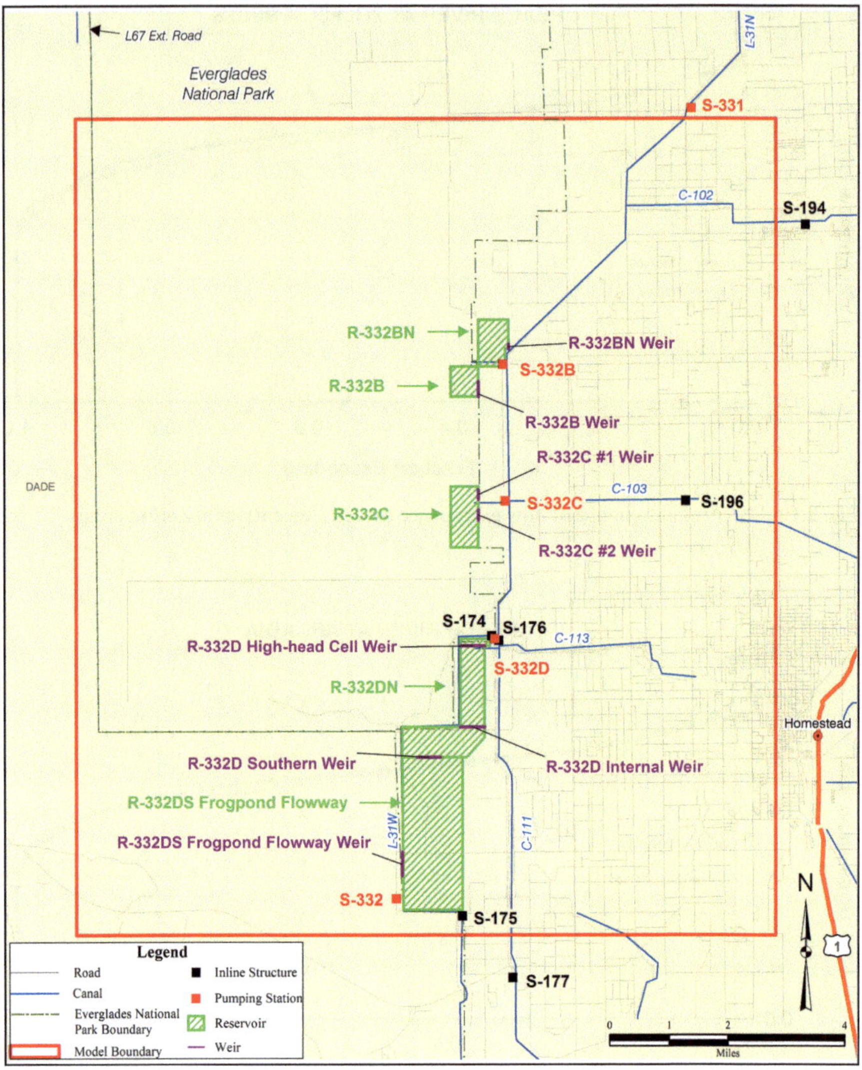

Fig. 2.21 Study area showing hydraulic structures, pumping stations, detention basins, and example observation locations

a geographic feature. As shown in Fig. 2.21, along the southeastern adjoining areas of ENP in the C-111 canal basin, a number of hydraulic structures have been constructed to help regulate the groundwater flow pattern within ENP. The C-111 canal separates ENP from highly productive subtropical agricultural lands to the east. In this area, surface water is highly interactive with groundwater because it is underlain by the Biscayne Aquifer, a highly transmissive aquifer. Owing to the extreme permeability of the Biscayne Aquifer in this area, the canals have a direct impact on water levels in adjacent areas. The C-111 canal project is intended to provide both a flood control function and an ecological restoration function. It aims at minimizing drainage of the adjacent wetlands by pumping water from the L-31 N canal into detention basins west of the L-31 N canal to maintain water levels that reduce the west-to-east gradient draining the Rocky Glades region of the ENP (Fig. 2.21). Phosphorus distribution and transport are also of concern in this area because the Everglades are fairly sensitive to phosphorus. Possible sources of phosphorus include water-borne phosphorus from the canals and from construction activities unearthing phosphorus in the soil from past agricultural uses [71]. At the study site, phosphorus is present in the form of dissolved and organic phosphates which constitute total phosphorus (TP). The main influx of TP is through the L-31 N canal from which water-borne TP gets pumped into the five detention basins west of the canal.

A model was developed for the National Park Service [14] to simulate integrated surface-water/groundwater flow as well as fate and transport of TP in the detention basins within the study area. A summary of the model is presented below.

4.2.2 Site Description

The study area is located in south Miami Dade County along the L-31 N canal (Fig. 2.21). The model covers approximately 157 square miles (400 km^2) in area. The northern boundary of the model is 1100 ft. (335 m) south of the S-331 pumping station and the southern boundary is 3650 ft. (1112 m) north of the S-177 pumping station for a total north-south extent of approximately 13.4 miles (21.4 km). The domain extends 7.1 miles (11.4 km) west and 4.6 miles (7.4 km) east of the L-31 N canal. Part of the subsurface system is drained by the L-31 W and C-111 canals. The L-31 W canal traverses from the northern boundary to the southern boundary of the S-332D basin. TP from outside the study area is first introduced into the surface system via the S-331 structure and pumped into the detention basins.

The hydrostratigraphy of the model area consists of the surface sediments underlain by the Biscayne Aquifer. In the western portion of the model area, the Biscayne aquifer is underlain by the Gray Limestone Confining Unit, and the Gray Limestone Aquifer. To the east beyond the extent of the Gray Limestone, the Biscayne Aquifer is underlain by the Lower Clastics of the Tamiami Formation. The Biscayne aquifer is conceptualized as a double-porosity medium [32]. This conceptualization is consistent with that presented by Cunningham et al. [72].

Water in the basins is derived from precipitation and water pumped from the canal. This water infiltrates downward through the basin bottoms (which can be the topsoil if present or exposed limestone) to the limestone below. Carried by the infiltration water is the TP from the canal, from wet atmospheric deposition, and from the leaching of the topsoil. It is also possible that the TP detected in groundwater may be associated with eroded sediment and flocculent material as TP concentration spikes were observed to follow heavy rainfall events. Around the detention basins, the water table is usually a few feet below the topsoil–limestone contact. When the basins are inundated by water from the canal or by heavy precipitation, the groundwater elevation in the limestone immediately below rises well above the topsoil–limestone contact, thereby allowing the groundwater to reach chemical equilibrium with the topsoil (where present). Once the TP enters the groundwater, part of it is transported back to the canal and reemerges into the canal downstream.

4.2.3 Model Development

A three-dimensional integrated surface-water/groundwater flow and transport model was developed to investigate the groundwater flow pattern and water quality in and around the detention areas in response to the operations of the system of canals, levees, and pumping stations, as well as climatic conditions. The model was discretized into four subsurface layers with 97 rows and 97 columns. Vertical and horizontal discretization of the groundwater grid of the model was based on the Pennsuco-Dade-Monroe (P-D-M) model [73]. The stratification information of subsurface layers and the associated hydrogeologic properties for the model were imported from the P-D-M model. The topmost subsurface model layer represents the surface sediments. Layer 2 represents the Biscayne Aquifer. The approximate eastern extent of the Gray Limestone Aquifer [74] passes through the model domain, and layers 3 and 4 represent the Gray Limestone Confining Unit and the Gray Limestone Aquifer, respectively, west of this line. To the east of the approximate eastern extent of the Gray Limestone Aquifer, layer 3 represents the lower portions of the Biscayne Aquifer, and layer 4 represents the Lower Clastics of the Tamiami Formation.

Surface-water flow is calculated on the overland grid which coincides areally with the subsurface grid. Grid resolution varies in the horizontal from 240 to 2400 ft. (73.15 to 731.5 m) and in the vertical from 0.1 to 120 ft. (0.03 to 36.6 m). The main canal feature present within the study area is the L-31 N canal flowing into the domain from the north. This canal branches out into canals C-102, C-103, and C-113 before it splits into C-111 and L-31 W canals. The canal network in the model domain was defined by 325 channel nodes.

Hydrostratigraphic and hydraulic information in subsurface layers were obtained from the P-D-M model. The information includes: subsurface layer geometry (thickness and elevation), hydraulic properties (hydraulic conductivity, leakance, and storativity). There are three groundwater pumping wells, east of the C-111 canal, within the model domain of this study (Fig. 2.21).

Manning's coefficient distribution for flow along the overland domain within the wetland areas was obtained from the Southern Inland and Coastal Systems (SICS) study of [75]. For the cultivated soil to the east of the ENP, Manning's coefficient for flow along the overland domain within the agricultural areas was obtained from the USDA database [14, 76]. The rill storage height was set according to the land use types (marl prairie and citrus/row crops) as defined by the South Florida Water Management Model (SFWMM) and P-D-M model. The SFWMM developed by the South Florida Water Management District [77] is a regional-scale hydrologic model used as a planning tool for system-wide evaluations in south Florida. Data for the channel segments, (Manning's coefficient, leakance, and channel geometry) were extracted from the SICS study [70], the SFWMM model, and the P-D-M model.

Water level information is available on a daily basis from January 1, 2000, through December 31, 2007, at 36 observation locations (wells, staff gages, or combinations of both) within and around the model domain. The water level information was also used to provide initial and boundary conditions on the overland domain and in the subsurface. The spatially interpolated head distribution of January 1, 2000, was used as initial conditions for a preliminary simulation which involved stabilizing the solution to steady-state conditions from which transient simulations were initiated.

Boundary conditions for the L-31 N canal running across the model domain include a flux boundary at the upstream reaches where water enters the domain. Daily flow observations from the S-331 monitoring station (see Fig. 2.21) were used as the inflow boundary condition at L-31 N canal for the model. This information was available for the period from January 1, 2000, through December 31, 2007.

The reservoirs or detention areas that exist within the model domain are shown in Fig. 2.21. There are four detention basins within the model domain—S-332BN, S-332B, S-332 C, and S-332D. The detention basins are typically surrounded by 6 ft. high berms which prevent the flow of water across these basins. There are four pumping stations in the L-31 N canal within the model area and one (S-331) just north of the domain. The pumping stations (Fig. 2.21) move water to their respective detention basins. There are also six inline structures (gated weirs and spillways) existing within the model area as located in Fig. 2.21. Various structures are present along the berms of the detention basins to let water flow in and out of the basins. These structures are shown in Fig. 2.21 and generally include weirs or culverts cut into the berm walls of the detention basins.

Data from eight rainfall gaging stations were used to apply average daily rainfall over the model area. This information was from January 1, 2000, through December 31, 2007 and was available on an hourly basis. The hourly data were summed to create daily inputs. The distribution of rainfall was estimated by interpolating between the rain gages.

Evapotranspiration (ET) was simulated according to a parametric model for water removal which describes the ET flux as a function of depth. The maximum ET flux occurs at or above a user defined maximum ET surface elevation, which linearly reduces to zero at or below the extinction surface elevation. The maximum ET surface was set at the shallow root zone and the extinction surface at the deep root zone. The maximum ET flux, which varied on a daily basis, was calculated as an average of the daily values over the simulation time period, at two ENP ET sites located just north and south of the model area.

Groundwater TP concentration data between 2000 and 2004 were collected by the US Army Corps of Engineers (USACE) and provided by the National Park Service (NPS) [76]. There are ten monitoring wells, seven of which are shallow and screened in layer 2 of the model. The remaining three wells are screened in layer 3 of the model. These ten wells are located within or around the S-332B basin. Concentration data from these ten wells were used to establish the transport model's initial conditions. Groundwater TP concentration data between 2005 and 2007 were collected and provided by Geiser et al. [78]. There are 21 monitoring wells, all of which are shallow and screened in layer 2 of the model. The depth of these wells varies from 4 to 14 ft. (1.2 to 4.3 m). There are eight, three, nine, and one observation well(s) in and around Basins S-332BN and BW, S-332 C, S-332DN, and S-332DS, respectively. Concentration data from these wells were used in the calibration of the transport model.

For some periods between 2000 and 2007, water with water-borne phosphorous was pumped from the L-31 N canal to the four detention basins. Daily TP concentration and pumping data related to all S-332 basins were obtained in 2008 [14, 76]. The daily TP concentration and pumping rate data were used as input conditions for the model to ensure that the estimation of the TP flux through the basins was as accurate as possible. The background TP concentration in the canal was typically low, on the order of 8–10 μg/L. Atmospheric wet TP deposition was based on the data across Florida compiled by Pollman et al. [79]. At the ENP study site, the volumetric weighted mean TP concentration for wet deposition was found to be 8 μg/L. This level of concentration is comparable to the typical TP concentration in the canal.

Data relating to topsoil thickness, TP distributions in soil and limestone, and physicochemical properties of the soils and limestone were obtained from several sources. For additional details, the reader is referred to Refs. [76, 80].

4.2.4 Model Calibration

The calibration strategy was designed to produce a model that is capable of simulating the general surface-water and groundwater flow patterns and regional trends in TP concentration. The calibration metrics used in the calibration were: observed groundwater and surface-water elevations, observed flow rates through inline structures, and observed TP in surface water and groundwater. Standard calibration metrics (see Sect. 3.3.3) were used. Calibration criteria are given in Refs. [76, 80]. Below is a summary of the calibration approach.

A steady-state flow condition was first simulated by the model to represent January 1, 2000 conditions, which provides a reproducible starting state from which transients of the model were simulated. The transient simulation was performed for a period of 8 years from January 1, 2000, through December 31, 2007. Data between 2000 and 2003 were used for calibration, and data from the remaining period were used for verification. Once the flow model had been calibrated, the transport component was calibrated. The model was calibrated using an expert interactive (manual) calibration approach. Initial calibration simulations were first conducted to probe the sensitivity of the model to critical parameters, to note cause-and-effect for determining alterations from the original dataset, necessary to achieve calibration. Following this, the model parameters were altered systematically to achieve calibration goals.

For the flow component of the model, excellent agreement between observed and simulated water elevations was obtained. Examples of comparisons between observed and simulated groundwater elevations at well RG4 and observed and simulated surface-water stages at the S-174 canal inline structure are shown in

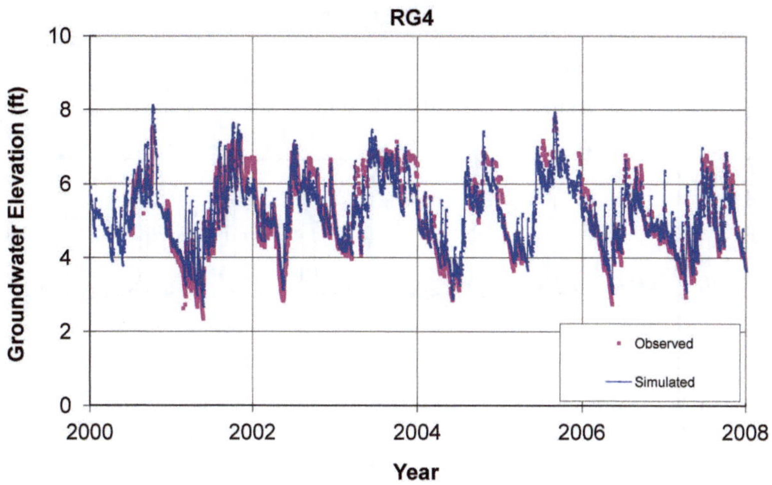

Fig. 2.22 Groundwater elevation at well RG4 versus time

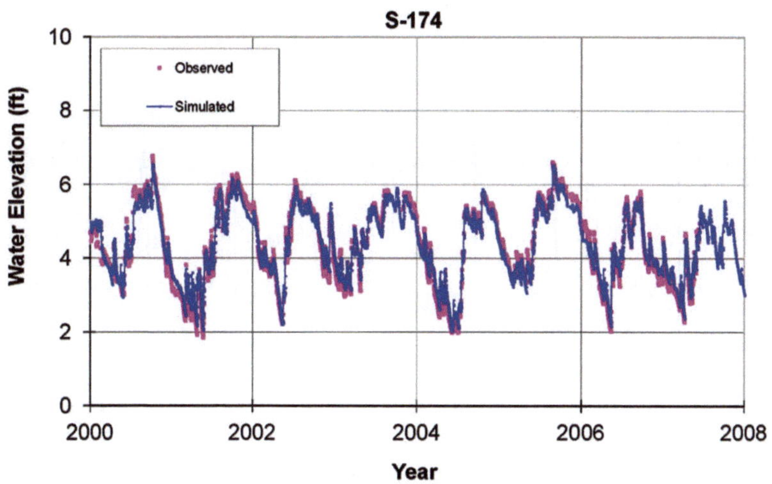

Fig. 2.23 Stage at inline structure S-174 versus time

Figs. 2.22 and 2.23, respectively. For the transport component, the model was able to capture the variation of TP concentration between 2005 and 2007 reasonably well around the detention basins. An example for well MW38 is shown in Fig. 2.24. Note that the peak TP concentrations predicted by the model correspond to peak precipitation but the observed high concentrations tend to lag behind the peak precipitation. However, some of the spikes do not correspond to peak precipitations, suggesting relatively complex transport processes of sediment and flocculent material

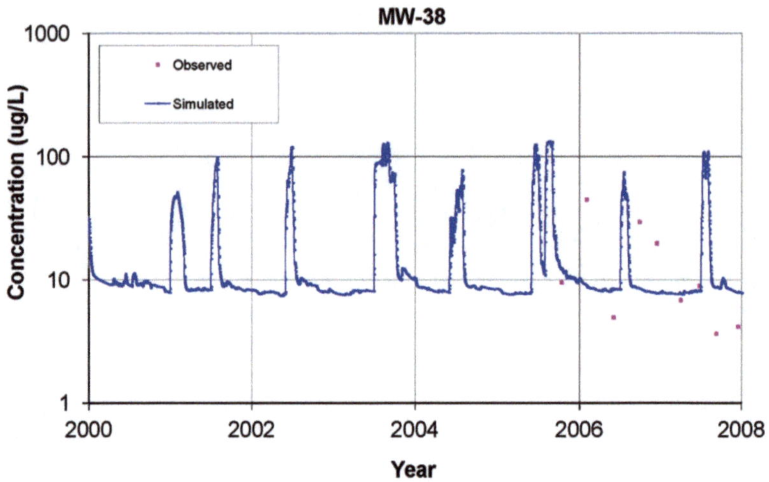

Fig. 2.24 Total phosphorus concentration versus time: Well MW38

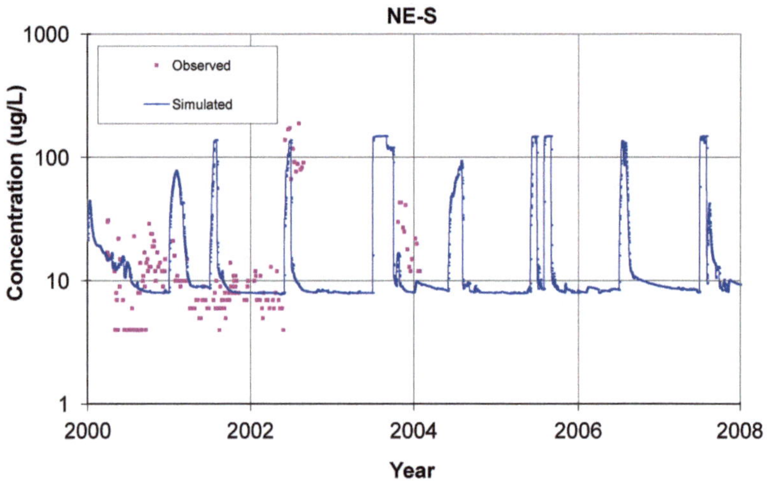

Fig. 2.25 Total phosphorus concentration versus time: Well NE-S

at the surface, as well as complex preferential flow paths in the aquifer below. In Fig. 2.25, it can be seen that at well NE-S, the model is able to simulate the decay of the elevated TP concentration mound that appeared prior to 2000 and the peaks that appear in 2002 and the end of 2003 reasonably well. A representative TP concentration in canal versus time curve near the Basin B pumping station is shown in Fig. 2.26. This figure shows that the TP concentration in the channel as predicted by the model agrees reasonably well with the observed data. All of the above-mentioned observation locations are shown in Fig. 2.21.

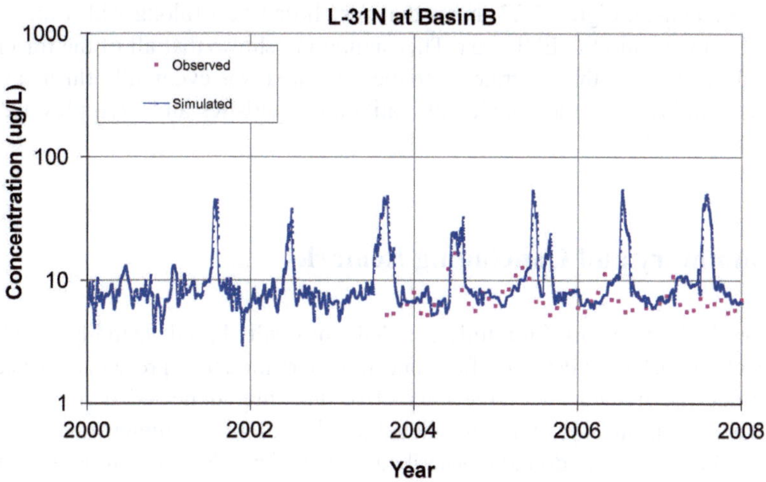

Fig. 2.26 Total phosphorus concentration versus time: L-31 N Canal at Basin B

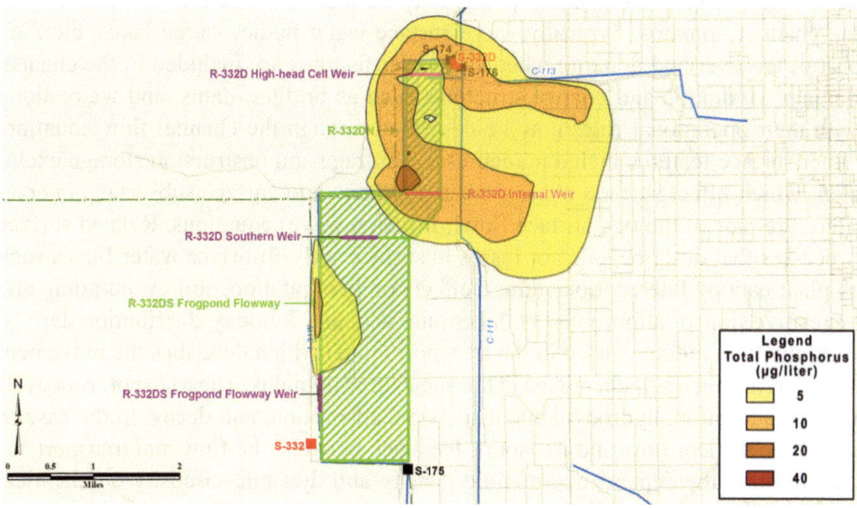

Fig. 2.27 Tracer distribution below the S-322D basin in the Biscayne aquifer (concentration values are in µg/L)

4.2.5 Model's Predictive Analysis Example

The developed model has been used as an analytical and management tool to investigate the operations and the impacts due to the construction of levees, canals, pumping stations, and detention basins on the groundwater flow pattern and water quality as part of the Mod Waters C-111 project. An example, shown in Fig. 2.27, demonstrates the movement of pumped-in water through a tracer (simulated) in the S-332D basin. Figure 2.27 shows that very little tracer (along with water) from Basin D moves into the ENP area. This simulation shows that all of the tracer and water from Basin D that migrated into the ENP area was eventually flushed out to the east. Additional details including calibration statistics and examples are provided in Refs. [14, 76].

5 Summary and Concluding Remarks

A theoretical framework for a fully coupled, physically based, spatially distributed conjunctive surface/subsurface flow and transport model is presented in Sect. 2. The model is divided into three interactive and interconnected domains: three-dimensional subsurface; two-dimensional overland; and one-dimensional channel. Flow in the subsurface domain, described by the Richards equation, is fully three-dimensional and variably saturated. Surface flow, in the overland and channel domains, is described by the two-dimensional and one-dimensional diffusion wave equations, respectively. Interactions between the three domains are described by hydraulic conductance at domain interfaces. Surface water bodies such as reservoirs, lakes, rivers, canals, and ponds may be included in either the overland or the channel domains. Typically, large surface water bodies (large lakes, etc.) are part of the overland domain, whereas smaller features are included in the channel domain. Hydraulic and control structures such as bridges, dams, and weirs along with their operational rules may be described through the channel flow equation. Other surface features including depression storage and obstruction storage exclusion, which affect surface-water flow and storage and surface/subsurface interactions, are part of the two surface flow (diffusion wave) equations. Related surface processes that describe gains or losses in surface and subsurface water fluxes such as plant canopy interception storage effects on precipitation, soil evaporation, and vegetative transpiration, must be taken into account. Velocity distribution derived from the flow model is used in the transport model which describes the movement of different species of chemicals in the same three domains. The transport processes include advection, hydrodynamic dispersion, adsorption, and decay. In the case of densitydependent flow and transport, feedback between the flow and transport occurs through the dependency of fluid density and dynamic viscosity on chemical concentrations.

A domain discretization scheme for finite-difference models is described in Sect. 2.6. The formation of the system equation that describes conjunctive flow and transport in the surface and subsurface domains along with solution techniques are presented in Sect. 2.7. Available simulators are discussed in Sect. 2.8. Model development and calibration are presented and discussed in Sect. 3. The discussion includes, model design, parameterization, calibration, and calibration criteria. Two application examples are provided in Sect. 4. These examples demonstrate the development and calibration of flow and transport models for two watersheds in Florida.

The framework presented in this chapter is very general. It can be extended to include other processes such as thermal transport, snowmelt, sediment transport, and flow and transport in fractured media.

References

1. Rosegrant, M. W., & Cai, X. (2002). Global water demand and supply projections part 2: Results and prospects to 2025. *Water International, 27*(2), 170–182.
2. Loucks, D. P. (1996). Surface water resource systems. In L. W. Mays (Ed.), *Water resources handbook* (pp. 15.3-15.44). New York: McGraw-Hill.
3. Freeze, R. A., & Harlan, R. L. (1969). Blueprint of a physically-based, digitally simulated hydrologic response model. *Journal of Hydrology, 9,* 237–258.
4. Spanoudaki, K., Nanou, A., Stamou, A. I., Christodoulou, G., Sparks, T., Bockelmann, B., & Falconer, R. A. (2005). Integrated surface water-groundwater modelling. *Global NEST Journal, 7*(3), 281–295.
5. Freeze, R. A. (1972). Role of subsurface flow in generating surface runoff: 1. Base flow contribution to channel flow. *Water Resources Research, 8*(3), 609–623.
6. Langevin, C., Swain, E., & Melinda, W. (2005). Simulation of integrated surface-water/ground-water flow and salinity for a coastal wetland and adjacent estuary. *Journal of Hydrology, 314,* 212–234.
7. Swain, E. D., & Wexler, E. J. (1996). A coupled surface-water and ground-water flow model (Modbranch) for simulation of stream-aquifer interaction. Techniques of Water-Resources Investigations of the United States Geological Survey Book 6, Chapter A6.
8. Bradford, S. F., & Katopodes, N. D. (1998). Nonhydrostatic model for surface irrigation. *Journal of Irrigation and Drainage Engineering, 124*(4), 200–212.
9. VanderKwaak, J. E. (1999). *Numerical simulation of flow and chemical transport in integrated surface-subsurface hydrologic systems*. PhD thesis, University of Waterloo, Waterloo, Ontario, Canada.
10. Panday, S., & Huyakorn, P. S. (2004). A fully coupled physically-based spatially distributed model for evaluating surface/subsurface flow. *Advances in Water Resources, 27,* 361–382.
11. Kumar, M., Duffy, C. J., & Salvage, K. M. (2009). A second order accurate, finite volume based, integrated hydrologic modeling (FIHM) framework for simulation of surface and subsurface flow. *Vadose Zone Journal.* doi:10.2136/vzj2009.0014.
12. Khambhammettu, P., Kool, J., Tsou, M.-S., Huyakorn, P. S., Guvanasen, V., & Beach, M. (2009). Modeling the integrated surface-water groundwater interactions in West-Central Florida, USA. The International Symposium on Efficient Groundwater Resources Management. *The Challenge of Quality and Quantity for Sustainable Future*, Bangkok, Thailand, February 16-21, 2009.
13. Barr, A., & Barron, O. (2009). *Application of a coupled surface water-groundwater model to evaluate environmental conditions in the Southern River catchment.* Commonwealth Scientific and Industrial Research Organisation: Water for a Healthy Country National Research Flagship, Australia.
14. Guvanasen, V., Wei, X. Y., Huang, D., Shinde, D., & Price, R. (2011). *Application of MODHMS to simulate integrated water flow and phosphorous transport in a highly interactive surface water groundwater system along the eastern boundary of the Everglades National Park, Florida*. MODFLOW and More 2011: Integrated Hydrologic Modeling Conference. The Colorado School of Mines, Golden, Colorado, USA. June 5-8, 2011.
15. Huang, G., & Yeh, G.-T. (2012). Integrated modeling of groundwater and surface water interactions in a manmade wetland. *Terrestrial, Atmospheric and Oceanic Sciences, 23*(5), 501–511.

16. Panday, S., & Huyakorn, P. S. (2008). MODFLOW SURFACT: A state-of-the-art use of vadose zone flow and transport equation and numerical techniques for environmental evaluations. *Vadose Zone Journal, 7*(2), 610–631.
17. Zhang, Q., & Werner, A. D. (2012). Integrated surface-subsurface modeling of Fuxianhu Lake catchment, Southwest China. *Water Resources Management, 23*(11), 2189–2204.
18. Bear, J. (1972). *Dynamics of fluids in porous media* (p. 764). New York: Elsevier.
19. Bear, J. (1979). *Hydraulics of groundwater* (p. 569). New York: McGraw-Hill.
20. Eagleson, P. S. (1969). *Dynamic hydrology* (p. 462). New York: McGraw-Hill.
21. Viessman, W., & Lewis, G. (1996). *Introduction to hydrology* (p. 760). New York: Harper-Collins.
22. Guvanasen, V., & Chan, T. (2000). A three-dimensional numerical model for thermohydromechanical deformation with hysteresis in fractured rock mass. *International Journal of Rock Mechanics and Mining Sciences, 37,* 89–106.
23. van Genuchten, M. Th. (1980). A closed-form equation for predicting the hydraulic conductivity of unsaturated soils. *Soil Science Society of America Journal, 44,* 892–898.
24. Guymon, G. L. (1994). *Unsaturated zone hydrology* (p. 210). New Jersey: Prentice Hall.
25. Wu, Y. S., Huyakorn, P. S., & Park, N. S. (1994). A vertical equilibrium model for assessing nonaqueous phase liquid contamination and remediation of groundwater systems. *Water Resources Research, 30,* 903–912.
26. Huyakorn, P. S., Panday, S., & Wu, Y. S. (1994). A three-dimensional multiphase flow model for assessing NAPL contamination in porous and fractured media: 1. Formulation. *Journal of Contaminant Hydrology, 16,* 109–130.
27. Gottardi, G., & Venutelli, M. (1993). A control-volume finite-element model for two-dimensional overland flow. *Advances in Water Resources, 16,* 277–284.
28. Chow, V. T., Maidment, D. R., & Mays, L. W. (1988). *Applied hydrology* (p. 572). New York: McGraw-Hill.
29. Chow, V. T. (1959). *Open-channel hydraulics* (p. 680). New York: McGraw-Hill.
30. Scheidegger, A. E. (1961). General theory of dispersion in porous media. *Journal of Geophysical Research, 66,* 3273–3278.
31. Roberson, J. A., & Crowe, C. T. (1985). *Engineering fluid mechanics* (3rd edn., p. 503). Boston: Houghton Mifflin.
32. HydroGeoLogic, Inc. (2012). *MODHMS-A MODFLOW-based hydrologic modeling system. Documentation and user's guide*. Reston, Virginia: HydroGeoLogic, Inc.
33. Kristensen, K. J., & Jensen, S. E. (1975). A model for estimating actual evapotranspiration from potential evapotranspiration. *Nordic Hydrology, 6,* 170–188.
34. Wigmosta, M. S., Vail, L. W., & Lettenmaier, D. P. (1994). A distributed hydrology-vegetation model for complex terrain. *Water Resources Research, 30*(6), 1665–1679.
35. Monteith, J. L. (1981). Evaporation and surface temperature. *Quarterly Journal of the Royal Meteorological Society, 107,* 1–27.
36. Senarath, S. U. S., Ogden, F. L., Downer, C. W., & Sharif, H. O. (2000). On the calibration and verification of two-dimensional, distributed, hortonian, continuous watershed models. *Water Resources Research, 36*(6), 1495–1510.
37. Feddes, R. A., Kowalik, P. J., & Zaradny, H. (1978). *Simulation of field water use and crop yield*. New York: Wiley.
38. Woolhiser, D. A., Smith, R. E., & Giraldez, J.-V. (1997). Effects of spatial variability of saturated hydraulic conductivity on hortonian overland flow. *Water Resources Research, 32*(3), 671–678.
39. McDonald, M. G., & Harbaugh, A. W. (1988). *A modular three-dimensional finite-difference ground water flow model*. United States Geological Survey Open File Report 83-875. Washington, DC.
40. Graham, N., & Refsgaard, A. (2001). MIKE SHE: A distributed, physically based modeling system for surface water/groundwater interactions. In *Proceedings of "MODLFOW 2001 and other modeling Odysseys,"* Golden, Colorado, USA. pp. 321–327.

41. Panday, P., Brown, N., Foreman, T., Bedekar, V., Kaur, J., & Huyakorn, P.S. (2009). Simulating dynamic water supply systems in a fully integrated surface-subsurface flow and transport model. *Vadose Zone Journal.* doi:10.2136/vzj2009.0020.
42. Celia, M. A., Bouloutas, E. T., & Zarba, R. L. (1990). A general mass-conservative numerical solution for the unsaturated flow equation. *Water Resource Research, 27*(7), 1483–1496.
43. Huyakorn, P. S., & Pinder, G. F. (1983). *Computational methods in subsurface flow* (p. 473). London: Academic.
44. Huyakorn, P. S., Springer, E. P., Guvanasen, V., & Wadsworth, T. D. (1986). A three dimensional finite element model for simulating water flow in variably saturated porous media. *Water Resources Research, 22*(12), 1790–1808.
45. Vinsome, P. K. W. (1976). Orthomin, an iterative method for solving sparse sets of simultaneous linear equations. In *Proceedings of Society of Petroleum Engineers Symposium on Numerical Simulation of Reservoir Performance*, Los Angeles, California, USA.
46. van der Vorst, H. A. (1992). Bi-CGSTAB: A fast and smoothly converging variant of Bi-CG for the solution of nonsymmetric linear systems. *SIAM Journal on Scientific and Statistical Computing, 13,* 631–644.
47. Forsyth, P. A. (1993). *MATB user's guide iterative sparse matrix solver for block matrices.* Canada: Department of Computer Science University of Waterloo.
48. Yeh, G. T., & Huang, G. (2003). *A numerical model to simulate water flow in watershed systems of 1-D stream-river network, 2-D overland regime, and 3-D subsurface media (WASH123D: Version 1.5).* Orlando, Florida: Department of Civil and Environmental Engineering, University of Central Florida.
49. Yeh, G. T., Huang, G., Cheng, H. P., Zhang, F., Lin, H. C., Edris, E., & Richards, D. (2006). A first principle, physics-based watershed model: WASH123D. In: V. P. Singh & D. K. Frevert (Eds.), *Watershed models.* Boca Raton: CRC.
50. Therrien, R., McLaren, R. G., & Sudicky, E. A. (2007). *Hydrogeosphere-a three-dimensional numerical model describing fully integrated subsurface and surface flow and solute transport.* Canada: Groundwater Simulations Group, University of Waterloo.
51. Kollet, S. J., & Maxwell, R. M. (2006). Integrated surface-groundwater flow modeling: A free-surface overland flow boundary condition in a parallel groundwater flow model. *Advances in Water Resources, 29,* 945–958.
52. Anderson, M. P., & Woessner, W. W. (1992). *Applied groundwater modeling* (p. 381). San Diego: Academic.
53. Poeter, E. P., & Hill, M. C. (1998). *Documentation of UCODE: A computer code for universal inverse modeling.* United States Geological Survey Water Resources Investigations Report 98-4080, Washington, DC.
54. Doherty, J. (2000). *PEST, model-independent parameter estimation.* Australia: Watermark Numerical Computing.
55. Duan, Q., Gupta, V. K., & Sorooshian, S. (1992). Effective and efficient global optimization for conceptual rainfall-runoff models. *Water Resources Research, 28,* 1015–1031.
56. Vrugt, J. A., Gupta, H. V., Bouten, W., & Sorooshian, S. (2003). A shuffled complex evolution metropolis algorithm for optimization and uncertainty assessment of hydrological model parameters. *Water Resources Research, 39,* 1201–1218.
57. Schoups, G., Lee Addams, C., & Gorelick, S. M. (2005). Multi-objective calibration of a surface water-groundwater flow model in an irrigated agricultural region: Yaqui Valley, Sonora, Mexico. *Hydrology and Earth System Sciences, 9,* 549–568.
58. Vrugt, J. A., Gupta, H. V., Bastidas, L. A., Bouten, W., & Sorooshian, S. (2003). Effective and efficient algorithm for multiobjective optimization of hydrologic models. *Water Resources Research, 39,* 1214–1232.
59. Duan, Q., Gupta, H. V., Sorooshian, S., Rousseau, A. N., & Turcotte, R. (Eds.). (2003). *Calibration of watershed models.* Water science and application 6 (p. 345). Washington, DC: American Geophysical Union.

60. Vrugt, J. A., Nuallain, B. O., Robinson, B. A., Bouten, W., Dekker, S. C., & Sloot, P. M. A. (2006). Application of parallel computing to stochastic parameter estimation in environmental models. *Computers & Geosciences, 32,* 1139–1155.
61. Schreuder, W. A. (2009). Running BeoPEST. In *Proceedings of the 1st PEST conference*, Potomac, Maryland, USA, November 1-3, 2009.
62. Hunt, R. J., Luchette, J., Schreuder, W. A., Rumbaugh, J. O., Doherty, J., Tonkin, M. J., & Rumbaugh, D. B. (2010). Using a cloud to replenish parched groundwater modeling efforts. *Ground Water, 48*(3), 360–365.
63. Hill, M. C., & Tiedeman, C. R. (2007). *Effective groundwater model calibration* (p. 455). New Jersey: Wiley.
64. Nash, J. E., & Sutcliffe. J. V. (1970). River flow forecasting through conceptual models part I-a discussion of principles. *Journal of Hydrology, 10*(3), 282–290.
65. Smith, M. B., Laurine, D. B., Koren, V. I., Reed, S. M., & Zhang, Z. (2003). Hydrologic model calibration strategy accounting for model structure. In: Q. Duan, H. V. Gupta, S. Sorooshian, A. N. Rousseau, & R. Turcotte (Eds.), *Calibration of watershed models*, Water science and application 6 (p. 133–152). Washington, DC: American Geophysical Union.
66. Beach, M. H. (2006). *Southern district ground-water flow model, version 2.0.* Hydrologic Evaluation Section, Resource Conservation and Development Department, Southwest Florida Water Management District, Brooksville, Florida.
67. Environmental Simulations, Inc. (2004). *Development of the district wide regulation model.* Report submitted to the Southwest Florida water Management District, Brooksville, Florida.
68. HydroGeoLogic, Inc. (2011). *Peace river integrated modeling project (PRIM) phase IV: Basin-wide model.* Report submitted to the Southwest Florida Water Management District, Brooksville, Florida.
69. HydroGeoLogic, Inc. (2009). *Peace river integrated modeling project (PRIM) phase III: Saddle creek basin integrated model.* Report submitted to the Southwest Florida Water Management District, Brooksville, Florida.
70. HydroGeoLogic, Inc. (2012). *Peace river integrated modeling project (PRIM) phase V: Predictive model simulations.* Report submitted to The Southwest Florida Water Management District.
71. U.S. Army Corps of Engineers (USACE). (2002). *Central and Southern Florida Project, Canal-111 South Dade County, Florida-S-332D detention area pre-operations and start-up monitoring data review report.* U.S. Army Corps of Engineers, Jacksonville District, Jacksonville, Florida.
72. Cunningham, K. J., Sukop, M. C., Huang, H., Alvarez, P. F., Curran, H. A., Renken, R. A., & Dixon, J. F. (2009). Prominence of ichnologically influenced macroporosity in the karst system of biscayne aquifer: Stratiform "Super-K" zones. *Geological Society of America Bulletin, 121*(1/2), 164–180.
73. Evans, R. A. (2000). *Calibration and verification of the MODBRANCH numerical model of South Dade County, Florida.* U. S. Army Corps of Engineers, Jacksonville District, February, 2000.
74. Fish, J. E., & Stewart, M. (1991). *Hydrogeology of the surficial aquifer system, Dade County, Florida.* U.S. Geological Survey Water-Resources Investigations Report 90-4108, 56 pp.
75. Swain, E. D., Wolfert, M. A., Bales, J. D., & Goodwin, C. R. (2004). *Two-dimensional hydrodynamic simulation of surface-water flow and transport to Florida Bay through the Southern Inland and Coastal Systems (SICS).* U.S. Geological Survey Water-Resources Investigations Report 03-4287. 56 pp.
76. HydroGeoLogic, Inc. (2010) *Surface water groundwater flow and transport model development for the eastern boundary of Everglades National Park.* Report submitted to The Florida International University, Miami, Florida, and National Park Services, Homestead, Florida.
77. South Florida Water Management District. (1997). *Documentation for the South Florida water management model.* Hydrologic Systems Modeling Division, Planning Department, South Florida Water Management District, West Palm Beach, Florida.

78. Geiser, E., Price, R., Scinto, L., & Trexler, J. (2008). *Phosphorus retention and subsurface movement through the S-332 detention basins on the eastern boundary of the Everglades National Park*. Florida International University, Report submitted to National Park Services, Homestead, Florida.
79. Pollman, C. D., Landing, W. M., Perry, J. J., & Fitzpatrick, T. (2002). Wet deposition of phosphorus in Florida. *Atmospheric Environment, 36,* 2309–2318.
80. HydroGeoLogic, Inc. (2006). *Conceptual and numerical model development using MODHMS for marsh driven operations at S −332 detention basins, prepared for: South Florida Ecosystem Office Everglades National Park*. Report submitted to The Florida International University, Miami, Florida, and National Park Services, Homestead, Florida.

Chapter 3
River Channel Stabilization with Submerged Vanes

A. Jacob Odgaard

Contents

Abstract Submerged vanes are an unobtrusive and cost-effective way for river engineers to address many problems associated with river channel stability and river management in general. The vanes are small flow-training structures designed and installed on the riverbed to modify the near-bed flow pattern and redistribute flow and sediment transport within the channel cross section. The structures are laid out so they create and maintain a flow and bed topography that is consistent with that of a stable channel creating optimum conditions for managing the river. A relatively new technology, submerged vanes are a low-impact method for restoring riverbanks,

A. J. Odgaard (✉)
IIHR-Hydroscience and Engineering, University of Iowa, 4114 Seamans Center for the Engineering Arts and Sciences, Iowa City, IA 52242, USA
e-mail: jacob-odgaard@uiowa.edu

C. T. Yang, L. K. Wang (eds.), *Advances in Water Resources Engineering*,
Handbook of Environmental Engineering, Volume 14, DOI 10.1007/978-3-319-11023-3_3

stabilizing or re-meandering river reaches previously modified (straightened) by humans, increasing flood flow capacity, reducing sediment deposits, and for helping maintain or enhance the ecosystem in and around rivers. Following laboratory research and feedback from field installations, guidelines are now available for designs that are effective and sustainable. These guidelines are described in the book by Odgaard *River Training and Sediment Management by Submerged Vanes,* ASCE Press, 2009. Following a brief summary of the theory with illustrations from the ASCE book (reprinted with permission of ASCE), this chapter presents the latest feedback from field installations and suggestions for future applications.

Keywords Submerged vanes · River restoration · Rivers · Re-meandering · River realignment · Sediment management

1 Introduction

Rivers play an important role in society. They provide water for irrigation, water supply, power generation, and many other uses. They also cause disasters primarily during floods when they inundate portions of the floodplain and destroy property and infrastructure. The sediment they transport from the watershed to the ocean often interferes with both navigation in the rivers and infrastructure along them. One of the river engineer's major tasks is to help facilitate optimum usage of this resource and at the same time provide protection against disasters. River training is the common solution. Traditionally, river training consists of construction of revetments, dikes, wing dams, weirs, and by dredging; and in some cases, construction of bypass channels. These techniques function by adjusting bank resistance and/or bank erodibility and/or flow and bed topography.

Revetments are structures that are aligned parallel to the flow. They are used most often to protect eroding banks and to form a smooth bank line. Types range from simple riprap embankments to revetments made with rock mattresses (e.g., gabions) and articulated concrete mattresses to revetments made with wooden piles and stone-filled trenches. Bio-mattresses are also used to promote vegetation on the banks. Dikes, wing dams, submerged vanes, and weirs (e.g., bend-way weirs) are structures placed at an angle to the flow. They are typically used for: (a) faring out sharp bends to a larger radius of curvature to provide a more desirable channel alignment (and thus stabilize concave banks); (b) closing off secondary channels and old bend ways; (c) redistributing flow within a channel cross section (for example, to constrict a channel to increase depth in certain areas or to transform a braided river into a single channel); and (d) protecting bridges, utility crossings, and structures along the bank. Most dikes are made with stone fill, but other materials are also used. The so-called bend-way weirs are made with rocks; they function like dikes, and are oriented upstream at an angle of, typically, 60–80° with the bank. Finally, dredging is the process of moving material from one part of a channel to another or to a disposal site on land. Dredging is used most often for deepening or widening navigation channels or for land reclamation.

All of the aforementioned techniques have been perfected over many years, and the experience and design guidelines are well documented in the literature (numerous reports by the US Army Corps of Engineers).

The submerged-vane technique is a relatively new and not-so-well-documented technique whose promise is becoming more evident every day. Both laboratory and field tests suggest that this technique has a broad range of applications. Vanes have already been installed in many rivers throughout the world, including the Nile River, Egypt; Waikato River, New Zealand; Kosi River, Nepal; Kuro River, Japan; Missouri River, the USA; and a number of smaller rivers in the Eastern and Midwest states of the USA. Feedback from these and other sites have resulted in an improved understanding of the functioning of vanes and improved design basis [1]. Following a brief summary of the theory with illustrations from [1], reprinted with permission of ASCE, this chapter presents the latest feedback from field installations and suggestions for future applications.

1.1 Concept

Submerged vanes are small flow-training structures (foils) designed to modify the near-bed flow pattern and redistribute flow and sediment transport within the channel cross section. The structures are installed at an angle of attack of, typically, 10–20° with the flow, and their initial height is 0.2–0.4 times the local water depth at design stage.

The vanes function by generating secondary circulation in the flow. The circulation alters magnitude and direction of the bed shear stresses and causes a change in the distribution of velocity, depth, and sediment transport in the area affected by the vanes. As a result, the riverbed aggrades in one portion of the channel cross section and degrades in another.

Typically, vanes are installed in arrays along one side (or both sides) of a river channel over a reach long enough to create a desired flow redistribution. Their advantage over traditional training structures, such as dikes and groins, is that they can produce a given redistribution of flow at less resistance to the flow and at less cost. Vanes produce flow redistribution by vorticity, whereas, groins and dikes, which are usually placed normal to the flow, produce flow redistribution by simple continuity and drag force. Because they are nearly aligned with the flow, the associated drag force is relatively small. Their alignment also eliminates the problem of structural stability associated with local scour, which is often a concern with the traditional structures. An important point in regard to flow resistance is that the resulting lower velocity within the vane field not only causes a reduction in flow depth but also results in dunes in the vane field that are smaller than prior to the installation of vanes. So, in the vane field, the increase in flow resistance due to vane-induced drag is partially outweighed by the decrease of flow resistance due to smaller size of bedforms. As a result, the overall change in water-surface slope is often negligible. This feature makes the vanes ideally suited for sustainable adjustments of flow in the river.

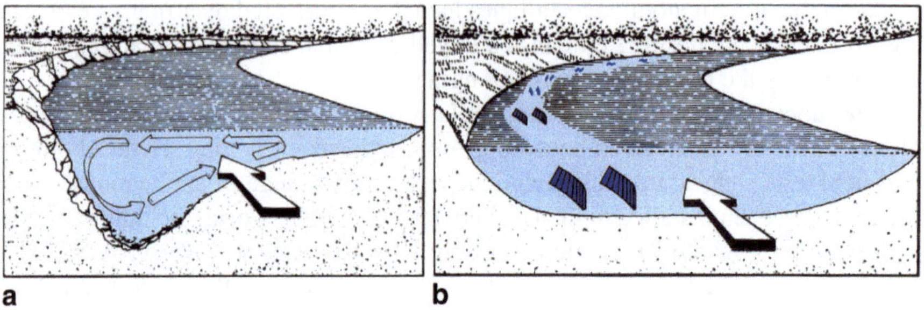

Fig. 3.1 Submerged vanes for mitigating stream bank erosion, **a** naturally occurring secondary current in river bend, **b** vane-induced secondary current eliminates the naturally occurring secondary current and stabilizes riverbank. (Source: Odgaard [1], with permission from ASCE)

1.2 Developments

The first known attempts to develop a theoretical design basis are by Odgaard and Kennedy [2] and Odgaard and Spoljaric [3]. Odgaard's and Kennedy's efforts are aimed at designing a system of vanes to stop or reduce bank erosion in river curves. In such an application, the vanes are laid out so that the vane-generated secondary current eliminates the centrifugal-induced secondary current, which is the root cause of bank undermining (Fig. 3.1). The centrifugally induced secondary current in river bends, also known as the transverse circulation or helical motion, results from the difference in centrifugal acceleration along a vertical line in the flow because of the nonuniform vertical profile of the velocity. The secondary current forces high-velocity surface current outward and low-velocity near-bed current inward (Fig. 3.1a). The increase in velocity at the outer bank increases the erosive attack on the bank, causing it to fail. By directing the near-bed current toward the outer bank, the submerged vanes counter the centrifugally induced secondary current, thereby inhibiting bank erosion. The vanes stabilize the toe of the bank (Fig. 3.1b).

Both laboratory and field tests have shown [1] that vane systems can be configured to stabilize meandering channels and channels with shoaling problems. Tests have shown that significant changes in depth and velocity distributions can be achieved without causing significant changes in cross-sectional area, energy slope, and downstream sediment transport. The changes in cross-sectional average parameters are small because the vane-induced secondary current changes the direction of the bed shear stresses by only a small amount. The vanes' effectiveness in sediment redistribution is particularly useful at diversions and water intakes where shoaling problems are notorious. Strategically placed, the vanes have been shown [1] to effectively prevent bed load from entering the diversion or water intake while maintaining the stability of the main channel.

Most applications so far have been with a simple, flat-panel design, where the vanes are made of either reinforced concrete (Fig. 3.2) or sheet piles, preferably flat-panel sheet piles like those used recently [4] at the Greenville Utilities Commission

Fig. 3.2 Precast concrete vane panels being placed between H-pile supports. Placement guides extend temporarily above H-columns. (Source: Odgaard [1], with permission from ASCE)

Fig. 3.3 Flat-panel sheet pile vane ready for installation at the Greenville Utilities Commission water supply intake on Tar River, North Carolina, 2012. Only the topmost 1.5–2.0 ft will be above the current bed level. (Courtesy of the Greenville Utilities Commission)

water supply intake on Tar River, North Carolina (Fig. 3.3). Laboratory studies [1] have shown that vane efficiency can be improved by making the vane double-curved as shown in Fig. 3.4.

Finally, vanes may be used effectively in conjunction with traditional river training strategies. For example, protecting a riverbank using a system of submerged vanes together with a moderate toe protection with riprap is, in many cases, less expensive than, and equally effective as a full-height riprap embankment. The vane–riprap solution is also more environmentally attractive because both vanes and riprap will be submerged most of the time, allowing the upper bank to maintain

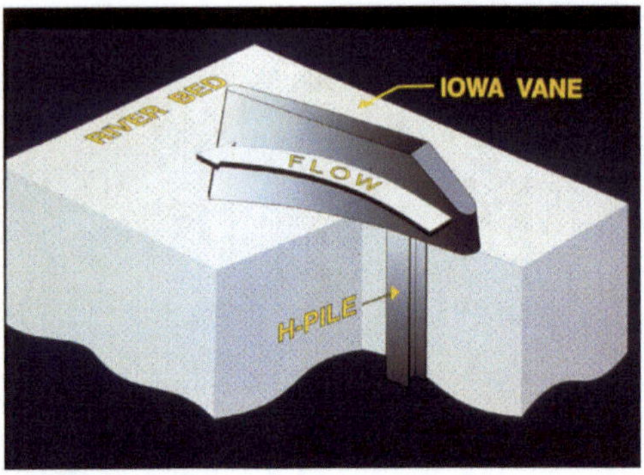

Fig. 3.4 Sketch showing improved final design. (Source: Odgaard [1] with permission from ASCE)

its natural structures and ecosystem. As will be demonstrated later, vanes also work well in combination with dikes and wing dams. In this case, the dikes and wing dams are installed upstream to stabilize a river segment and provide optimum approach-flow conditions for the vane system.

1.3 Sustainability

As suggested earlier, vanes are ideally suited for sustainable adjustments of the flow in a river channel. This is because the adjustments are made without causing significant changes in the variable that has probably the greatest effect on channel stability, namely, water-surface slope or energy expenditure per unit length. By preserving slope and rate of energy expenditure of the river flow, vanes also help preserve the ecosystem of the river and its environment. However, at the same time, it is important to recognize that the vane-induced adjustments are static. River channels are dynamic. Many river channels maintain their dynamic equilibrium by the natural process of meandering, that is, a certain amount of bank-line migration may be part of the natural process of maintaining dynamic equilibrium. Therefore, in certain river environments, vane design must be preceded by a channel stability analysis so the vanes can be laid out to anticipate migration and provide long-term stability.

A channel stability analysis is also important when new channel alignments are being designed, for example, in channel restoration projects and in re-meandering of previously straightened channels. An example of a stability analysis is given later in this chapter. The example shows how a stable meander planform is obtained by perturbation stability analysis, and how vanes may be used to preserve the stable planform. However, channel restoration is still an area of much research, and a stability analysis based on a single-thread meander planform may not always be sufficient for a sustainable solution.

2 Theory

2.1 Airfoil Analogy

A submerged vane at a small angle of attack with the flow induces a horizontal circulation in the flow downstream. The circulation arises because the vertical pressure gradients on the two surfaces of the vane cause the fluid flowing along the high-pressure (upstream) side to acquire an upward velocity component, whereas on the low-pressure (downstream) side there is a downward velocity component. The resulting vortices (vortex sheet) at the trailing edge of the vane roll up to form a large vortex springing from a position near the top of the vane. This vortex is carried with the flow downstream, where it gives rise to a secondary or helical motion of the flow and associated changes in bed shear stress and bed topography. These changes can be calculated [1].

The area of streambed affected by a single vane is limited. To generate a larger, coherent vortex that affects the flow pattern over a wider area of the channel cross section, several vanes must be employed. If the vanes are arranged in an array as shown in Figs. 3.5 and 3.6, the width of the affected area is increased. Calculations and experiments show that a vane spacing of two to three times vane height is appropriate for the vortices induced by a vane pair to merge into a distinct, common vortex with relatively large transverse velocity in the region between the vane axes and not too large reduction of circulation per vane due to vortex interference.

To sustain a certain induced circulation and induced bed shear stress downstream, the vane array must be repeated at intervals in the downstream direction. The distance between the arrays depends on the design objective, which must stipulate lower limits on induced stresses. Formulas for estimating induced bed shear stresses as well as the impact on a channel's energy slope are presented in [1].

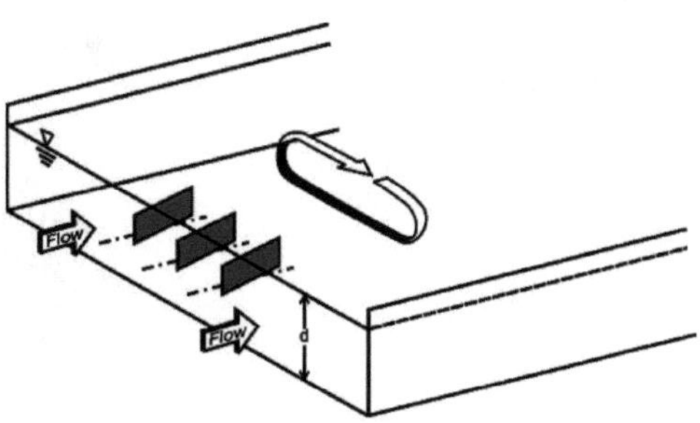

Fig. 3.5 Schematic showing circulation induced by array of three vanes. (Source: Odgaard [1] with permission from ASCE)

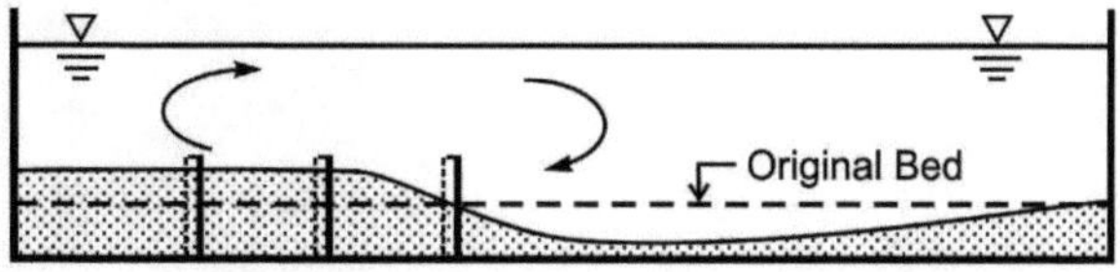

Fig. 3.6 Schematic showing change in bed profile induced by array of three vanes. (Source: Odgaard [1] with permission from ASCE)

2.2 Flow Equations and Solutions

To calculate the effect of a submerged-vane system on flow and bed topography in an alluvial channel, the vane-induced stress distribution is introduced into the equations for conservation of mass (water and sediment) and momentum. The resulting equations are reduced using order of magnitude analysis and applying a stability criterion for sediment particles on the streambed [5].

Odgaard and Wang [5] solve the governing equations using a finite difference scheme. The boundary condition is obtained from the continuity equation. The computation is carried out starting at the bank farthest away from the vanes. In a river curve, this is normally the inner bank. Initially, the flow depth at the starting point is set equal to the pre-vane flow depth, and the cross-sectional distributions are calculated. If these distributions do not satisfy the boundary condition, a new starting depth is selected. The process is repeated until the boundary condition is fulfilled.

2.3 Laboratory Validation Tests

A series of laboratory tests were conducted to validate the theory, one series of proof-of concept tests conducted in a rigid-bed channel and several test series conducted in movable-bed channels [1, 6]. Figure 3.7 is from a test in straight movable-bed channel; it shows the bed topography after draining most of the water from the flume.

Fig. 3.7 Upstream view of a nearly drained, straight channel with vanes. Before the water was drained from the flume, flow depth was about 18.2 cm; discharge 0.154 m^3/s; and water-surface slope 0.00064. The vanes reduced the depth near the right bank by about 50 %; this caused the depth near the left bank to increase by 20–30 %

One of the most important observations made in the tests in both the curved and straight flumes is that the vane-induced changes occurred without causing significant changes in the area of the cross sections and of the longitudinal slope of the water surface. The changes in slope were less than 8 %. This observation is important because it implies that the vanes will not cause any changes of the stream's sediment-transport capacity upstream and downstream from the vane field and, therefore, should not alter the overall characteristics of the stream.

Another notable observation is that the vane-induced redistribution of sediment within a cross section is an irreversible process in the sense that a reduction of discharge does not lead to recovery of original distributions. A reduction in discharge does not result in a reduction in the volume of sediment accumulated in the vane field because, at the lower discharge, the sediment-transport capacity in the vane field is too low to remove the sediment that accumulated at the higher discharge.

3 Field Tests and Experience

3.1 Stabilization of River Channel Alignment

As mentioned, many river channel stabilization projects include realignment of the channel, for example re-meandering of a reach previously straightened. In such cases, design is preceded by a channel stability analysis. Such an analysis consists of either a review of historical, sequential aerial photos of stable upstream and downstream channel reaches, or a formal perturbation stability analysis or both [7].

The following is a stabilization project in which a vane system was installed to ameliorate a channel instability problem caused by channel straightening. The river is the West Fork Cedar River, Iowa, USA [1]. The channel was straightened and widened at the time of construction of a new bridge (1970) to allow a 100-year flood flow to pass through. Figure 3.8 shows the excavation plan for the project. It is a 150-m-long, 9-m-wide, six-span, I-beam bridge with the road surface about 5 m above the low-flow streambed. The top width and bank-full depth of the river upstream of the excavation are 30–40 m and 1.9–2.1 m, respectively. The bed material is sand with a median particle diameter of about 0.5 mm. Annual mean flow in the river is about 14 m^3/s and bank-full flow about 100 m^3/s. By 1984, a considerable portion of the excavation upstream from the bridge had filled in and become vegetated. Figure 3.9a shows the 1984 bank line and a sandbar that subsequently developed along the left bank. The sandbar occupied four of the six spans, causing the flow to be thrown toward the right bridge abutment, where it undermined and eroded the bank. Annual dredging became necessary because the sandbar grew in size after each storm. It was clear from aerial photos that the bar formed as part of the river's adjustment to the 1970 channel straightening, which essentially eliminated two meanders and shortened the channel segment by 482 m (from 1189 to 707 m). The straightening resulted in a 69 % increase in the local channel slope, from 0.00047 to 0.00083. This increase in slope caused the channel reach to transition from a meandering regime toward a braided regime.

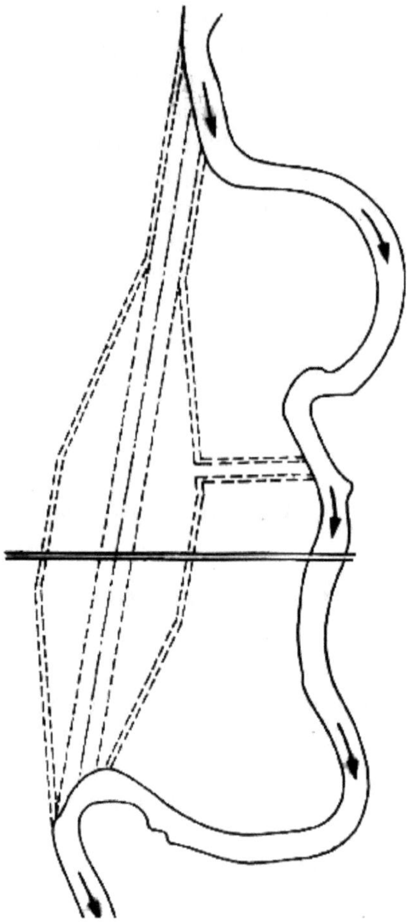

Fig. 3.8 Excavation plan for West Fork Cedar River channel straightening

A system of 12 vanes (four arrays with three vanes in each array) was installed in the summer of 1984. The layout is shown in Fig. 3.9b. Each vane consists of vertical sheet piles driven into the streambed and aligned at a 30° angle with the main channel. With this angle, the vanes are at about 20° with the 1984 mean flow direction, which is indicated by the arrows in Fig. 3.9a. Each sheet piling is 3.7 m long, and its top elevation is 0.6 m above the streambed. The vane system was designed to cause flow depth and velocity to decrease along the right bank and increase along the centerline. As seen in Fig. 3.10, the system has accomplished this. A permanent, protective berm now is seen along the bank that was previously eroding (along a reach of about 450 m). In fact, the vanes are now maintaining a cross-sectional bed profile similar to that designed when the bridge was constructed. Nine of the 12 vanes are now permanently covered with vegetation. Maintenance has not been necessary.

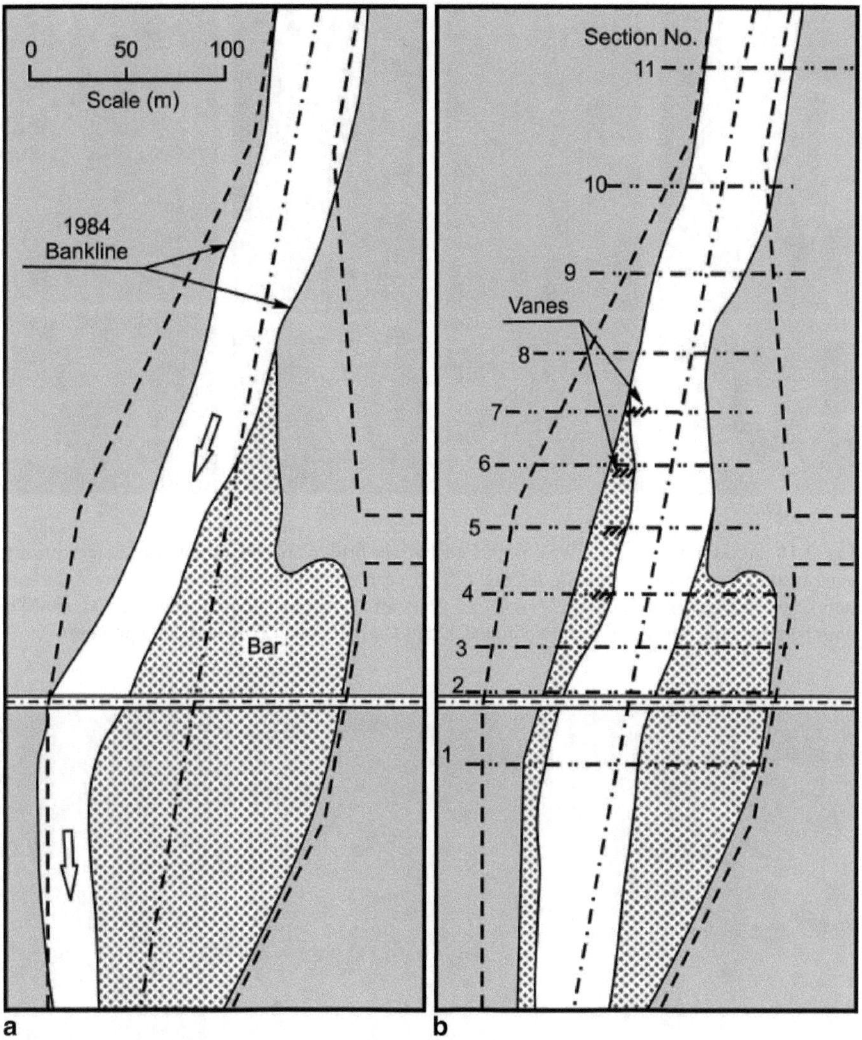

Fig. 3.9 Plan of West Fork Cedar River bridge crossing, **a** prior to vane installation in 1984, and **b** 5 years after vane installation. (Source: Odgaard [1], with permission from ASCE)

A major flood occurred in 2008 which by several accounts exceeded the 100-year flood. Portions of the right bank were undermined and washed away. The right photo in Fig. 3.10 shows the low flow channel the following year 2009. As seen, the "damages" were minor. The right bank did not come even close to the bridge abutment. As seen in Fig. 3.11, the channel is recovering, and in 2011 the right bank is essentially back to where it was prior to the 2008 flood.

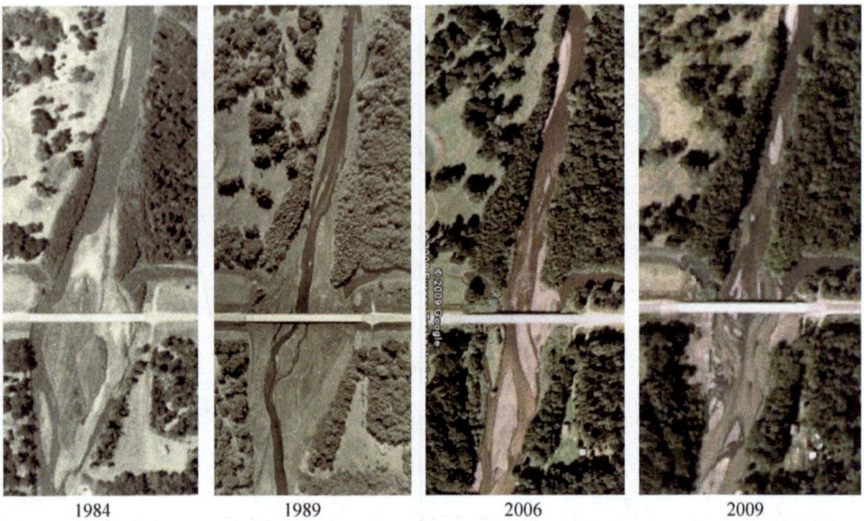

Fig. 3.10 Aerial photos of the West Fork Cedar River bridge crossing at low flow, (*left*) prior to vane installation in 1984, (*middle left*) in 1989 5 years after vane installation (along right bank only), (*middle right*) in 2006, and (*right*) 25 years after vane installation. (Source: Odgaard [1], with permission from ASCE (left two images), and DigitalGlobe (2006 and 2009 photos))

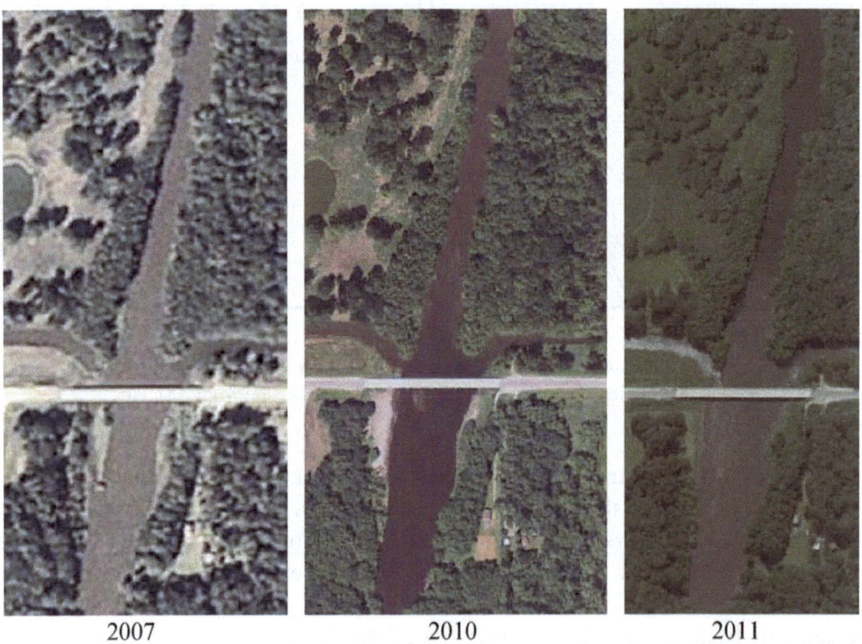

Fig. 3.11 Aerial photos of the West Fork Cedar River bridge crossing at bank-full flow, (*left*) in 2007, (*middle*) in 2010, and (*right*) in 2011, 27 years after vane installation. (Source: DigitalGlobe)

3.2 Stabilization of Riverbanks

Often, stabilization of a river alignment also brings vegetation back to the riverbanks and restores the banks' natural ecology. However, adjustment of alignment is not always possible. Infrastructure such as highways and bridges are often in the way, and a more localized approach is needed. In such cases, submerged vanes are also an effective measure for restoring and stabilizing the banks, in some cases aided by riprap along the toe of the bank. Typically, the most unstable reaches of a river are the bends. In bends, the interaction between the vertical gradient of the velocity and the curvature of the flow generates a so-called secondary or spiraling flow. The secondary flow moves high-velocity, near-surface current outward and the low-velocity, near-bed current inward, thereby producing larger depths and velocities near the outer banks. The deepening of the channel diminishes the toe support of the bank and the larger velocities attack it, setting the stage for bank erosion. As indicated in Fig. 3.2, the vanes are laid out so that the vane-generated secondary current eliminates all or part of the centrifugal-induced secondary current. The vanes stabilize the toe of the bank.

Figure 3.12 shows a bend of Wapsipinicon River, USA, that was restored in 1988. The bank was eroding at a rate of 3 m per year toward a county road and was endangering a bridge structure. The bank height is 3.5 m and bank-full flow about 600 m^3/s. A system of 28 vanes was installed along approximately 100 m of the bend in May 1988. The vanes were fabricated of reinforced concrete and each mounted on an H-pile, which was driven approximately 4.6 m into the streambed. See the schematic in Fig. 3.4. The vanes were oriented at approximately 20° with the direction of flow at bank-full flow. At around 2000, the toe of the bank downstream of the vane installation was provided with a riprap protection. Figure 3.12 shows aerial views of the river, in 1988 just before vane installation, and in 2009, 19 years after installation. It is seen that the vane system aided by the riprap protection not only helped to restore the bank but also helped to create a more favorable approach-flow condition for the bridge opening.

Fig. 3.12 Aerial view of Wapsipinicon River in 1988 (*left*) and in 2009 (*right*). (Courtesy of Robert DeWitt, River Engineering International (*left photo*) and DigitalGlobe (*right photo*))

3.3 Stabilization of River Channel at Diversions and Water Intakes

Channel stability is often a challenge when water is withdrawn from the channel through a diversion or water intake. The withdrawal of water reduces the downstream flow velocity and, hence, the downstream sediment-transport capacity in the vicinity of the diversion or intake. As a result, the bed level may increase in the area around the intake (sometimes even blocking the flow into the intake) upsetting the balance between flow and sediment transport and the natural channel stability. A submerged vane system is typically designed to fulfill two purposes: (a) prevent bed load from being withdrawn into the diversion or intake and (b) maintain stability of the flow in the main channel. In this case, the vane system must (a) intercept the bed load heading for the intake and deflect it around the intake and (b) increase flow velocity in the main channel enough that the deflected sediment stays in motion and continues downstream without causing unwanted degradation.

The design scenario is depicted in Fig. 3.13. Figure 3.13a shows the channel section with elevated bed level due to the water withdrawal. Figure 3.13b shows the bed-level changes (at bank-full flow) that the vanes need to accomplish to (a) get sediment-free water into the intake and (b) increase bed level and flow velocity outside the intake enough to keep sediment in motion past the intake. Figure 3.13c shows the bed level the vane system needs to maintain at low flow. The width of the vane field, δ_v, must be sufficient that the induced aggradation within the field results in a channel along the intake of sufficient width, depth, and velocity to accommodate the flow into the intake as well as maintain a sufficiently high flow velocity past it. Channel stability is achieved and maintained when the slope of the energy grade line (energy dissipation per unit length along the channel) and main-flow

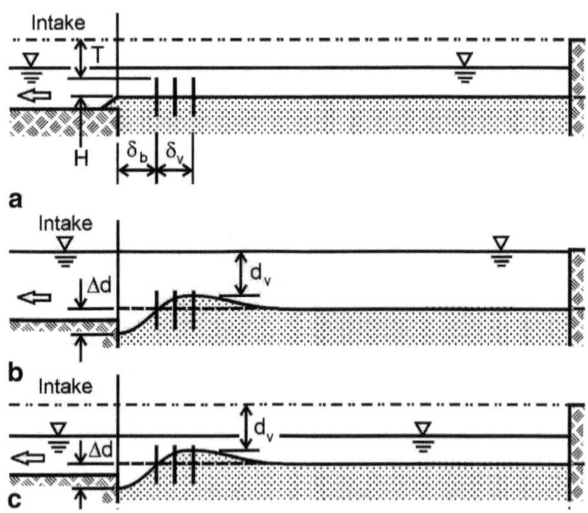

Fig. 3.13 Schematic showing design environment and variables for a vane system at a water intake or diversion

curvature are maintained. As mentioned earlier, this is a feasible objective because of the vanes' relatively small angle of attack. With a relatively small angle of attack, the increase in flow resistance caused by the drag force on the vanes is modest and not more than can be compensated for by the decrease in bed-form drag due to smaller dunes within the vane field. The design challenge is to maintain stability at all flow rates and water-surface elevations at the site; vanes are static while the channel flow is dynamic. The following case studies demonstrate both the challenge and the solution.

The vane system at *Duane Arnold Energy Center* (DAEC) was the first (1991) designed with the aforementioned dual purpose: (a) prevent bed load from entering the intake and (b) maintain channel stability. As seen in Fig. 3.14, the solution consisted of the installation of a guide wall and nine submerged vanes. The guide wall is attached to the upstream corner of the intake and extends upstream as it tapers into the bank line, thus smoothing the approach flow past the intake structure. The vanes intercept the approaching bed load and divert it away from the face of the intake. Installation details are described in [1].

As described by Odgaard [1], one of the challenges with the DAEC project was the gradual change in flow split between meander curve and cutoff. By 2005, the flow through the meander curve had decreased to less than 10%. As a result, the channel upstream from the intake had incurred changes that affected the flow

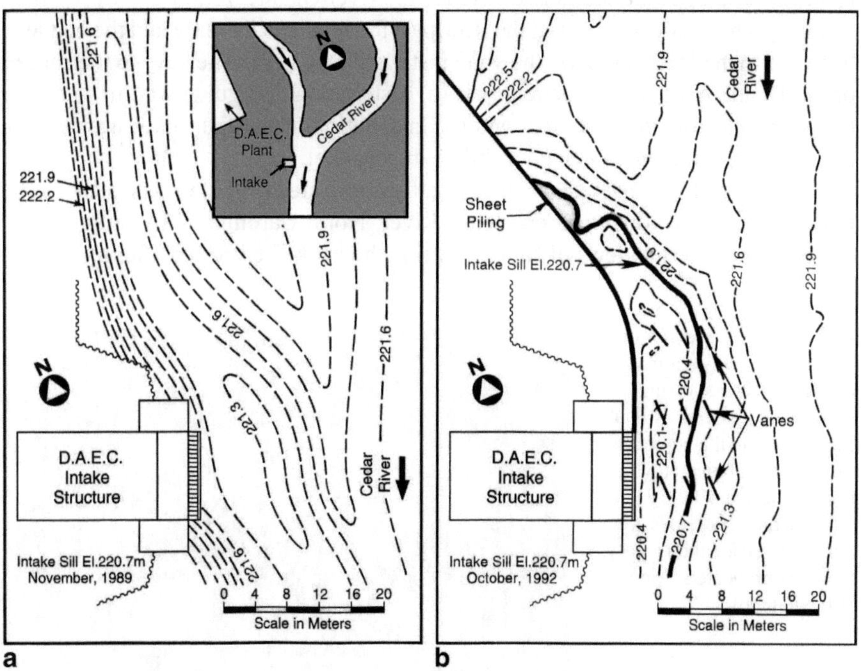

Fig. 3.14 Bed-level contours in Cedar River at the DAEC intake structure, **a** in 1989, and **b** in 1992. (Source: Odgaard [1], with permission from ASCE)

Fig. 3.15 2008 view of Goldsboro raw water intake on Neuse River, North Carolina. (Source: DitigalGlobe)

approaching the vane field and intake. Between 1990 and 2005, a considerable amount of bank erosion occurred along the right bank about 200 m upstream from the intake in the upstream segment of the cutoff, and an even larger amount along the left bank just upstream of the intake. By 2005, the width of the channel just upstream of the intake had become nearly twice the width in 1980, and shoaling occurred in the channel. A decision was subsequently made to build four spur dikes upstream and opposite the intake to make the river channel return to its 1980 planform [1]. The spur dikes caused the flow past the intake to increase and the bed level to return to the level just after the vane installation in 1991. The keys to the success of the Duane Arnold vane system was (a) installation of a guide wall to smoothen the approach flow to the system and (b) installation of the spur dikes upstream on the opposite bank which helped stabilize the channel.

A guide wall was also used at the recently completed (2012) vane system for the *Goldsboro raw water intake* on Neuse River, North Carolina, USA. Figures 3.15 and 3.16 show aerial views of Neuse River at the intake before and after installation

Fig. 3.16 2012 view of Goldsboro raw water intake on Neuse River showing guide wall upstream of intake for smoothing the approach flow to the submerged vane system located off the end of the structure; six buoys are installed outside the vane system to warn boaters. (Source: DitigalGlobe)

of vanes and guide wall. The Goldsboro design was aided by Hydrologic Engineering Centers River Analysis System (HEC-RAS) calculations to ensure the design would preserve the energy grade line through the reach. So far, the design has met expectations.

The vane system at the *Byron Station intake* on Rock River, Illinois, completed in 1998 was the first dual-purpose system at an intake located on the inside of a river curve [1]. In this case, the bed-load deflection challenge is exacerbated by the river's natural tendency to build a point bar at the intake (Fig. 3.17). The intake draws 1.8 m^3/s of water from the river for make up for the cooling system of the plant. Note the location of the spur dikes.

The design approach at the Byron Station intake was later (1993) used at the *Huntly Power Station,* New Zealand [1]. However, at Huntly, the intake was located on the outer bank of the meander curve relatively close to the crossover from the upstream meander curve. The upstream spur dikes or weirs were designed to stabilize the crossover. Because of the near-braided regime of the river at this location, the crossover was unstable and sandbars tended to form at the intake. These weirs, together with the vanes, now maintain a stable, relatively deep channel along the face of the intake.

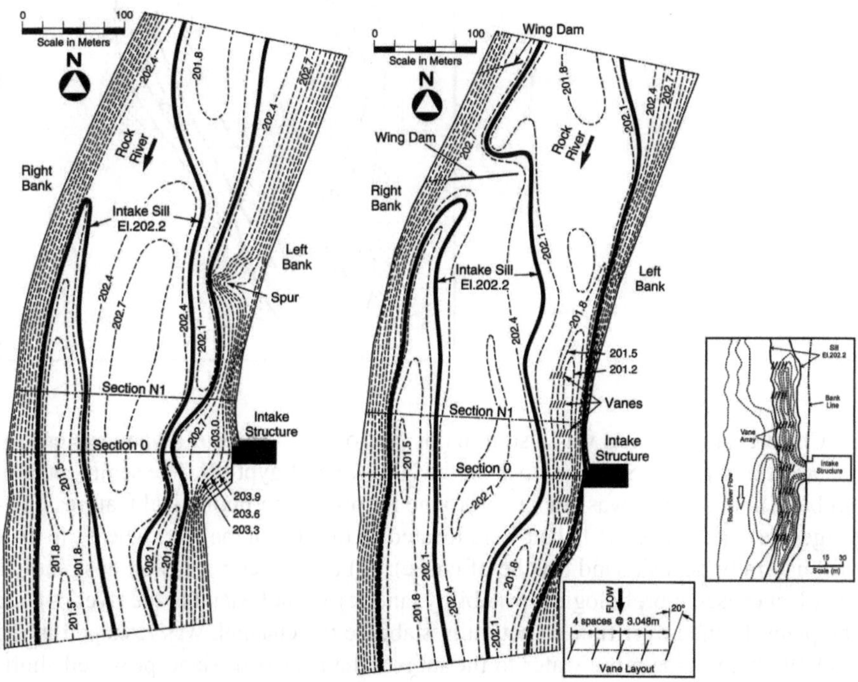

a b c

Fig. 3.17 Bed-level contours in Rock River at the Byron Station (Illinois) intake structure, **a** in 1990, **b** in 1994, and **c** in 2007. **c** is based on survey data used with permission of Exelon Corporation, all rights reserved. (**a**, **b**, and **c** are adapted from Odgaard [1] with permission from ASCE)

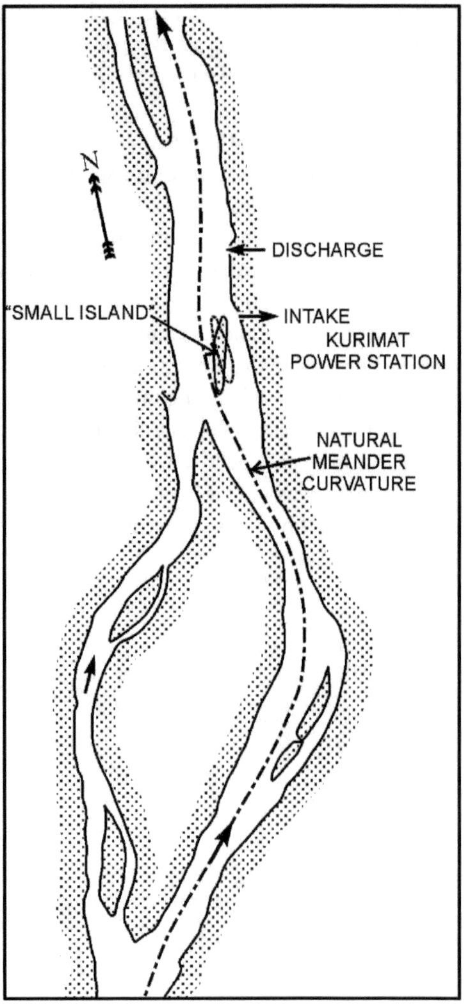

Fig. 3.18 Plan of the Nile River at Kurimat Power Station

Channel stabilization was also a major part of the design of a vane system for the *Kuraimat Power Station intake* on the Nile River, Egypt [8]. The sedimentation problem at this intake was due in part to the following morphological features/challenges: (a) The intakes and plant were located on the side of the river where the river will naturally deposit sand (inside of curve); (b) a channel expansion upstream of the plant caused morphological instability; and (c) a small island in the river outside the plant, formed over time to naturally stabilize the channel, was reshaped in the mid-1990s to guide more water to the intake; the reshaped island provided short-term improvement but not long-term improvement, and the channel reach became less stable. Figure 3.18 shows the alignment of the river as it flows past the intake. The meander curve (half-wavelength) is approximately 2.6 km long. The solution consisted of (a) construction of two flow deflectors upstream the small island to

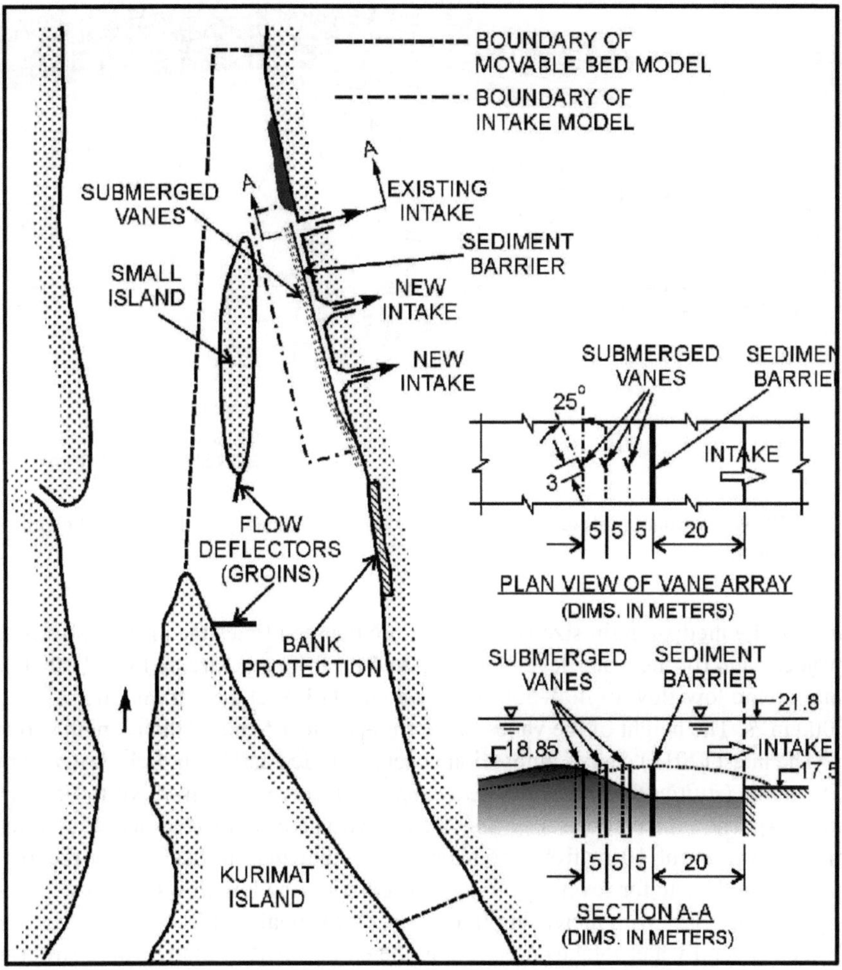

Fig. 3.19 Flow and sediment management measures, and model boundaries

adjust the flow split around it and (b) installation of a sediment barrier and a series of submerged vanes near the intake to intercept and deflect sediment that would otherwise enter the intake. A major challenge was to not only preserve but also enhance channel stability at this location; this was a major challenge because bringing the island shape back to "normal" was out of the question. The solution was developed using a combination of physical and numerical modeling as well as perturbation stability analysis. Two physical model studies were included, the boundaries of which are indicated in Fig. 3.19.

The intake for the *Sunsary Morang Irrigation Project* on Kosi River, Nepal, suffered a similar sedimentation problem, also triggered by channel instability (braiding). See Odgaard [1]. What distinguishes the Nepal project from other projects

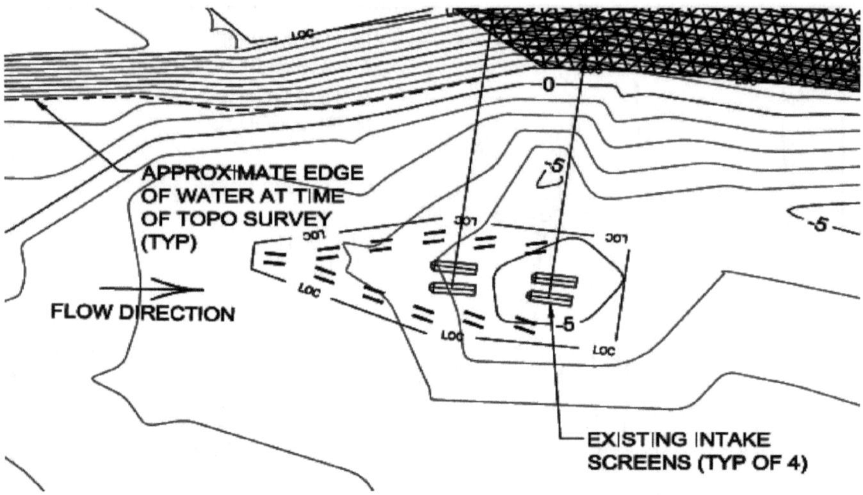

Fig. 3.20 Vane layout at intake screens in Tar River, N.C. (Courtesy of the Greenville Utilities Commission)

are: (a) the median grain size is 30 mm, more than ten times the size at most other projects, and (b) the water level varies up to 4 m, from an extreme low of 109 m at an extreme low flow of 400–500 m^3/s to about 113 m at an extreme high flow of 9000 m^3/s. The height of the vanes range from 1.2 to 1.5 m and their length is 6 m.

The latest (2012) design innovation is demonstrated and being field-tested at the *Greenville Utilities Commission's water supply intake on Tar River*, North Carolina, USA [4]. The intake consists of four wedge-wire screens located near the bottom of the river about 10 m from the shoreline. Sedimentation has been a recurring problem at the site for several years, in particular during low flows, and back flushing was necessary with increasing frequency. Occasional shutdowns have occurred. In fact, several water intakes on rivers in the eastern part of the USA are suffering from similar sedimentation problems which seem to correlate with unprecedented, extended periods of droughts.

The new vane system design is a "wedge-plow" design where the vanes are oriented such that they split the oncoming bed load evenly around the intake screens much like the way the wedge snow plow on a locomotive pushes the snow off railroad tracks. The layout is shown in Fig. 3.20. Each vane is approximately four [4] ft long and stands 1.5–2.0 ft above the river bottom. They are made from flat-panel sheet piling driven 6–8 ft into the riverbed. Installed in pairs, the vanes are fully submerged during all but the very lowest river stages. Figure 3.21 shows the template that was used to guide installation, and Fig. 3.22 shows the installation process with all vanes still exposed above the water surface. After installation, the frequency of back flushing has been significantly reduced, and so far the plant has not experienced any shutdowns due to sediment infiltration. After several rain events, a sandbar developed in proximity of the vanes. Figure 3.23 shows how the vanes

Fig. 3.21 Template used for guiding vane installation at intake screens in Tar River, N.C. (Courtesy of the Greenville Utilities Commission)

Fig. 3.22 Vanes being installed around intake screens in Tar River, N.C. (Courtesy of the Greenville Utilities)

Fig. 3.23 Vane system deflecting bed load around intake screens in Tar River, N.C. (Courtesy of the Greenville Utilities Commission)

created an angular face of the sandbar indicating that the vanes deflected it around the intake and accomplished their purpose.

Obviously, it is too early to assess the long-term performance of the Greenville system. Although the system was designed to solve the sedimentation problem at the intake more so than solve a channel stability problem, the observations at the site confirm that a low-impact (low-resistance) vane system can be designed to have a significant effect on sediment movement through a channel without affecting the expenditure of flow energy—an observation of significance to many channel stabilization projects.

4 Proposed Applications

4.1 Creating a Stable Meander Planform

The following example illustrates how the experience from the aforementioned case studies is applied to a typical channel re-meandering project. The data used in the example are adapted from an ongoing project on a channel reach that is unstable and has braiding tendencies. Figure 3.24 is a plan view of the reach, which is 2 km long. The objective is to stabilize the reach as a single-thread channel allowing the ecology to be restored and the area to better accommodate the needs of the nearby community. First, the alignment of the new channel is determined such that it is consistent with that of a stable meander. Alternatives 1 and 2 in Fig. 3.25 are two possible solutions for channel alignment. The most stable alignment is determined by a stability analysis. Using data from the field for the situation depicted in Fig. 3.24, the upstream stable reach has average width, depth, slope, and grain size of 75 m, 3 m, 0.0009, and 0.6 mm, respectively. Bank-full (channel forming) discharge is 423 m^3/s. At this discharge, the Darcy–Weisbach friction factor is $f=0.06$ and average velocity 1.88 m/s. According to Leopold and Wolman [9], these values

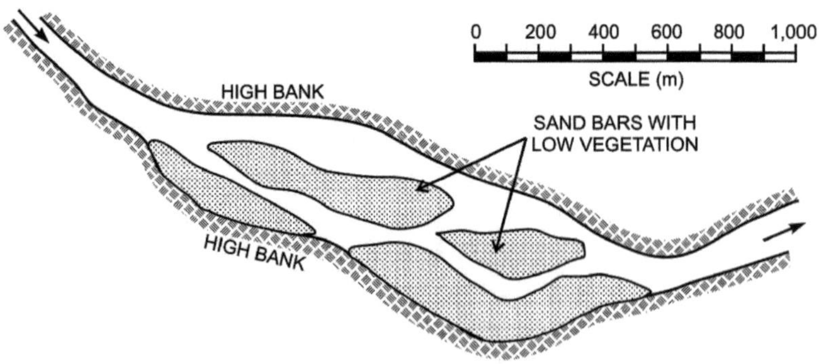

Fig. 3.24 Channel reach to be stabilized. (Source: Odgaard [1], with permission from ASCE)

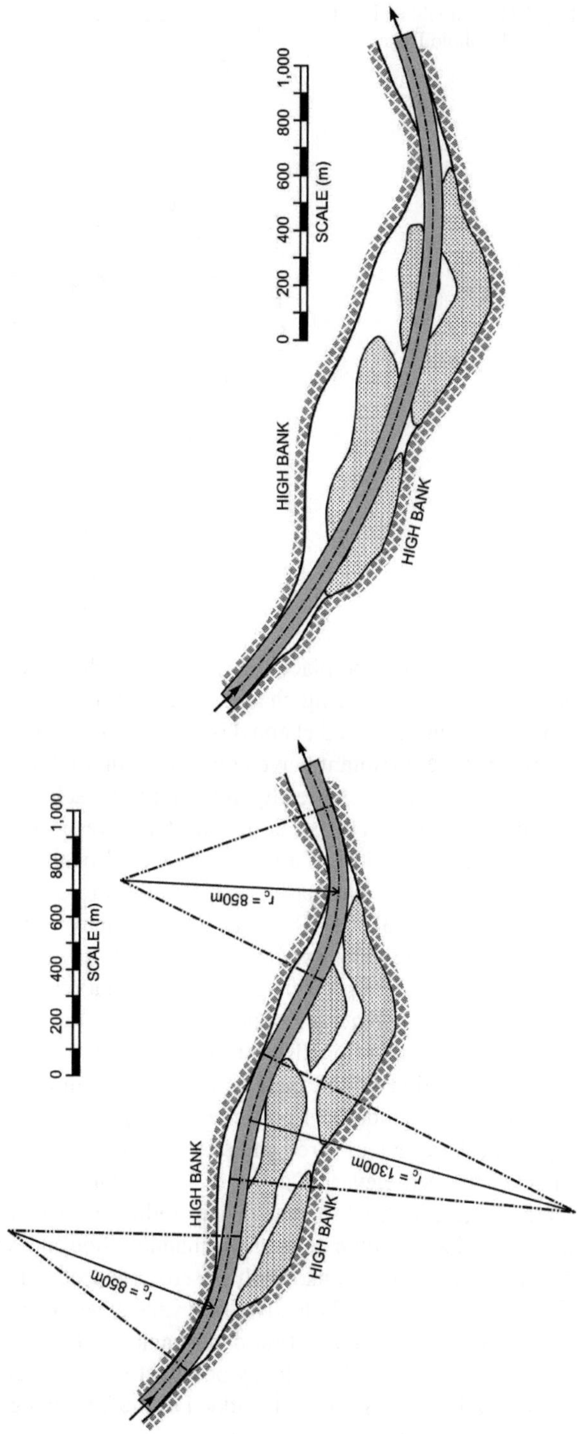

Fig. 3.25 Alternative channel alignments through the reach, Alternative 1 (*left*) and Alternative 2 (*right*). (Source: Odgaard [1], with permission from ASCE)

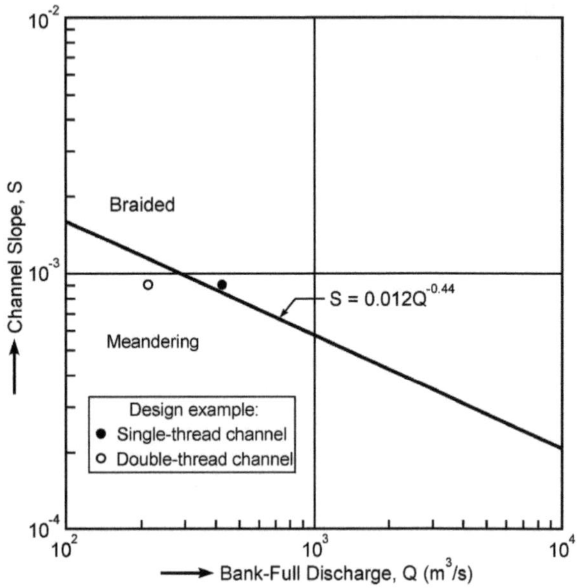

Fig. 3.26 Leopold and Wolman's Threshold Relation

of discharge and slope place this channel reach at the transition between a braided channel and a meandering channel (Fig. 3.26), close enough to the meandering regime that a single-thread channel is sustainable. A perturbation stability analysis [1] shows that the dominant wavelength is about 1600 m. That means Alternative 1 in Fig. 3.25 is the most stable alignment for the reach. The alignment consists of three circular arcs of radii 850, 1300, and 850 m, respectively, connected with straight-channel segments. Using a two-dimensional bend-flow model [1, 10], maximum scour depth in this alternative is estimated to be 4.3 m and maximum near-bank velocity about 2.3 m/s. The near-bank velocity is about 20 % larger than cross-sectional average velocity. Considering that the channel is a constructed channel, the banks have to be protected to allow vegetation to get established.

A system of submerged vanes is proposed for stabilizing the alignment, much like in the aforementioned West Fork Cedar River case. Vanes would be installed along the portion of the bank with greater-than-average near-bank velocity. Using the design guidelines presented in [1], a system of 1.34-m-tall vanes would be placed in arrays of three and spaced longitudinally at a distance of 20–25 m to increase the bed elevation along the outer bank to approximately the elevation of the average bed. Such a system will reduce near-bank velocity to cross-sectional average velocity with no change in channel slope and will thus prevent erosion from occurring along the bank. With no erosion occurring along the bank, vegetation should be able to take hold. The first vane array of each system would be installed at the point of estimated "first outer-bank erosion occurrence," which in this case is calculated to be at the halfway point between the beginning of the bend and bend apex. Lead-in vanes would be added at the beginning of the reach to promote the development of the first meander curve.

In general, the selection of channel alignment must be, to some extent, based on existing topography and high-bank alignment. When the calculated wavelength differs from that fitting in naturally with the existing high-bank alignment, the selected wavelength should be adapted to fit the existing high-bank alignment as much as possible. The alignment shown is the most appropriate for the given high-bank alignment and the calculated wavelength. Had the calculated wavelength been twice that calculated above (if, say, channel width and grain size had had different values), a more appropriate channel alignment fitting with the given high-bank alignment could have been Alternative 2.

4.2 Controlling a Braided River Channel

The aforementioned example suggests that vanes can also be a useful tool for stabilizing reaches of braided rivers. By redistributing flow and sediment transport in strategic sections of a braided river, for example at nodal points, vanes may help stabilize not only channel alignment but also flow distribution between each channel of a braided channel system. By redistributing flow among a select number of (parallel) channels, and stabilizing each channel using the aforementioned guidelines, vanes have the potential for helping reduce the lateral extent of a braided river segment. Indeed, vanes have been used in the past to successfully close off secondary branches of a river [11]. Thus, a braided reach could be narrowed by closing off the outside branches of the reach.

Referring to the aforementioned design example, had bank-full discharge Q been significantly larger (and slope unchanged), the reach would be well within the braided regime. The data point in Fig. 3.26 would be further to the right. In this case, stabilizing the reach with a single-thread meandering channel would be difficult if not impossible. However, splitting the flow into two parallel channels could potentially make both become stable meandering channels. The flow split could be an even split, or it could be a split as indicated in Fig. 3.27, in which the design flow (channel forming discharge) of the upper channel is smaller than that of the lower channel. If total flow rate were 423 m^3/s, as in the previous example, and the flow were split evenly between the two channels, the flow point for each channel would be that circled to the left of the threshold line in Fig. 3.26. In other words, each channel would be well within the meandering regime.

The design process would include determining sets of corresponding discharges and meander wavelengths and selecting the sets for which the meander wavelengths

Fig. 3.27 Stabilization by channel split

have the closest match to those upstream and downstream from the reach. The amplitudes of the meanders would be determined based on natural lateral constraints and on the slope of the channel immediately upstream and downstream of the reach. Submerged vanes appropriately placed at the bifurcation and at strategic points along the channel may control the flow split between the two channels.

In braided channels, one of the management challenges is to limit the lateral expansion of the flow area. In some large rivers like the Kosi River in Nepal and the Brahmaputra River in India, the braiding causes the flow area to be several kilometers wide, often leading to unpredictable shifts in the braiding pattern from year to year. These shifts can be devastating to the communities living around the river. One approach to help prevent this behavior is to identify so-called nodal points along the river, points where the lateral expansion is limited and relatively stable. Often, because of inherent flow instability, a small flow adjustment immediately downstream from such a nodal point can affect flow patterns over a large distance downstream. A limited number of submerged vanes appropriately placed at the exit from such nodal points can create and maintain a desired flow pattern for many kilometers downstream possibly all the way to the subsequent nodal point and thus help restrict the lateral expansion tendencies of the river in between the nodal points. It should be noted that submerged vanes have already proved successful in closing off secondary branches of a river channel [11]. By orienting the vanes such that they intercept and deflect sediment into the entrance region of the secondary branch, as shown schematically in Fig. 3.28, velocities decrease and a gradual channel aggradation occurs in the entrance section that eventually closes off the entrance.

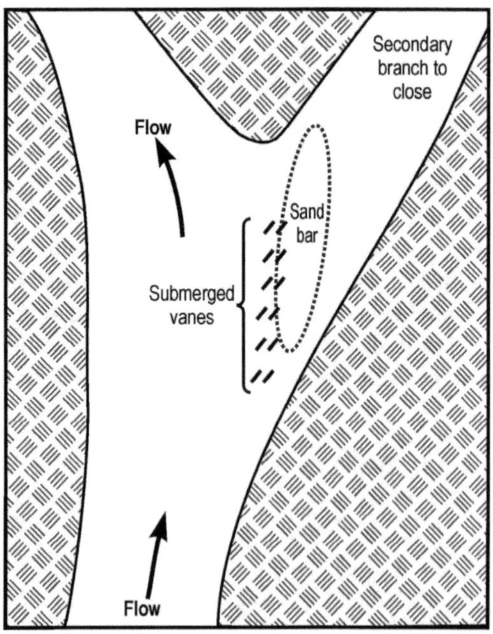

Fig. 3.28 Schematic showing how submerged vanes could help close off a secondary branch

5 Conclusions

The case studies described show that submerged vanes are an effective method for stabilizing river channels. They can be used for realigning or re-meandering a river and for stabilization of the riverbanks. The design process includes (a) site inspection and surveys, (b) review of sequential aerial photos, and (c) a formal perturbation stability analysis. The process typically provides dominant wavelength and amplitude of channel alignment, most stable channel cross section, flow and depth distribution throughout channel, and bank erosion patterns. Feedback from field installations, like the ones described above, and laboratory results have provided guidelines for the development of specifications for vane layouts. The specifications include vane height, length, orientation, number of vanes, and array configuration, longitudinal, and lateral spacing, distance to bank, location of first and last vane in system, and staging. The experience to date suggests that vanes may be effective in managing braided rivers; they may be used to close off select secondary branches and reduce the lateral extent of a braided channel system.

Glossary

Aggradation the process by which streambeds, floodplains, and the bottoms of other water bodies are raised in elevation by the deposition of material eroded and transported from other areas. It is the opposite of degradation.

Alluvial pertains to alluvium deposited by a stream or flowing water (usually sand).

Alluvial stream a stream whose channel boundary is composed of alluvium, and which generally changes its cross section and bedform due to the interaction of the flow and mobile boundary adjustment.

Bank migration lateral shifting of the banks of a stream course.

Bedforms wave-like irregularities found on the bottom (bed) of a stream that are related to flow characteristics. They are given names such as "dunes," "ripples," and "anti-dunes." They are related to the transport of sediment and they interact with the flow because they change the roughness of the streambed. An analog to streambed forms is desert sand dunes.

Bed load material moving on or near the streambed by rolling, sliding, and sometimes making brief excursions into the flow a few diameters above the bed, i.e., jumping. The term "saltation" is sometimes used in place of "jumping." Bed load is bed material that moves in continuous contact with the bed; contrast with suspended load.

Bed material the sediment mixture of which the bed is composed. In alluvial streams, bed-material particles are liable to be moved at any moment or during some future flow condition. Bed material may include grain sizes that travel both as bed load and as suspended load.

Boundary conditions definitions or statements of conditions or phenomena at spatial or temporal boundaries of a model. Water levels, flows, sediment concentrations, etc. that are specified at the boundaries of the area being modeled.

A specified tailwater elevation and incoming upstream discharge are typical boundary conditions.

Braided channel a stream that is characterized by random interconnected channels divided by islands or bars. Bars that divide the stream into separate channels at low flow are often submerged at high flow.

Channel a natural or artificial waterway that periodically or continuously contains moving water.

Conveyance a measure of the flow capacity of a channel section. Flow is directly proportional to conveyance for steady uniform flow.

Cross section the shape of the channel in which a stream flows on a line perpendicular to the flow or banks.

Cross-sectional area the wetted area of a cross section perpendicular to the direction of flow.

Degradation the process by which streambeds, floodplains, and the bottoms of other water bodies are lowered in elevation by erosion of material. It is the opposite of aggradation.

Deposition the mechanical or chemical processes through which sediments accumulate in a (temporary) resting place.

Depth of flow the vertical distance from the bed of a stream to the water surface.

Discharge the volume of a fluid or solid passing a cross section of a stream per unit time.

Dunes bedforms with triangular profile that advance downstream due to net deposition of particles on the steep downstream slope. Dunes move downstream at velocities that are small relative to the streamflow velocity.

Erosion the wearing away of the land surface or stream boundaries by detachment and movement of soil and rock fragments through the action of moving water or other geological agents.

Floodplain normally dryland adjacent to a body of water which is susceptible to periodic inundation by floodwaters.

Fluvial in this chapter, pertaining to streams.

Fluvial sediment particles derived from rocks or biological materials that are transported by, suspended in, or deposited by streams.

Grain size see Particle size.

Local scour erosion caused by an abrupt change in flow direction or velocity. Examples include erosion around bridge piers, downstream of stilling basins, at the end of dikes or vanes, and near snags.

Meandering a planform (alluvial) process that generates a series of bends of alternate curvature connected by straight reaches.

Meandering stream an alluvial stream characterized in planform by a sequence of alternating bends. The bends are usually a result of alluvial processes rather than the nature of the terrain.

Mean velocity the discharge divided by the wetted area of a cross section.

Particle size a linear dimension, usually designated as "diameter," used to characterize the size of a particle.

Planform the shape and size of channel and overbank features as viewed form directly above.

Point bar deposits of sediment that occur on the convex side or inside of channel bends. Their shape may vary with changing flow conditions, but they do not move significantly relative to the bends. However, the general magnitude and location of the bars vary with discharge and sediment load.

River training enhancements made to the banks of a river to constrain the river to a desired course; the enhancements are usually in the form of realignment or regrading of the banks, and installation of bank protection structures such as rock revetments (riprap), dikes, weirs, or submerged vanes.

Scour concentrated erosive action by water. The enlargement of a flow section or creation of a depression by the removal of bed material through the action of moving water.

Secondary currents (or flow) the movement of water particles normal to the principal direction of flow.

Sediment naturally occurring material that is broken down by processes of weathering and erosion and is subsequently transported by the action of wind, water, or ice, and/or by the force of gravity acting on the particle itself. In this chapter, sediment is material (sand) transported by water.

Sedimentation refers to the gravitational settling of suspended particles that are heavier than water.

Sediment discharge the mass or volume of sediment (usually mass) passing a stream cross section in a unit of time. The term may be qualified, for example, as suspended sediment discharge, bed-load discharge, or total sediment discharge. See Sediment load.

Sediment load a general term that refers to material in suspension and/or in transport. It is not necessarily synonymous with either discharge or concentration. It may also refer to a particular type of load; for example, total, suspended, bed, or bed material load.

Sediment particle solid fragments of mineral material in either a singular or an aggregate state.

Sediment transport (rate) see Sediment discharge.

Shear stress (boundary shear stress) frictional force per unit or area exerted on a channel boundary by the flowing water. An important factor in the movement of bed material.

Stable channel a stream channel that does not change in planform, cross section, Wor bed profile during a particular period of time (but may over longer periods of time).

Stream bank erosion the removal of bank material primarily by hydraulic action.

Top width the width of a stream section at the water surface; it varies with stage in most natural channels.

Validation/Verification check of the behavior of a model against a set of prototype conditions that was not used for calibration.

References

1. Odgaard, A. J. (2009). *River training and sediment management with submerged vanes*. Reston: ASCE Press.
2. Odgaard, A. J., & Kennedy, J. F. (1983). River-bend bank protection by submerged vanes. *Journal of Hydraulic Engineering, 109*(8), 1161–1173.
3. Odgaard, A. J., & Spoljaric, A. (1986). Sediment control by submerged vanes. *Journal of Hydraulic Engineering, 112*(12), 1164–1181.
4. Personal communication with Randy Emory and Anthony Whitehead of the Greenville Utilities Commission. (2013). Greenville, North Carolina, U.S.A.
5. Odgaard, A. J., & Wang, Y. (1991). Sediment management with submerged vanes. I: Theory. *Journal of Hydraulic Engineering, ASCE, 117*(3), 267–283.
6. Wang, Y., & Odgaard, A. J. (1993). Flow control with vorticity. *Journal of Hydraulic Research, IAHR, 31*(4), 549–562.
7. Odgaard, A. J., & Abad, J. D. (2007). River meandering and channel stability, sedimentation engineering: Processes, measurements, modeling, and practice. In M. Garcia (ed.), *ASCE Manuals and Reports on Engineering* (Practice No. 110). Reston: ASCE Press.
8. Odgaard, A. J., Abdel-Fattah, S., Ali, A. M., El Ghorab, D. A. S., & Alsaffar, A. (2006). *Sediment management at a water intake on the Nile River*, Egypt. Conference proceedings, an international perspective on environmental and water resources, ASCE/EWRI, New Delhi, India, December 18–20.
9. Leopold, L. B., & Wolman, M. G. (1957). *River channel patterns—braided, meandering and straight*. U.S. Geological Survey Professional Paper 282B, U.S. Geological Survey, Washington, D.C.
10. Odgaard, A. J. (1989). River-meander model. I: Development. *Journal of Hydraulic Engineering, ASCE, 115*(11), 1433–1450.
11. Chabert, J., Remillieux, M., & Spitz, I. (1961). Application de la circulation transversal a la correction des rivieres et a la protection des prises d'eau [in French]. *Ninth Convention* (pp. 1216–1223). Dubrovnik: IAHR.

Chapter 4
Mathematic Modelling of Non-Equilibrium Suspended Load Transport, Reservoir Sedimentation, and Fluvial Processes

Qiwei Han and Mingmin He

Contents

Abstract This chapter consists of two parts: mechanism of non-equilibrium transport of non-uniform suspended load and its application to mathematical modelling. Based on a stochastic approach of sediment transport proposed by the authors, a 1D equation of non-equilibrium transport for each size group of non-uniform sediment is developed. The equations to predict the change of sediment concentration and the corresponding size distribution of suspended load and bed material are also derived. The concept that changes in size distribution are interrelated to sediment-carrying capacity is explored. These results reveal the essence of sediment transport of non-uniform sediment. In the second part, a mathematical model incorporating

M. He (✉) · Q. Han
Sediment Research Department, China Institute of Water Resources and Hydroelectric Power Research, 20, Chegongzhuang West Road, 100044 Beijing, China
e-mail: mingmin@iwhr.com; mingmin377@163.com

C. T. Yang, L. K. Wang (eds.), *Advances in Water Resources Engineering*,
Handbook of Environmental Engineering, Volume 14, DOI 10.1007/978-3-319-11023-3_4

the mentioned equations to compute deposition and scouring in reservoirs as well as the fluvial processes of river channels has been developed. Verification of the model agrees well with field data.

Keywords Bed material · Sediment-carrying capacity · Consolidation of deposit · Deposition · Exchange intensity · Fluvial process · Mathematical model · Non-equilibrium sediment transport · Reservoir sedimentation · Scouring · Size distribution · Suspended load

Nomenclature

$1-\varepsilon_{0\cdot l}$	Probability of lth size group being stopped from motion
α_l	Coefficient of saturation recovery
β_l	Probability of incipient suspension of lth size group
γ_s	Specific weight of sediment (kg/m^3)
γ'_s	Specific weight of deposits or bed material (kg/m^3)
ε_y	Diffusion coefficient of flow in vertical direction (m^2/s)
λ	Percentage of deposition (kg/m^{2}s)
$\lambda_{1\cdot 4\cdot l}$	Exchange intensity of sediment from rest at bed surface to suspension of lth size group (kg/m^{2}s)
$\lambda_{4\cdot 1\cdot l}$	Exchange intensity of sediment from suspension to rest at bed surface of lth size group (kg/m^{2}s)
λ^*	Percentage of scouring of bed material
ω	Mean settling velocity of suspended load (m/s)
ω_l	Settling velocity of lth size group (m/s)
ω_m	Median value of settling velocity during deposition (m/s)
ω^*	Mean settling velocity of sediment-carrying capacity (m/s)
ω^*_m	Median value of settling velocity during scouring (m/s)
Δa	Scoured or silted area at Δx in Δt (m^2)
Δh	Depth of deposition during Δt (m)
ΔS	Concentration of sediment supplement from bed material (kg/m^3)
ΔS_h	Sediment concentration corresponding to disturbed thickness of bed material taken part in scouring and sorted, but not scoured (kg/m^3)
ΔS_m	Sediment concentration corresponding to the amount of disturbed bed material (kg/m^3)
Δt	Time increment(s)
Δt_i	Time increment from instant t_{i-1} to t_i (s)
$\Delta V_{i.j}$	Volume of deposit at interval Δx_j during Δt_i (m^3)
Δx	Space interval along flow direction (m)
Δx_j	Space interval along flow direction from x_{j-1} to x_j (m)
a	Scoured or silted area (m^2)
$a(x, t)$	Equation of total area of erosion or deposition in period t (m^2)
A	Cross-sectional area of flow (m^2)
$A(x,z,t)$	Equation of cross-sectional area (m^2)
B	Cross-sectional width of flow (m)
$B(x,z,t)$	Equation of cross-sectional width (m)

B_k	Stable width of cross section (m)
D_{50}	Median size of deposit or bed material (mm)
D_l	Particle size of lth size group (mm)
g	Gravitational acceleration (m/s^2)
h	Average depth of flow (m)
H	Water level (m)
$h_0(x)$	Equation of the water surface (m)
$h_1(x)$	Equation of the bed surface (m)
i	Subscript, which is the parameter that indicates mean value from instant t_{i-1} to t_i
j	Subscript, which is the parameter indicates mean value from x_{j-1} to x_j or at position x_j
J_f	Energy slope
K	Coefficient of sediment-carrying capacity
L	Distance from inlet section to outlet section (m)
$L_{4 \cdot l}$	Mean step length of suspended particle (m)
n	Manning's coefficient of roughness (s/m$^{1/3}$)
m	Coefficient of sediment-carrying capacity
m_l	Total number of size groups of sediment
$P_{1 \cdot l}$	Size distribution of bed material
$P_{4 \cdot l}$	Size distribution of suspended load
$P_{1 \cdot l \cdot 0}$	Size distribution of bed material at initial instant
$P_{4 \cdot l \cdot 0}$	Size distribution of suspended load at inlet section
$P^*_{4 \cdot l}$	Size distribution of sediment-carrying capacity
$P^*_{4 \cdot l \cdot 0}$	Size distribution of sediment-carrying capacity at inlet section
$\tilde{P}^*_{4 \cdot l}$	Size distribution of sediment supplement
$\tilde{P}^*_{4 \cdot l \cdot 0}$	Size distribution of sediment supplement when $\lambda^* = 1$
q	Flow discharge of unit width (m^2/s)
Q	Flow discharge (m^3/s)
$q_{4 \cdot l}$	Sediment discharge of unit width (kg/ms)
S	Sediment concentration or sediment concentration at outlet section (kg/m^3)
S_0	Sediment concentration at inlet section (kg/m^3)
S_l	Sediment concentration of lth size group (kg/m^3)
S^*	Sediment-carrying capacity (kg/m^3)
S^*_0	Sediment-carrying capacity at inlet section (kg/m^3)
S^*_l	Sediment-carrying capacity at outlet section (kg/m^3)
$\bar{S}$	Mean concentration along vertical direction of total suspended load (kg/m^3)
$\bar{S}_l$	Mean concentration along vertical direction of lth size group (kg/m^3)
$S(x,y)$	Concentration of total suspended load at point (x, y) (kg/m^3)
$S_l(x,y)$	Concentration of lth size group at point (x, y), kg/m^3
$S^*(l)$	Sediment-carrying capacity of uniform sand with particle size, D_l (kg/m^3)

$t_{4\cdot0\cdot l}$	Time period for incipient suspension of a rest particle, i.e. time period for a particle transformed from rest to suspension to get off the bed surface (s)
V	Average velocity of flow (m/s)
V_x, V_y	Mean flow velocity in x and y direction, respectively (m/s)
W	Total weight of deposit (kg)
W_l	Deposited weight of lth size group (kg)
W_m	Amount of total bed material before scouring (kg)
$W_{m\cdot l}$	Amount of lth size group of bed material before scouring (kg)
W'	Weight of deposits in the river channel (kg)
W''	Weight of deposits on the flood plain (kg)
z	Elevation (m)
z_k	Elevation corresponding to the stable width of section (m)

1 Introduction

Most sediment transport models applied to engineering practice are based on equilibrium transport approach, i.e. sediment-carrying capacity is replaced by the actual sediment concentration [1–9]. Sediment-carrying capacity, in general, is not equal to sediment concentration, especially in the case of reservoir sedimentation and/or in the scouring process of river channel downstream a reservoir. Many researchers have worked on 2D diffusion equations of non-equilibrium transport [10–24]. Most of them deal with uniform sediment. Some 1D non-equilibrium transport equations have been published for uniform sediment [25–28]. Thus, further studies of generalized non-equilibrium transport for non-uniform sediment are needed.

Based on stochastic theory of sediment transport proposed by the authors, a 1D equation system of non-equilibrium transport for each size group of non-uniform sediment is presented. A mathematical model incorporating this equation system to compute deposition and scouring in reservoirs as well as the fluvial processes of river channels has been developed. Verifications of the model are given in this chapter.

2 Mechanism of Non-Equilibrium Transport| of Non-Uniform Suspended Load

2.1 General Description of Deposition and Scouring Associated with Changes of Size Distribution of Suspended Load and Bed Material

The deposition and/or scouring process caused by the change of flow intensity is associated with the changes of size distribution of suspended load and bed material as a self-adjustment process. Han [29] showed the deposition in the desilting

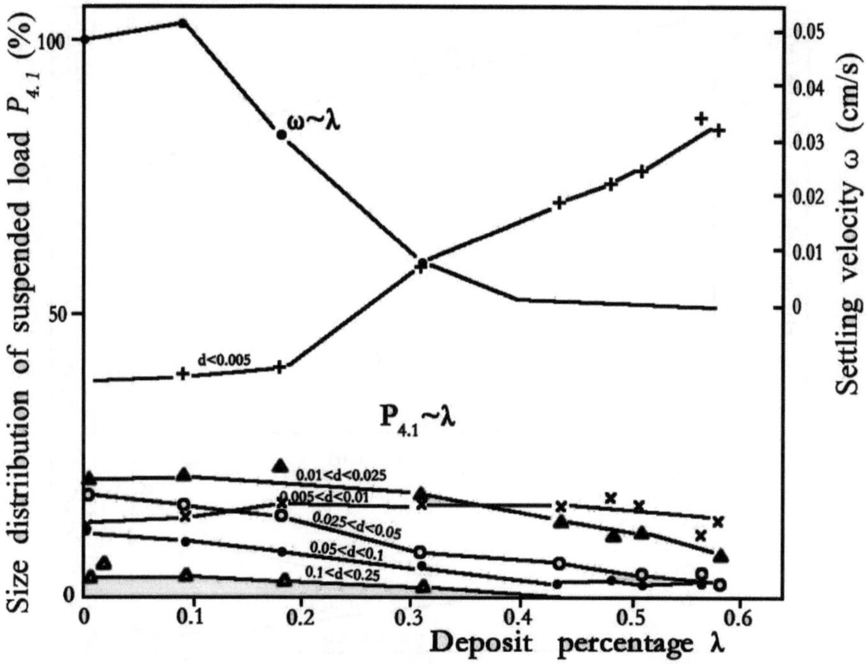

Fig. 4.1 Changes in size distribution of suspended load and average settling velocity during deposition in Wotousi desilting canal

canal at Wotousi. It is found that for different size groups the changes in sediment concentration are different. At the exit section, sediments with the particle size greater than 0.3 mm are deposited, while the sediment concentration of the finest size ($d<0.004$ mm) remained unchanged. In Fig. 4.1, the sorting curve of the data of Wotousi is given. It shows that the size distribution of suspended load varied greatly. The average settling velocity changes from 0.1089 cm/s at the entrance to 0.00325 cm/s at the exit, decreasing by 32.5 times. In Fig. 4.1,

$$\lambda=\frac{S_0-S}{S_0}, \tag{4.1}$$

Where

λ is the percentage of deposition
S_0 is the concentration at entrance, and
S is the concentration at the exit

The sediment-carrying capacity depends exclusively on flow intensity is generally different from the actual sediment concentration. It was expressed by Han [30, 31] as

$$S^* = K\gamma_s\left(\frac{V^3}{gh\omega}\right)^m, \tag{4.2}$$

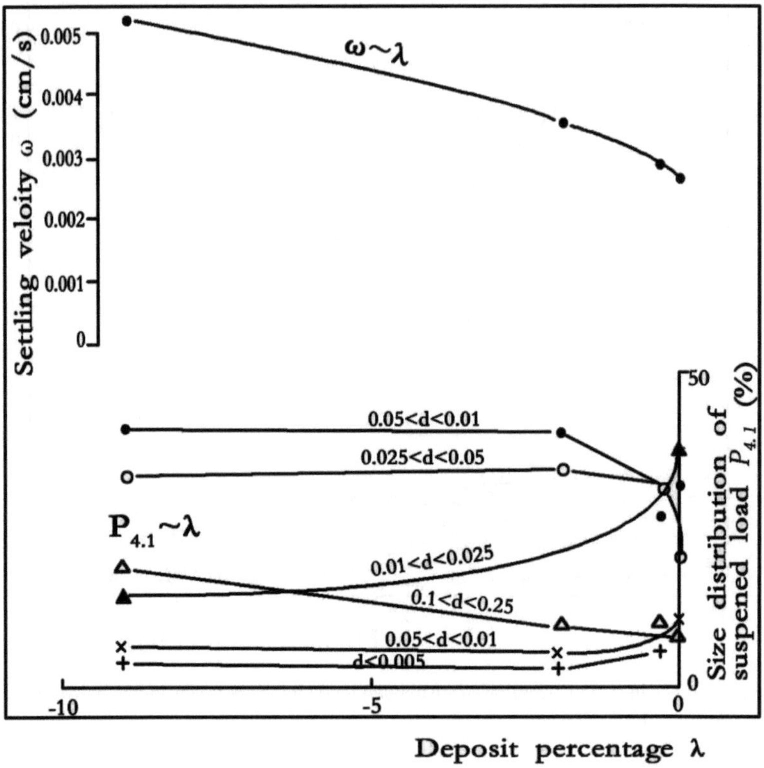

Fig. 4.2 Changes in size distribution of suspended load $P4\cdot_l$ and average settling velocity during scouring in the Sanshenggong Reservoir

Where

S^* is the sediment-carrying capacity
V is the average flow velocity
h is the mean water depth
g is the gravitational acceleration
γ_s is the unit weight of sediment
ω is the settling velocity of particle, and
$K = 0.927 \times 10^{-4}$ and $m = 0.92$

It can be deduced from this equation that under the same flow condition when settling velocity decreases 32.5 times, sediment-carrying capacity increases 25.3 times. However, in the case of the desilting canal at Wotousi, the concentration reduces only by 1.37 times.

Figure 4.2 shows the scouring at the Sanshenggong reservoir in Inner Mongolia Autonomous Region of China. During scouring process, concentration of the coarser size groups of suspended load picked up from the river bed increased faster than the finer size groups. As a result, the average settling velocity increased and the sediment-carrying-capacity decreased.

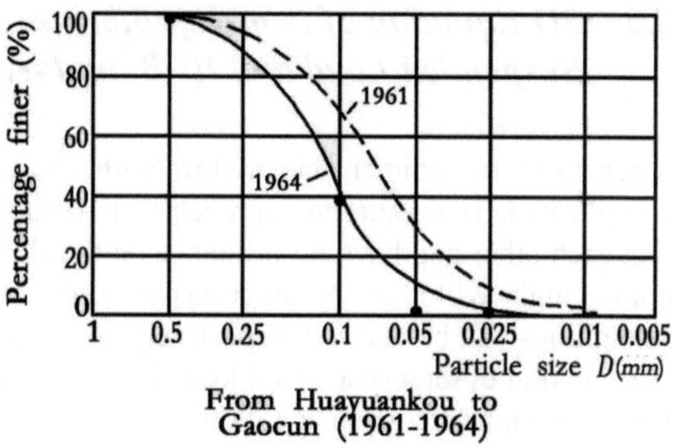

Fig. 4.3 Changes in size distribution of bed material during scouring in the lower Yellow River

Equation 4.2 has been extended to the case of hyper-concentration and applied to the Yellow River sedimentation studies. The studies show that the equation is applicable to low-, median-, and hyper-concentrated flows [32, 33].

The change in size distribution of suspended load during the process of scouring or deposition will continue to a certain extent; because coarse particles settle first in the deposition process, suspended load becomes finer, the fall velocity becomes smaller, and hence the sediment-carrying capacity becomes larger. Thus, the decrease of sediment concentration is eventually balanced by the increase of sediment-carrying capacity. During scouring process, particles picked up from bed material are larger than those in the original suspended load. Consequently, suspended load becomes coarser, its settling velocity increases, the sediment-carrying capacity decreases, and the rate of increasing of sediment concentration slows down. In other words, the development of scouring attenuates by itself.

During erosion and deposition, the change in size distribution of bed material takes place simultaneously. In Fig. 4.3, the bed materials coarsened in the lower Yellow River and downstream of the Sanmenxia reservoir from Huayuankou to Gaocun, when large amount of sediment was deposited in the reservoir and the clear water scoured the downstream channel. During the deposition process, sediment settled on the river bed is finer than the original bed material and the size distribution of bed material becomes finer. Consequently, in the deposition process, the flow velocity tends to increase and the size distribution of suspended load gets finer significantly. These processes will lead to the increasing of sediment-carrying capacity and reduce further deposition. In the process of scouring, the coarsening of bed material results in a decrease of the flow velocity. The suspended load becomes coarser because of supplement of coarse particles from bed material. Thus, the sediment-carrying capacity is reduced. Therefore, the tendency of increase in both the sediment-carrying capacity and the sediment concentration will be reduced. In other words, the development of scouring will also be reduced eventually by itself.

The examples mentioned above indicate that a close interrelationship exists among sediment concentration, size distribution of suspended load, and bed material. It suggests that in non-equilibrium transport of non-uniform sediment, the study on sediment concentration has to take into account the sorting process of suspended load and the armouring or getting finer of bed material (Fig. 4.2).

2.2 2D Equation of Non-Equilibrium Transport of Non-uniform Suspended Load and Its Boundary Conditions

Mechanism of sediment-laden flow with non-uniform suspended load is rather complicated. The common approach is to assume that moving particles do not affect each other and have no influence on the flow if the sediment concentration is not too high. With these assumptions, the sediment transport equations for different size groups can be derived respectively, the summation of these equations will be the equation of total suspended load. For 2D steady and non-equilibrium transport, these equations are given as follows:

$$\frac{\partial}{\partial x}(V_x S_l)+\frac{\partial}{\partial y}(V_y S_l)-\frac{\partial}{\partial y}\left(\varepsilon_y \frac{\partial S_l}{\partial y}-\omega_l S_l\right)=0 \qquad (l=1,2,\ldots,m_l), \tag{4.3}$$

Where

V_x and V_y are the mean flow velocity in longitudinal and vertical direction, respectively

S is the sediment concentration

ω is the settling velocity

m_l is the total number of size groups

ε_y is the diffusion coefficient in vertical direction, and the parameters with subscript l represent lth size group.

Various boundary conditions on the river bed for the diffusion equation have been used by different authors [10–24]. The concept of exchange intensity [34–37] is applied by the authors. Take a unit area near bed surface into consideration. Based on the theory of turbulent diffusion, the weight of sediment particle falling on a unit area of bed surface is

$$-\varepsilon_y \frac{\partial S_l}{\partial y}+\omega_l S_l \qquad (l=1,2,\ldots,m_l). \tag{4.4}$$

Following the principle of exchange intensity, the sediment particle picked up from bed surface is [36, 38]

$$(\lambda_{1\cdot 4\cdot l}-\lambda_{4\cdot 1\cdot l})\frac{\pi}{6}\gamma_s D_l^3, \tag{4.5}$$

where $\lambda_{1\cdot 4\cdot l}$ and $\lambda_{4\cdot 1\cdot l}$ are the exchange intensities from rest to suspension and that from suspension to rest, respectively,

$$\lambda_{1\cdot 4\cdot l}=\frac{4m_0 P_{1\cdot l}}{\pi D_l^2}\frac{\beta_l}{t_{4\cdot 0\cdot l}}, \tag{4.6}$$

and

$$\lambda_{4\cdot1\cdot l} = (1-\beta_l)(1-\varepsilon_{0\cdot l})K_{4\cdot l}\mu_{4\cdot l} = \frac{(1-\beta_l)(1-\varepsilon_{0\cdot l})}{L_{4\cdot l}} \frac{P_{4\cdot l}S}{\frac{\pi}{b}\gamma_s D_l^3}, \tag{4.7}$$

Where

$P_{1\cdot l}$ and $P_{4\cdot l}$ are the size distribution of bed material and suspended load, respectively

$K_{4\cdot l} = \dfrac{P_{4\cdot l}S}{\frac{\pi}{b}\gamma_s D_l^3}$ is the concentration of suspended load counted as particle number

S is the mean concentration of total suspended load

β_l is the probability of incipient suspension

$1-\varepsilon_{0\cdot l}$ is the ceasing probability

$t_{4\cdot0\cdot l}$ is the time period for incipient suspension of a rest particle, i.e. time period for a particle transformed from rest to suspension to get off the bed surface

$L_{4\cdot l}$ is the mean step length of suspended particle

β_l, $\varepsilon_{0\cdot l}$, $t_{4\cdot0\cdot l}$, and $L_{4\cdot l}$ all are functions of flow parameters and the particle size

$L_{4\cdot l}$ is also related to the distribution of sediment concentration in the vertical direction. Therefore, the boundary condition at the river bed is Eq. (4.34)

$$\varepsilon_y \frac{\partial S_l}{\partial y} - \omega S_l \bigg|_{y=h_1(x)} = (\lambda_{1\cdot4\cdot l} - \lambda_{4\cdot1\cdot l})\frac{\pi}{6}\gamma_s D_l^3. \tag{4.8}$$

The total sediment concentration is

$$S(x,y) = \sum_{l=1}^{m_l} S_l(x,y) \tag{4.9}$$

and the size distribution of suspended load is

$$P_{4.l} = \frac{\bar{S}_l(x)}{\bar{S}(x)}, \tag{4.10}$$

where $\bar{S}(x)$ and $\bar{S}_l(x)$ are the mean value of the total concentration $S(x,y)$ and lth size group concentration $S_l(x,y)$ along the vertical cross section, respectively.

The boundary condition at the water surface is

$$\left(\varepsilon_y \frac{\partial S_l}{\partial y} - \omega S_l\right)\bigg|_{y=h_0(x)} = 0. \tag{4.11}$$

2.3 1D Equation of Non-Equilibrium Transport

1D equation of non-equilibrium transport can be derived from Eqs. (4.3), (4.8), and (4.11). Integrating Eq. (4.3) from $h_0(x)$ to $h_1(x)$ yields

$$\int_{h_0(x)}^{h_1(x)}\left[\frac{\partial}{\partial x}(V_x S_l)+\frac{\partial}{\partial y}\left(V_y S_l\right)\right]\mathrm{d}y=\int_{h_0(x)}^{h_1(x)}\frac{\partial}{\partial y}\left(\varepsilon_{y.l}\frac{\partial S_l}{\partial y}-\omega_l S_l\right)\mathrm{d}y. \tag{4.12}$$

The transport rate of suspended load can be expressed by

$$q_{4\cdot l}=\int_{h_0(x)}^{h_1(x)}V_x S_l \mathrm{d}x. \tag{4.13}$$

The geometric relations are

$$\begin{aligned} V_y\Big|_{y=h_1(x)}&=V_x\Big|_{y=h_1(x)}t_g\theta_1=V_x\Big|_{y=h_1(x)}h_1'(x)\\ V_y\Big|_{y=h_0(x)}&=V_x\Big|_{y=h_0(x)}t_g\theta_0=V_x\Big|_{y=h_0(x)}h_0'(x). \end{aligned} \tag{4.14}$$

Therefore, the term on the left side of Eq. (4.12) equals to $\mathrm{d}q_{4\cdot l}/\mathrm{d}x$. In fact, from Eqs. (4.13) and (4.14)

$$\begin{aligned} \frac{\mathrm{d}q_{4\cdot l}}{\mathrm{d}x}&=\frac{\mathrm{d}}{\mathrm{d}x}\int_{h_0(x)}^{h_1(x)}V_x S_l \mathrm{d}x\\ &=\int_{h_0(x)}^{h_1(x)}\frac{\partial}{\partial x}(V_x S_l)\mathrm{d}x+V_x S_l\Big|_{h_1(x)}h_1'x-V_x S_l\Big|_{h_0(x)}h_0'(x)\\ &=\int_{h_0(x)}^{h_1(x)}\frac{\partial}{\partial x}(V_x S_l)\mathrm{d}x+V_y S_l\Big|_{h_0(x)}^{h_1(x)}\\ &=\int_{h_0(x)}^{h_1(x)}\left[\frac{\partial}{\partial x}(V_x S_l)+\frac{\partial}{\partial y}\left(V_y S_l\right)\right]\mathrm{d}x. \end{aligned} \tag{4.15}$$

From Eq. (4.8), the term on the right side of Eq. (4.12) can be expressed as

$$\begin{aligned} \int_{h_0(x)}^{h_1(x)}\frac{\partial}{\partial y}\left(\varepsilon_y\frac{\partial S_l}{\partial y}-\omega_l S_l\right)\mathrm{d}y&=-\left(\varepsilon_y\frac{\partial S_l}{\partial y}-\omega_l S_l\right)\Bigg|_{y=h_1(x)}\\ &=-(\lambda_{1\cdot 4\cdot l}-\lambda_{4\cdot 1\cdot l})\frac{\pi}{6}\gamma_s D_l^3. \end{aligned} \tag{4.16}$$

Then Eq. (4.12) can be rewritten as

$$\frac{\mathrm{d}q_{4\cdot l}}{\mathrm{d}x}=-(\lambda_{1\cdot 4\cdot l}-\lambda_{4\cdot 1\cdot l})\frac{\pi}{6}\gamma_s D_l^3 \tag{4.17}$$

and the transport rate can be represented by

$$q_{4\cdot l} = qS_l, \tag{4.18}$$

where q is the flow discharge of unit width. For simplicity, $\overline{S}_l$ and $\overline{S}$ are replaced by S_l and S for the mean value of sediment concentration of the lth size group and total concentration, respectively. Substituting Eqs. (4.6) and (4.7) into Eq. (4.17) yields

$$\frac{\mathrm{d}S_l}{\mathrm{d}x} = -\frac{(1-\varepsilon_{0\cdot A})(1-\beta_l)}{L_{4\cdot l}}\left[P_{4\cdot l}S - \frac{2}{3}\frac{m_0\gamma_s}{q}\frac{P_{1\cdot l}D_l\beta_l}{t_{4\cdot 0\cdot l}}\frac{L_{4\cdot l}}{(1-\varepsilon_{0\cdot l})(1-\beta_l)}\right] \quad (l = 1, 2, \ldots, m_l). \tag{4.19}$$

Define sediment-carrying capacity of lth size group as

$$S_l^* = \frac{2}{3}m_0\frac{\gamma_s}{q}\frac{P_{1\cdot l}D_l\beta_l}{t_{4\cdot 0\cdot l}}\frac{L_{4\cdot l}}{(1-\varepsilon_{0\cdot l})(1-\beta_l)} \quad (l = 1, 2, \ldots, m_l). \tag{4.20}$$

The total sediment-carrying capacity S^* and the size distribution of sediment-carrying capacity $P_{4\cdot l}^*$ is related to S_l^*,[31, 36, 39]

$$S_l^* = P_{4\cdot l}^* S^*. \tag{4.21}$$

Thus, Eq. (4.19) becomes

$$\frac{\mathrm{d}S_l}{\mathrm{d}x} = -\frac{(1-q_{0\cdot l})(1-\beta_l)}{L_{4\cdot l}}(P_{4\cdot l}S - P_{4\cdot l}^*S^*) \quad (l = 1, 2, \ldots, m_l), \tag{4.22}$$

which is the fundamental equation system of the 1D non-equilibrium sediment transport.

The capacity for carrying uniform sand with particle size D_l can be expressed as

$$S^*(l) = \frac{2}{3}m_0\frac{\gamma_s}{q}\frac{D_l\beta_l}{t_{4\cdot 0\cdot l}}\frac{L_{4\cdot l}}{(1-\varepsilon_{0\cdot l})(1-\beta_l)} \quad (l = 1, 2, \ldots, m_l). \tag{4.23}$$

Obviously,

$$P_{4\cdot l}^* S^* = P_{1\cdot l}S^*(l) \quad (l = 1, 2, \ldots, m_l). \tag{4.24}$$

Because

$$\sum_{l=1}^{m} P_{4\cdot l}^* = 1, \tag{4.25}$$

then

$$P_{4\cdot l}^{*}=\frac{P_{1\cdot l}D_l\beta_l L_{4\cdot l}}{(1-\varepsilon_{0\cdot l})(1-\beta_l)t_{4\cdot 0\cdot l}}\Bigg/\sum_{l=1}^{m_l}\frac{P_{1\cdot l}D_l\beta_l L_{4\cdot l}}{(1-\varepsilon_{0\cdot l})(1-\beta_l)t_{4\cdot 0\cdot l}}\qquad (l=1,2,\ldots,m_l). \quad (4.26)$$

Thus, the size distribution for a specific sediment-carrying capacity depends on the size distribution of bed material rather than the size distribution of the suspended load. This is different from what is commonly believed at present time. The concept of exchange between bed material and suspended load suggests that sediment-carrying capacity represents the weight of sediment eroded from the river bed and hence is related to the size distribution of bed material [39]. It must be emphasized that the size distribution of bed material is also subject to change with respect to time because of the exchanges between bed material and suspended load. In general, when the exchanges are considered, the size distribution of sediment-carrying capacity is also related to suspended load and $P_{1\cdot l}$ is the distribution of effective bed material [39–41].

Equation (4.26) is available for all the case of deposition and scouring. Especially, in the case of intensive deposition or scouring, there are $P_{4\cdot l}^{*}=P_{4\cdot l}$, $(l=1,\ 2,\ldots,m_l)$.

Sediment transport is in intensive equilibrium, if the sediment transport of each size group is in equilibrium, i.e.

$$\frac{\mathrm{d}S_l}{\mathrm{d}x}=0\qquad (l=1,2,\ldots,m_l). \quad (4.27)$$

In this case,

$$\begin{aligned} &S=S^{*}\\ &P_{4\cdot l}=P_{4\cdot l}^{*}\qquad (l=1,2,\ldots,m_l). \end{aligned} \quad (4.28)$$

For engineering application, the following equation is recommended.

$$S^{*}(l)=K\gamma_{\mathrm{s}}\left(\frac{V_3}{gh\omega_l}\right)^{m}. \quad (4.29)$$

From Eqs. (4.24) and (4.29), S^{*} can be expressed as

$$S^{*}=1\Bigg/\sum_{l=1}^{m_l}\frac{P_{4.l}^{*}}{K\gamma_{\mathrm{s}}\left(\frac{V^3}{gh\omega_l}\right)^{m}}=K\gamma_{\mathrm{s}}\left(\frac{V^3}{gh}\right)^{m}\frac{1}{\sum_{l=1}^{m_l}P_{4\cdot l}^{*}\omega_l^{m}}=K\gamma_{\mathrm{s}}\left(\frac{V^3}{gh\omega^{*}}\right)^{m}, \quad (4.30)$$

where

$$\omega^{*m}=\sum_{l=1}^{m_l}P_{4\cdot l}^{*}\omega_l^{m}. \quad (4.31)$$

Equation (4.30) shows that the capacity for carrying non-uniform sediment is assumed to have the same form as uniform sediment, provided that the corresponding mean value of the terminal velocity ω^* is defined by Eq. (4.31).

Assume that [31, 36]

$$\frac{(1-\varepsilon_{0\cdot l})(1-\beta_l)}{L_{4\cdot l}} = \alpha_l \frac{\omega_l}{q} \qquad (l = 1, 2, \ldots, m_l), \tag{4.32}$$

then, Eq (4.22) becomes

$$\frac{\mathrm{d}S_l}{\mathrm{d}x} = -\frac{\alpha_l \omega_l}{q}(P_{4\cdot l}S - P^*_{4\cdot l}S^*) \qquad (l = 1, 2, \ldots, m_l), \tag{4.33}$$

where α_l is the coefficient of saturation recovery. It is generally a function of settling velocity and flow parameters. If shear velocity U_* is not too small, for example $U_* \geq 0.01$ m/s, then according to Eq. (4.32) the mean value of α_l for different size group of suspended load is about 0.5 for equilibrium sediment transport. Considering the difference between equilibrium and non-equilibrium sediment transport, α_l can be taken as 0.25 for reservoir sedimentation and 1.0 for scouring during flushing of reservoir and in river channel with fine bed material. However, in the case of coarse bed material in river channel, α_l has different values for different size groups. In the paper of Han and Chen [42], the coefficient of saturation recovery for concentration and sediment-carrying capacity $\tilde{\alpha}_l$, α^* has been introduced, respectively. The calculated results show little difference among them. Therefore, the experience values can be used in practice.

The boundary condition and the non-equilibrium equation based on the stochastic theory have been applied in some models and have checked by a lot of field data [10, 11, 43, 44]. Some simulated boundary conditions and transport equations have been introduced not strictly based on theory, and experiential coefficients of saturation recovery were adopted [45–48].

2.4 Change of Sediment Concentration along River Course

The change of sediment concentration along a river course can be obtained from Eq. 4.33. Take $x=0$ at the entrance of a river course and $x=L$ at the exit. If L is not too large, then as an approximation the sediment-carrying capacity at the reach between $x=0$ and $x=L$ can be assumed to change linearly

$$\frac{\mathrm{d}S^*}{\mathrm{d}x} = -\frac{S^*_0 - S^*}{L}. \tag{4.34}$$

Substituting Eq. (4.34) into Eq. (4.33), integrating and summing up all lth equations yields

Table 4.1 Comparison of concentration at Dayuzhang Wotousi desilting canal

No. of cross section	S^*	$\left(S_0 - S_0^*\right)\times e^{-\frac{\alpha L}{q}}$ Equation (4.36)	$\left(S_0^* - S^*\right)\times\frac{1}{\alpha L}\left(1-e^{-\frac{\alpha L}{q}}\right)$ Equation (4.36)	S calculated (kg/m³)	S observed (kg/m³)
2	26.2	5.42	−8.22	23.4	30.8
3	23.4	−1.64	2.24	24.0	27.8
4	13.3	0.33	7.57	21.2	23.5
5	9.77	5.64	3.09	18.5	19.2
6	6.97	8.73	2.50	18.2	17.6
7	7.46	9.56	−0.32	16.7	16.7
8	9.33	9.24	−1.77	16.8	14.8
9	24.6	7.06	−14.8	16.9	14.3

$$S = S^* + \left[S_0 \sum_{l=1}^{m_l} P_{4\cdot l\cdot 0} \exp\left(-\frac{\alpha_l \omega_l L}{q}\right) - S_0^* \sum_{l=1}^{m_l} P_{4\cdot l\cdot 0}^* \exp\left(-\frac{\alpha_l \omega_l L}{q}\right)\right]$$
$$+\left[S_0^* \sum_{l=1}^{m_l} P_{4\cdot l\cdot 0}^* \frac{q}{\alpha_l \omega_l L}\left\{1-\exp\left(-\frac{\alpha_l \omega_l L}{q}\right)\right\} - S^* \sum_{l=1}^{m_l} P_{4\cdot l}^* \frac{q}{\alpha_l \omega_l L}\left\{1-\exp\left(-\frac{\alpha_l \omega_l L}{q}\right)\right\}\right], \quad (4.35)$$

where subscript '0' represents values at the entrance, while those without subscript denote values at the exit. Equation (4.35) shows that sediment concentration at the exit consists of three parts. The first part is the sediment-carrying capacity at the exit. The second part stands for the effect of excess of sediment load, i.e. the amount beyond its sediment-carrying capacity at the entrance section. The third part is the effect of changes of the sediment-carrying capacity along the river course.

In the case of uniform sediment, Eq. (4.35) can be simplified to

$$S = S^* + (S_0 - S_0^*)e^{-\frac{\alpha\omega L}{q}} + \frac{q}{\alpha\omega L}\left(1-e^{-\frac{\alpha\omega L}{q}}\right)(S_0^* - S^*). \quad (4.36)$$

Equation (4.36) can be used for non-uniform sediment, provided that (a) the change of sediment size distribution is small, or (b) the relative distance $\frac{\omega L}{q}$, which can be interpreted as the ratio of L to settling distance from water surface to bed surface, is small. Then, ω has to be modified by

$$\omega = \sum_{l=1}^{m_l} P_{4.l}\,\omega_l \quad (4.37)$$

and ω^* modified by $\omega^{*m} = \sum_{l=1}^{m_l} P_{4.l}^*\,\omega_l$ as shown in Eq. (4.31).

The verifications of Eqs. (4.35) and (4.36) are shown in Tables 4.1, 4.2, Figs. 4.5, and 4.6. Computed values agree well with field data. Computed values in Table 4.1 and Fig. 4.5 are obtained from Eq. (4.36), while those in Table 4.2 and Fig. 4.6 are obtained from Eq. (4.35). From Table 4.1, it can be seen that none of the three terms

Table 4.2 Difference of calculation between Eqs. (4.35) and (4.36)

No. of cross section	Observed S(kg/m³)	Calculated S(kg/m³)		$\frac{\omega L}{q}$
		Eq. (4.35)	Eq. (4.36)	
D1	1.66			0.828
D4	1.63	1.66	1.65	
				1.19
D7	1.40	1.11	1.23	
				23.4
D9	0.931	0.313	0.600	
				8.52
D11	0.214	0.027	0.198	

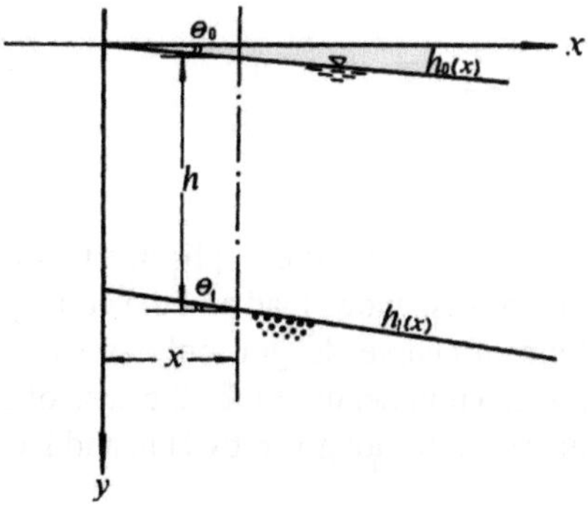

Fig. 4.4 Sketch of 2D flow in vertical direction

in Eq. (4.36) can be neglected. From Table 4.2, if $\frac{\omega L}{q} < 1$, then Eq. (4.36) can give satisfactory result. However, when $\frac{\omega L}{q} > 1$, only Eq. (4.35) can be used.

2.5 Equations of Size Distribution of Suspended Load

Considering different features of the changes of size distribution associated with deposition and scouring processes, three categories are specified: intensive deposition, intensive scouring, and slight scouring of finer particles with slight deposition of coarse particles.

Intensive deposition is defined in the case where deposition is predominant. No particles of the original bed material will be involved in the process of exchange. Intensive scouring is defined in the case where scouring is predominant. No particle from incoming suspended load can rest on the bed, while the original bed is subject to continuing scouring in the process of exchange. The slight scouring accompanied with slight deposition is defined in the case that the scouring of finer particles is accompanied simultaneously with deposition of coarse particles.

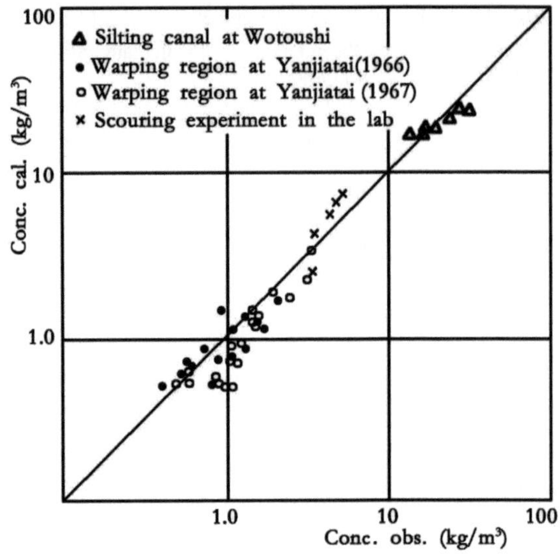

Fig. 4.5 Verification of concentration (using the mean settling velocity)

Equation (4.35) is applicable to all three cases. However, size distribution equation of suspended load and the corresponding bed material are different. Due to the limit of chapter length, only cases of intensive deposition and intensive scouring are given in detail. As for the case of slight scouring accompanied with deposition, more details are given by Han and He [39–41].

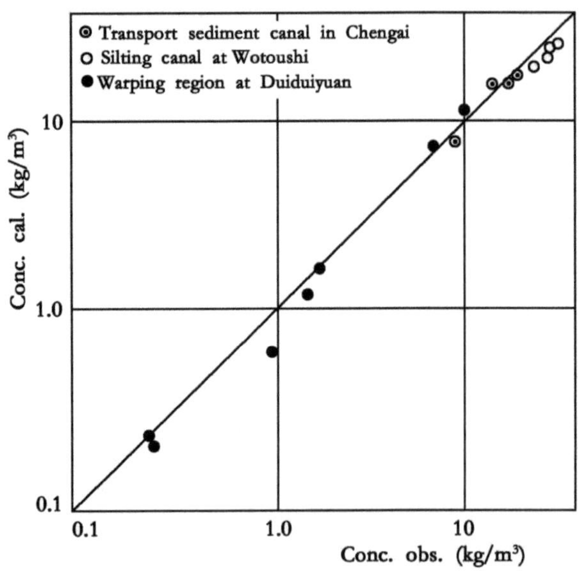

Fig. 4.6 Verification of concentration (using the summation of concentrations of different size groups)

In Eq. (4.33), $P_{4\cdot l}$ and $P_{4\cdot l}^*$ are unknown, so the equation system is not closed. In the case of intensive deposition, it is easy to demonstrate that [39]

$$P_{4\cdot l} = P_{4\cdot l}^*. \tag{4.38}$$

Then, from Eq. (4.33),

$$\frac{\mathrm{d}S_l}{\mathrm{d}x} = -\alpha\frac{\omega_l}{q}P_{4\cdot l}(S - S^*) \qquad (l = 1, 2, \ldots, m_l). \tag{4.39}$$

Summing up all these equations yields

$$\frac{\mathrm{d}S}{\mathrm{d}x} = -\alpha\frac{\omega}{q}(S - S^*), \tag{4.40}$$

where ω is determined by Eq. (4.37). Dividing Eq. (4.39) by Eq. (4.40) and taking S as independent variable come to

$$\frac{\mathrm{d}P_{4\cdot l}S}{\mathrm{d}S} = \frac{\omega_l}{\omega_0}P_{4\cdot l} \qquad (l = 1, 2, \ldots, m_l). \tag{4.41}$$

After integrating

$$P_{4\cdot l} = P_{4\cdot l\cdot 0}\exp\left[\int_{S_0}^{S}\left(\frac{\omega_l}{\omega} - 1\right)\frac{\mathrm{d}s}{s}\right] \qquad (l = 1, 2, \ldots, m_l). \tag{4.42}$$

From the median theorem of integration

$$\int_{S_0}^{S}\left(\frac{\omega_l}{\omega} - 1\right)\frac{\mathrm{d}s}{s} = \left(\frac{\omega_l}{\omega_m} - 1\right)\ln\frac{S}{S_0} \qquad (l = 1, 2, \ldots, m_l), \tag{4.43}$$

Where $\omega(S_0) > \omega_m > \omega(S)$. The percentage of deposition is defined as

$$\lambda = \frac{S_0 - S}{S_0}. \tag{4.44}$$

Then, Eq. (4.42) becomes

$$P_{4\cdot l} = P_{4\cdot l\cdot 0}(1 - \lambda)^{\left(\frac{\omega_l}{\omega_m} - 1\right)} \qquad (l = 1, 2, \ldots, m_l), \tag{4.45}$$

where ω_m is determined by

$$\sum_{l=1}^{m_l} P_{4\cdot l\cdot 0}(1 - \lambda)^{\left(\frac{\omega_l}{\omega_m} - 1\right)} = 1. \tag{4.46}$$

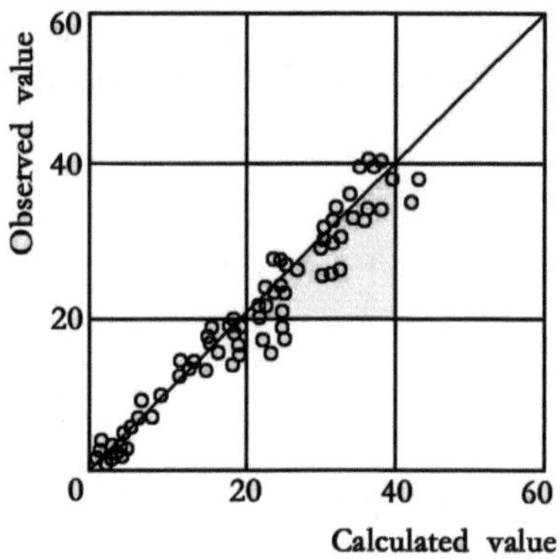

Fig. 4.7 Comparison of distribution at Yanjiatai warping region

It is shown in Eq. (4.45) that the size distribution at the exit section is related to the size distribution of suspended load at entrance section, i.e. the deposit percentage, the settling velocity of *l*th size group, and the median of settling velocity. Equation (4.45) also indicates that the size distribution is getting finer in deposition process. For coarser particles, $\omega_l > \omega_m$, hence $P_{4\cdot l} < P_{4\cdot l\cdot 0}$, while for the finer particles, $\omega_l < \omega_m$, $P_{4\cdot l} > P_{4\cdot l\cdot 0}$. Figure 4.7 shows that field data and computed results for Yanjiatai warping area are in good agreement. Similar comparisons at Wotousi, Diudiuyuan, and Danjiangkou reservoir are shown in Figs. 4.8, 4.9 and 4.10, respectively. Tables 4.3 and 4.4 summarize the comparisons between computed and measured results from Sanmenxia and Danjiangkou reservoirs. Because the mean values are used, the agreement is better than those given in Figs. 4.7, 4.8, 4.9 and 4.10.

For intensive scouring, the concentration and size distribution of suspended load exclusively eroded from bed surface are defined by the following equations [30, 31]

$$SP_{4\cdot l} = S_0 P_{4\cdot l\cdot 0} + \Delta S \tilde{P}^*_{4\cdot l}, \tag{4.47}$$

where

$$S^* P^*_{4\cdot l} = S_0 P_{4\cdot l\cdot 0} + \Delta S^* \tilde{P}^*_{4\cdot l}, \tag{4.48}$$

$$\Delta S = S - S_0, \tag{4.49}$$

$$\Delta S^* = S^* - S_0, \tag{4.50}$$

and $\tilde{P}^*_{4\cdot l}$ is the size distribution of sediment supplement. Summing up these equations and substituting ΔS for *x* into Eq. (4.33) yields

$$\frac{\mathrm{d}\tilde{P}^*_{4\cdot l}\Delta S}{\mathrm{d}x} = \alpha \frac{\omega_l}{q} \tilde{P}^*_{4\cdot l}(\Delta S - \Delta S^*).$$

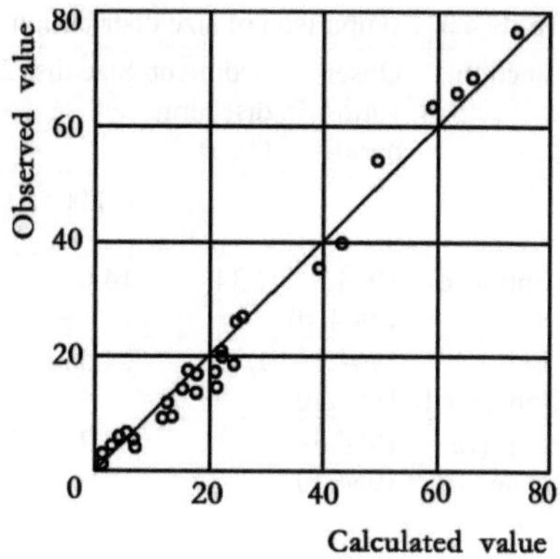

Fig. 4.8 Comparison of size distribution at Wotousi desilting canal

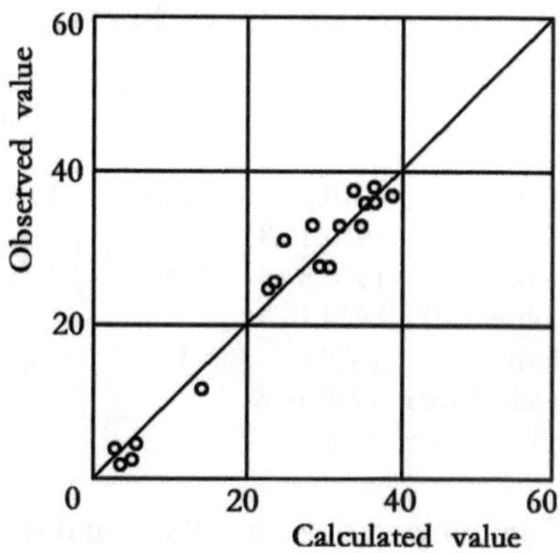

Fig. 4.9 Comparison of size distribution at Diudiuyuan warping region

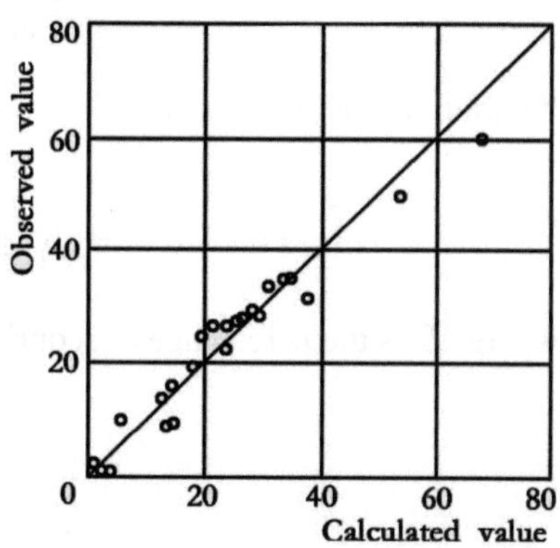

Fig. 4.10 Comparison of size distribution in Danjiangkou reservoir

Table 4.3 Comparison of size distribution of suspended load in Sanmenxia reservoir

Section	Observation period	Sediment discharge (10^9t)	Size distribution (mm)							
			<0.005	0.005 ~ 0.01	0.01 ~1560.025	0.225 ~ 0.05	0.05 ~ 0.10	0.10 ~ 0.25	0.25 ~ 0.50	0.50 ~ 1.0
Entrance	1964.3–1.964.10	2.34	14.0	9.9	17.5	25.4	28.2	4.3	0.6	0.1
Exit (observed)	1964.3–1964.10	1.34	24.5	13.9	21.1	25.3	13.7	1.3	0.2	
Exit (calculated)	1964.3–1964.10		23.2	15.2	22.7	23.6	14.6	0.7		

Table 4.4 Comparison of size distribution of suspended load in Danjiankou reservoir

Section	Observation period	Sediment discharge (10^6t)	Size distribution (mm)						
			<0.01	0.01 ~ 0.025	0.025 ~ 0.05	0.05 ~ 0.10	0.10 ~ 0.25	0.25 ~ 0.50	0.50 ~ 1.0
Entrance	1970.8.17–1970.10.8	12.05	19.8	20.9	24.8	26.6	6.1	1.8	
Exit (observed)	1970.8.17–1970.10.8	5.89	34.5	28.7	22.9	13.5	0.4		
Exit (calculated)	1970.8.17–1970.10.8		35.6	29.3	22.7	11.2	1.1	0.1	

Summing up these equations and substituting ΔS for x comes to

$$\frac{\mathrm{d}\tilde{P}^*_{4\cdot l}\Delta S}{\mathrm{d}\Delta S} = \frac{\omega_l}{\omega^*}\tilde{P}^*_{4\cdot l}. \tag{4.51}$$

Integrating Eq. (4.51) gives

$$\tilde{P}^*_{4\cdot l} = \tilde{P}^*_{4\cdot l\cdot 0}\lambda^{*\left(\frac{\omega_l}{\omega^*_m}-1\right)}, \tag{4.52}$$

where λ^* is the percentage of scouring which can be expressed as

$$\lambda^* = \frac{\Delta S}{\Delta S_m}, \tag{4.53}$$

ΔS_m is the sediment concentration corresponding to the amount of bed material taken part in scouring and becomes sorted. Taking the effect of sand wave into account, it can be found that ΔS_m is usually larger than ΔS by a certain amount of ΔS_h, i.e.

$$\Delta S_m = \Delta S + \Delta S_h, \tag{4.54}$$

where ΔS_h is the sediment concentration corresponding to disturbed sediment amount, which is taken part in scouring and becomes sorted, but is not the scoured amount of sediment. When $\lambda^* = 1$, it is obvious that

$$\tilde{P}^*_{4\cdot l} = \tilde{P}^*_{4\cdot l\cdot 0} = P_{1\cdot l\cdot 0}. \tag{4.55}$$

Equation (4.55) means that the size distribution of initial supplement of sediment is that of bed material. Substituting $\tilde{P}^*_{4\cdot l}$ into Eq. (4.46) comes to

$$P_{4\cdot l} = \frac{1}{1-\lambda}\left[P_{4\cdot l\cdot 0} - \lambda P_{1\cdot l\cdot 0}\lambda^{*\left(\frac{\omega_l}{\omega^*_m}-1\right)}\right] \quad (l = 1, 2, \ldots, m_l). \tag{4.56}$$

In scouring process, λ defined in Eq. (4.44) is negative. Superimposing these m_l equations leads to the equation for ω^*_m as

$$\sum_{l=1}^{m_l} P_{1\cdot l\cdot 0}\lambda^{*\left(\frac{\omega_l}{\omega^*_m}-1\right)} = 1. \tag{4.57}$$

Equation (4.56) shows that the size distribution of suspended load at exit of the reach are functions of size distribution of suspended load at entrance, percentage of scouring, settling velocity of its size groups, the median of settling velocity, and the initial size distribution of bed material.

The verification of Eq. (4.56) is shown in Figs. 4.11 and 4.12. Those field data are collected from gauging stations of Yellow River downstream Sanmenxia reservoir during the period of degradation.

2.6 Change of Size Distribution of Bed Material

When intensive deposition takes place, the amount of deposits of the lth size group between entrance and exit section in time interval Δt is

$$W_l = (QS_0 P_{4\cdot l\cdot 0} - QSP_{4\cdot l})\Delta t. \tag{4.58}$$

Then, the total amount of deposit is

$$W = (QS_0 - QS)\Delta t. \tag{4.59}$$

The size distribution of the deposited layer is [29, 30]

$$P_{1\cdot l} = \frac{W_l}{W} = \frac{1}{\lambda}\left[P_{4\cdot l\cdot 0} - (1-\lambda)P_{4\cdot l}\right] = \frac{P_{4\cdot l\cdot 0}}{\lambda}\left[1 - (1-\lambda)^{\frac{\omega_l}{\omega_m}}\right]. \tag{4.60}$$

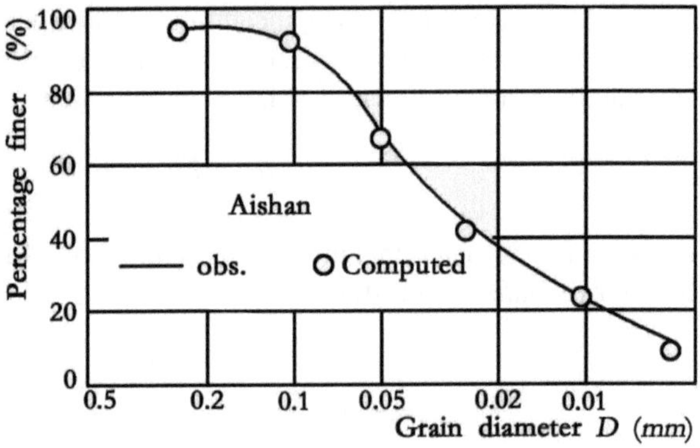

Fig. 4.11 Verification of cumulative curve of grain size at Aishan station

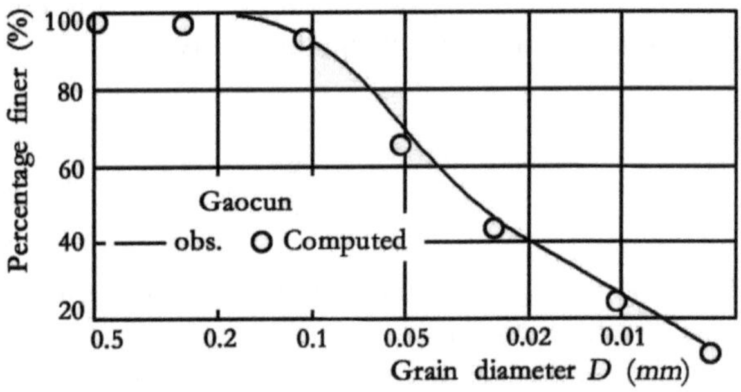

Fig. 4.12 Verification of cumulative curve of grain size at Gaocun station

When intensive scouring takes place, the residual bed material size distribution can be expressed by the following equation [30, 31]

$$P_{1\cdot l} = \frac{W_{m\cdot l} - W_l}{W_m - W} = \frac{W_{m\cdot l} - \dfrac{W_l}{W}\dfrac{W}{W_m}}{1 - \dfrac{W}{W_m}} = \frac{P_{1\cdot l\cdot 0} - \tilde{P}^*_{4\cdot l}\lambda^*}{1 - \lambda^*} = P_{1.l.0}\frac{1 - \lambda^{*\frac{\omega_l}{\dot{\omega}_m}}}{1 - \lambda^*}, \tag{4.61}$$

where W_m and $W_{m\bullet l}$ are the amount of the total and the lth size group of bed material before scouring, respectively. It can be seen that as λ^* increases, the bed material tends to get coarser.

Verification of Eqs. (4.60) and (4.61) are shown in Figs. 4.13, 4.14, 4.15 and 4.16. Field data plotted in Fig. 4.12 are taken from the Sanmenxia reservoir for the period from March to October 1964. Data in Fig. 4.14 are taken from the Danjiangkou reservoir for the period from 17 August to 8 October 1970. Field data plotted in Figs. 4.15 and 4.16 are taken from the lower Yellow River at reaches from Gaocun to Aishan, and from Huayuankou to Gaocun, respectively, in 1961–1964. During this period, strong degradation took place in the whole river course downstream of the Sammenxia reservoir.

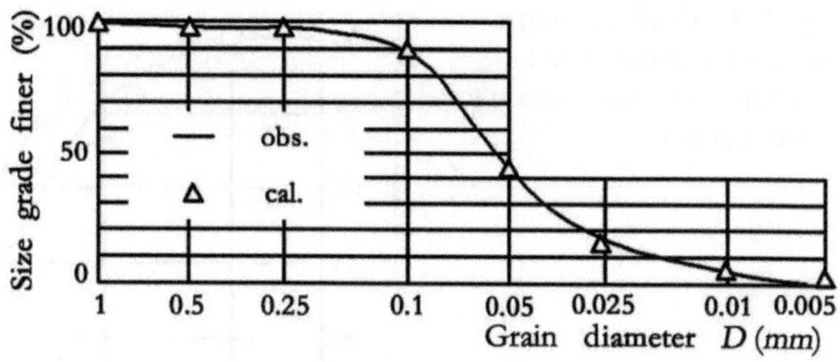

Fig. 4.13 Verification of cumulative curve of grain size in Sanmenxia reservoir

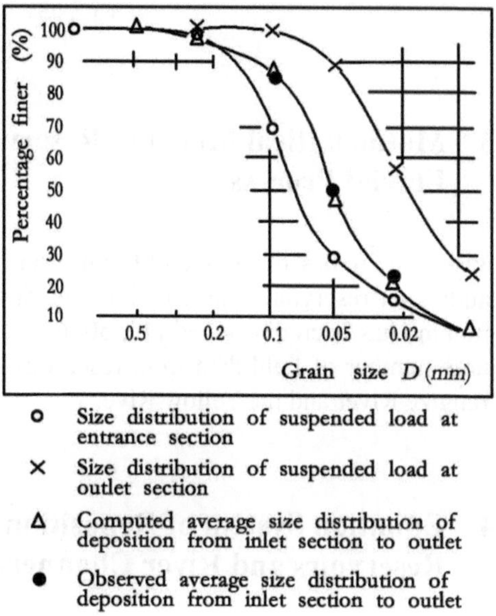

Fig. 4.14 Verification of cumulative curve of grain size in Danjiangkou reservoir

Equations (4.35), (4.45), (4.56), (4.60), and (4.61) have been checked with a large number of field data by Han et al. [29, 30, 49, 50] Because of limited space available, only a small part of verification is given here.

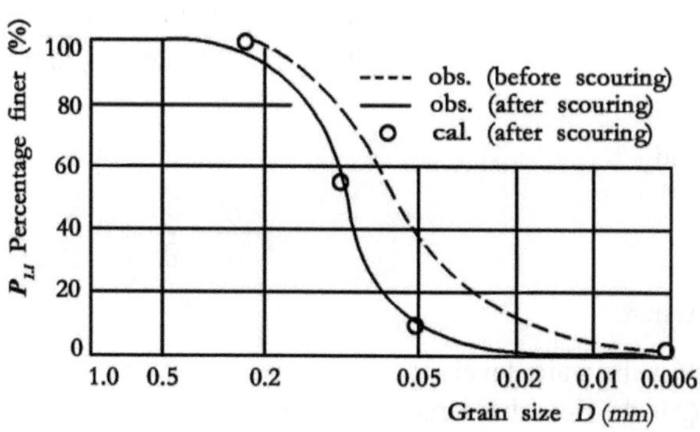

Fig. 4.15 Verification of cumulative curve of bed material from Gaocun to Aishan stations

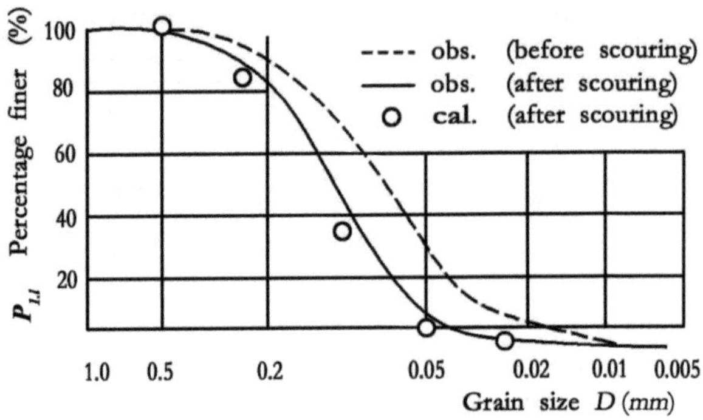

Fig. 4.16 Verification of cumulative curve of bed material from Huayuankou to Gaocun stations

3 Mathematical Model of Reservoir Sedimentation and Fluvial Process

Based on studies of non-equilibrium transport of non-uniform sediment and the studies on reservoir sedimentation and fluvial process [49–64], a computer programme has been developed [49, 50, 65]. This programme has been verified by a large number of field data from reservoirs and downstream river channels of the Yangtze River and the Yellow River.

4 Equation System of Deposition and Scouring in Reservoirs and River Channels

The system of 1D equations of the non-equilibrium transport of non-uniform sediment for reservoir sedimentation and fluvial process consists of momentum equation of flow, continuity equation, equation of morphology, equation of sediment balance, and equation of sediment concentration.

Momentum equation of flow and continuity are

$$\frac{\partial H}{\partial x} + j_f + \frac{1}{2g}\frac{\partial V^2}{\partial x} + \frac{1}{g}\frac{\partial V}{\partial t} = 0 \tag{4.62}$$

and

$$\frac{\partial H}{\partial x} + \frac{\partial A}{\partial t} = 0, \tag{4.63}$$

Where

H is the water level
Q is the flow discharge

A is the cross-sectional area of flow
g is the gravitational acceleration
$V=Q/A$ is the average velocity of flow, and
j_f is the energy slope

$$j_f = \frac{n^2 Q^2 B^{4/3}}{A^{10/3}}, \tag{4.64}$$

Where

n is the Manning's roughness coefficient and
B is the flow width of cross section.

For a rigid boundary, the area and width of cross section are functions of x and H. For a movable bed, those are functions of time also. Three types of bed deformation have been summarized, i.e. the deformation of wide section, narrow section, and gradually enlarged section. Only the deformation of wide cross section is discussed in this chapter. It is assumed that the deposit is uniformly distributed along the whole wetted perimeter during deposition. In the case of scouring, it is assumed that the depth of scouring is also uniformly distributed, but limited to the stable width of section determined by the morphological equation. If Δh is the depth of scouring or deposition in time increment Δt, the equation of area and width of cross section can be expressed as

$$A(x,z,t) = \begin{cases} A(x,z-\Delta h,t-\Delta t) & (z \in A_1) \\ A(x,z,t-\Delta t) & (z \in A_2) \end{cases}, \tag{4.65}$$

$$B(x,z,t) = \begin{cases} A(x,z-\Delta h,t-\Delta t) & (z \in A_1) \\ B(x,z,t-\Delta t) & (z \in A_2) \end{cases}, \tag{4.66}$$

Where

z is the bed elevation
Δh is the depth of deposition with a positive value during deposition and a negative value during scouring, and
Δa is the scoured or silted area in Δt, i.e.

$$\Delta a = a(x,t) - a(x,t-\Delta t). \tag{4.67}$$

Where

$A(x,z,t)$ is the area of cross section
$B(x,z,t)$ is the width of cross section
$a\,(x,\,t)$ is the total area of erosion or deposition in period t, and A_1 and A_2 are determined by

$$A_1 = \{z : z < H(x,t), \text{ if } \Delta a \geq 0; \text{ or } z < z_k \text{ and } z < H(x,t), \text{ if } \Delta a < 0\}$$
$$A_2 = \{z : z > H(x,t), \text{ if } \Delta a \geq 0; \text{ or } z \geq z_k \text{ or } z \geq H(x,t), \text{ if } \Delta a < 0\}, \tag{4.68}$$

where z_k is the elevation corresponding to the stable width of section determined by morphological equation. The equation of sediment balance is

$$\frac{\partial(QS)}{\partial x} + \frac{\partial(AS)}{\partial t} + \frac{\partial(\gamma_s' a)}{\partial t} = 0, \tag{4.69}$$

Where

S is the total sediment concentration
QS is the discharge of suspended load
AS is the sediment weight in water column with unit length in longitudinal direction, and
γ_s' is the density of deposits

If the time span for deposition in a reservoir is not too large, consolidation of deposits can be neglected and the initial density of deposits is used. The formulae for the initial density of deposits of uniform and non-uniform sediment settled uniformly or randomly have been derived [56]. For a long-term sedimentation dealing with consolidation process, an empirical formula describing the change of density should be used [56].

The equation describing the change of sediment concentration along the river course is

$$\frac{\partial QS_l}{\partial x} + \frac{\partial (AS_l)}{\partial t} = \alpha_l \omega_l (S_l - S_l^*) \qquad (l = 1, 2, \ldots, m_l), \tag{4.70}$$

Where

S_l is the concentration of lth size group and
S_l^* is the sediment-carrying capacity of that group
In the case of intensive deposition and scouring, a relationship of $P_{4\cdot l}^*$ in Eq. (4.35) is given by Eq. (4.38) as follows:

$$P_{4\cdot l}^* = P_{4\cdot l}$$

For the case of intensive deposition, Eq. (4.38) is valid. However, for intensive scouring, it is an approximation. Thus, in the cases of intensive scouring and deposition

$$S_l^* = P_{4\cdot l}^* S^* = P_{4\cdot l} K \frac{Q^{3m} B^m}{A_m^{4m} \sum_{l=1}^{m_l} P_{4\cdot l}\, \omega_l^m} \qquad (l = 1, 2, \ldots, m_l). \tag{4.71}$$

The total sediment-carrying capacity is

$$S^* = \sum S_l^* = K\frac{Q^{3m}B^m}{A_m^{4m}\sum_{l=1}^{m_l} P_{4\cdot l}\omega_l^m} \qquad (l = 1,2,\ldots,m_l). \tag{4.72}$$

There are five variables (Q, H, A, B, a) and $m_l + 1$ (S_1, S_2... S_{m_l}, S) unknowns in m_l+6 Eqs. (4.62), (4.63), (4.65), (4.66), (4.69), (4.70), and

$$S = \sum_{l=1}^{m_l} S_l. \tag{4.73}$$

In the computation of S^*, Eq. (4.72) should be used.

Equation (4.61) is used to compute the size distribution of residual bed material during scouring and Eq. (4.60) is suitable for size distribution of deposited material during deposition.

Thus, the system of 1D equations of transport of non-uniform sediment is closed. Its boundary conditions are sediment concentration and size distribution at the entrance section, the flow discharge at the entrance and exit sections, and the water level at the exit section. The initial conditions are the depth and size distribution of the bed material at the initial moment. Dividing time period $(0,T)$ by $\Delta t_i = t_i - t_{i-1}$ and integrating Eqs. (4.62), (4.63), (4.69), and (4.70) from t_{i-1} to t_i, the partial differential equations can be changed into ordinary differential equations. By separating the reach under consideration $(0, L)$ into space intervals (x_{j-1}, x_j), ($j=1, 2,\ldots,m_j$, $x_0=0$, $x_{m_j} = L$) and approximating the differential to difference, a system of finite difference equations can be derived. With some terms omitted, Eqs. (4.62), (4.63), and (4.69) become [49, 50]

$$H_{i\cdot j-1} = H_{i\cdot j} + \frac{1}{2}\frac{n_{j-1}^2 \Delta x_{j-1}}{2}\left(\frac{Q_{i\cdot j-1}^2 B_{i\cdot j-1}^{4/3}}{A_{i\cdot j-1}^{10/3}} + \frac{Q_{i\cdot j}^2 B_{i\cdot j}^{4/3}}{A_{i\cdot j}^{10/3}}\right), \tag{4.74}$$

$$Q_{i\cdot j} = Q_{i-1} + \frac{Q_{i\cdot m} - Q_{i\cdot j}}{L}x_j, \tag{4.75}$$

$$\Delta a_{i\cdot j} = \frac{Q_{i\cdot j-1}S_{i\cdot j-1} - Q_{i\cdot j}S_{i\cdot j}}{\Delta x_j \gamma_{s\cdot i\cdot j}}, \tag{4.76}$$

where the variable with subscripts i and j indicate the mean value of the variable in time interval (t_{i-1} to t_i) and at position x_j or in the space interval (x_{j-1}, x_j), respectively, $\Delta x_j = x_j - x_{j-1}$ is the interval length. $\Delta a_{i.j}$ is the mean deposit area at section (x_{j-1}, x_j) from instant t_{i-1} to t_i,

$$\bar{\gamma}_{s\cdot i\cdot j}\Delta a_{i\cdot j} = \frac{\gamma_{s\cdot i\cdot j}a_{i\cdot j} - \gamma_{s\cdot i-1\cdot j}a_{i-1\cdot j}}{\Delta t_i}. \tag{4.77}$$

$Q_{i\cdot 0}$ and $Q_{i\cdot m_j}$ are the discharge at entrance and exit section, respectively.

The sediment-carrying capacity S^* is assumed to be linearly changed between (x_j-1, x_j) and the unsteady term can be ignored. Then total concentration can be expressed as

$$S_{i\cdot j} = S^*_{i\cdot j} + (S_{i\cdot j-1} - S^*_{i\cdot j-1})\sum_{l=1}^{m_l} P_{4\cdot l\cdot i\cdot j-1}\mu_{l\cdot i\cdot j} + S^*_{i\cdot j-1}\sum_{l=1}^{m_l} P_{4\cdot l\cdot i\cdot j-1}\beta_{l\cdot i\cdot j} - S^*_{i\cdot j}\sum_{l=1}^{m_l} P_{4\cdot l\cdot i\cdot j}\beta_{l\cdot i\cdot j}, \tag{4.78}$$

where

$$\mu_{l\cdot i\cdot j} = \exp\left(-\alpha\omega_l \frac{B_{i\cdot j-1} + B_{i\cdot j}}{Q_{i\cdot j-1} + Q_{i\cdot j}}\right), \tag{4.79}$$

$$\beta_{l\cdot i\cdot j} = \frac{Q_{i\cdot j-1} + Q_{i\cdot j}}{\alpha\omega_l (B_{i\cdot j-1} + B_{i\cdot j})\Delta x_j}(1 - \mu_{l\cdot i\cdot j}), \tag{4.80}$$

$$S^*_{i\cdot j} = K\frac{Q^{3m}_{i\cdot j} B^m_{i\cdot j}}{(A^{4m}_{i\cdot j}\,\omega^m_{i\cdot j})}, \tag{4.81}$$

$$\omega^m_{i\cdot j} = \sum_{l=1}^{m_l} P_{4\cdot l\cdot i\cdot j}\,\omega^m_l. \tag{4.82}$$

The size distribution of suspended load during intensive deposition is

$$P_{4\cdot l\cdot i\cdot j} = P_{4\cdot l\cdot i\cdot j-1}\frac{(1-\lambda_{i\cdot j})^{\left(\frac{\omega_l}{\omega_{m\cdot i\cdot j}}\right)}}{1-\lambda_{i\cdot j}} \qquad (l = 1, 2, \ldots, m_l), \tag{4.83}$$

where $\lambda_{i\cdot j}$ is the deposition percentage

$$\lambda_{i\cdot j} = \frac{Q_{i\cdot j-1}S_{i\cdot j-1} - Q_{i\cdot j}S_{i\cdot j}}{Q_{i\cdot j-1}S_{i\cdot j-1}}, \tag{4.84}$$

$\omega_{m\cdot i\cdot j}$ is given by

$$\sum_{l=1}^{m_l} P_{4\cdot l\cdot i\cdot j-1}\frac{(1-\lambda_{i\cdot j})^{\left(\frac{\omega_l}{\omega_{m\cdot i\cdot j}}\right)}}{1-\lambda_{i\cdot j}} = 1. \tag{4.85}$$

The size distribution of suspended load during intensive scouring is

$$P_{4\cdot l\cdot i\cdot j} = \frac{1}{1-\lambda_{i\cdot j}}\left[P_{4\cdot l\cdot i\cdot j-1} - \lambda_{i\cdot j} P_{1\cdot l\cdot i-1\cdot j} \lambda_{i\cdot j}^{*\left(\frac{\omega_l}{\omega_{m\cdot i\cdot j}^*}-1\right)}\right], \tag{4.86}$$

where $\omega_{m\cdot i\cdot j}^*$ satisfies the equation

$$\sum_{l=1}^{m_l} P_{1\cdot l\cdot i-1\cdot j} \lambda_{i\cdot j}^{*\left(\frac{\omega_l}{\omega_{m\cdot i\cdot j}^*}\right)-1} = 1, \tag{4.87}$$

$\lambda_{i\cdot j}^*$ is the percentage of scouring

$$\lambda_{i\cdot j}^* = \frac{\Delta h'_{i\cdot j}}{\Delta h'_{i\cdot j} + \Delta h'_0} \tag{4.88}$$

as only the main channel is subject to scouring, $\Delta h'_{i\bullet j}$ is the eroded weight corresponding to the eroded depth in main channel, $\Delta h'_0$ is the weight of disturbed sediment related to the sand wave. $P_{1\bullet l\bullet i-1\bullet j}$ is the size distribution of bed material within the depth of $\Delta h'_0 + h'_{i\bullet j}$ before scouring.

From Eqs. (4.65), (4.66), and (4.68), the width and area of cross section at *j*th section from the instant t_{i-1} to t_i expressed in Eqs. (4.89, 4.90 and 4.91) are

$$A_{i\cdot j}(z) = \begin{cases} A_{i-1,j}(z - \Delta h_{1\cdot j}), & (z \in A_1) \\ A_{i\cdot j}(z) - \Delta\alpha_{i\cdot j}, & (z \in A_2) \end{cases}, \tag{4.89}$$

$$B_{i\cdot j}(z) = \begin{cases} B_{i-1,j}(z - \Delta h_{1.j}), & (z \in A_1) \\ B_{i\cdot j}(z), & (z \in A_2) \end{cases}, \tag{4.90}$$

and

$$\begin{aligned} A_1 &= \{z : z < H_{i\cdot j},\ \text{if}\ \Delta\alpha_{i\cdot j} > 0;\ \text{or}\ \ z < z_k\ \ \text{and}\ \ z < H_{i\cdot j},\ \text{if}\ \Delta\alpha_{i\cdot j} < 0\} \\ A_2 &= \{z : z \geq H_{i\cdot j},\ \text{if}\ \Delta\alpha_{i\cdot j} \geq 0;\ \text{or}\ \ z \geq z_k\ \ \text{or}\ \ z \geq H_{i\cdot j},\ \text{if}\ \Delta\alpha_{i\cdot j} < 0\}. \end{aligned} \tag{4.91}$$

4.1 Computation Procedure

The computation proceeds step by step. Each step corresponds to a certain time interval Δt_i, within which the sediment transport can be considered steady and in equilibrium with movable bed. Thus, flow factors, sediment transport rate, the sediment eroded or deposited, and the deformation of river bed are all interrelated with each other. Consequently, the iterated method is used for the computation. The computation consists of four aspects: (1) water level, (2) concentration and size distribution of suspended load, (3) weight and area of scouring and deposition and thickness and size distribution of each layer of bed material, and (4) deformation of cross section. At the first step of integration at a time interval, the geometrical parameters of cross section of the previous time interval is used to compute water level. After calculating the deformation of cross section, water level should be re-calculated with the new area and the new width of cross section, and then, the other steps can be repeated. The iterative steps are repeated until the required accuracy is reached.

In the case of dealing with high sediment concentration, if the time interval is not too short (not less than several days) the above-mentioned steps must be followed. In this case, neither the influence of the changed area on the flow velocity nor the influence on the concentration within a step can be neglected. In the case dealing with low sediment concentration in longer time interval, for example in one tenth of a year, or a river with high concentration in short time interval (less than a couple of days), it may not exert much influence on the water level, only its effect on the flow velocity and the concentration should be taken into consideration. In the case dealing with low concentration, the bed deformation in a time interval less than one fiftieth of a year is rather small and little influence on the water level and flow velocity may be caused, so the iterative cycle of computation can be ignored.

4.2 Computation of Size Distribution of Suspended Load and Sediment Concentration

The computation of size distribution of suspended load and sediment concentration is explored herein.

In the case of intensive deposition, concentration and the size distribution of suspended load are given in Eqs. (4.78, 4.79, 4.80, 4.81, 4.82, 4.83, 4.84 and 4.85), and $P_{4 \cdot l \cdot i \cdot j}$, $S_{i \cdot j}$, and $\omega_{i \cdot j}$ can all be expressed as functions of $S_{i \cdot j}$ through Eqs. (4.79, 4.80, 4.81, 4.82, 4.83, 4.84 and 4.85). Terms in the right side in Eq. (4.78) are function of $S_{i \cdot j}$, i.e.

$$F(S_{i \cdot j}) = S^{*}_{i \cdot j} + (S_{i \cdot j-1} - S^{*}_{i \cdot j-1}) \sum_{l=1}^{m_l} P_{4 \cdot l \cdot i \cdot j-1} \mu_{l \cdot i \cdot j} + S^{*}_{i \cdot j-1} \sum_{l=1}^{m_l} P_{4 \cdot l \cdot i \cdot j-1} \beta_{l \cdot i \cdot j} - S^{*}_{i \cdot j} \sum_{l=1}^{m_l} P_{4 \cdot l \cdot i \cdot j} \beta_{l \cdot i \cdot j}.$$

Equation (4.78) can then be expressed as

$$S_{i \cdot j} = F(S_{i \cdot j}). \tag{4.92}$$

It can be seen that $F(S_{i\cdot j})$ is a decreasing function of $S_{i\cdot j}$, thus the unique solution of Eq. (4.92) exists and the method for separation of root is used for its solution. In this case, a great amount of work must be done in calculating $P_{4\cdot l\cdot i\cdot j}$ and $\omega_{i\cdot j}$. For simplicity, a sorting curve for size distribution of suspended load is served. Assuming that

$$\eta_v = (1-\lambda_v)^{1/\omega_v} \quad (v = 1, 2, \ldots, m_v) \tag{4.93}$$

from Eq. (4.83),

$$P_{4\cdot l\cdot i\cdot j\cdot v} = P_{4\cdot l\cdot i\cdot j-1}\eta_v^{\omega_l} / (1-\lambda_{i\cdot j\cdot v}). \tag{4.94}$$

It is obvious that $0 \le \eta_v \le 1$. From Eqs. (4.93) and (4.94), the percentage of deposition can be calculated by

$$\lambda_{i\cdot j\cdot v} = 1 - \sum_{l=1}^{m_l} P_{4\cdot l\cdot i\cdot j-1}(1-\lambda_{i\cdot j})^{\left(\frac{\omega_l}{\omega_v}\right)} = 1 - \sum_{l=1}^{m_l} P_{4\cdot l\cdot i\cdot j-1}\eta_v^{\omega_l}. \tag{4.95}$$

Then, $P_{4\cdot l\cdot i\cdot j\cdot v}$ can be obtained from Eq. (4.94). The relationship between $\lambda_{i\cdot j\cdot v}$ and $P_{4\cdot l\cdot i\cdot j\cdot v}$ is defined by sorting curves. With the help of these curves, the computation work will be greatly reduced.

In the case of intensive scouring, the suspended load at certain position x along the river course depends on not only the size distribution of suspended load at the position x_{j-1} but also on the size distribution of bed material changed due to scouring. This part of bed material consists of the eroded bed material with a height $\Delta h'$ and the disturbed part with a height of Δh_0. During the scouring process, the top layer of bed material becomes coarser, while the deeper layers remain unchanged. Thus in the computation, the bed material should be divided into several layers with a constant depth Δh, except for the top layer and the lowest layer. The top layer sometimes has a height less than Δh and the lowest layer may has a much larger value. The size distribution of each layer can be obtained. From these data of bed material, total concentration and the size distribution of suspended load can be solved similarly.

From the concentration and size distribution of suspended load, the weight and area of deposits can be obtained. The deposit weight of each size group is

$$W_{l\cdot i\cdot j} = (Q_{i\cdot j-1}P_{4\cdot l\cdot i\cdot j-1}S_{i\cdot j-1} - Q_{i\cdot j}P_{4\cdot l\cdot i\cdot j}S_{i\cdot j})\Delta t_i. \tag{4.96}$$

The total weight of deposits is

$$W_{i\cdot j} = \sum_{l=1}^{m_l} W_{l\cdot i\cdot j},$$

where the weight of deposits in the river channel is

$$W'_{i\cdot j} = \begin{cases} W_{i\cdot j}\dfrac{B_k}{B_{i\cdot j}} & (\text{if } S_{i\cdot j} < S_{i\cdot j-1} \text{ and } B_{i\cdot j} > B_k) \\ W_{i\cdot j} & (\text{otherwise}) \end{cases} \tag{4.97}$$

and that on the flood plain is

$$W''_{i\bullet j} = W_{i\bullet j} - W'_{i\bullet j}, \tag{4.98}$$

B_k is the stable width of cross section. The size distribution of deposits is

$$r_{l\bullet i\bullet j} = \frac{W_{l\bullet i\bullet j}}{W_{i\bullet j}}. \tag{4.99}$$

The median size of deposit is

$$D_{50\cdot i\cdot j} = D_{m-1} + \frac{D_m - D_{m-1}}{r_{m\cdot i\cdot j}}\left(0.5 - \sum_{k=1}^{m-1} r_{l\cdot i\cdot j}\right), \tag{4.100}$$

where m is determined by

$$\sum_{l=1}^{m-1} r_{l\bullet i\bullet j} \le 0.5 < \sum_{l=1}^{m} r_{l\bullet i\bullet j}, \tag{4.101}$$

where D_m is the diameter of the m-th size group. Density of deposits $\gamma_{s\cdot i\cdot j}$ can be found according to the table in the paper of Han et al. [56]. The volume of deposit is

$$\Delta V_{i\cdot j} = \frac{W_{i\cdot j}}{\gamma_{s\cdot i\cdot j}}. \tag{4.102}$$

The deposit area is

$$\Delta a_{i\cdot j} = \frac{\Delta V_{i\cdot j}}{\Delta x_j}. \tag{4.103}$$

As scouring occurs in the main channel only, it is necessary to calculate the thickness and the size distribution of the surface layer of the flood plain. For the main channel, the number of the layers, the thickness, and the size distribution of each layer should be regulated according to whether it is under scouring or deposition, respectively.

Fig. 4.17 Comparison of accumulative deposits in Sanmenxia reservoir from Tongguan to Sanmenxia (March 1964–October 1964)

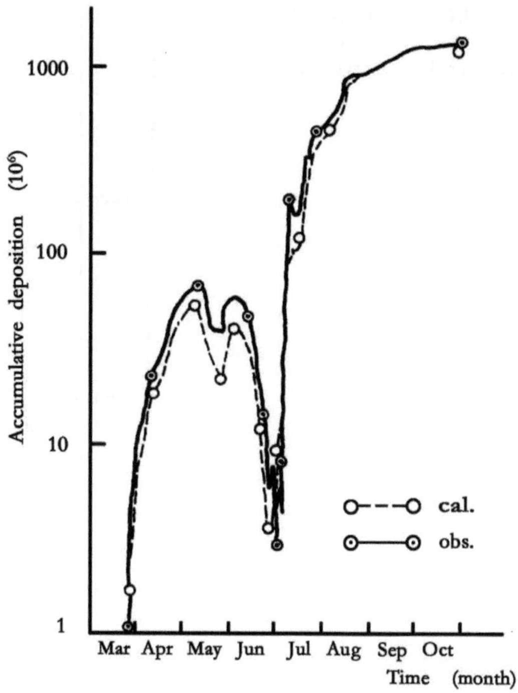

Some results are given here to show the effectiveness of the model of non-equilibrium transport. Figure 4.17 illustrates the accumulative deposition from Tongguan to Sanmenxia in Sanmenxia reservoir during the period from March to October in 1964. Figure 4.18 indicates that the accumulative deposit in Danjiangkou reservoir from 1967 to 1969. Figure 4.19 shows the comparison of accumulative deposit between measured and computed results in upstream of the Danjiangkou reservoir in 1970. The corresponding sediment concentration and its size distribution at the exit are listed in Tables 4.4 and 4.5. Figure 4.20 gives the comparison of deposition process in the Yanjiatai Warping region in 1966. Verification of accumulative deposits along the river course and size distribution of suspended load and deposits and water stage for Yanjiatai Warping region is given in Figs. 4.21, 4.22 and 4.23 and Table 4.6. Figure 4.24 demonstrates the deposition and the scouring in Zhongzhouzi cut-off project in the Yangtze River. In Fig. 4.24 curve (a) is the sediment discharge at the exit section of the new river course of Jianli; (b) is the flow discharge percentage in the new river course; and (c) is the deposited and eroded sediment in the old and new river course, respectively. The corresponding verification of the concentration of new and old river at entrance and exit section is shown in Fig. 4.25. Figures 4.26 and 4.27 show the accumulative sedimentation at Chouyanji in the Yangtze River and the calculations conform well to the observed data and the experiment made by Tsinghua University. Figure 4.28 shows the process of the delta formation in a model reservoir with a rectangular section

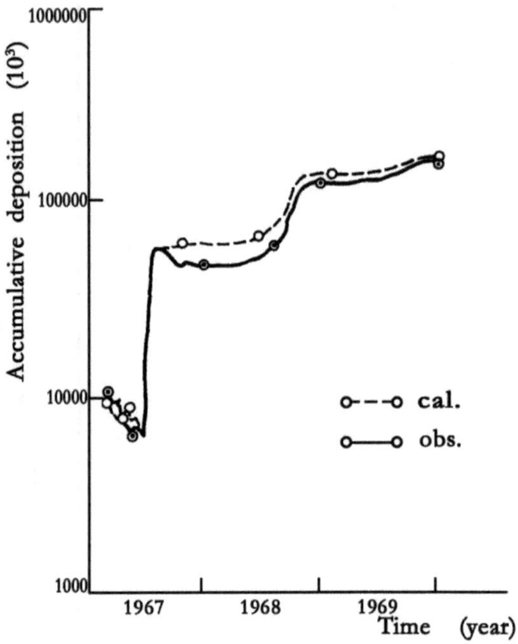

Fig. 4.18 Comparison of accumulative deposits in Danjiangkou reservoir from 1967 to 1968

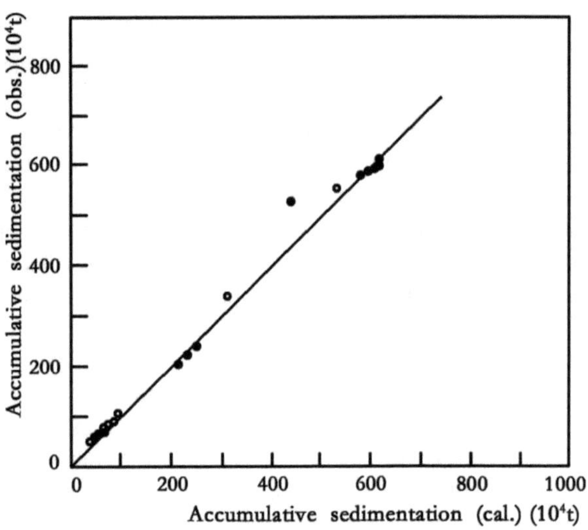

Fig. 4.19 Comparison of accumulative deposits of upstream reach of Danjiangkou reservoir in 1970

Table 4.5 Comparison of concentration at outlet section of upstream reach of Danjiangkou reservoir in 1970

No. of time interval	No. of days	Concentration at outlet section (kg/m³)	Concentration at inlet section (kg/m³)		Note
			Observed	Calculated	
1	16	0.066	0.028	0.025	Omitting flood peak
2	1	3.52	0.034	0.824	
3	19	0.469	0.048	0.133	
4	3	1.19	0.297	0.490	
5	1	4.56	2.29	1.98	
6	3	2.83	1.59	1.72	
7	4	0.971	0.628	0.521	
8	6	0.112	0.014	0.051	
total	52	1.54	0.759	0.794	

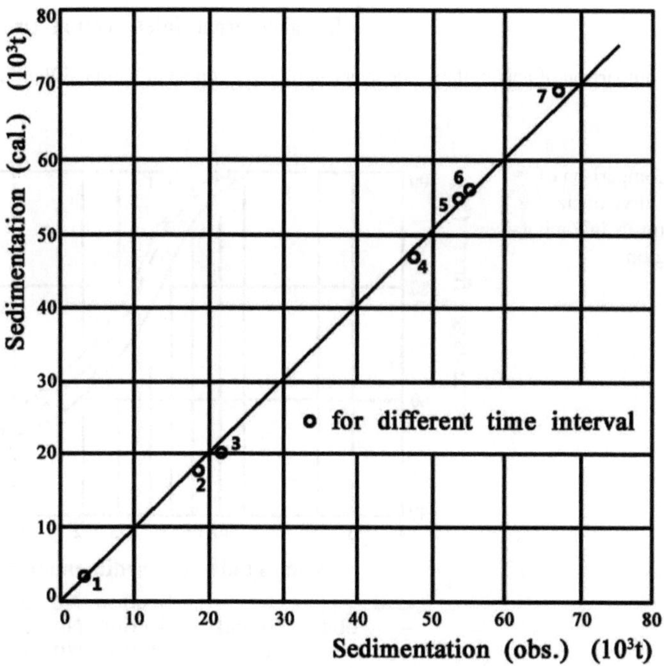

Fig. 4.20 Comparison of deposition process for different time interval in Yanjiatai Warping region

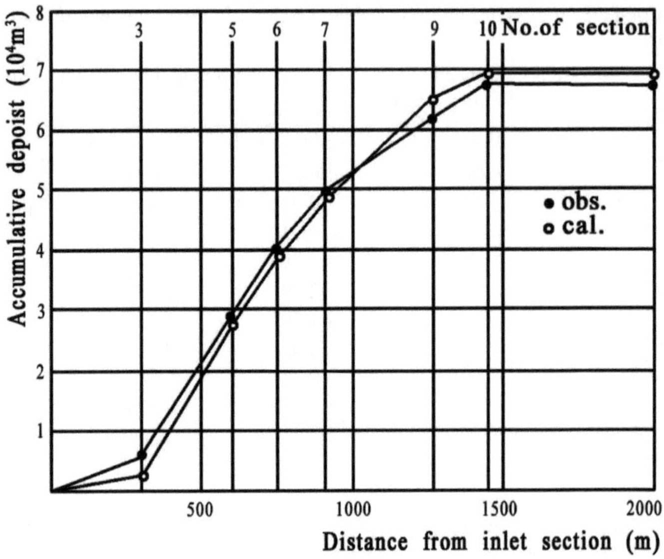

Fig. 4.21 Comparison of accumulative deposits along river course in Yanjiatai Warping region

Fig. 4.22 Comparison of cumulative curve of size grade of deposits in Yanjiatai Warping region

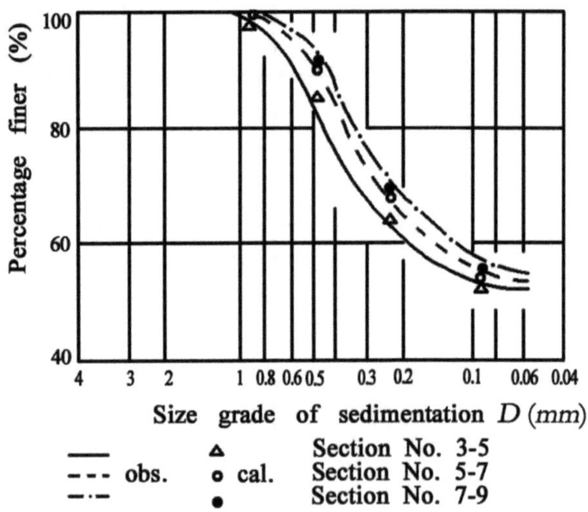

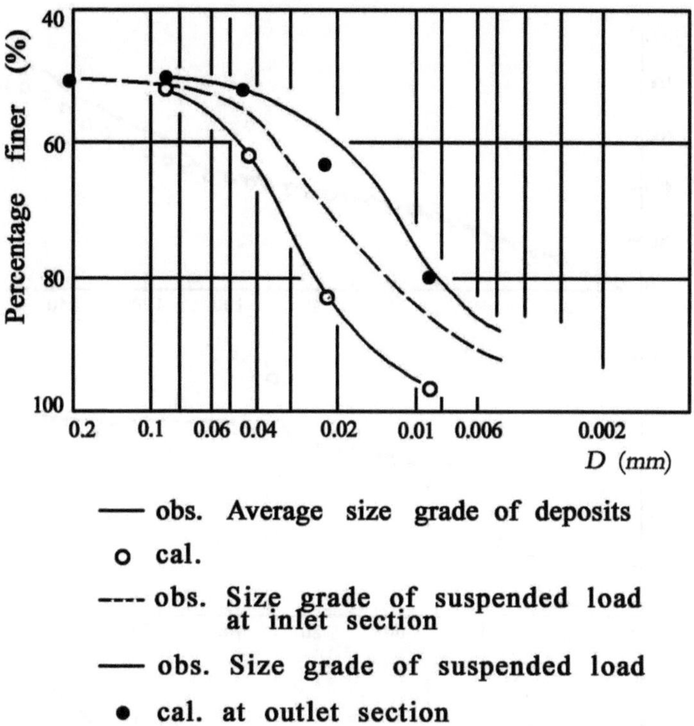

Fig. 4.23 Comparison of cumulative curve of size grade of suspended load and deposits in Yanjiatai Warping region

Table 4.6 Comparison of water stage

No. of time interval	Water stage (m)	
	Observed	Calculated
1	30.20	30.22
2	31.03	31.01
3	30.50	30.48
4	30.81	30.83
5	30.07	30.15
6	30.33	30.37
7	30.93	30.95

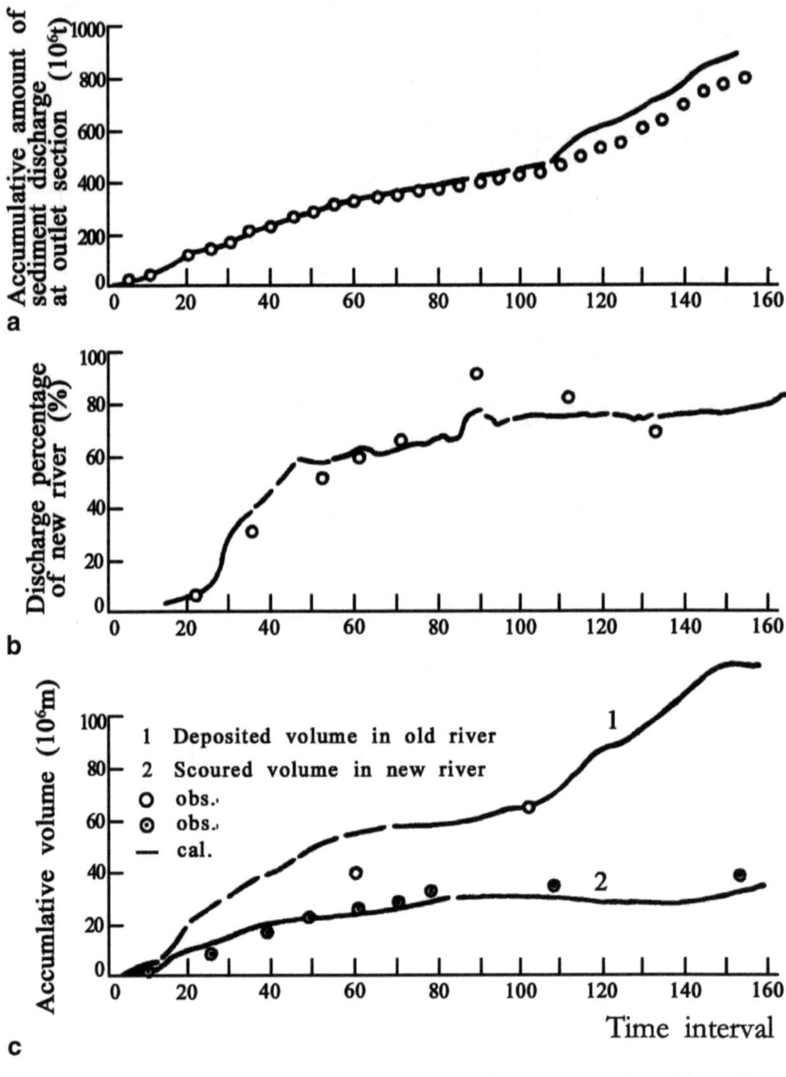

Fig. 4.24 **a** Verification of total amount of sediment discharge at the outlet of Cut-off Project at Zhongzhouzi of the Yangtzer River from May 1967 to December 1968. **b** Verification of diversion ratio into the new channel of Cut-off Project at Zhongzhouzi of the Yangtze River from May 1967 to December 1968. **c** Verification of deposition and scouring in the old and new channel of Cut-off Project at Zhongzhouzi of the Yangtze River from May 1967 to December 1968

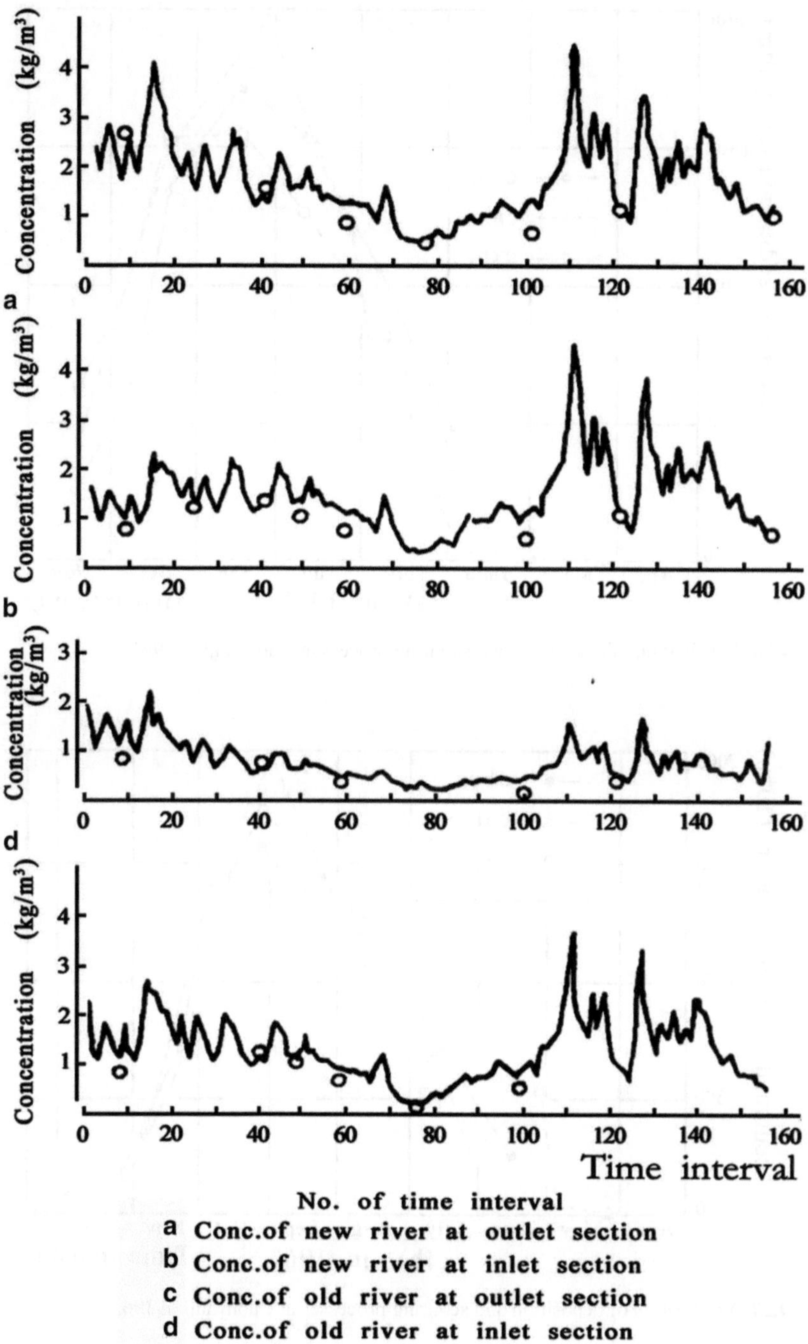

Fig. 4.25 Verification of the hydrograph of concentration at inlet and outlet sections of old and new channels of Cut-off Project at Zhongzhouzi

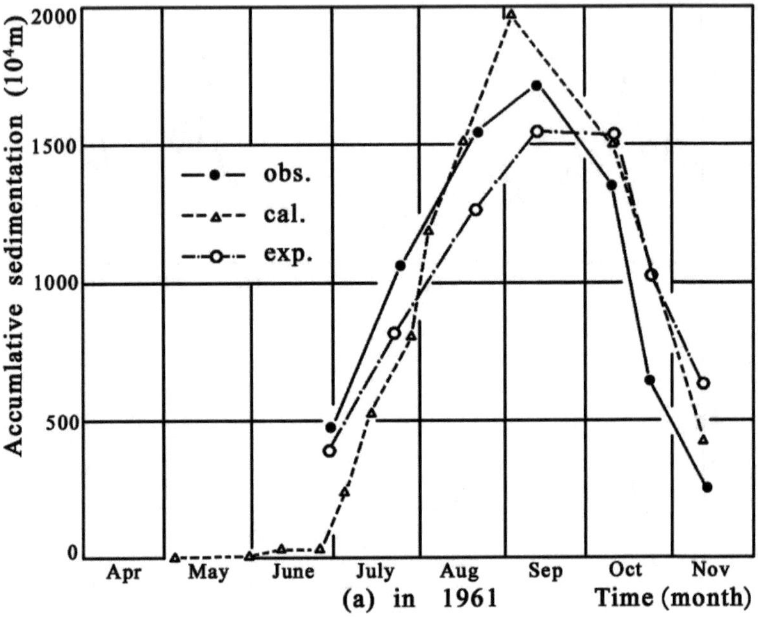

Fig. 4.26 Verification of deposition and scouring process at Chouyanji in 1961

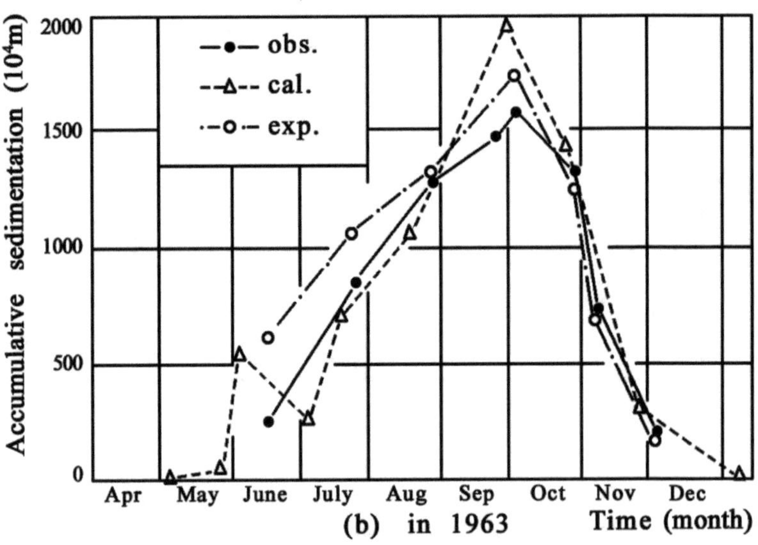

Fig. 4.27 Verification of deposition and scouring processes at Chouyanji in 1962

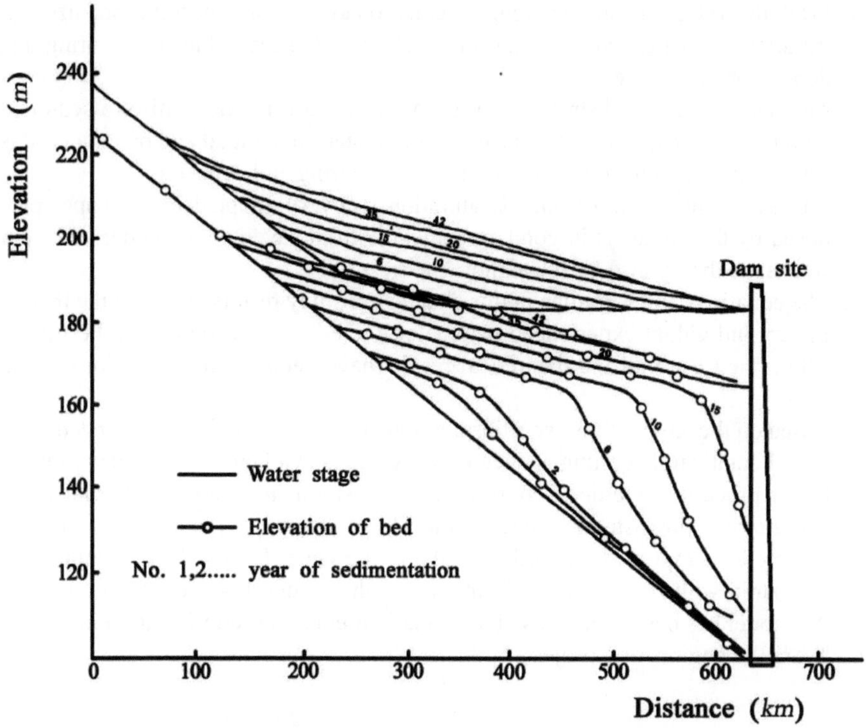

Fig. 4.28 Example of computation of delta formation process in a reservoir

and a slope of 2×10^{-4}. The another example is that from 2003 to 2005 the deposit amount in the Three-Gorges reservoir is 4.341×10^{8} t, close to the calculated value 4.490×10^{8} t [66, 67].

More data are published by Han et al. [49, 50].

The above-mentioned mathematical model is currently used in China for many important engineering projects, including the Three-Gorges Project [68–70].

5 Conclusions

1. The study results dealing with important aspects of sedimentation provide a complete theoretical description of non-equilibrium transport of non-uniform sediment.
2. Based on stochastic theory of exchange intensity, the boundary condition of 2D sediment diffusion equation is introduced and the 1D diffusion equation is also derived. This boundary condition is strict in theory and covers most boundary conditions presented previously.

3. From the 1D diffusion equation, the equation system of concentration, size distribution of suspended load, and their relation with the values of scouring and deposition are derived.
4. The relation of size distribution of capacity for carrying non-uniform sediment with the size distribution of concentration of suspended load and bed material is presented under the condition of intensive scouring and deposition.
5. The experienced coefficient of saturation recovery adopted in the paper proposed by the author is in good agreement with the average value derived from stochastic theory also by the author.
6. The equations and formulae in this chapter are study results based on the theory. Except individual experienced parameter, there is no coefficients to be further determined. And all the theoretical results have been verified by a lot of field data.
7. Based on the study of non-equilibrium transport of non-uniform suspended load, a mathematical modelling is developed, which is available for various sedimentation process, including scouring and deposition in reservoir, alluvial river, and the river downstream the reservoir. The model is in common used, since its parameters adopted in the model are almost the same. The results of computation compared with field data justify the use of the model. For more than 40 years, the model has been widely used in China to predict the sedimentation in a lot of hydraulic engineering projects.

References

1. U. S. Army Corps of Engineers. (1976). HEC-6 Scour and Deposition in Rivers and Reservoir, User's Manual.
2. United States Department of the Interior, Geological Survey. (1982). *Modeling of transient stream-bed in the Rio Grande, Cochiti Dam to near Alluquerque*, New Mexico, by Robort C. Mengis, Denver, Colorado.
3. Krishnappan, B. G. (1981). *Users manual: Unsteady, non-uniform, mobile boundary flow Model—MOBED* (CCIW (p. 107)). Burlington: Hydraulic Division, National Water Research Institute.
4. Stand, R. I., & Pemberton, E. L. (1982). *Reservoir sedimentation*. Denver: Technical Guideline for Bureau of Reclamation, Sedimentation and River Hydraulics Section, Hydrology Branch, Division of Planning Technical Services, Engineering and Research Center.
5. Karim, F., & Kennedy, J. F. (1983). *Missouri River computer-based predictors for sediment discharges friction factors of alluvial streams*. US Army Corps of Engineers Missouri River Division, MRD Sediment Series No. 19.
6. Zhang, Q. S., et al. (1983). *A mathematical model for the prediction of the sedimentation process in rivers*. Proceedings. of the 2nd International Symposium on River Sedimentation, Water Resources and Electric Power Press, Beijing, China (pp. 106–116) (In Chinese).
7. Chang, H. H. (1986). *Generalized Computer Program Fluvial—12 Mathematical Model for Erodible Channel* (Users Manual). San Diego.
8. Mahmood, K. (1986). Reservoir sedimentation: Impact, extent and mitigation. Agriculture and Rural Development Department. Washington, DC: World Bank.

9. Molinas, A., & Yang, C. T. (1986). *Generalized Stream Tube Model (GSTARS) for Alluvial River Simulation*. Denver: U. S. Department of the Interior, Bureau of Reclamation, Engineering and Research Center.
10. Yang, C. T., & Simoes, F. J. M. (2002). *User's Manual for GSTARS3* (Generalized sediment transport model for alluvial river simulation version 3.0), Denver: U.S. Department of the Interior, Bureau of Reclamation Technical Service Center.
11. Yang, C. T., & Ahn, J. (2011). *User's manual for GSTARS4* (Generalized sediment transport model for alluvial river simulation version 4.0). Fort Collins: Hydroscience and training center, Colorado State University.
12. Dobbins, W. E. (1944). Effect of turbulence on sedimentation. *Transaction of the ASCE, 109*, 619–658.
13. Rossinski, K. I., & Kuzmin, T. A. (1955). *The method of solving diffusion equation for suspended load*. USSR: Fluvial Processes Publishing House of Academy of Sciences. (in Russian)
14. Mei, C. C. (1969). Non-uniform diffusion of suspended sediment. *Journal of the Hydraulic Division ASCE, 95*(Hyl), 581–594.
15. Jobson, H. E., & Sayre, W. W. (1970). Predicting concentration profiles in open channels. *Journal of the Hydraulic Division ASCE, 96*(Hy10), 1983–1966.
16. Helmfelt, A. T., & Lenau, C. W. (1970). Non-equilibrium transport of suspended sediment. *Journal of the Hydraulic Division ASCE, 96*(Hy7), 1567–1586.
17. Sarikaya, H. Z. (1973). *Numerical calculation of the removed ratio of suspended matter in settling basins*. Proceedings of the 15th Congress of IAHR, Istanbul, pp. 207–214.
18. Sumer, B. M. (1977). Settlement of solid particles in open channel flow. *Journal of the Hydraulic Division ASCE, 103*(Hy11), 1323–1337.
19. Kerssons, P. J. M., Prins, A., & Leo, C. V. R. (1979). Model for suspended sediment transport. *Journal of the Hydraulic Division ASCE, 105*(Hy5), 461–476.
20. Ashida, K. (1980). *How to predict resevoir sedimentation*. Proceedings of the International Symposium on River Sedimentation (Vol. 2, pp. 821–850), Guanghua Press.
21. Stefan, H. G. (1980). *Suspended sediment mixing and settling in reservoirs*. Proceedings of the International Symposium on River Sedimentation (Vol. 2, pp. 865–896), Guanghua Press.
22. Zhang, Q. S. (1980). Diffusion process of sediment in open channel and its application. [In Chinese] *Journal of Sediment Research, 1*, 37–52.
23. Bechteler, W., & Schrimpt, W. (1983). *Mathematical model of non-uniform suspended sediment transport*. Proceedings. of the 2nd International Symposium on River Sedimentation (pp. 390–401). Beijing: Water Resources and Electric Power Press.
24. Parthaniades, E. (1984). *The present status of knowledge and needs for research on cohesive sediment dynamic*. Proceedings of the 3rd International Symposium on River Sedimentation (pp. 3–25). School of Engineering, University of Mississippi, USA.
25. Miheev, M. V., & Unerich, A. P. (1959). *River channel regulation for irrigation and drainage purpose*. Selhozyaizdat (in Russian).
26. Karausherv, A. V. (1960). Dynamics problems of natural water courses. Gidrometeoizdat (in Russian).
27. Dou, G. R. (1963). Suspended sediment movement by tidal currents and bed deformation calculations. *Journal of Hydraulic Engineering,* No 4. 1964 pp.
28. Lin, B. N., et al. (1981). Mathematical model of suspended sediment transport by tidal flow in the Qiantang Estuary. *Journal of Sediment Research, Beijing, 2*, 16–29.
29. Han, Q. W. (1973). *Preliminary studies of the non-equilibrium transport of sediment in reservoir*. Report on Sedimentation in Reservoir, Task Group on Yellow River Sedimentation Research (pp. 145–168) [In Chinese].
30. Han, Q. W. (1980). *A study on the non-equilibrium transportation of suspended load*. Proceedings of the International Symposium on River Sedimentation (Vol. 2, pp. 793–801), Guanghua Press [In Chinese].
31. Han, Q. W. (1979). A Study of the non-equilibrium transportation of non-uniform suspended load. [In Chinese] *Chinese Science Bulletin, 24*(17), 804–808.

32. Liu, Y. L., & Yu, X. (2005). Calculation of capacity for carrying non-uniform suspended load of non-equilibrium transport in Yellow River. *Journal of Sediment Research, 1,* 79–81. (In Chinese).
33. Zheng, W., Guo, Q. C., & Lu, Q. (2011). Review of basic theories of hyperconcentrated flows. *Journal of Sediment Research, 2,* 75–80. (In Chinese).
34. Han, Q. W., Wang, Y. C., & Xiang, X. L. (1982). Erosion and recovery of sediment concentration in the river channel downstream from Danjiangkou Reservoir. Proceedings of the Exeter Symposium, IAHS, pp. 145–152.
35. Han, Q. W. (1983). A discussion for analysis and calculation of deposits of suspended load in the reservoir. *Journal of Sediment Research, 2,* 79–81. (In Chinese).
36. Han, Q. W., & He, M. M. (1984). *Stochastic theory of sediment motion.* [In Chinese] Beijing: Science Publishing House.
37. Han, Q. W., & He, M. M. (1984). *Stochastic characters of sediment exchange and its application.* Proceedings of the 4th IAHR International Symposium on Stochastic Hydraulics (pp. 100–114), University of Illinois, USA.
38. He, M. M., & Han, Q. W. (1986). Mechanism and characteristics of non-uniform sediment transport. Proceedings of the 3rd International Symposium on River Sedimentation (pp. 845–852), University of Mississippi, USA.
39. He, M. M., & Han, Q. W. (1989). Concepts about grain size distribution of carrying capacity and effective bed material. *Journal of Hydraulic Engineering, 3,* 17–26. (In Chinese).
40. He, M. M., & Han, Q. W. (1990). The determination of the size distributions of carrying capacity and effective bed material. *Journal of Hydraulic Engineering, 3,* 1–12. (In Chinese).
41. Han, Q. W. (2008). *Size distribution of carrying-sediment capacity.* Proceedings of 6th National Symposium on Basic Theory of Sedimentation (Vol. I, pp. 1–20), Water Conservancy of Yellow River Press.
42. Han, Q. W., & Chen, X. J. (2008). Theory and calculation method of coefficient of saturation recovery. *Journal of Sediment Research, 6,* 8–16. (In Chinese).
43. Othman, K. I., & Wang, D. (2004). *Application of GSTARS 2.1 model for degradation in alluvial channels.* Proceedings of 9th International Symposium on River Sedimentation (Vol. III, pp. 1532–1537). Tsinghua University Press.
44. Zhang, W., Yan, Y. X., Zhu, Y. L., & Zheng, J. H. (2004). *1-D Numerical model of flow motion and suspended-sediment transportation in river networks of the Pearl River delta.* Proceedings of 9th International Symposium on River Sedimentation (Vol. III, pp. 1538–1543). Tsinghua University Press.
45. Margvelashvili, V., Herzfeld, M., & Wester, I. T. (2006). Modelling of fine -sediment transport in the Fitzoy Estuary and Keppel Bay, Australia Technical Report 39, June 2006, Cooperative Research Center for Coastal Zone, Estuary & Waterway Management.
46. LI, R. J., Luo, F., Zhou, H. M., & Qing, Y. (2010). Analysis on sediment settling probability and erosion coefficient. *Journal of Sediment Research, 1,* 63–66.
47. Zhou, C. J.,Chaolun,B., Zhang, H. W.,Peng, Y., & Zhang,Y. (2004). *2-D numerical modeling unsteady flow and sediment transport in Chongqing Reach of the Yangtze River.* Proceedings of 9th International Symposium on River Sedimentation (Vol. III, pp. 1560–1564). Tsinghua University Press.
48. Han, Q. L., & Zhang, Q. W. (2004). *1-D mathematical model for the Lower Yellow River.* Proceedings of 9th International Symposium on River Sedimentation (Vol. III, pp. 1550–1554), Tsinghua University Press.
49. Han, Q. W., & Huang, Y. L. (1974). *Methods of computation of sedimentation and erosion in the reservoir and utilization of digital computer.* Selected Papers on Yangtze Water Conservancy and Hydroelectric Power Research, Yangtze Water Conservancy and Hydroelectric Power Research Institute (No. 1, pp. 45–85). (In Chinese).
50. Han, Q. W., & He, M. M. (1987). Mathematical model of non-equilibrium transport of non-uniform sediment (MINENS)—A model of fluvial process and reservoir sedimentation. Lecture Note of Advanced Course on Mathematical Modeling of Alluvial River (Vol. 2, pp. IX–2–1–X–2–40). Beijing: IRTCES Publication.

51. Han, Q. W. (1980). *Long-term operation of the reservoir*. Selected Paper of Yangtze Water Conservancy and Hydroelectric Power Research (No. 6, pp. 47–56). Yangtze Water Conservancy and Hydroelectric Power Research Institute (In Chinese).
52. Han, Q. W. (1978). Studies on equilibrium state and topographic modification of long-term operating reservoir. Yangtze River (No. 2, pp. 18–35). (In Chinese).
53. Han, Q. W., & Xiang, X. L. (1981). *The characteristics of sediment transport in density current*. Yangtze River (No. 4 pp. 76–81). (In Chinese).
54. Han, Q. W., & Liang, Q. R. (1981). Discussion on the methods of composition for wetted perimeter of different roughness. *Journal of Hydraulic Engineering, 4,* 64–67. (In Chinese).
55. He, M. M., & Han, Q. W. (1982). Stochastic model of incipient sediment motion. *Journal of the Hydraulic Division ASCE, 108*(HY2), 211–224
56. Han, Q. W., Wang, Y. C., & Xiang, X. L. (1981). Initial specific weight of deposits. *Journal of Sediment Research, 1,* 1–13. (In Chinese).
57. Han, Q. W., Xiang, X. L., & Wang, Y. C. (1983). *A study on armoring of bed material*. Proceedings of the 2nd International Symposium on River Sedimentation (pp. 366–375). Water Resource and Electric Power Press (In Chinese).
58. Tong, Z. J., & Han, Q. W. (1983). *Deformation of river bed downstream from reservoir affected by changing of flow and sediment*. Proceedings of the 2nd International Symposium on River Sedimentation (pp. 673–681). WREP Press (In Chinese).
59. Han, Q. W., & Tong, Z. J. (1982). *The mechanism and characteristics of fluvial process downstream from Danjiangkou Reservoir*. Selected Papers on Fluvial Process Downstream from Danjioangkou Reservoir, Hydrologic Dept. of YVPO (pp. 17–39). (In Chinese).
60. Han, Q. W., & Shen, X. G. (1984). A study on reservoir sedimentation with conical profile. *Journal of Sediment Research, 2,* 33–51. (In Chinese).
61. Han, Q. W., He, M. M., Tong, Z. J., Wang, Y. C., Yang, K. C., & Pu, Y. X. (1986) Bed load aggradations in reservoir and deposition in variable backwater region. *Journal of Sediment Research, 1,* 1–16. (In Chinese)
62. Han, Q. W., He, M. M., & Wang, Y. C. (1989). *A discussion on distinction between wash load and bed material load*. Proceedings of the 4th International Symposium on River Sedimentation (Vol. 1, pp 506–513). China Ocean Press.
63. He, M. M., & Han, Q. W. (1989). *Raising of backwater elevation during Reservoir sedimentation*. Proceedings of the 4th International Symposium on River Sedimentation (vol. II, pp. 999–1006).
64. Han, Q. W., & He, M. M. (1988). Some problems of sedimentation calculation in the mathematical models. *Journal of Hydraulic Engineering, 5,* 16–25 (in Chinese).
65. Han, Q. W., & He, M. M. (1987). Mathematical modeling of reservoir sedimentation and fluvial process. *Journal of Sediment Research, 3,* 1–13. (In Chinese).
66. Experts Group of Sedimentation Engineering of Three Gorges Reservoir, Chinese Academy of Engineering. (2008). Appraisal Report of Sedimentation Engineering of Three Gorges Reservoir.
67. China Institute of Water Resources and Hydroelectric Power Research et al. (2011). Summarize report of recent study on sedimentation calculation of Three Gorges Reservoir.
68. Han, Q. W., He, M. M., & Sun, W. D. (1987). *The calculation and analysis of reservoir sedimentation of suspended load*. Collected Research Papers IWHR (Vol. 26, pp. 83–95). Water Resources and Electric Power Press (In Chinese).
69. Han, Q. W., He, M. M., & Sun, W. D. (1988). *The calculation and analysis of reservoir sedimentation of suspended load*. Collected Research Results on Engineering Sedimentation for the Three Gorges Project (160–180 planned Scheme), Bureau of Science and Technology, the Ministry of Water Resources and Electric Power (pp. 29–98). (In Chinese).
70. Han, Q. W. (1988). *Analysis of reliability of computed results on the reservoir sedimentation for the Three Gorges Project*. Collected Research Results on Engineering Sedimentation for the Three-Gorges Project (Planned Scheme: 160–180 m Control Elevation), Bureau of Science and Technology, the Ministry of Water Resources and Electric Power (pp. 144–161). (In Chinese).

Chapter 5
Minimum Energy Dissipation Rate Theory and Its Applications for Water Resources Engineering

Guobin B. Xu, Chih Ted Yang and Lina N. Zhao

Contents

G. B. Xu (✉) · L. N. Zhao
State Key Laboratory of Hydraulic Engineering Simulation and Safety, Tianjin University, 92 Weijin Road, Tianjin 300072, China
e-mail: xuguob@sina.com

L. N. Zhao
e-mail: neimengguzhaolina@163.com

C. T. Yang
Department of Civil and Environmental Engineering, Colorado State University, Fort Collins, CO 80523, USA
e-mail: ctyang@engr.colostate.edu; ctyang23@gmail.com

C. T. Yang, L. K. Wang (eds.), *Advances in Water Resources Engineering*,
Handbook of Environmental Engineering, Volume 14, DOI 10.1007/978-3-319-11023-3_5

Abstract Minimum energy dissipation rate principle can be derived from minimum entropy production principle. Minimum entropy production principle is equivalent to the minimum energy dissipation rate principle. The concept of minimum energy dissipation rate principle is that, when an open system is at a steady nonequilibrium state, the energy dissipation rate is at its minimum value. The minimum value depends on the constraints applied to the system. If the system deviates from the steady nonequilibrium state, it will adjust itself to reach a steady nonequilibrium state. The energy dissipation rate will reach a minimum value again. In order to verify the fluid motion following minimum energy dissipation rate principle, re-normalisation group (RNG) $k-\varepsilon$ turbulence model and general moving object (GMO) model of Flow-3D were applied to simulate fluid motion in a straight rectangular flume. The results show that fluid motion satisfies the minimum energy dissipation rate principle. Variations of energy dissipation rate of alluvial rivers have been verified with field data. When a river system is at a relative equilibrium state, the value of its energy dissipation rate is at minimum. The minimum value depends on the constraints applied to the river system. However, due to the dynamic nature of a river, the minimum value may vary around its average value. When a river system evolves from a relative state of equilibrium to another state, the process is very complicated. The energy dissipation rate does not necessarily decrease monotonically with respect to time. When a system is at a new relative state of equilibrium, the energy dissipation rate must be at a minimum value compatible with the constraints applied to the system. Hydraulic geometry relationships can be derived from the minimum energy dissipation rate principle. Combining the minimum energy dissipation rate principle with optimization technology as the objective function under the given constraints, the optimum design mathematical models can be developed for a diversion headwork bend structure and stable channel design.

Keywords Diversion bend · Fluvial process · Hydraulic geometry · Minimum energy dissipation rate principle · Minimum entropy production principle · Stable channel design · Stream power · Unit stream power

Nomenclature

A	Cross-sectional area of flow (m^2)
A_k	Chemical affinity (J/mol)
B	River width (m)
b	Channel bottom width (m)
C	Chezy coefficient ($m^{1/2}/s$)
C_H	Nonsilting sediment concentration (kg/m^3)
C_{max}	Maximum permissible sediment concentration (kg/m^3)
C_r	Criterion of circulation intensity (dimensionless)
C_s	Sediment concentration (kg/m^3)
C_V	Sediment concentration by volume (dimensionless)
C_*	Sediment transport capacity (kg/m^3)
d_{50}	Sediment median diameter (m or mm)
$\mathrm{d}_e E / \mathrm{d}t$	The entropy flux (W/K)
$\mathrm{d}_i E / \mathrm{d}t$	The entropy production (W/K)
E	Entropy (J/K)
E_V	Local entropy, also known as unit volume entropy or entropy density ($J/(K\cdot m^3)$)
e	Internal energy (J/kg)
$\boldsymbol{F}$	Mass force acting on a unit of fluid mass (N/kg)
G_b	Cross-sectional rate of bed-load transport (kg/s)
g	Acceleration of gravity (m/s^2)
h	Average water depth (m)
$\boldsymbol{J}_i$	Generalized flows (no unique units)
L_{kl}	Phenomenological coefficients (dimensionless)
m	Mass or bankside slope (kg or dimensionless)
n	Roughness ($s/m^{1/3}$)
$\boldsymbol{n}$	The outward unit vector (dimensionless)
$\boldsymbol{P}$	Second-order stress tensor (Pa)
P	Entropy production (W/K)
p	Pressure (Pa)
Q	Water discharge (kg/m^3)
$\boldsymbol{q}_\lambda$	Thermal transport vector (W/m^2)
q_R	Thermal radiation per unit mass (W/kg)
R	Hydraulic radius (m)
S	Slope (dimensionless)
T	Absolute temperature (K)
t	Time (s)
U	Velocity (m/s)
U_c	Incipient velocity (m/s)
$\boldsymbol{u}$	Velocity vector (m/s)
u_i	Component of velocity (m/s)
V	Volume (m^3)
$\boldsymbol{X}_i$	Generalized forces (no unique units)

z_b	Elevation at the bottom of cross section (m)
Γ	Permissible ratio (dimensionless)
γ	Specific weight of water (N/m^3)
$\boldsymbol{\delta}$	Second-order unit tensor (dimensionless)
ν	Molecular viscosity (m^2/s)
ν_t	Turbulent viscosity (m^2/s)
$\boldsymbol{\Pi}$	Tangential stress tensor (Pa)
ρ	Concentration or density (kg/m^3)
ρ_s'	Dry density of sediment (kg/m^3)
σ	Local entropy production (W/(K·m^3))
Φ	Energy dissipation rate per unit length (W/m)
Φ_V	Energy dissipation rate per unit fluid volume (W/m^3)
ϕ	Energy dissipation function per unit volume of energy in unit time (W/m^3)
χ	Wetted perimeter (m)
ω	Sediment particle fall velocity (m/s or mm/s)

1 Introduction

Minimum energy dissipation rate principle was proposed in 1868 by the German physicist Helmholtz. It has been applied to irrotational uniform flow of clear water within solid walls. Due to the lack of rigorous theoretical proof, it has been called the rate of energy dissipation extremum hypothesis. In the early 1950s, this principle was applied to movable-bed sediment-laden flow by Velikanov (Великанов). After the 1970s, Yang and Chang et al. made significant progresses in the study of minimum energy dissipation rate principle and its applications [1−24].

Yang [1, 3] assumed that there is an analogy between a thermo and a river system. The concept of entropy was introduced by him to the study of river system. He believed that the only useful energy in the river system is its potential energy. He further assumed that potential energy and elevation of a river system are equivalent to thermal energy and absolute temperature, respectively, of a heat system. Based on this analogy and the direct application of entropy concept in thermodynamics, it can be shown that

$$\frac{\mathrm{d}y}{\mathrm{d}t} = \frac{\mathrm{d}x}{\mathrm{d}t} \cdot \frac{\mathrm{d}y}{\mathrm{d}x} = US = \text{a minimum}, \tag{5.1}$$

where

y is the potential energy per unit weight of water in a river system (J/N)
t is the time (s),
x is the channel reach length (m)
U is the average flow velocity (m/s) and
S is the energy slope (dimensionless)

Yang [3, 7] defined the *US* product as the unit stream power. Based on the concept of unit stream power, the minimum energy dissipation rate principle can be written as:

$$\Phi = \gamma \iiint u_i S_i \, dxdydz = \gamma QS = \text{a minimum}, \tag{5.2}$$

where

Φ is the energy dissipation rate per unit length (W/m)
γ is the specific weight of water (N/m^3)
u_i is the longitudinal local velocity (m/s)
S_i is the local energy slope (dimensionless)
x, y, z are the longitudinal, lateral, and vertical coordinates, respectively (m)
Q is the average water discharge (m^3/s) and
S is the average energy slope (the discharge-slope product is called stream power; dimensionless).

2 Minimum Entropy Production Principle and Minimum Energy Dissipation Rate Principle

2.1 *Minimum Energy Dissipation Rate Principle*

The minimum energy dissipation rate principle was introduced by the Belgian physicist and chemist Prigogine in 1945. After decades of continuous development and improvement, the principle has become one of the basic theories of nonequilibrium thermodynamics [25–28]. The study of nonequilibrium thermodynamics theory is for an open system. Open system means that it can have exchange of energy and matter with the surrounding world.

2.1.1 Entropy Change in an Open System

According to the theory of classical thermodynamics, the state of a macroscopic system can be expressed by the state function, namely, entropy E. Prigogine divided the entropy change dE into the sum of two contributions [25], i.e.,

$$\frac{\mathrm{d}E}{\mathrm{d}t} = \frac{\mathrm{d}_e E}{\mathrm{d}t} + \frac{\mathrm{d}_i E}{\mathrm{d}t}, \tag{5.3}$$

where

$\mathrm{d}_e E / \mathrm{d}t$ is the entropy flux due to exchanges (of energy or matter) with the environment (W/K) and

$d_i E / dt$ is the entropy production due to irreversible processes inside the system, represented by the symbol P (W/K)

The second law of thermodynamics states that $d_i E / dt$ is positive, i.e.,

$$P = \frac{d_i E}{dt} \geq 0. \tag{5.4}$$

2.1.2 Assumption of Local Equilibrium

Classical thermodynamics method cannot be applied to nonequilibrium state open systems directly. The assumption of local equilibrium introduced to the study of nonequilibrium thermodynamics was one of the important contributions of the Brussels School headed by Prigogine.

Although a system that may consist of many small local subsystems is nonequilibrium, each subsystem can be considered as in equilibrium. From a macro point of view, the subsystem may be small, but it contains a sufficient number of microscopic particles inside, and still satisfies the classical macroscopic thermodynamic system conditions. So the classical thermodynamics function can still be used to describe the subsystem. One equilibrium state of subsystem may be different from the other, but the entire system is still nonequilibrium. If a nonequilibrium system satisfies the local equilibrium assumption and a thermodynamic function is used to describe the state of the local subsystem, then the thermodynamic function of the entire system is the sum of each of the local thermodynamic function.

Consider that a nonequilibrium open system of n constituents is isothermal and isobaric, but contains chemical reaction. When the system satisfies the local equilibrium assumption, entropy E can be used to describe the state of the system as the thermodynamic function. The local entropy expression of the system E_V can be defined as:

$$E_V = E_V\left(\left\{\rho_j(t,\boldsymbol{r})\right\}\right), \tag{5.5}$$

where

E_V is the local entropy (J/(K·m^3)), also known as unit volume entropy or entropy density, and

$\rho_j(t,\boldsymbol{r})$ is the concentration of each component in the system, which can vary with respect to space $\boldsymbol{r}$ and time t (kg/m^3)

Taking time derivatives of Eq. (5.5) yields:

$$\frac{\partial E_V}{\partial t} = \sum_{j=1}^{n}\left(\frac{\partial E_V}{\partial \rho_j}\right)\left(\frac{\partial \rho_j}{\partial t}\right). \tag{5.6}$$

2.1.3 Mass Conservation Equation

Assume an open system of n components simultaneously containing chemical reaction and diffusion movement. At a certain moment, the masses of n matters are $m_1, m_2, \ldots, m_j, \ldots, m_n$, respectively. Component j participates in r kind of chemical reactions. w_k is the chemical reaction rate per unit volume of chemical reaction k, v_{jk} is the number of grams of component j produced per gram of reaction k. Mass conservation equations of nonequilibrium systems containing chemical reaction and diffusion movement are

$$\frac{\partial \rho_j}{\partial t} = -\mathrm{div}\boldsymbol{J}_j + \sum_{k=1}^{r} v_{jk} w_k, \quad (j = 1, 2, \ldots, n), \tag{5.7}$$

where

ρ_j is the partial mass density of the component j (kg/m^3) and
$\boldsymbol{J}_j$ is the mass current density of the component j (kg/(m$^2 \cdot$ s)).

Equation (5.7) is the reaction–diffusion equation consists of nonlinear partial differential equations.

2.1.4 Entropy Balance Equation

According to the Gibbs equation in thermodynamics [28], Gibbs equation for unit volume can be written as:

$$T\mathrm{d}E_V = \rho \mathrm{d}e + \rho p \mathrm{d}V_m - \mu \mathrm{d}\rho, \tag{5.8}$$

where

E_V is the unit volume entropy (local entropy) (J/(K·m^3))
T is the absolute temperature (K)
ρ is the concentration or density (kg/m^3)
e is the internal energy (J/kg)
p is the pressure (Pa)
μ is the chemical potential (J/kg) and
V_m is the volume per unit mass (m^3/kg), i.e.,

$$V_m = \frac{V}{m} = \frac{1}{\rho}, \tag{5.9}$$

where

V is the volume (m^3) and
m is the mass (kg)

Assuming that there is no external force under mechanical equilibrium conditions for a local subsystem, Eq. (5.8) becomes [25]:

$$\frac{\partial E_V}{\partial \rho_j} = -\frac{\mu_j}{T}. \tag{5.10}$$

Substitution of Eq. (5.10) into Eq. (5.6) yields:

$$\frac{\partial E_V}{\partial t} = -\sum_{j=1}^{n} \frac{\mu_j}{T} \frac{\partial \rho_j}{\partial t}. \tag{5.11}$$

Substitution of mass conservation Eq. (5.7) into Eq. (5.11) yields:

$$\frac{\partial E_V}{\partial t} = \nabla \cdot \left(\sum_{j=1}^{n} \frac{\mu_j}{T} \boldsymbol{J}_j \right) - \sum_{j=1}^{n} \boldsymbol{J}_j \cdot \nabla \frac{\mu_j}{T} + \sum_{k=1}^{r} \left(-\sum_{j=1}^{n} \frac{\mu_j}{T} \nu_{jk} \right) w_k. \tag{5.12}$$

Let

$$\left. \begin{aligned} -\mathrm{div} \boldsymbol{J}_s &= \nabla \cdot \left(\sum_{j=1}^{n} \frac{\mu_j}{T} \boldsymbol{J}_j \right), \quad (s = 1, 2, \ldots, n) \\ \sigma &= -\sum_{j=1}^{n} \boldsymbol{J}_j \cdot \nabla \frac{\mu_j}{T} + \sum_{k=1}^{r} \frac{A_k}{T} w_k \end{aligned} \right\}, \tag{5.13}$$

where

$\boldsymbol{J}_s$ is the entropy flux due to diffusion (W/(K·m^2))
σ is the local entropy production due to chemical reaction (W/(K·m^3)) and
A_k is the chemical affinity (J/mol), $A_k = -\sum_{j=1}^{n} \nu_{jk} \mu_j$

Substitution of Eq. (5.13) into Eq. (5.12) yields:

$$\frac{\partial E_V}{\partial t} = -\mathrm{div} \boldsymbol{J}_s + \sigma. \tag{5.14}$$

Equation (5.14) is the local entropy balance equation. Integration of Eq. (5.14) over the volume yields:

$$\frac{\mathrm{d}E}{\mathrm{d}t} = \frac{\partial}{\partial t} \iiint_V E_V \mathrm{d}V = -\iiint_V \mathrm{div} \boldsymbol{J}_s \mathrm{d}V + \iiint_V \sigma \mathrm{d}V = -\iint_\Omega \boldsymbol{n} \cdot \boldsymbol{J}_s \mathrm{d}\Omega + \iiint_V \sigma \mathrm{d}V, \tag{5.15}$$

where $\boldsymbol{n}$ is the outward unit vector to the surface element dΩ (dimensionless).

Comparing Eq. (5.3) with Eq. (5.15),

$$\frac{\mathrm{d}_e E}{\mathrm{d}t} = -\iint_\Omega \boldsymbol{n} \cdot \boldsymbol{J}_s \mathrm{d}\Omega, \tag{5.16}$$

$$\frac{\mathrm{d}_i E}{\mathrm{d}t} = \iiint_V \sigma \mathrm{d}V. \tag{5.17}$$

The local entropy production can be written in the following form [25, 26]:

$$\sigma = \sum_{i=1}^{m} \boldsymbol{J}_i \cdot \boldsymbol{X}_i, \tag{5.18}$$

where

$\boldsymbol{J}_i$ is the generalized flow (no unique units) and
$\boldsymbol{X}_i$ is the generalized force (no unique units). Just requiring the units of $\boldsymbol{J}_i \cdot \boldsymbol{X}_i$ is $\mathrm{W/(K \cdot m^3)}$

There are certain rules to follow in choosing generalized flows and generalized forces in Eq. (5.18) [26, 27].

2.1.5 Lyapounov Stability Theory

Let the nonlinear differential equations be defined as:

$$\frac{\mathrm{d}x_i}{\mathrm{d}t} = f\left(t, \{x_i\}\right) \quad (i = 1, 2, \ldots, n). \tag{5.19}$$

The stability theorems are:

Theorem 1: If it is possible to find a function $V(t, \{x_i\})$ with a fixed sign for differential Eq. (5.19), and $V\frac{\mathrm{d}V}{\mathrm{d}t} \le 0$ within the neighborhood *D*, then the solution of Eq. (5.19) is stable.

Theorem 2: If it is possible to find a function $V(t, \{x_i\})$ with a fixed sign for differential Eq. (5.19), and $V\frac{\mathrm{d}V}{\mathrm{d}t} < 0$ within the neighborhood *D* except for the origin of the coordinate, then the solution of Eq. (5.19) approaches stable asymptotically.

Theorem 3: If it is possible to find a function $V(t, \{x_i\})$ with a fixed sign for differential Eq. (5.19), and $V\frac{\mathrm{d}V}{\mathrm{d}t} > 0$ within the neighborhood *D* except for the origin of the coordinate, then the solution of Eq. (5.19) is unstable.

The function having the above properties is called the Lyapounov function, which is also known as the stability criterion. Therefore, the key of study of differential equations' stability is to find a Lyapounov function. It is very difficult, in some cases, to find this function for most of the complex differential equations. Fortunately, Prigogine found a Lyapounov function (i.e., entropy production *P*) in the near-equilibrium range. The Lyapounov function can be used to determine a system's stability in the near-equilibrium range.

2.1.6 Phenomenological Equations and Onsager Reciprocal Relations

Generalized flows and generalized forces of a system vanish at thermodynamic equilibrium. Once a system deviates from the equilibrium state, the generalized forces immediately generate and trigger the generalized flows. Functional relationship between generalized flows and generalized forces is $\boldsymbol{J}_k = \boldsymbol{J}_k(\{\boldsymbol{X}_l\})$. One might expand generalized flows $\boldsymbol{J}_k$ in power series of the generalized forces $\boldsymbol{X}_l$ near equilibrium [25]:

$$\boldsymbol{J}_k\left(\{\boldsymbol{X}_l\}\right) = \boldsymbol{J}_k(0) + \sum_{l=1}^{m}\left(\frac{\partial \boldsymbol{J}_k}{\partial \boldsymbol{X}_l}\right)_0 \boldsymbol{X}_l + \frac{1}{2}\sum_{l=1}^{m}\sum_{s=1}^{m}\left(\frac{\partial^2 \boldsymbol{J}_k}{\partial \boldsymbol{X}_l \partial \boldsymbol{X}_s}\right)_0 \boldsymbol{X}_l \boldsymbol{X}_s + \cdots. \tag{5.20}$$

The first term of Eq. (5.20) vanishes at the equilibrium state. Because the impacts of the generalized forces are weak, the contributions of the third and subsequent terms can be neglected, provided that the system is near equilibrium. The remaining terms yield:

$$\boldsymbol{J}_k = \sum_{l=1}^{m} L_{kl} \boldsymbol{X}_l, \tag{5.21}$$

where

$$L_{kl} = \left(\frac{\partial \boldsymbol{J}_k}{\partial \boldsymbol{X}_l}\right)_0. \tag{5.22}$$

Equation (5.21) is the phenomenological equation. This equation shows that the relationship between generalized forces and generalized flows is linear. Therefore, the near-equilibrium range is also known as linear nonequilibrium range ("linear range" for short). Coefficients L_{kl} are known as the phenomenological coefficients and they are constants at linear range. The matrix of phenomenological coefficient is symmetric in the linear range of the irreversible process

$$L_{kl} = L_{lk}. \tag{5.23}$$

Equation (5.23) is the Onsager reciprocal relations. In other words, the increase of flux $\boldsymbol{J}_k$ arising from the increase of force $\boldsymbol{X}_l$ is equal to the increase of flux $\boldsymbol{J}_l$ arising from the increase of $\boldsymbol{X}_k$.

2.1.7 Minimum Energy Dissipation Rate Principle and Stability of Steady Nonequilibrium State

A system is either at equilibrium or nonequilibrium state. If the state parameters of a system do not change with respect to time, then this system is at a steady state. Steady state is not necessarily the equilibrium state, and it may also be at nonequi-

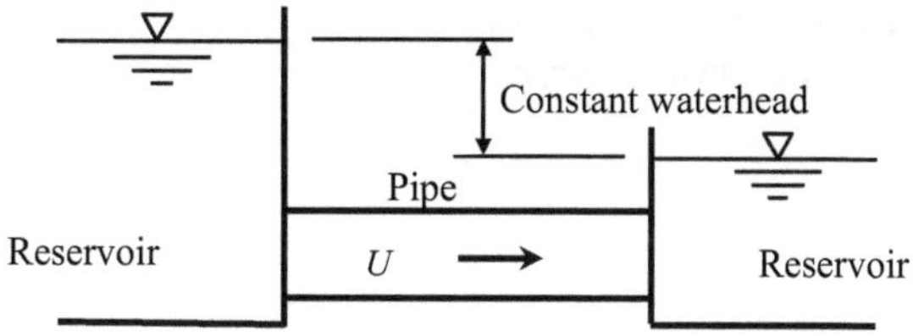

Fig. 5.1 Steady flow (steady nonequilibrium state)

librium state. Although the state parameters of a system do not change with respect to time at steady nonequilibrium states, macroscopic flow of physical property can still occur within the system. The exchange of internal and external properties makes the state parameters of the system remain constant. For instance, Fig. 5.1 [16] shows that the two end stations of a pipe are connected with two reservoirs of infinite volumes to maintain a constant head between the two end stations of the pipe. In this case, a flow in the pipeline is constant and all state parameters do not change with respect to time. It means that the flow reaches the steady state, and this steady state is the nonequilibrium rather than equilibrium because water still flows from high to low level in a macro-flow process. Figure 5.1 was also used by Yang and Song [16] to illustrate that an open system of pipe flow can be treated as a closed system including two reservoirs with a constant head difference.

According to the second law of thermodynamics, the total entropy production $P = \mathrm{d}_i E / \mathrm{d}t$ of a system is positive definite in the entire range of nonequilibrium thermodynamics. Using the linear relationship between linear generalized flow and generalized force and the Onsager reciprocal relations in the linear range, Prigogine made a series of complex derivation [25], to prove that

$$\frac{\mathrm{d}P}{\mathrm{d}t} \le 0. \tag{5.24}$$

The derivative of entropy production is negative definite in the linear range of non-equilibrium. Therefore, the entropy production is a function with a fixed sign and $P\frac{\mathrm{d}P}{\mathrm{d}t} \le 0$. So it can be a Lyapunov function of a system in the linear range.

$\frac{\mathrm{d}P}{\mathrm{d}t} = 0$ means the system is at steady state. $\frac{\mathrm{d}P}{\mathrm{d}t} < 0$ means the system is away from the steady state. Equation (5.24) shows that, when boundary conditions are constant, in the linear range of the nonequilibrium state, the irreversible evolution processes inside an open system always move toward the direction of reducing the entropy production until its minimum value is attained. At equilibrium, the state of the system no longer varies with respect to time (see Fig. 5.2). At this point, the system is compatible with the external constraints of the steady nonequilibrium state. This conclusion is called the principle of minimum entropy production. The principle of minimum entropy production guarantees the stability of steady nonequilibrium state. Once the system reaches a steady nonequilibrium state, in the absence of outside influence, it will not deviate from the steady state [24].

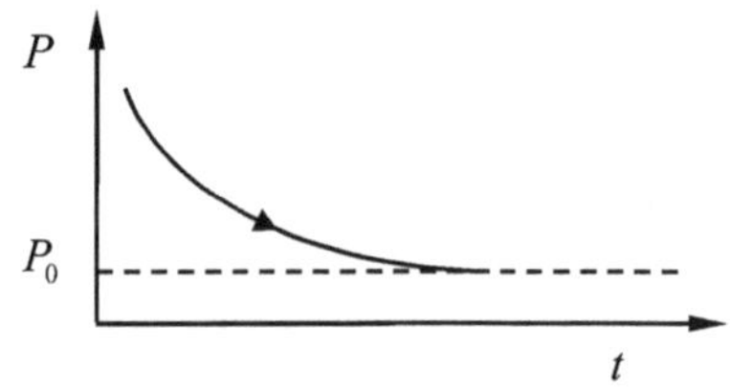

Fig. 5.2 Variation of entropy production in linear range

Prigogine derived the principle of minimum entropy production, with a series of assumptions. These assumptions include that the boundary conditions of the system are independent of time. Some people suspect that the principle of minimum entropy production is only applicable to open systems with constant boundary conditions, but not applicable to open systems with the boundary conditions evolving in time. Actually, the principle of minimum entropy production is applicable to any open system, regardless of the boundary condition to be constant or not. To reach this conclusion, we can examine the entropy change in Eq. (5.3) for open systems. The system is at steady nonequilibrium state if $\mathrm{d}E/\mathrm{d}t = \mathrm{d}_e E/\mathrm{d}t + \mathrm{d}_i E/\mathrm{d}t = 0$. If the external constraints (including boundary conditions) change, it will lead to the state of the system parameters change. Then, the system will deviate from the original steady state and evolve to a new steady state compatible with the new external constraints. In this process, the entropy production $P = \mathrm{d}_i E/\mathrm{d}t$ of the system is not necessarily monotonically decreasing. In other words, it may increase or reduce with the change of the entropy flux $\mathrm{d}_e E/\mathrm{d}t$. When the system evolves to a new steady state, the entropy generation must be the minimum value compatible with the new external constraints (see Fig. 5.3). Numerical flume simulation and river field data can confirm this viewpoint.

2.2 Minimum Entropy Production Principle Equivalence to Minimum Energy Dissipation Rate Principle

Prigogine derived the principle of minimum entropy production from local entropy production. The same conclusion can also be obtained by using the energy dissipation function.

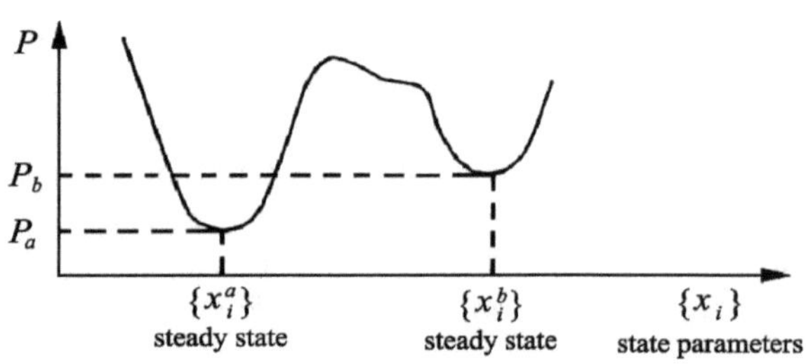

Fig. 5.3 Principle of minimum entropy production and stability of the steady state

The relationship between the local entropy production and the energy dissipation function [26, 27] is

$$\phi = T\sigma \tag{5.25}$$

where

T is the absolute temperature (K) and
ϕ is the energy dissipation function per unit volume of energy in unit time (W/m^3)

This is caused by the system's irreversible process, representing the dissipated energy in unit time per unit volume.

Similar to local entropy generation, energy dissipation function can also be written as the sum of the product of generalized forces and generalized flow (5.26):

$$\phi = \sum_{j=1}^{m} \boldsymbol{J}_j \cdot \boldsymbol{X}_j. \tag{5.26}$$

The principle for generalized forces and generalized flow in Eq. (5.26) is the same for generalized forces and generalized flows of local entropy production in Eq. (5.18). Generalized forces and generalized flow product must have the dimension of energy dissipation function.

According to Eq. (5.25), the energy dissipation rate and local entropy production P satisfy the relation

$$\Phi = \iiint_V \phi \mathrm{d}V = T\iiint_V \sigma \mathrm{d}V = TP, \tag{5.27}$$

where P can be replaced by the energy dissipation rate Φ to represent the principle of minimum entropy production

$$\frac{\mathrm{d}\Phi}{\mathrm{d}t} \le 0. \tag{5.28}$$

Principle of minimum entropy production is equivalent to the minimum energy dissipation rate principle in the linear range. When boundary conditions are constant in the linear range of nonequilibrium, the irreversible processes inside an open system always move toward the direction of reducing the entropy production until the entropy production reaches the minimum value. In this case, the system no longer varies with respect to time, and the system is compatible with the external constraints of steady nonequilibrium state.

2.3 *Minimum Energy Dissipation Rate Principle of Fluid [28–30]*

Fluid dynamic equations include the equation of motion and continuity equation. For viscous fluid, the Navier–Stokes equation of motion vector expression is

$$\boldsymbol{F} + \frac{1}{\rho}\nabla\cdot\boldsymbol{P} = \frac{\mathrm{d}\boldsymbol{u}}{\mathrm{d}t} \tag{5.29}$$

or

$$\boldsymbol{F} + \frac{1}{\rho}\nabla\cdot\boldsymbol{P} = \frac{\partial \boldsymbol{u}}{\partial t} + (\boldsymbol{u}\cdot\nabla)\boldsymbol{u}. \tag{5.30}$$

Equation (5.30) can be written in the tensor form

$$F_i + \frac{1}{\rho}\frac{\partial p_{ij}}{\partial x_j} = \frac{\partial u_i}{\partial t} + u_j\frac{\partial u_i}{\partial x_j}, \tag{5.31}$$

where

$\boldsymbol{F}$ is the mass force acting on a unit of fluid mass (F_i is the component of mass force tensor; N/kg)
ρ Is the density of fluid (kg/m^3)
$\boldsymbol{P}$ is the second-order stress tensor (p_{ij} is the component of stress tensor; Pa)
$\boldsymbol{u}$ is the velocity vector (u_i is the component of velocity; m/s)
$\mathrm{d}\boldsymbol{u}/\mathrm{d}t$ is the inertial acceleration (m/s^2)
$\partial\boldsymbol{u}/\partial t$ is the time variation of acceleration ($\partial u_i/\partial t$ is the component of variation of acceleration, m/s^2)
$(\boldsymbol{u}\cdot\nabla)\boldsymbol{u}$ is the variation of displacement acceleration and
$(u_j(\partial u_i/\partial x_j)$ is the component of variation of displacement acceleration, m/s^2)

The stress tensor is defined as

$$\boldsymbol{P} = -\boldsymbol{\delta} p + \boldsymbol{\Pi} \tag{5.32}$$

or

$$p_{ij} = -\delta_{ij} p + \tau_{ij} \tag{5.33}$$

where

$\boldsymbol{\delta}$ is the second-order unit tensor (δ_{ij} is the component of second-order unit tensor; dimensionless)

$\boldsymbol{\Pi}$ is the tangential stress tensor (τ_{ij} is the component of tangential stress tensor; Pa), and
p is the average hydrodynamic pressure along the normal direction (Pa)

Fluid continuity equation in vector form is

$$\frac{\mathrm{d}\rho}{\mathrm{d}t}+\rho\nabla\cdot\boldsymbol{u}=0. \tag{5.34}$$

For an incompressible fluid, ρ is a constant. Therefore, the continuity equation can be written as:

$$\nabla\cdot\boldsymbol{u}=0. \tag{5.35}$$

Equation (5.35) can also be written in tensor form

$$\frac{\partial u_i}{\partial x_i}=0. \tag{5.36}$$

Regardless of whether fluid movement in nature is at dynamic equilibrium or non-equilibrium state, its flow velocity distribution must satisfy both the equation of motion and continuity equation. Only when the entropy production or energy dissipation rate is at its minimum value, the velocity distribution satisfies the solution of the equation of motion and the continuity equation at the steady state (dynamic equilibrium). According to the principle of minimum entropy production, the entropy production or energy dissipation rate is a Lyapunov function which is related to the steady-state solution of fluid equation of motion and the continuity equation.

Using the assumption of local equilibrium during the derivation of minimum entropy production principle, Prigogine deduced the local entropy balance equation of the system under no external force. Considering an external force acting on the viscous fluid, time derivative on both sides of the unit volume Gibbs Eq. (5.8) yields:

$$\frac{\mathrm{d}E_V}{\mathrm{d}t}=\frac{\rho}{T}\frac{\mathrm{d}e}{\mathrm{d}t}+\frac{\rho p}{T}\frac{\mathrm{d}V_m}{\mathrm{d}t}-\frac{\mu}{T}\frac{\mathrm{d}\rho}{\mathrm{d}t}, \tag{5.37}$$

where

$$\frac{\mathrm{d}V_m}{\mathrm{d}t}=\frac{\mathrm{d}}{\mathrm{d}t}\left(\frac{1}{\rho}\right)=-\frac{1}{\rho^2}\frac{\mathrm{d}\rho}{\mathrm{d}t}=\frac{1}{\rho}\nabla\cdot\boldsymbol{u}. \tag{5.38}$$

In the derivation of Eq. (5.38), the continuity Eq. (5.34) is used. Vector expression of the fluid energy equation is:

$$\rho\frac{\mathrm{d}}{\mathrm{d}t}\left(e+\frac{u^2}{2}+gh\right)=\nabla\cdot(\boldsymbol{P}\cdot\boldsymbol{u})+\nabla\cdot\boldsymbol{q}_\lambda+\rho q_R, \tag{5.39}$$

where

e is the internal energy of unit mass of the fluid (J/kg)
$u^2/2$ is the kinetic energy of flow per unit mass (J/kg)
gh is the potential energy of flow per unit mass (J/kg), g is the gravitational acceleration (m/s^2), and h is the elevation (only considering gravitational field; m)
$\boldsymbol{q}_\lambda$ is the thermal transport vector through surface into a differential volume in a unit time by heat conduction (W/m^2)
q_R is the thermal radiation per unit mass (W/kg), $\nabla\cdot(\boldsymbol{P}\cdot\boldsymbol{u})$ is the power exerted by surface forces (W/m^3).

It can be written in the following form [31]:

$$\nabla\cdot(\boldsymbol{P}\cdot\boldsymbol{u}) = \boldsymbol{u}\cdot(\nabla\cdot\boldsymbol{P}) - p\nabla\cdot\boldsymbol{u} + \boldsymbol{\Pi}:\nabla\boldsymbol{u}. \tag{5.40}$$

Multiply both sides of Eq. (5.29) by the velocity vector yields:

$$\rho\boldsymbol{F}\cdot\boldsymbol{u} + \boldsymbol{u}\cdot(\nabla\cdot\boldsymbol{P}) = \rho\frac{\mathrm{d}}{\mathrm{d}t}\left(\frac{u^2}{2}\right). \tag{5.41}$$

If only the gravity force is considered

$$\boldsymbol{F}\cdot\boldsymbol{u} = -g\frac{\mathrm{d}h}{\mathrm{d}t}. \tag{5.42}$$

Substituting Eq. (5.41) into Eq. (5.39) and taking Eqs. (5.40) and (5.42) into consideration yield:

$$\rho\frac{\mathrm{d}e}{\mathrm{d}t} = -p\nabla\cdot\boldsymbol{u} + \boldsymbol{\Pi}:\nabla\boldsymbol{u} + \nabla\cdot\boldsymbol{q}_\lambda + \rho q_R. \tag{5.43}$$

Equation (5.43) is the time derivative of the internal energy of the fluid or another form of expression of energy equation. If the temperature of flow is uniform in the flow field, there is no heat exchange, i.e., $\boldsymbol{q}_\lambda = q_R = 0$. Eq. (5.43) can be reduced to

$$\rho\frac{\mathrm{d}e}{\mathrm{d}t} = -p\nabla\cdot\boldsymbol{u} + \boldsymbol{\Pi}:\nabla\boldsymbol{u}. \tag{5.44}$$

Substituting Eqs. (5.34), (5.38), and (5.44) into Eq. (5.37), the local entropy equilibrium equation of fluid becomes

$$\frac{\mathrm{d}E_V}{\mathrm{d}t} = \nabla\cdot\left(\frac{\mu\rho}{T}\boldsymbol{u}\right) - \boldsymbol{u}\cdot\frac{\mu}{T}\nabla\rho + \frac{1}{T}\boldsymbol{\Pi}:\nabla\boldsymbol{u}, \tag{5.45}$$

where

$$\left.\begin{aligned} -\mathrm{div}\boldsymbol{J} &= \nabla\cdot\left(\frac{\mu\rho}{T}\boldsymbol{u}\right) \\ \sigma &= -\boldsymbol{u}\cdot\frac{\mu}{T}\nabla\rho+\frac{1}{T}\boldsymbol{\Pi}:\nabla\boldsymbol{u} \end{aligned}\right\}. \tag{5.46}$$

For incompressible fluid, ρ is a constant, so $\nabla\rho = 0$. The local entropy production of fluid σ can be expressed as:

$$\sigma = \frac{1}{T}\boldsymbol{\Pi}:\nabla u. \tag{5.47}$$

According to the relationship between local entropy production σ and energy dissipation function Eq. (5.25), the energy dissipation function of fluid can be written as:

$$\phi = T\sigma = \boldsymbol{\Pi}:\nabla\boldsymbol{u}. \tag{5.48}$$

The tensor form of Eq. (5.48) is

$$\phi = \tau_{ij}\frac{\partial u_i}{\partial x_j}. \tag{5.49}$$

Integrating ϕ, the rate of energy dissipation in unit time Φ of fluid is:

$$\Phi = \iiint_V \phi \mathrm{d}V. \tag{5.50}$$

Substituting the expression of energy dissipation rate function Eq. (5.49) into Eq. (5.50), the rate of energy dissipation of fluid becomes

$$\begin{aligned} \Phi &= \iiint_V \phi \mathrm{d}V = \iiint_V \tau_{ij}\frac{\partial u_i}{\partial x_j}\mathrm{d}V \\ &= \iiint_V \frac{\partial}{\partial x_j}(\tau_{ij}u_i)\mathrm{d}V - \iiint_V u_i\frac{\partial \tau_{ij}}{\partial x_j}\mathrm{d}V \\ &= \iint_\Omega \tau_{ij}u_i n_i \mathrm{d}\Omega - \iiint_V u_i\frac{\partial \tau_{ij}}{\partial x_j}\mathrm{d}V, \end{aligned} \tag{5.51}$$

where n_i is the component of the outward unit vector $\boldsymbol{n}$ normal to the surface $\mathrm{d}\Omega$ (dimensionless).

The partial integration and the Gauss formula are used for the derivation of Eq. (5.51). The boundary conditions are constant if the flow is at steady state. In this case, the first surface integral term of Eq. (5.51) vanishes. Thus,

$$\Phi = -\iiint_V u_i\frac{\partial \tau_{ij}}{\partial x_j}\mathrm{d}V. \tag{5.52}$$

For steady flows, the time-variation acceleration vanishes in the equation of flow motion Eq. (5.31), i.e., $\frac{\partial u_i}{\partial t} = 0$. Based on Eq. (5.33), the equation of flow motion Eq. (5.31) can be written as:

$$\frac{\partial \tau_{ij}}{\partial x_j} = -\rho F_i + \frac{\partial p}{\partial x_i} + \rho u_j \frac{\partial u_i}{\partial x_j}. \tag{5.53}$$

Submitting Eq. (5.53) into Eq. (5.52) yields:

$$\Phi = -\iiint_V u_i \frac{\partial \tau_{ij}}{\partial x_j} \mathrm{d}V = -\iiint_V u_i \left(-\rho F_i + \frac{\partial p}{\partial x_i} \right) \mathrm{d}V - \rho \iiint_V u_i u_j \frac{\partial u_i}{\partial x_j} \mathrm{d}V. \tag{5.54}$$

According to the field theory,

$$(\boldsymbol{a} \cdot \nabla)\boldsymbol{a} = \mathbf{grad}\frac{a^2}{2} - \boldsymbol{a} \times \mathbf{rot}\boldsymbol{a}. \tag{5.55}$$

For irrotational flow, Eq. (5.55) becomes

$$(\boldsymbol{a} \cdot \nabla)\boldsymbol{a} = \mathbf{grad}\frac{a^2}{2}. \tag{5.56}$$

Taking Eq. (5.56) into consideration, the second term of the right side of Eq. (5.54) can be written as:

$$\iiint_V u_i u_j \frac{\partial u_i}{\partial x_j} \mathrm{d}V = \iiint_V u_i (\boldsymbol{u} \cdot \nabla)\boldsymbol{u} \mathrm{d}V = \iiint_V u_i \mathbf{grad}\frac{u^2}{2} \mathrm{d}V = \iiint_V u_i \frac{\partial \left(\frac{u^2}{2} \right)}{\partial x_i} \mathrm{d}V.$$

Therefore,

$$\begin{aligned} \Phi &= -\iiint_V u_i \left(-\rho F_i + \frac{\partial p}{\partial x_i} \right) \mathrm{d}V - \rho \iiint_V u_i \frac{\partial \left(\frac{u^2}{2} \right)}{\partial x_i} \mathrm{d}V, \\ &= -\iiint_V u_i \left(-\rho F_i + \frac{\partial p}{\partial x_i} + \rho \frac{\partial \left(\frac{u^2}{2} \right)}{\partial x_i} \right) \mathrm{d}V \end{aligned} \tag{5.57}$$

If only the gravitational force F_i is considered in the direction of h,

$$F_i = -g \frac{\partial h}{\partial x_i} \tag{5.58}$$

Substitution of Eq. (5.58) into Eq. (5.57) yields:

$$\Phi = -\iiint_V u_i \left(\rho g \frac{\partial h}{\partial x_i} + \frac{\partial p}{\partial x_i} + \rho \frac{\partial \left(\frac{u^2}{2} \right)}{\partial x_i} \right) \mathrm{d}V = -\gamma \iiint_V u_i \frac{\partial}{\partial x_i} \left(h + \frac{p}{\gamma} + \frac{u^2}{2g} \right) \mathrm{d}V, \quad (5.59)$$

where

$$-\frac{\partial}{\partial x_i}\left(h + \frac{p}{\gamma} + \frac{u^2}{2g} \right) = S_i.$$

Equation (5.59) can now be written as:

$$\Phi = \gamma \iiint_V u_i S_i \mathrm{d}V, \quad (5.60)$$

where

γ is the specific weight of water (N/m^3)
S_i is the energy gradient of flow. The velocity-slope product is the unit stream power (dimensionless).

For one-dimensional flow, Eq. (5.60) can be simplified to:

$$\Phi = \gamma \iiint_V u_x S_x \mathrm{d}V = \gamma \int_L S_x \mathrm{d}x \iint_A u_x \mathrm{d}A, \quad (5.61)$$

where A is the cross-sectional area of the flow (m^2).

Taking unit length $L=1$ along the direction of the flow and the assumption that $S_x = S$ is a constant per unit length yield:

$$\Phi = \gamma S \iint_A u_x \mathrm{d}A = \gamma QS. \quad (5.62)$$

The discharge-slope product is the stream power. Equation (5.62) is an expression of total rate of energy dissipation per unit length.

Equation (5.62) was derived from using the continuity equation, the N–S equation of motion, and the energy equation applicable to the movement of viscous fluid. They are valid for laminar flow and turbulent flow. Equation (5.62) can be applied to laminar flow as well as to turbulent flow for solving practical problems.

According to the principle of minimum entropy generation of nonequilibrium thermodynamics or the principle of minimum energy dissipation rate, when boundary conditions are constant, the evolution processes always move toward the direction of reducing the entropy production or the energy dissipation rate until the system reaches a steady nonequilibrium state which is compatible with the external constraints. At this point, the minimum value is attained. For fluid, taking the expression of the energy dissipation rate into account, the mathematical expression of the principle of minimum energy dissipation rate or minimum stream power is

$$\Phi = \gamma QS = \text{a minimum.} \tag{5.63}$$

The principle of minimum entropy production, the principle of minimum energy dissipation rate, the principle of minimum stream power, or minimum unit stream power indicate that the direction of evolution of fluid motion is to minimize its energy dissipation rate, stream power, or unit stream power with a minimum value compatible with the external constraints. The value of entropy production or energy dissipation rate reaches a minimum value at steady state. Equation (5.63) was deduced based on the equation of flow motion Eq. (5.53) in which the time-variant acceleration in the inertia terms was negligible. Therefore, Eq. (5.63) can be applied to the following three cases: (1) any open systems of fluid with steady boundary conditions, for example, rivers; (2) steady nonuniform flow or uniform flow; and (3) laminar flow or turbulent flow.

2.4 *Interim Summary*

The principle of minimum entropy production, minimum energy dissipate rate, minimum stream power, and minimum unit stream power are the basic principles of nonequilibrium thermodynamics and fluid mechanics. These principles can be applicable to an open system. A river system is an open system that exchanges matter and energy with the outside world in the process of movement. The minimum energy dissipation rate principle or its simplified minimum stream power or minimum unit stream power is proven with sound theoretical basis from the minimum entropy production principle. The minimum entropy production principle is equivalent to the minimum energy dissipation rate principle.

The principle of minimum entropy production or the principle of minimum energy dissipation rate reflects the behavior of steady nonequilibrium state. If an open system is at steady state and the boundary conditions of the system are constant, the system may deviate from this steady state because of the changing external conditions or constraints. Once the system deviates from the steady state, the entropy production or the energy dissipatation rate of the system may become greater than that at the steady state. At this point, the entropy production or the energy dissipation rate of the system will decrease with respect to time. Eventually, the system is restored to a steady state. Once the system reaches a steady nonequilibrium state, in the absence of outside influence, it will not vary from the steady state. The boundary condition of the system maintains constant at steady state. If the boundary condition of the system keeps constant and once the system deviates from the steady state, the system will be restored to the original steady state. If the boundary conditions change, the system will deviate from the original steady state and reach another new steady state which is compatible with the new boundary conditions. When a system changes from one steady state to another, it may go through a series of nonequilibrium evolution. Eventually, it will reach a new steady state. The value of entropy production or the energy dissipation rate must be at minimum at the new steady state, but it does not necessarily decrease monotonically with time in nonequilibrium evolution. Therefore, the minimum energy dissipation rate or the principle of minimum entropy production can be considered as a stability criterion for the evolution process of an open system.

3 Numerical Simulation of Fluid Motion in Flume Based on Minimum Energy Dissipation Rate Principle [32]

In order to verify that fluid motion satisfies minimum energy dissipation rate principle, the re-normalisation group (RNG) $k-\varepsilon$ turbulence model combined with general moving object (GMO) model is used in the numerical simulation study on fluid motion in a straight rectangular flume. The GMO model in Flow-3D can simulate rigid body motion, which is either user-prescribed (prescribed motion) or dynamically coupled with fluid flow (coupled motion).

By selecting the fluid between different cross sections as the research system, variation of energy dissipation rate per unit volume of water was calculated for the system. The simulation results show that the energy dissipation rate of the steady flow can reach a stable constant minimum value. The energy dissipation rate of an unstable flow has no definite minimum value, and its variation can increase or decrease.

3.1 Mathematical Model

The RNG $k-\varepsilon$ turbulence model was selected to simulate fluid motion. This model consists of the following equations:

Continuity equation:

$$\frac{\partial \rho}{\partial t}+\frac{\partial(\rho u_i)}{\partial x_i}=0. \tag{5.64}$$

Momentum equation:

$$\frac{\partial u_i}{\partial t}+\frac{\partial}{\partial x_j}(u_i u_j)=-\frac{\partial p}{\rho \partial x_i}+\frac{\partial}{\partial x_j}\left[(\nu+\nu_t)\left(\frac{\partial u_i}{\partial x_j}+\frac{\partial u_j}{\partial x_i}\right)\right], \tag{5.65}$$

where

ρ is the fluid density (kg/m^3)
t is the time (s)
u_i, u_j are the flow velocities in the i and j direction, respectively, ($i, j = 1, 2, 3$; m/s)
x_i, x_j are the displacement components in the i and j direction, respectively, ($i, j = 1, 2, 3$; m)
p is the pressure (Pa)
v is the molecular viscosity (m^2/s), and
v_t is the turbulent viscosity (m^2/s), in the following form:

$$v_t = C_\mu \frac{k^2}{\varepsilon}, \tag{5.66}$$

with

$$k = \frac{1}{2}\overline{u_i' u_i'}; \quad \varepsilon = v \overline{\frac{\partial u_i' \partial u'_i}{\partial x_j \partial x_j}}$$

where

k is the turbulent kinetic energy (J/kg)
ε is the turbulent dissipation rate (W/kg), and
C_μ is the dimensionless constant of 0.0845 in the RNG $k-\varepsilon$ turbulence model (dimensionless)

Equation of k:

$$\frac{\partial k}{\partial t} + \frac{\partial (k u_i)}{\partial x_i} = G_k + \varepsilon + \frac{\partial}{\partial x_j}\left[\alpha_k (v + v_t)\frac{\partial k}{\partial x_j}\right]. \tag{5.67}$$

Equation of ε:

$$\frac{\partial \varepsilon}{\partial t} + \frac{\partial (\varepsilon u_i)}{\partial x_i} = C_{\varepsilon 1}^* \frac{\varepsilon}{k} G_k - C_{\varepsilon 2}\frac{\varepsilon^2}{k} + \frac{\partial}{\partial x_j}\left[\alpha_\varepsilon (v + v_t)\frac{\partial \varepsilon}{\partial x_j}\right], \tag{5.68}$$

where G_k is the production term of the turbulent kinetic energy caused by average velocity gradient (m^2/s^3), i.e.,

$$G_k = v_t\left(\frac{\partial u_i}{\partial x_j} + \frac{\partial u_j}{\partial x_i}\right)\frac{\partial u_i}{\partial x_j}. \tag{5.69}$$

In Eqs. (5.67) and (5.68), there are four coefficients, i.e., $C_{\varepsilon 1}^*$, $C_{\varepsilon 2}$, α_k, α_ε. They can be represented by

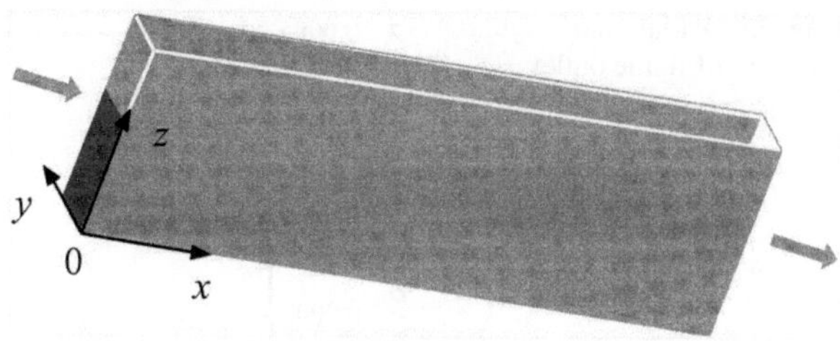

Fig. 5.4 Illustration of numerical flume installation

$$C_{\varepsilon 1}^{*} = C_{\varepsilon 1} - \frac{\eta(1-\eta/\eta_0)}{1+\beta\eta^3}; \quad C_{\varepsilon 2} = 1.68; \quad \alpha_k = \alpha_\varepsilon = 1.39,$$

where

$C_{\varepsilon 1} = 1.42$ (dimensionless)
$\eta = Sk/\varepsilon$ (dimensionless)
$S = \sqrt{2S_{ij}S_{ij}}$ (1/s), and

$$S_{ij} = \frac{1}{2}\left(\frac{\partial u_i}{\partial x_j} + \frac{\partial u_j}{\partial x_i}\right) \text{ (1/s)}$$

$\beta = 0.012$ (dimensionless)
$\eta_0 = 4.377$, dimensionless

3.2 Description of Numerical Flume

The numerical flume is a rectangular water flume with length, width, and height of 10, 1, and 1.5 m, respectively. The inlet of the flume is fixed, and the outlet is rotatable around the *y* axis of inlet in the *x–z* plane at a certain angular velocity to adjust the flume bottom slope (see Fig. 5.4).

Using the rectangular numerical flume, the simulation of the variation of the energy dissipation rate under three different calculation conditions is given below.

3.3 Boundary Conditions and Initial Conditions of Calculation

1. *Calculation condition one*: The flume is tilted with bottom slope of 0.01. First, water is injected from the inlet with a constant velocity until a steady nonuniform flow is reached. Second, the outlet begins to rotate until the bottom slope changes from 0.01 to 0.011. The flow becomes unsteady in this condition. After a period of time, the flow reaches a steady nonuniform condition again. The energy dissipation rate change of the whole process is from a steady to an unsteady flow condition, and then to a steady flow again.

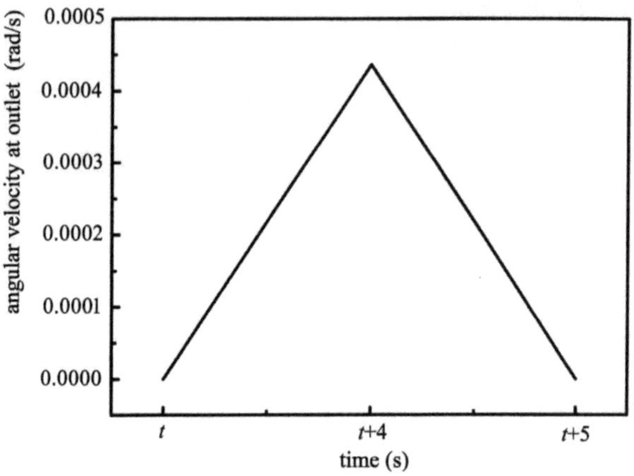

Fig. 5.5 Variation of angular velocity of flume outlet

Boundary conditions are: (1) inflow boundary condition: inlet velocity $u=5.5$ m/s, water depth $h=0.5$ m; (2) outflow boundary condition: free flow; (3) sidewall condition: wall roughness coefficient $n=0.009$.
Initial conditions: water is immobile in the flume, upstream water depth $h=0.4$ m, downstream water depth $h=0.5$ m.

2. *Calculation condition two*: Compared with calculation condition one, only flow velocity of the upstream cross-section is changed.
   Boundary conditions are: (1) Inflow boundary condition: inlet velocity $u=5.06$ m/s, water depth $h=0.5$ m; (2) outflow boundary condition: free flow; (3) sidewall condition: wall roughness coefficient $n=0.009$.
   Initial conditions: water is immobile in the flume, upstream water depth $h=0.4$ m, downstream water depth $h=0.5$ m.
3. *Calculation condition three*: The calculation condition two is the one with bottom slope 0.011 at steady flow condition as the initial state and then change the bottom slope from 0.011 to 0.01. Other conditions are the same as those of calculation condition one.

When bottom slope is changed, the angular velocity of the outlet accelerates to the maximum linearly and then decelerates to zero linearly as shown in Fig. 5.5. The angular velocity direction of calculation condition three is opposite to calculation conditions of one and two.

3.4 Calculation Process

Longitudinal profile of the flume is shown in Fig. 5.6a. The fluid between upstream section 1 and downstream section 2 is the study system. The length of the reach between the two sections is ΔL. Start from section 1, the system is divided into several rectangular units, and each unit size is 0.1 m (height) $\times$ 0.1 m (lateral) $\times$ 0.1 m (longitudinal). The system is shown in Fig. 5.6b.

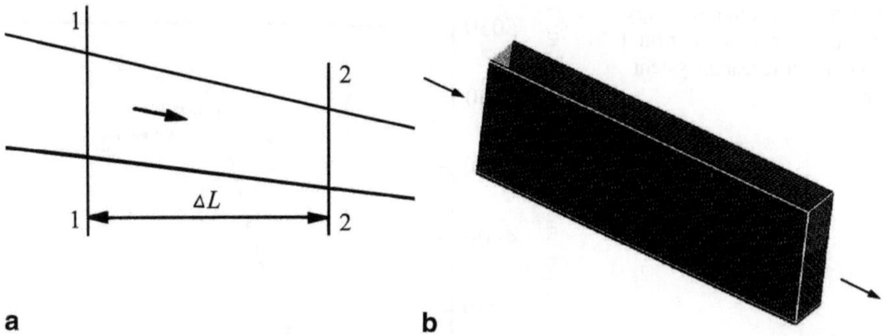

Fig. 5.6 Calculation illustration of flume and unit division of calculation region. **a** Calculation illustration of flume. **b** Unit division of calculation region

Energy dissipation is due to the viscosity of fluid. Therefore, the mechanical energy is irreversibly transformed into heat and dissipated to overcome the viscous force and the sidewall friction. The energy of fluid inside of the system does not change with respect to time for a steady flow. If there is no energy loss, the energy input from the upstream cross section shown in Fig. 5.6a should be the same as the energy output at the downstream cross section. If there is energy dissipation, the energy output at downstream cross section will be less than that of the input energy at the upstream section. The difference between the upstream cross section and the downstream cross section is the energy dissipation of the study system. For unsteady flow, the fluid volume in the system varies with respect to time. In the middle of the system, water surface may have some slight fluctuations, and the energy in the system varies with respect to time. The energy difference between the upper and lower cross sections includes the energy dissipation and energy stored in the system. Through the simulation of the whole variation process from steady flow to unsteady flow, the energy dissipation rate of the unit volume of water can be calculated. By doing so, whether the entire variation process of fluid motion satisfies the principle of minimum energy dissipation rate can be examed. Energy dissipation rate of unit volume of water is the energy losses of the unit volume of water in unit time.

In the system shown in Fig. 5.6a, the energy dissipation rate of unit volume of water of the steady and unsteady flows can be calculated. Assume that energy of the system is e_t at time t ($t = t_0 + \sum \Delta t$), where t_0 = initial time of the entire calculation period), and fluid volume of the system is V_t. After interval Δt, due to the energy loss, the energy of system becomes $e_{t+\Delta t}$ and the fluid volume of system becomes $V_{t+\Delta t}$. During interval Δt, the input energy at the upstream section is e_u, and the output energy at the second downstream section is e_d. If there is no energy loss, then $e'_{t+\Delta t} = e_t + e_u - e_d$ at time $t + \Delta t$. Therefore, the energy dissipation rate of system is $(e'_{t+\Delta t} - e_{t+\Delta t}) / \Delta t$.

For each calculation unit, it is assumed that water discharge is Q_i; flow velocity is u_i; volume of water is V_i; and the corresponding water level of the grid center point is z_i. Taking the interval from the first steady flow transition to the second steady

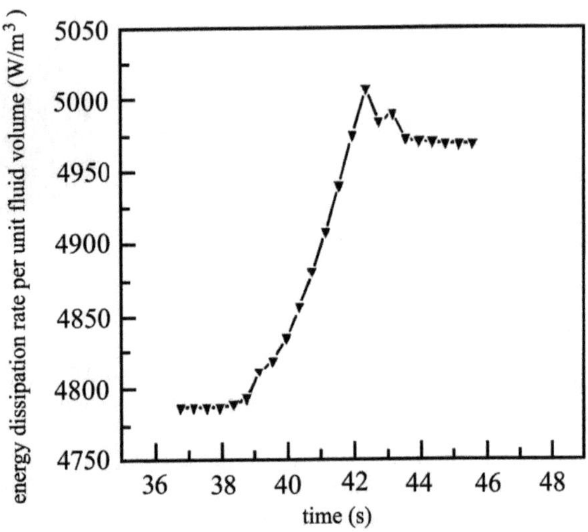

Fig. 5.7 Variation of energy dissipation rate per unit fluid volume for research system one

flow as calculation period of time, the energy dissipation rate per unit fluid volume during interval Δt can be calculated as:

$$\Phi_V = \frac{(e_t + e_u - e_d) - e_{t+\Delta t}}{\frac{1}{2}(V_t + V_{t+\Delta t})\Delta t} \tag{5.70}$$

where $V_t = V_{t+\Delta t} = \sum V_i$ (m^3), $e_t = e_{t+\Delta t} = \sum e_i$ (J), $e_i = \left(z_i + \frac{u_i^2}{2g}\right)\rho g V_i$ (J), $e_u = e_d = \sum e_j$ (J), and $e_j = \left(z_i + \frac{u_i^2}{2g}\right)\rho g Q_i \Delta t$ (J).

3.5 *Results and Analysis*

The energy dissipation rate per unit fluid volume in each system was calculated according to Eq. (5.70) and the results are shown in Figs. 5.7, 5.8, 5.9, 5.10, and 5.11. The fluid in the system from 2 to 2.5 m (the flume inlet is the origin of coordinates) and from 2 to 3 m are the two study systems, respectively. Under calculation condition one, the former is defined as the research system one, and the latter as the research system two. Under calculation condition two, the former is system three, and the latter is system four. Under the third calculation condition, the former is system five.

As shown in Figs. 5.7, 5.8, 5.9, 5.10, and 5.11, the energy dissipation rate per unit fluid volume is a constant when the flow reaches steady flow initially. When the flume outlet starts to rotate, the flow in flume begins to deviate from the original steady flow state and becomes unsteady. The energy dissipation rate also begins to

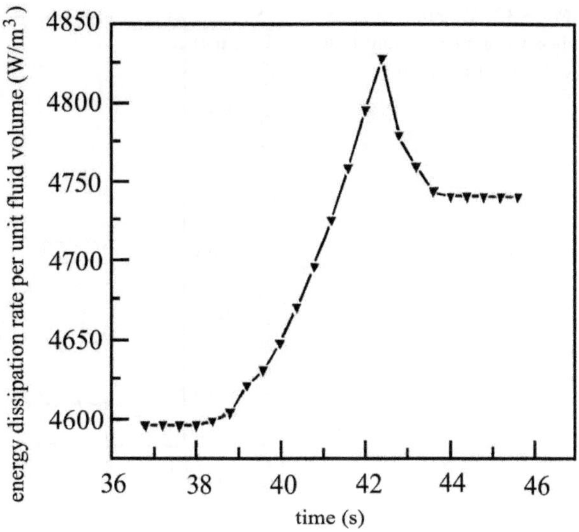

Fig. 5.8 Variation of energy dissipation rate per unit fluid volume for research system two

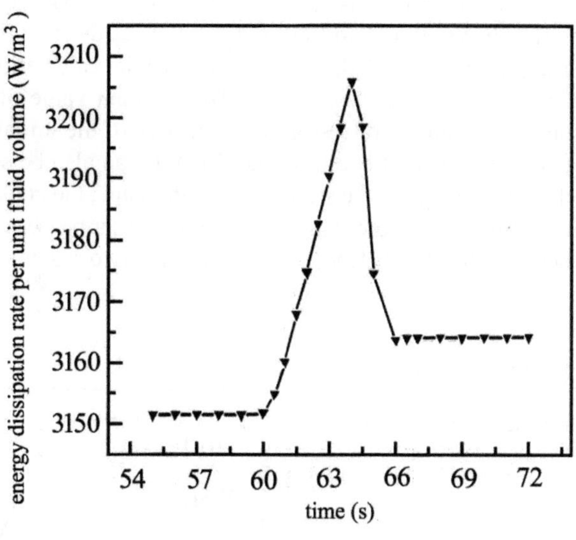

Fig. 5.9 Variation of energy dissipation rate per unit fluid volume for research system three

vary with respect to time. When the flume outlet stops rotation, the flow will automatically adjust the hydraulic elements to make it reach a steady flow state gradually. The energy dissipation rate per unit fluid volume will eventually reach a constant minimum value. Due to the different constraint conditions applied to the first and the second steady flow conditions, the constant values of energy dissipation rate of fluid are not the same. The energy dissipation rate can be expressed by $\Phi = \gamma QS$. Let Q be a constant. The energy dissipation rate Φ increases with increasing slope

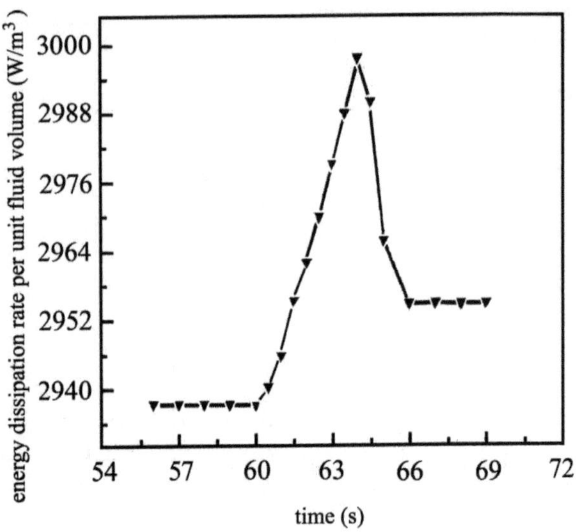

Fig. 5.10 Variation of energy dissipation rate per unit fluid volume for research system four

S. Under the calculation conditions of one and two, the flume bottom slope increases gradually. The constant value of energy dissipation rate per unit fluid volume of research system one to four in the initial steady state is less than that of the second one (see Figs. 5.7, 5.8, 5.9, and 5.10). Under calculation condition three, the flume bottom slope slowly decreases. The constant value of energy dissipation rate per unit fluid volume of research system five of the initial steady flow is greater than that of the second one (see Fig. 5.11). It can also be seen from Figs. 5.9 and 5.11 that, if the slope of the steady flow is the same, the constant value of energy dissipation rate is the same. However, the transition process is different from each other because the initial steady flow state of the two systems is different.

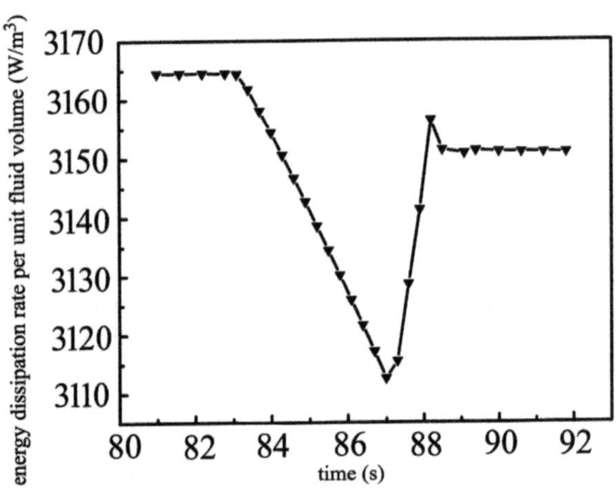

Fig. 5.11 Variation of energy dissipation rate per unit fluid volume for research system five

3.6 *Interim Summary*

1. The energy dissipation rate can reach a minimum constant value when the flow is steady. This conclusion is consistent with the minimum energy dissipation rate principle. The energy dissipation rate of unsteady flow is not definite. Its variation is complex, and may increase or decrease. The variation of energy dissipation rate in an alluvial river is consistent with this conclusion. More detailed discussions will be made later.
2. If the external constraints of a system vary significantly, the flow deviates from the original state through unsteady flow transition processes to reach a new steady flow condition compatible with the new external constraints. Although the energy dissipation rate of each steady flow can reach a minimum value, different values may be obtained with different constraints.

4 Adjustment of Energy Dissipation Rate in Alluvial Rivers

4.1 *Self-adjustment Function of Alluvial Rivers*

Alluvial rivers can adjust their physical conditions to restore equilibrium for different water, sediment, and boundary conditions [33].

Alluvial rivers' bed consists of loose sediments, so riverbed can be adjusted with different water and sediment conditions. Riverbed geometry of alluvial rivers is formed by the interaction among water, sediment, and riverbed material. On the one hand, the action of water and sediments on riverbed changes riverbed geometry. On the other hand, the variation of riverbed geometry also affects stream movement.

Due to the dynamic feedback mechanism of alluvial rivers, alluvial rivers can adjust automatically to a certain extent. In the process of adjustment, alluvial river adjusts its physical quantities by the interaction of stream and riverbed to reach equilibrium between sediment transport capacity and incoming sediment. The adjustments gradually decrease in strength and range, and tend to reach a relative equilibrium state. This relative equilibrium state is a dynamic equilibrium. There are two kinds of physical quantities that can be adjusted in the process: (1) physical quantities related to boundary conditions, such as river cross-sectional shape, roughness, longitudinal profile (including slope, riffle, pool, etc.), river pattern, etc. and (2) physical quantities related to flow characteristics, such as the depth of water, distribution of flow velocity and sediment concentration, and turbulence characteristics.

There are short- and long-term adjustments in the process of automatic adjustment of alluvial rivers. The short-term adjustment means a river system adjusts its physical quantities continuously in response to the variation of incoming water and sediment, and it will not stop even when the river is at a state of relative equilibrium. The long-term adjustment is a long evolution process to reach a relative equilibrium state.

4.2 Energy Dissipation Rate of Alluvial River at Relative Equilibrium State

A river is a complicated open system. Its evolution process satisfies the nonequilibrium thermodynamic theory. Consequently, automatic adjustments of a river not only tend to maintain a relative equilibrium state (steady nonequilibrium state) but also satisfy the minimum entropy production principle or the minimum energy dissipation rate principle. In the process, the entropy production or energy dissipation rate of river system tends to be at its minimum value compatible with the constraints applied to the system [34]. There are four kinds of external constraints applied to alluvial rivers: (1) meteorological hydrological conditions, such as rainfall, runoff, water temperature, etc.; (2) characteristics of bed materials, such as erodibility, stability, gradation of sediment diameter, roughness, etc.; (3) geologic and topographic conditions, such as river slope, width, and strength of river bedrock; and (4) artificial constraints on water conservancy projects, such as dam and riverbank.

When a river is at a state of relative equilibrium, external constraint conditions are basically stable or vary slightly with respect to time. Once the river system deviates from the relative equilibrium state, it will restore the original relative equilibrium state by reducing entropy production or energy dissipation rate. If external constraint conditions vary significantly, the river will deviate from the original state of relative equilibrium and reach a new state compatible with new constraints. The process that a river system evolves from a relative equilibrium state to another is very complex. In this process, the energy dissipation rate does not necessarily decrease monotonically with respect to time. When a system is at a new state of relative equilibrium, the energy dissipation rate will be at a new minimum value as shown in Fig. 5.12.

Figure 5.12 was presented by Yang to show that the unit stream power varied with respect to time at gaging station Halls on the South Fork Deer River [9, 16]. Artificial dredging near the gaging station was carried out between 1964 and 1966. As shown in Fig. 5.12, unit stream power was at a minimum value before dredging. This means that the reach is at a relative equilibrium state. After dredging, the external constraint conditions changed. Consequently, the river deviated from the relative equilibrium state and began to self-adjust. In the process of self-adjustments, unit stream power deviated from the original minimum value, went through increase and decrease processes, and reached a new stable condition compatible with new constraints. In the early stage of self-adjustment, bed boundary varies significantly due to a large magnitude of scouring and sedimentation deformation. This means that external constraint conditions no longer remain constant. In this case, unit stream power does not necessarily decrease with respect to time and is

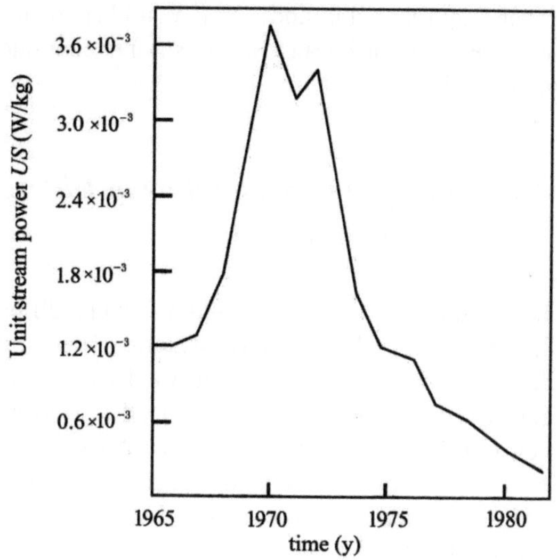

Fig. 5.12 Variation of unit stream power at gaging station Halls on the South Fork Deer River, Tennessee

likely to increase first and then decrease. After 1970, riverbed boundary tends to be stable gradually with decreasing scour and sedimentation. When external constraints begin to remain constant, river adjustments tend to decrease unit stream power to a new minimum value in 1982. This new minimum value is not the same as that before dredging in 1964 [35].

It should be pointed out that the entropy variation related to isolated systems in classical thermodynamics may not be applied to the river systems directly. For example, Leopold et al. [36] in 1962 considered that, based on the second law of thermodynamics (namely maximum entropy principle), when alluvial river is at the relative equilibrium state, the entropy E of alluvial river reaches the maximum value. Although Leopold's viewpoint was popular, it has the following shortcomings [37]:

First, a river is an open system rather than an isolated system. An isolated thermodynamic system is the one not influenced by external factors. There is no exchange of matter or energy with external environment. Strictly speaking, there is no absolute isolated system in nature. On the basis of the second law of thermodynamics, the entropy of an isolated system increases monotonically due to the irreversible process inside the system, i.e., $\mathrm{d}E = \mathrm{d}_i E > 0$. This monotonic increase cannot cease until the system reaches thermodynamic equilibrium state. When entropy reaches maximum, $\mathrm{d}E = 0$. However, the entropy change of an open system is $\mathrm{d}E = \mathrm{d}_e E + \mathrm{d}_i E$, and $\mathrm{d}E$ may be equal to, greater or less than zero with the variation of entropy flux term $\mathrm{d}_e E$. When $\mathrm{d}E = \mathrm{d}_e E + \mathrm{d}_i E = 0, \mathrm{d}_e E = -\mathrm{d}_i E$, a river is at a state of relative equilibrium (steady state).

Second, the relative equilibrium state of a river system is at dynamic equilibrium which is equivalent to a linear, steady nonequilibrium state of the open-system thermodynamics. The equilibrium state of thermodynamics means a static equilibrium state. If a river is at this state, there will not be any flow and sediment transport. The entropy of a river tends to be maximum, that is to say, the adjustment of the river may reach thermodynamic equilibrium state eventually. However, this state may never occur because its constraints continuously change with respect to time and location.

4.3 Energy Dissipation Rate of Different River Patterns and Its Variations

In the adjustment process of alluvial rivers, different external constraint conditions lead to forming different river patterns. There are four kinds of river patterns: meandering, straight, braided, and wandering rivers [33, 38]. A river often presents several river patterns from river head to estuary due to the variation of external constraint conditions along the river. A river reduces its entropy production or energy dissipation rate by adjusting its pattern to be compatible with the variation of external constraint conditions. There are several ways for a river to reduce its entropy production or energy dissipation rate, such as by increasing river width or by decreasing slope. Depending on the external constraint conditions, different adjustments may result in forming different river patterns. For example, if a river reduces entropy production or energy dissipation rate by decreasing slope, it may evolve to a meandering river. If river width is increased, it may evolve to a braided river or wandering river.

External constraints include incoming water and sediment conditions and riverbed boundary conditions. External constraint conditions for different river patterns are [39]:

1. *Meandering river*. Composition of river bank material is more stable than river bottom material, the runoff fluctuation is small, and the river keeps sediment transport equilibrium in a long period of time.
2. *Straight river*. Composition of river bank material such as clay and silt clay can resist erosion. As a result, river bank is hard to be scoured and transverse deformation of the river is limited. Composition of river bottom material is medium sand and fine sand.
3. *Braided river*. Composition of riverbed material is nonhomogeneous, and there are stable nodes in the upstream and downstream reaches. The runoff fluctuation is small, and the sediment concentration is low. This channel pattern is basically at a state of equilibrium sediment transport.
4. *Wandering river*. Composition of river bank and river bottom material is fine sediment with less clay. The scouring resistance of the bank matterial is poor, so it is easy to be silted and eroded. As a result, river bottom and banks are easy to deform. Besides, annual runoff fluctuation is large. The difference of runoff between flood period and dry period is very large. Flood rises and falls rapidly; and incoming sediment and sediment concentration are high.

According to the field data of 12 reaches of the Yellow River and Yangtze River, Huang [40] in 2008 used multiple linear regression method to analyze river patterns

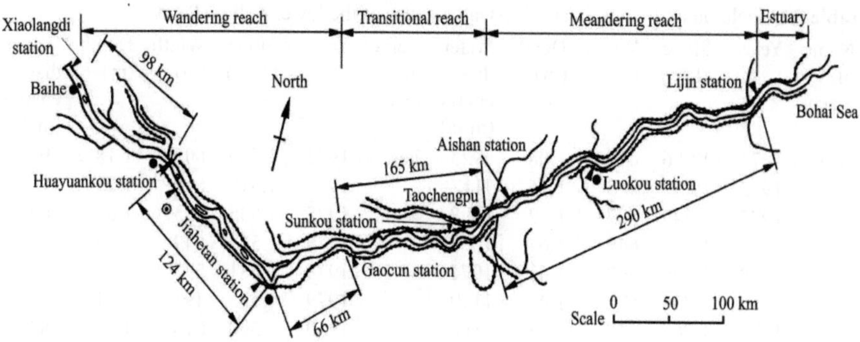

Fig. 5.13 Location of the seven hydrological stations along the lower Yellow River, and channel patterns of different reaches

and concluded that the influence of riverbed boundary condition on river patterns is greater than that of incoming water and sediment conditions. Based on the field data of seven hydrologic stations in Huayuankou, Jiahetan, Gaocun, Sunkou, Aishan, Luokou, and Lijin of the lower Yellow River, Xu and Zhao [41] in 2013 calculated the weights of all influencing factors on fluvial process by adopting information entropy. They obtained a similar conclusion that the influence of bed boundary condition on fluvial process of those reaches is stronger than that of incoming water and sediment conditions by comparing weights.

Figure 5.13 shows that the reach of the lower Yellow River between Baihe and Gaocun is a wandering river, Gaocun–Taochengpu reach is a transitional channel from a wandering river to a meandering river, Taochengpu–Lijin reach is a meandering river, and the lower Lijin reach is an estuary reach. There are seven hydrologic stations in the lower Yellow River (Huayuankou, Jiahetan, Gaocun, Sunkou, Aishan, Luokou, and Lijin). The six reaches represented by the stations are considered as six subsystems. Based on the monthly mean hydrological data of 21 years (1972, 1973, 1975–1980, 1982, 1985, 1987, 1988, and 1991–2000) of the seven hydrologic stations, the yearly mean hydrological data of each of the six reaches are summarized in Tables 5.1, 5.2, and 5.3. The variation of unit stream power *US* of each reach with respect to time [42] is shown in Fig. 5.14.

Figure 5.14 shows the order of unit stream powers for the six reaches in descending order: Huayuankou–Jiahetan reach, Jiahetan–Gaocun reach, Gaocun–Sunkou reach, Sunkou–Aishan reach, Aishan–Luokou reach, and Luokou–Lijin reach. Huayuankou–Jiahetan reach and Jiahetan–Gaocun reach are typical wandering reaches and their unit stream powers are the largest. Therefore, it can be concluded that the wandering reach is the most unstable. Aishan–Likou reach and Likou–Lijin reach are meandering reaches and unit stream powers of the two reaches are the smallest, so the meandering river is more stable than others. Gaocun–Sunkou reach and Sunkou–Aishan reach are the transitional reaches with double characteristics of the wandering river and the meandering river. Therefore, the value of unit stream power of the transitional reach is between that of meandering river and wandering river patterns.

Table 5.1 Field data of Hua–Jia and Jia–Gao reaches of the lower Yellow River

Name of reach	Year	Slope (10^{-4})	Width (m)	Depth (m)	Water discharge (m^3/s)	Name of reach	Year	Slope (10^{-4})	Width (m)	Depth (m)	Water discharge (m^3/s)
Hua-Jia	1972	1.796	697	1.08	923	Jia-Gao	1972	1.530	605	1.18	903
	1973	1.797	758	1.16	1117		1973	1.518	614	1.22	1111
	1975	1.749	827	1.42	1698		1975	1.535	585	1.63	1642
	1976	1.771	649	1.64	1635		1976	1.513	549	1.65	1589
	1978	1.699	562	1.41	1074		1978	1.581	595	1.26	1016
	1979	1.736	577	1.4	1149		1979	1.525	491	1.51	1112
	1980	1.768	648	1.13	868		1980	1.500	616	1.19	820
	1982	1.753	700	1.46	1325		1982	1.468	691	1.41	1277
	1985	1.768	614	1.67	1455		1985	1.474	549	1.86	1426
	1987	1.758	385	1.42	684		1987	1.495	327	1.59	616
	1988	1.774	429	1.62	1082		1988	1.495	380	1.68	995
	1991	1.758	374	1.45	722		1991	1.482	396	1.25	657
	1992	1.769	679	1.58	814		1992	1.496	401	1.34	759
	1993	1.770	401	1.64	934		1993	1.538	389	1.53	894
	1994	1.669	439	1.54	940		1994	1.642	400	1.49	902
	1995	1.588	387	1.25	723		1995	1.739	368	1.15	665
	1996	1.583	429	1.31	839		1996	1.723	504	0.99	776
	1997	1.587	330	1.03	419		1997	1.718	323	0.95	357
	1998	1.605	353	1.41	661		1998	1.682	370	1.19	605
	1999	1.606	311	1.4	647		1999	1.547	370	1.12	564
	2000	1.575	280	1.55	509		2000	1.649	356	1.16	464

It can be seen from Fig. 5.14 that unit stream power *US* of Huayuankou–Jiahetan reach and Jiahetan–Gaocun reach varies greatly during the 21 years. This indicates that although the wide, shallow, and disordered conditions of the wandering reaches have been improved to some extent by river regulation, the mainstream swing is still very large because the density and length of engineering are not enough to restrain the mainstream. Unit stream power of Gaocun–Sunkou and Sunkou–Aishan reaches is increasingly close to those of Aishan–Likou reach and Likou–Lijin reach. This shows that the transitional reach is gradually changing to a meandering reach and tends to reach a relative equilibrium state. During 1949–1960, the transitional reach was not regulated, and during 1979–1990, the reach was regulated. Before and after river regulation, the maximum amplitude of mainstream swing decreases from 5400 to 1850 m; the average of amplitude of mainstream swing drops by 58 % with a decrease from 1802 to 753 m; and the mean intensity of mainstream swing drops by 62 % with a decrease from 425 to 160 m/a. The above results show that the characteristics of the transitional reach are increasingly changing to those of a meandering reach with the wandering characteristics gradually disappeared [43]. The change of unit stream power of the 21 years of Aishan–Luokou reach and Luokou–Lijin reach becomes weaker and weaker. This means that the two reaches are gradually approaching to relative equilibrium state. The closure of Xiaolangdi on October 28, 1997, caused cutoffs in the Luokou–Lijin reach. This resulted in sudden increase

Table 5.2 Field data of Gao–Sun and Sun–Ai reaches of the lower Yellow River

Name of reach	Year	Slope (10^{-4})	Width (m)	Depth (m)	Water discharge (m^3/s)	Name of reach	Year	Slope (10^{-4})	Width (m)	Depth (m)	Water discharge (m^3/s)
Gao-Sun	1972	1.145	464	1.48	866	Sun-Ai	1972	1.200	383	1.81	832
	1973	1.143	603	1.48	1078		1973	1.196	496	1.88	1032
	1975	1.137	510	1.84	1605		1975	1.228	410	2.49	1595
	1976	1.127	494	1.74	1568		1976	1.163	423	2.40	1617
	1978	1.163	426	1.45	964		1978	1.199	346	2.04	933
	1979	1.160	414	1.59	1078		1979	1.182	368	2.02	1041
	1980	1.153	426	1.4	769		1980	1.177	342	2.04	731
	1982	1.168	513	1.58	1217		1982	1.176	397	2.08	1158
	1985	1.153	449	2.15	1376		1985	1.155	380	2.39	1307
	1987	1.160	363	1.54	571		1987	1.216	292	2.37	521
	1988	1.206	390	1.61	950		1988	1.218	329	2.12	866
	1991	1.160	388	1.44	617		1991	1.194	311	2	587
	1992	1.162	376	1.42	703		1992	1.111	276	2.12	648
	1993	1.102	410	1.75	867		1993	1.169	329	2.06	811
	1994	1.126	423	1.37	872		1994	1.218	361	1.86	850
	1995	1.126	392	1.18	627		1995	1.172	319	1.87	601
	1996	1.144	463	1.38	700		1996	1.207	300	1.93	648
	1997	1.155	279	1.21	300		1997	1.259	189	1.90	245
	1998	1.178	341	1.37	558		1998	1.224	242	2.08	523
	1999	1.165	340	1.35	510		1999	1.245	237	2.04	448
	2000	1.165	362	1.24	384		2000	1.235	259	1.99	362

of the unit stream power in 1997. After 1997, the unit stream power of the reach regained previous level, and the reach tended to be at a relative equilibrium state.

4.4 *Interim Summary*

1. If the external constraint conditions vary significantly, the river will deviate from the previous state of relative equilibrium to a new state compatible with new constraints. The river system evolution from a relative equilibrium state to another state is very complicated. In this process, the energy dissipation rate does not necessarily decrease monotonically with respect to time, and it is also likely to increase temporarily. However, when a system is at a new state of relative equilibrium, the value of energy dissipation rate must be at minimum compatible with the new constraint conditions.
2. When a system is at a relative equilibrium state, the value of energy dissipation rate must be at minimum subject to the constraint conditions. The relative equilibrium of a river is a dynamic equilibrium. As a result, energy dissipation rate fluctuates around its average when a river is at a relative equilibrium state.
3. The analysis of unit stream power of 21 years of record for the six reaches of three kinds of river patterns concludes that unit stream power of the wandering reach is the largest and that of the meandering reach is the smallest, and that of

Table 5.3 Field data of Ai–Luo and Luo–Li reaches of the lower Yellow River

Name of reach	Year	Slope (10^{-4})	Width (m)	Depth (m)	Water discharge (m^3/s)	Name of reach	Year	Slope (10^{-4})	Width (m)	Depth (m)	Water discharge (m^3/s)
Ai-Luo	1972	1.007	287	2.52	793	Luo-Li	1972	0.902	495	2.54	734
	1973	1.002	266	2.50	978		1973	0.907	245	2.46	916
	1975	0.983	285	3.35	1579		1975	0.917	284	3.11	1530
	1976	0.994	285	3.40	1500		1976	0.912	301	3.13	1445
	1978	1.017	496	2.72	904		1978	0.927	234	2.72	833
	1979	1.076	258	2.82	998		1979	0.908	254	2.68	911
	1980	0.997	245	2.66	698		1980	0.924	259	2.24	630
	1982	1.002	529	2.82	1085		1982	0.955	276	2.52	983
	1985	0.992	294	3.19	1296		1985	0.925	345	2.97	1268
	1987	1.028	196	2.82	449		1987	0.934	169	2.03	378
	1988	1.027	200	2.79	774		1988	0.939	220	2.18	671
	1991	0.999	204	2.75	536		1991	0.936	188	2.37	443
	1992	1.013	180	2.6	576		1992	0.987	163	2.17	475
	1993	1.002	225	2.65	723		1993	0.940	225	2.28	633
	1994	0.998	240	2.63	801		1994	0.983	223	2.37	721
	1995	1.212	201	2.44	551		1995	0.802	188	1.85	472
	1996	1.000	200	2.25	601		1996	0.992	183	2.24	527
	1997	1.442	127	1.96	178		1997	0.930	132	1.17	99
	1998	1.016	164	2.52	471		1998	0.957	182	1.83	439
	1999	1.012	170	2.22	371		1999	0.956	198	1.60	273
	2000	1.004	175	2.18	300		2000	0.975	185	1.44	209

the transitional reach is between that of the meandering reach and wandering reach. The stability of river decreases with increasing unit stream power.

5 Hydraulic Geometry Based on Minimum Energy Dissipation Rate Principle

5.1 Brief Introduction to Hydraulic Geometry

Alluvial rivers may reach a relative equilibrium state with a long-term interaction between flow and alluvial river boundaries. The hydraulic geometry relationships among hydraulic parameters and channel geometry and longitudinal profile can be written in the following mathematical form:

$$\left.\begin{aligned} B &= f_1(Q,G,D) \\ h &= f_2(Q,G,D) \\ S &= f_3(Q,G,D) \end{aligned}\right\}. \tag{5.71}$$

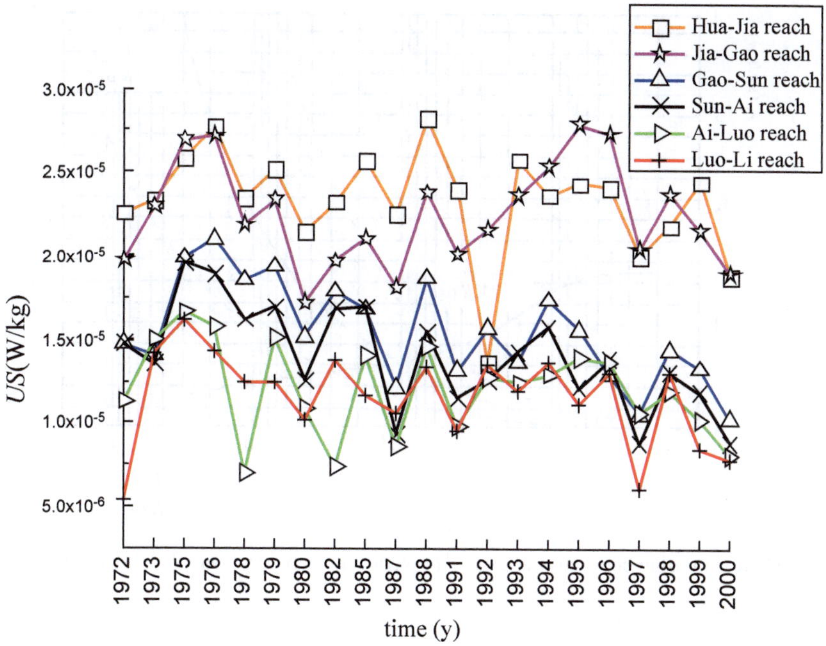

Fig. 5.14 Variation of unit stream power *US* of six reaches of the lower Yellow River

where

B is the river width (m)
h is the average water depth (m)
S is the longitudinal riverbed slope (dimensionless)
Q is the water discharge or channel-forming discharge (m^3/s)
G is the sediment discharge from upstream reach (kg/s), and
D is the riverbed boundary conditions (dimensionless).

In order to obtain the specific expressions for Eq. (5.71), the following equations are used:

Continuity equation for flow:

$$Q = AU. \tag{5.72}$$

Flow motion equation (Manning formula):

$$S = \frac{n^2 U^2}{R^{4/3}}. \tag{5.73}$$

Sediment load equation:

$$C_* = K\left(\frac{U^3}{gh\omega}\right)^m, \tag{5.74}$$

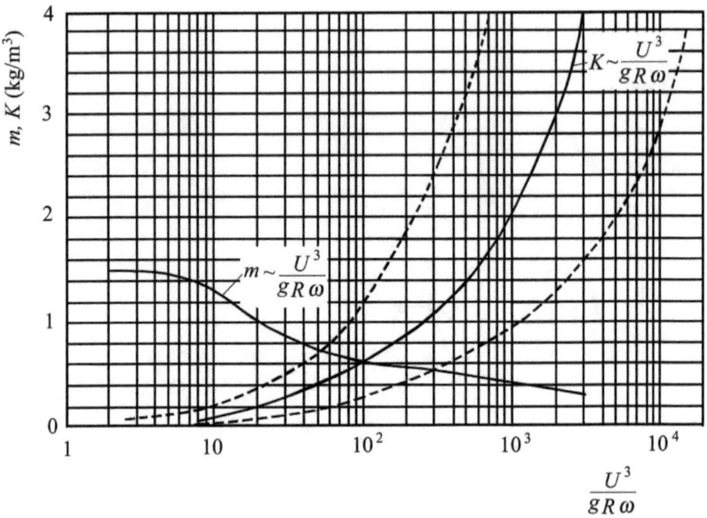

Fig. 5.15 Relationship between m and $U^3/(gR\omega)$, and between K and $U^3/(gR\omega)$

where

Q is the channel-forming discharge (m^3/s)
A is the cross-sectional area of flow (m^2)
U is the cross-sectional mean velocity (m/s)
R is the cross-sectional hydraulic radius, which can be replaced by water depth h (m)
n is the roughness ($s/m^{1/3}$)
C_* is the sediment transport capacity (kg/m^3)
ω is the sediment particle fall velocity (m/s)
K is the coefficient (kg/m^3), and
m is the index (dimensionless).
K and m can be determined from field data or Fig. 5.15 [44].

In Fig. 5.15, the solid line represents the equilibrium state. The upper and lower dashed lines represent the equilibrium state from silting and erosion processes, respectively.

It is important to note that Eq. (5.74) is a suspended load sediment-carrying capacity formula applicable to channel-forming process. Bed-load sediment-carrying capacity formula is applicable to the channel-forming process dominated by bed-load movement.

In the Eqs. (5.72), (5.73), and (5.74), there are four unknowns, i.e., river width B, average water depth h, longitudinal bed slope S, and flow velocity U. The number of unknowns is more than the equations available for solving these unknowns. Consequently, no solution can be obtained. In order to solve the equation set, another independent equation must be added into the equation set. The additional independent equation can be empirical or theoretical.

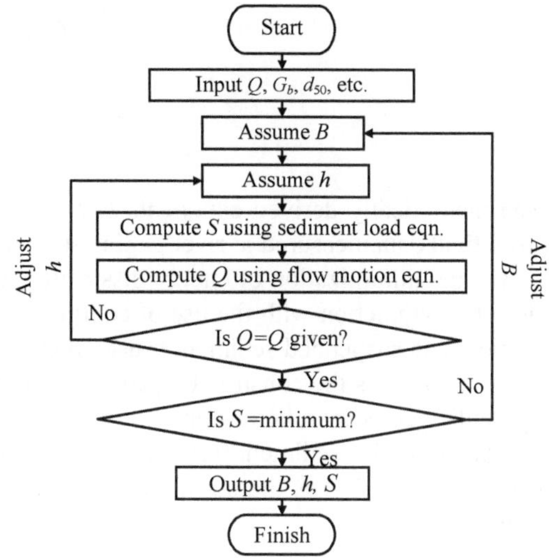

Fig. 5.16 Flow diagram showing major steps of computation

By statistical analysis of a large number of data, the relationship between channel geometry, hydraulic factors, and sediment can be established. Empirical hydraulic geometry relationships include longitudinal hydraulic geometry and cross-sectional hydraulic geometry. Theoretical hydraulic geometry relationships include those using the hypothesis of riverbed minimum activity, minimum variance theory, minimum energy dissipation rate principle, etc. Among the theoretical relationships, the principle of minimum energy dissipation rate is most promising. Chang, Yang, Yang, and Xu et al., among others, have made significant advancements in this area.

5.2 *Hydraulic Geometry Based on Minimum Energy Dissipation Rate Principle*

Chang [22] in 1980 developed a computer program by taking flow equation of motion and bed-load transport rate equation as constraint equations, and used them to determine channel geometry when the slope S is at the minimum. This is because, for a given Q, the minimum energy dissipation rate $\Phi = \gamma QS$ means minimum S. Figure 5.16 is the flow diagram of the computer program which can be used for stable channel design. Chang used Engelund and Hansen equation as the equation of motion, and DuBoys formula for bed-load movement. It should be pointed out that other flow equation of motion and bed-load transport rate formula can also be used.

Yang [11] in 1981 obtained an objective function by substituting Eq. (5.73) into $\Phi = \gamma QS$. He used his dimensionless unit stream power equation as the constraint equation, and constituted a Lagrange function. The theoretical exponents of the following equations can be obtained from the extremal solution of the Lagrange function:

$$B = aQ^b. \tag{5.75}$$

$$h = cQ^f. \tag{5.76}$$

$$U = kQ^m. \tag{5.77}$$

The theoretically derived exponents b, f, and m are very close to those obtained from field data measured at river gaging stations.

Let the minimization of energy dissipation rate $\Phi = \gamma QS$ be the objective function. In conjunction with the use of continuity equation for flow, flow equation of motion, and suspended sediment transport capacity equation (or bed-load transport rate equation) as the constraint equations, Xu [45] in 1993 obtained the following hydraulic geometry relationships.

For suspended sediment forming rivers, the use of suspended sediment transport capacity Eq. (5.74) as the constraint equation, the following equations can be derived:

$$\left.\begin{aligned} B &= 3.120\frac{K^{1/(7m)}\,Q^{3/7}}{g^{1/7}\,C_*^{1/(7m)}\,\omega^{1/7}} \\ h &= 0.426\frac{K^{1/(7m)}\,Q^{3/7}}{g^{1/7}\,C_*^{1/(7m)}\,\omega^{1/7}} \\ S &= 2.437\frac{g^{16/21}\,n^2 C_*^{16/(21m)}\,\omega^{16/21}}{K^{16/(21m)}\,Q^{2/7}} \\ U &= 0.752\frac{g^{2/7}\,C_*^{2/(7m)}\,\omega^{2/7}\,Q^{1/7}}{K^{2/(7m)}} \end{aligned}\right\}. \tag{5.78}$$

For bed-load-forming rivers, the following Zhang's bed-load transport rate equation can be used as the constraint equation [44], i.e.,

$$G_b = 0.00124\frac{\alpha\rho_s' U^4 B}{g^{3/2}\,h^{1/4}\,d_{50}^{1/4}}, \tag{5.79}$$

where

G_b is the cross-sectional rate of bed-load transport (kg/s)
ρ_s' is the dry density of sediment (kg/m^3)
d_{50} is the the median diameter of bed material (m), and
α is the sand wave shape coefficient (dimensionless), α=0.52–0.53.

For bed-load-forming rivers, the following hydraulic geometry relationships can be obtined:

$$\left.\begin{aligned} B &= 1.634\frac{\alpha^{4/29}\rho_s'^{4/29}Q^{16/29}}{g^{6/29}d_{50}^{1/29}G_b^{4/29}} \\ h &= 0.146\frac{\alpha^{4/29}\rho_s'^{4/29}Q^{16/29}}{g^{6/29}d_{50}^{1/29}G_b^{4/29}} \\ S &= 284.572\frac{g^{32/29}n^2 d_{50}^{16/87}G_b^{64/87}}{\alpha^{64/87}\rho_s'^{64/87}Q^{82/87}} \\ U &= 4.192\frac{g^{12/29}d_{50}^{2/29}G_b^{8/29}}{\alpha^{8/29}\rho_s'^{8/29}Q^{3/29}} \end{aligned}\right\}. \quad (5.80)$$

It is important to note that the selection of appropriate and accurate sediment transport capacity equation as the constraint equation is very important for the determination of hydraulic geometry relationships. Inapporpriate selection of sediment transport equation can lead to incorrect hydraulic geometry relationships.

5.3 *Interim Summary*

Hydraulic geometry relationships at the state of relative equilibrium are important from both theoretical and engineering points of view. When alluvial rivers reach the state of relative equilibrium, their energy dissipation rate must be a minimum value subject to external constraints. Hydraulic geometry relationships derived from the minimum energy dissipation rate principle have physical meanings, and the river geometry relationships thus obtained are in good agreement with the field data. The results have been verified by many scholars.

6 Optimum Design of Low-Head Water Diversion Project

The low-head diversion projects in this study include headworks of water diversion projects and stable channel designs. The arrangement of water diversion project must satisfy the requirements of water diversion and sediment prevention. The key

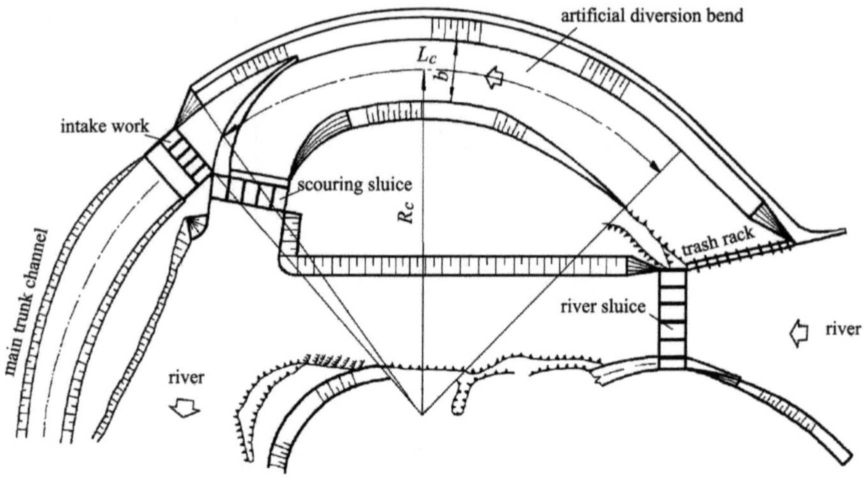

Fig. 5.17 Layout of diversion bend structure

problem that needs to be solved in the stable channel design is the scouring and sedimentation deformation of the channels. The minimum energy dissipation rate principle is used as the optimal design theory.

6.1 Optimum Design of Water Diversion Bend Structure [46]

A typical layout of a water diversion bend structure is shown in Fig. 5.17. Water diversion bend structure uses artificial bend to produce transverse circulation for the separation of water and sediment. The purpose of this structure is to discharge water to one side and the sediment flushing to the other side of the channel. This type of diversion structure, known as the Fergana headworks, was first built in the Fergana region of Soviet Central Asia. It was introduced to Xinjiang, China, in 1958. After continuous development and improvement, it has become one of the main types of headworks used in China. Water diversion bend structure is suitable for mountain rivers of medium to large size where the bed load is pebble. This kind of structure consists of debris guard, artificial diversion bend, intake work scouring sluice, etc.

The key design consideration of water diversion bend structure is to determine stable geometry of the artificial diversion bend to produce stable and strong transverse circulation. The purpose is to make the surface-water flow to the concave bank and make the bottom sediment-laden flow to convex bank through scouring sluice into the downstream river. The effect of water diversion and sediment flushing increases with increasing bend circulation [47, 48]. In the past, the design of diversion bend was mainly based on experience or physical model tests without much theoretical basis. Sometimes, due to the unreasonable design of the diversion bend, it may cause serious siltation and affects normal operation of the water diversion structure. In view of this, the following mathematical model is used for optimum diversion bend design.

6.1.1 Mathematical Model for Optimum Water Diversion Bend Design

The optimum design model includes the objective function, design variables, and constraint conditions.

(1) Objective function

The artificial diversion bend design should satisfy the minimum energy dissipation rate principle. The minimum stream power should be treated as the objective function for diversion bend design, i.e.,

$$\Phi = \gamma QS = \text{a minimum}, \tag{5.81}$$

where

Φ is the rate of energy dissipation on the unit river length (W/m)
γ is the specific weight of water (N/m^3)
Q is the water discharge (m^3/s), and
S is the longitudinal slope of the diversion bend (dimensionless).

For a specific diversion bend, water discharge Q and specific weight of water γ are given constants. Minimum energy dissipation rate Φ means minimum slope S, i.e.,

$$S = \text{a minimum}. \tag{5.82}$$

The slope S can be calculated using the Manning formula. Diversion bend cross section is designed as a trapezoidal cross section. Substitution of the trapezoidal cross-sectional hydraulic elements and flow continunity equation into Manning formula yields:

$$S = \frac{Q^2 n^2 (b + 2h\sqrt{1+m^2})^{4/3}}{(bh + mh^2)^{10/3}}, \tag{5.83}$$

where

b is the trapezoidal cross-sectional bottom width (m)
h is the average depth of cross section (m)
m is the bankside slope (dimensionless), and
n is the roughness ($s/m^{1/3}$)

(2) Decision variables

Water discharge Q, roughness n, sediment grain size d, and bankside slope m are given independent variables. The design is for the determination of bottom width b, sectional average water depth h, radius of bend center curvature R_c, length L_c of the bend center, longitudinal slope S, and cross-sectional longitudinal average velocity U. S and U are not independent variables, which can be obtained from Manning's equation and the equation of continuity of flow, after b and h are determined. So b, h, R_c, and L_c are the independent variables. These four variables constitute the design variables of the optimum design model, i.e.,

$$\boldsymbol{X} = \begin{bmatrix} b & h & R_c & L_c \end{bmatrix}^T.$$

(3) Constraint conditions

The geometry of diversion bend should satisfy the following conditions:

a. Circulation intensity conditions

Stable and strong transverse circulation is the key to guarantee normal operation of diversion structure with sediment transport. Zhang [49] took the ratio of transverse slope S_r and the longitudinal slope S as the criteria of circulation intensity, i.e.,

$$C_r = \frac{S_r}{S}. \tag{5.84}$$

Substituting

$$S_r = \frac{U^2}{gR_c},\ S = \frac{U^2 n^2}{R^{4/3}} \text{ and } R = \frac{bh + mh^2}{b + 2h\sqrt{1+m^2}}$$

into Eq. (5.84) yields:

$$C_r = \frac{(bh + mh^2)^{4/3}}{gn^2 R_c (b + 2h\sqrt{1+m^2})^{4/3}}. \tag{5.85}$$

Zhang [49] has analyzed more than ten water diversion bend structures in Xinjiang and discovered that, when criterion of circulation intensity C_r is greater than 1, the effect of the circulation flow is noticeable. When C_r is less than 1, the circulation flow is is not apparent; the operating condition is poor, and water diversion structures have serious siltation problems. Therefore, the value of C_r should satisfy the following relationship:

$$1 \le C_r \le 1.5. \tag{5.86}$$

Equation (5.86) can be written for the two constraints:

$$G_1(\boldsymbol{X}) = 1.5 - C_r \ge 0. \tag{5.87}$$

$$G_2(\boldsymbol{X}) = C_r - 1 \ge 0. \tag{5.88}$$

In order to ensure transverse circulation is fully developed, the diversion bend should have sufficient length. According to the completed project statistics [47, 50], the length of diversion bend center should satisfy the following relationships:

$$R_c \le L_c \le 1.4R_c, \tag{5.89}$$

or

$$G_3(\boldsymbol{X}) = 1.4R_c - L_c \ge 0. \tag{5.90}$$

$$G_4(\boldsymbol{X}) = L_c - R_c \ge 0. \tag{5.91}$$

b. Flushing conditions

In order to sluice sediment smoothly to the downstream river, bend longitudinal average flow velocity U should be greater than the required critical incipient flushing flow velocity, i.e.,

$$U \ge 1.3U_c, \tag{5.92}$$

where U_c is the incipient velocity (m/s). It can be calculated by Zhang's incipient velocity formula [44], i.e.,

$$U_c = 5.39h^{0.14}d_{50}^{0.36} \tag{5.93}$$

where

h is an average depth of bend (m) and
d_{50} is the sediment median diameter (m)

Subtituting Eq. (5.93) and the equation of continuity for flow into Eq. (5.92) yields:

$$\frac{Q}{bh+mh^2} \ge 7.01h^{0.14}d_{50}^{0.36}, \tag{5.94}$$

or

$$G_5(\boldsymbol{X}) = \frac{Q}{bh+mh^2} - 7.01h^{0.14}d_{50}^{0.36} \ge 0. \tag{5.95}$$

In summary, the optimum design consists of solving the following equations:

$$\left.\begin{array}{ll} \text{Solve} & \boldsymbol{X} = \begin{bmatrix} b & h & R_c & L_c \end{bmatrix}^T \\ \text{make} & \Phi(\boldsymbol{X}) \to \min \\ \text{subject to} & G_i(\boldsymbol{X}) \ge 0 \quad (i = 1,2,3,4,5) \end{array}\right\}. \tag{5.96}$$

Equation (5.96) can be used to solve the constrained nonlinear minimization problem. There are many methods for solving this kind of problems. The recommended method is the use of internal penalty function to change Eq. (5.96) into an unconstrained minimization problem. The specific form of the internal penalty function is:

$$F(\boldsymbol{X}, M_k) = \Phi(\boldsymbol{X}) + M_k \sum_{i=1}^{5} \frac{1}{G_i(\boldsymbol{X})} \tag{5.97}$$

where

$\Phi(\boldsymbol{X})$ is the original objective function (dimensionless)
$M_k \sum_{i=1}^{5} \frac{1}{G_i(\boldsymbol{X})}$ is the penalty term (dimensionless)
M_k is the penalty factor (dimensionless)
$F(\boldsymbol{X}, M_k)$ is the augmented objective function (dimensionless)

There are many ways to solve the unconstrained minimum value of Eq. (5.97). The Hooke–Jeeves pattern-search method [51, 52] is recommended because there is no special requirement for the function properties, and its adaptability is strong. The disadvantage of this method is that the rate of convergence is slow.

In order to avoid local minimum value of the inequality constraints in the feasible region, several initial values are substituted into Eq. (5.97) during the iteration process. The final optimum solution should be the one making the objective function a minimum.

6.1.2 Optimum Results and Verification

In order to vertify the correctness and reliability of the optimum design model, design results from three bend diversion structures at Kashi, Yeerqiang, and Jiangka in Xinjiang are used as examples. The project in Kashi is a completed project. The actual operation of many years shows that there is a strong circulation in the diversion bend and the effect of water diversion and sediment preventions is good. The two projects in Yeerqiang and Jiangka are used to test the model. The test shows that the effect of transverse circulation is less strong and there are sedimentations in the bend. Table 5.4 summarizes the basic data of the three diversion projects.

The optimum results of the three projects are shown in Table 5.5. The dimensions of calculated diversion bend are close to the actual dimensions. It can be concluded that the optimum design model based on the principle of minimum energy dissipation rate can be used for the design of water diversion bend.

Table 5.4 Basic data of water diversion bend structures

Name of structure	Name of river	Slope of riverbed	Q (m³/s)	d_{50} (m)	n	m	S	b (m)	R_c (m)	L_c (m)
Kashi structure	Kashi river	0.0067	300	0.04 ~0.05	0.035	1.5	0.0067	30	163.35	230.93
Yeerqiang structure	Yeer-qiang river	0.0040	330	0.066	0.035	1.75	0.0033	35	208.69	302.31
Jiangka structure	Jiangka river	0.0059 ~0.0067	120	0.037	0.035 ~0.038	1.5	0.0059	18	115	128.50

Table 5.5 Optimum results

Name of structure	S	b (m)	R_c (m)	L_c (m)	h (m)	U (m/s)
Kashi structure	0.0036	34.43	198.81	238.16	2.63	2.98
Yeerqiang structure	0.0042	26.91	221.12	265.15	3.01	3.41
Jiangka structure	0.0046	18.21	123.83	148.40	2.11	2.66

6.2 *Optimum Design of Stable Channel [53, 54]*

The concept of "stable channel" was proposed by Kennedy for unlined channel design in India. The definition of stable channel is that a channel can maintain a relative stable geometry in longitudinal and lateral directions. Longitudinally stable channel is neither eroding nor silting, or erosion and deposition are basically in balance for a given period of time. The lateral stability of the channel is that the cross-sectional shape of channel remains stable and satisfies certain hydraulic geometry relationships.

Stable channel is divided into two categories: one is without erosion and silting and the other one is an eroding–silting-equilibrium channel. The design methods of the two kinds of channels are different. Stable channel geometry is closely related to water discharge and sediment load variation and bed soil conditions of the channel. Different methods of regime channel design or the eroding–silting-equilibrium channel design can be used for different field conditions.

When sediment concentration of a river is at its minimum, the velocity of flow is the noneroding velocity. When sediment concentration of a river is at its maximum, the velocity of flow is the nonsilting velocity. Stable channel design should make the mean sectional velocity of channel between the noneroding velocity and the nonsilting velocity. In the northwest loess region of China, sediment concentration of rivers is extremely high in flood seasons but very low in drought seasons. The yearly sediment concentration variation is large and the nonsilting velocity is greater than the noneroding velocity. In this situation, the method of regime channel design without erosion and silting cannot be applied. The method of the eroding–silting-equilibrium channel design should be applied in this case. Channel materials of silt may be deposited or scoured. However, there should be a balance between scour and deposition in a certain period.

6.2.1 Optimium Design of Regime Channel

The traditional method of optimization design for regime channel is a trial-and-error method. Vertical and horizontal cross-sectional dimensions of the channels and the corresponding velocity are obtained using hydraulic geometry relationships, continuity equation, and Manning's equation first. Design will then be judged whether the velocity is between the noneroding velocity and the nonsilting velocity. If not, velocity will be recalculated. The trial-and-error method has two drawbacks. One is that emperical hydraulic geometry relationships may not be applicable to all regions. There is a large difference among different regions. Once unreasonable hydraulic geometry relationships are used, the velocity may lead to severe deformation of the channels. Another drawback is that different trial-and-error methods may reach different calculated results.

Due to these shortcomings of the trial-and-error method, the minimum energy dissipation rate principle should be used as the optimum design mathematical model for the design of regime channels. Optimum design mathematical model consists of the use of objective function, design variables, and constraints.

(1) Optimum design mathematical model

a. Objective function

For stable channel design, the objective is to obtain the optimum channel geometry under the given constraints. The objective function of stable channel design is the minimum energy dissipation rate or the minimum stream power principle:

$$QS = \text{a minimum.} \tag{5.98}$$

For a specific channel discharge, the minimization of the rate of total energy dissipation is equivalent to the minimization of slope S of channel, i.e.,

$$S = \text{a minimum.} \tag{5.99}$$

The slope can be calculated by Manning's equation for a traspzoid channel, i.e.,

$$S = \frac{U^2 n^2}{R^{4/3}}. \tag{5.100}$$

Sha's statistical analysis of stable channel shape of the loess areas in Northwest China indicates that a disk-shaped channel can be approximated by a tropizdal channel with a bankside slope $m=1$ [55] (see Fig. 5.18). Substituting continuity equation, Manning's equation, and trapezoidal cross-sectional geometry relationships into the Eq. (5.99) yields:

$$S = \frac{Q^2 n^2 (b + 2h\sqrt{1+m^2})^{4/3}}{(bh + mh^2)^{10/3}} = \text{a minimum,} \tag{5.101}$$

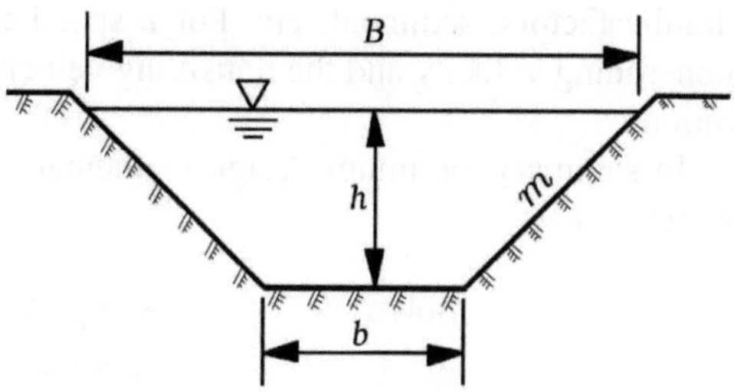

Fig. 5.18 Illustration of trapezoidal section

where

b is the channel bottom width (m)
h is the depth of flow (m)
m is the bankside slope (dimensionless), and
n is the roughness coefficient selected in the preliminary design (s/m$^{1/3}$)

b. Decision variables

Bankside slope m of the stable channel can be determined based on channel bed soil conditions, depth of water, construction conditions, etc. In the channel design process, water discharge, sediment concentration, sediment, and bed soil conditions are given parameters. Bottom width of channel b, depth of water h, longitudinal slope of channel S, and velocity of flow U are unknown parameters to be determined. Once b and h are determined, S and U can be expressed as functions of b and h using Manning's equation and flow continuity equation. b and h are considered as decision variables which can be expressed in the following vector form:

$$\boldsymbol{X} = \begin{bmatrix} b & h \end{bmatrix}^T.$$

c. Constraint conditions

The regime channel should satisfy the condition of noneroding and nonsilting velocity, i.e.,

$$U_{ns} < U < U_{ne}. \tag{5.102}$$

Substituting flow continuity equation into Eq. (5.102) yields:

$$G_1(\boldsymbol{X}) = \frac{Q}{bh + mh^2} - U_{ns} > 0, \tag{5.103}$$

$$G_2(\boldsymbol{X}) = U_{ne} - \frac{Q}{bh + mh^2} > 0. \tag{5.104}$$

The nonsilting velocity of the channel U_{ns} depends on channel sediment transport capacity. The noneroding velocity U_{ne} depends on channel bed soil conditions, hydraulic factors, sediment, etc. For a specific engineering design, the formula of noneroding velocity and the nonsilting velocity can be selected based on field conditions.

In summary, optimum design mathematical model for the regime channel can be written as:

$$\left.\begin{array}{ll} \text{Solve} & \boldsymbol{X} = [b\ h]^T \\ \text{make} & \Phi(\boldsymbol{X}) \to \min \\ \text{subject to} & G_i(\boldsymbol{X}) = \dfrac{Q}{bh + mh^2} - U_{ns} > 0 \\ & G_2(\boldsymbol{X}) = U_{ne} - \dfrac{Q}{bh + mh^2} > 0 \end{array}\right\}. \tag{5.105}$$

Equation (5.105) can be solved as an inequality constrained nonlinear minimization problem. Using inner point penalty function, Eq. (5.105) can be transformed into unconstrained minimization problem as:

$$\text{Min } F(\boldsymbol{X}, M_k) = \Phi(\boldsymbol{X}) + M_k \sum_{i=1}^{2} \frac{1}{G_i(\boldsymbol{X})}, \tag{5.106}$$

where

$F(\boldsymbol{X}, M_K)$ is the augmented objective function (dimensionless)
$\Phi(\boldsymbol{X})$ is the original objective function (dimensionless)
$M_k \sum_{i=1}^{2} \frac{1}{G_i(\boldsymbol{X})}$ is the penalty term (dimensionless), and
M_K is the penalty factor (dimensionless)

When M_K approaches zero from a positive value, the minimum values of the augmented objective function will approach the optimum solution of Eq. (5.105). The Hooke–Jeeves pattern-search method is used to solve the unconstrained minimum of Eq. (5.106). In order to avoid a local minimum value, different initial values for the iterative calculation in the feasible region should be selected to obtain several minimum values. The final choice of minimum value should make the objective function a minimum as the optimum solution of the objective function (i.e., slope).

The iterative steps of using the Hooke–Jeeves pattern-search methodto solve Eq. (5.106) are as follows:

i. Assign initial penalty factor $M_k > 0$, constant $c > 1$, and accuracy $\varepsilon > 0$.
ii. Select the initial point $\boldsymbol{X}^{(0)}$ in the feasible region and set $k = 0$.
iii. Taking $\boldsymbol{X}^{(k)}$ as the initial point and using the Hooke–Jeeves pattern-search method to solve the unconstrained minimum of Eq. (5.106), and obtain the optimum solution $\boldsymbol{X}^{(k+1)}$.

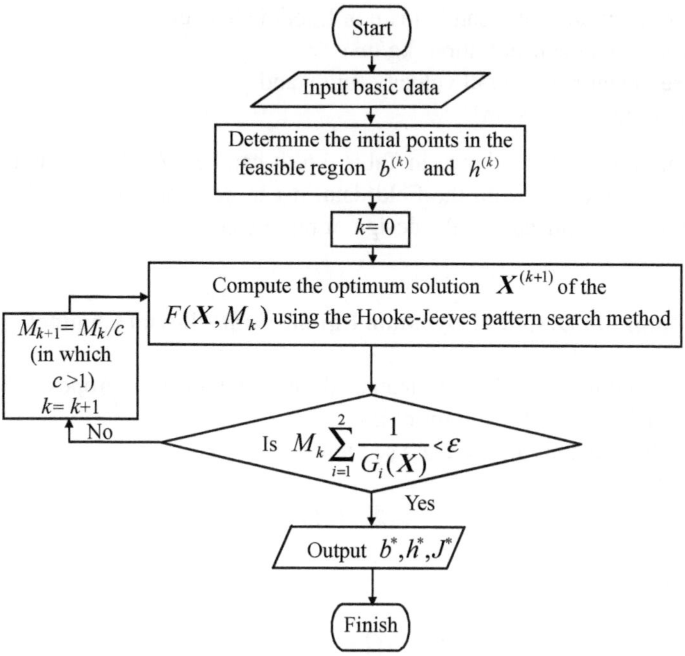

Fig. 5.19 Flow diagram of major computation steps

iv. If $M_k \sum_{i=1}^{2} \frac{1}{G_i(\boldsymbol{X}^{(k+1)})} < \varepsilon$, the optimum solution $\boldsymbol{X}^* = \boldsymbol{X}^{(k+1)}$ and stop the iterative calculation. If $M_K \sum_{i=1}^{2} \frac{1}{G_i(\boldsymbol{X}^{(k+1)})} > \varepsilon$, set $M_{k+1} = M_k / c$, use $k+1$ to replace k, then repeat step (iii) to continue the iteration. The calculated procedure is shown in Fig. 5.19.

(2) Verification of optimum design model

Field data of regime channels of Jinghui, Luohui, and Weihui in Shaanxi Province of China are used to calculate and verify optimum design model of the regime channel [56]. The following equations are used to compute the noneroding velocity and the nonsilting velocity [39]:

$$U_{\mathrm{ne}} = \alpha R^{0.4}, \tag{5.107}$$

$$U_{ns} = 2.72(C_s \omega)^{1/4} R^{3/8}. \tag{5.108}$$

where

α is coefficient = 0.96 for medium dense silty loam in Jinghui channel and Weihui channel

α=0.86 for medium dense sandy loam in Luohui channel
C_s is the sediment concentration (kg/m^3)
ω is the sediment particle fall velocity (m/s), and
R is the hydraulic radius (m)

The optimum results are shown in Table 5.6. It can be seen that calculated results are in close agreement with the field data. It can be concluded that the optimum design model is applicable to the design of regime channels.

6.2.2 Optimum Design of Dynamic Equilibrium Alluvial Channel

An eroding–silting-equilibrium channel design method was proposed by Sha in 1959 [55] using the following formulas:

Hydraulic geometry of the cross section:

$$\chi = kQ^{1/2}. \tag{5.109}$$

Continutity equation for flow:

$$Q = AU. \tag{5.110}$$

Flow equation of motion:

$$S = \frac{n^2U^2}{R^{4/3}}. \tag{5.111}$$

Sediment-carrying capacity formula (nonsilting sediment-carrying capacity):

$$C_H = \frac{0.176}{d_{50}^2}\frac{(U - 0.4d_{50}^{1/8}\,R^{1/6})^2}{R^{5/6}}. \tag{5.112}$$

where

χ is the wetted perimeter (m)
Q is the water discharge (m^3/s)
k is the coefficient of wetted perimeter. Sha studied many channels around the world, and concluded that the value of k should be between 2.3 and 7.5
A is the the area of the channel of cross section (m^2)
U is the velocity (m/s), R is the hydraulic radius (m)
d_{50} sediment median diameter (mm)
n is the roughness (s/m$^{1/3}$)
S is the slope (dimensionless), and
C_H is the nonsilting sediment concentration (kg/m^3), which can be computed by the following formula:

$$C_H = \frac{C_{\max}}{\Gamma}, \tag{5.113}$$

Table 5.6 Field data and calculated results of Jinghui, Luohui, and Weihui channel in Shaanxi Province

Irrigation districts	Labels of canals	Discharge (m^3/s)	Bank slope	Roughness	Saturation sediment concentration (kg/m^3)	Sediment median size (mm)	Depth (m)		Top width (m)		Slope (10^{-4})		Velocity (m/s)	
							Field	Calculated	Field	Calculated	Field	Calculated	Field	calculated
Jinghui irrigation district	1	14.46	1:1.28	0.025	40.3	0.0195	1.41	1.74	9.5	10.61	5.2	4.69	1.07	1
	2	4.5	1:1.00	0.025	40.3	0.0172	0.9	1.5	5.15	5.09	8.8	5.44	0.97	0.84
	3	2.76	1:0.88	0.025	40.3	0.019	0.76	0.8	4.45	5.3	7.75	6.92	0.82	0.76
	4	0.92	1:0.87	0.025	40.3	0.0212	0.58	0.89	2.4	2.3	8.6	8.79	0.67	0.68
	5	1..57	1:1.36	0.025	40.3	0.0128	0.66	0.8	3.25	4.13	8	6.28	0.73	0.65
	6	0.72	1:1.44	0.03	40.3	0.0112	0.42	0.5	3.1	3.38	9.2	10.74	0.55	0.55
	7	2.07	1:1.00	0.025	40.3	0.0155	0.76	0.8	3.95	4.5	5.9	6.52	0.7	0.7
Luohui irrigation district	1	4.01	1:1.08	0.03	23.6	0.0196	0.92	1	5.95	6.61	5.65	7.06	0.73	0.73
	2	2.49	1:0.67	0.02	23.6	0.0235	0.64	0.7	5.58	5.69	4.08	3.99	0.7	0.68
	3	0.27	1:0.88	0.03	14	0.0155	0.35	0.4	1.83	1.9	8.33	10.25	0.42	0.45
	4	0.96	1:0.83	0.0275	14	0.03	0.55	0.5	3.5	3.66	4.86	9.2	0.5	0.59
Weihui irrigation district	1	13.3	1:0.90	0.02	27.6	0.0225	1.41	1.5	8.6	10.59	4	3.04	1.1	0.96
	2	9.05	1:1.00	0.019	27.6	0.0097	1.03	2.47	8.85	6.97	4	1.82	0.99	0.82
	3	5.97	1:0.80	0.022	27.6	0.0235	1.04	1	6.85	7.91	4	4.64	0.84	0.84
	4	5.33	1:1.22	0.022	27.6	0.024	0.99	1	6.7	7.64	4	4.81	0.81	0.83
	5	1.8	1:0.63	0.0225	27.6	0.0222	0.79	0.74	3.4	3.91	5	5.93	0.67	0.71

where

Γ is the permissible ratio which is not a constant. The value can be determined by the comparison of similarity channels (dimensionless)

C_{max} is the maximum permissible sediment concentration (kg/m^3)

Sha assumed that muddy water has a silting rate greater than that of clear water. If the reasonable choice of permissible ratio Γ and the maximum permissible sediment concentration are made, the balance between siltation and erosion of channels can be maintained.

There are several shortcomings in Sha's design method. First, the value of coefficient of wetted perimeter k and permissible ratio Γ are uncertain. Therefore, designers rely on their experiences. Second, when water discharge is given, as sediment concentration increases, the cross-sectional shape becomes broader and shallower. Field data show that the shape of regime channels is narrow and deep [57, 58].

(1) Optimum design mathematical model

a. Objective function: The objective function of regime channel and the eroding–silting-equilibrium channel is to minimize energy dissipation rate.
b. Decision variables:The decision variables of the regime channel and the eroding–silting-equilibrium channel are bottom width b and depth h.
c. Constraint conditions: The concept of eroding-silting-equilibrium refers to in a certain period (such as a design represents years or flood) both siltation and erosion occurring, but the amount of sediment is almost equal to the amount of erosion. The concept as constraint conditions can be written to noneroding and nonsilting or eroding-silting equilibrium.

(i) The method of noneroding and nonsilting constraint conditions

The design velocity of flow U is greater than the noneroding velocity U_K corresponding to the minimum sediment concentration C_{min}, but less than the nonsilting velocity U_H corresponding to the maximum permissible sediment concentration C_{max} (see Fig. 5.20), i.e.,

$$U_H > U > U_K. \tag{5.114}$$

Equation (5.114) shows that channel may silt or erode, but silting and eroding are in balance by choosing reasonable maximum permissible sediment concentration C_{max}. According to the field data and specific operating conditions of channels, the maximum permissible sediment concentration C_{max} can be determined from Fig. 5.20.

Substitution of flow continuity equation and trapezoidal cross-sectional area relationship into Eq. (5.114) yields the following two constraint conditions:

$$G_1(\boldsymbol{X}) = U_H - \frac{Q}{bh + mh^2} > 0, \tag{5.115}$$

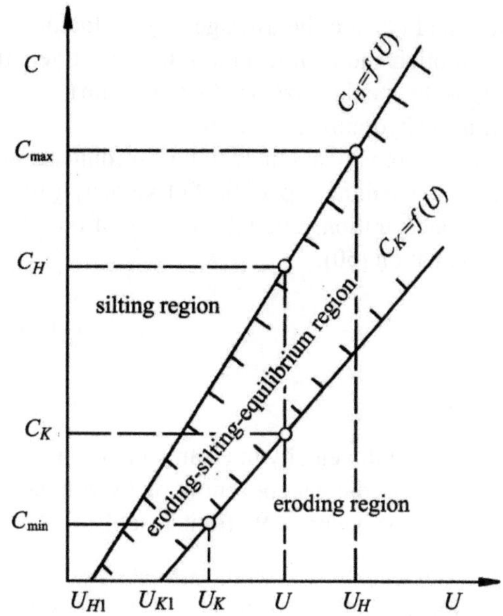

Fig. 5.20 Relationship between maximum sediment concentration (or minimum permissible sediment concentration) and noneroding velocity (or the nonsilting velocity)

$$G_2(X) = \frac{Q}{bh + mh^2} - U_K > 0. \tag{5.116}$$

According to the Sha's sediment-carrying capacity formula [59]:

$$C_* = \frac{K\mathrm{d}_{50}}{\omega^{4/3}}\left(\frac{U - U_0}{\sqrt{R}}\right)^2. \tag{5.117}$$

The following nonsilting velocity and noneroding velocity can be obtained:

$$U_H = \frac{C_{\max}^{1/2}\omega^{2/3}R^{1/2}}{K_H^{1/2}d_{50}^{1/2}} + U_{H_1}R^{0.2}, \tag{5.118}$$

$$U_K = \frac{C_{\min}^{1/2}\omega^{2/3}R^{1/2}}{K_K^{1/2}d_{50}^{1/2}} + U_{K_1}R^{0.2}. \tag{5.119}$$

where

$$U_{H_1} = (0.011 + 0.39d_{50}^{3/4})^{1/2}$$

$$U_{K_1} = \left[1.1\frac{(0.7-\varepsilon)^4}{d_{50}} + 0.43d_{50}^{3/4}\right]^{1/2}$$

where

K_H and K_K are the nonsilting sediment-laden coefficient and noneroding sediment-laden coefficient, respectively. For loess channel, K_H=3200, K_K=1100 (dimensionless)

U_{H_1} and U_{K_1} are the average flow velocities of unit hydraulic radius at the end of motion and incipient motion, respectively (m/s)
d_{50} is the median size of sediment (mm)
R is the hydraulic radius (m)
ε is the porosity assumed to be 0.4 (dimensionless)
ω is the sediment particle fall velocity (mm/s), which is influenced by sediment concentration. The fall velocity of muddy water is calculated by the following equation [60]:

$$\omega = \omega_0 (1 - C_V)^m, \tag{5.120}$$

where

ω_0 is the fall velocity in clear water (mm/s)
C_V is the sediment concentration by volume (dimensionless), and
m is the exponent=4.91 (dimensionless).

(ii) The method of eroding–silting equilibrium as constraint conditions

Water and sediment transportation mathematical model can be used to calculate yearly cumulative erosion and deposition thickness $\Delta H_i (i = 1, 2, \ldots, n. n = \text{the number of segment})$ in a segment of the channel. According to the concept of eroding–silting equilibrium, cumulative erosion, and deposition thickness of all the channel segments should be zero. In fact, cumulative erosion is not equal to cumulative deposition accurately. Therefore, a permissible error ε_H is set. If ΔH_i is less than ε_H, the channel in eroding–silting equilibrium can be expressed by

$$\Delta H_i \le \varepsilon_H \quad (i = 1, 2, \ldots, n). \tag{5.121}$$

Equation (5.121) is written to satisfy the constraint condition:

$$g(X) = \varepsilon_H - \Delta H_i \ge 0 \quad (i = 1, 2, \ldots, n). \tag{5.122}$$

A one-dimentional water and sediment transportation mathematical model can be used to calculate cumulative erosion and deposition thickness ΔH_i of each segment of channel based on the following equations:

Continuity equationfor flow:

$$\frac{\partial Q}{\partial x} + B\frac{\partial h}{\partial t} = 0. \tag{5.123}$$

Flow equation of motion:

$$\frac{\partial Q}{\partial t} + \frac{\partial}{\partial x}\left(\frac{Q^2}{A}\right) + gA\frac{\partial (z_b + h)}{\partial x} + g\frac{Q^2}{C^2 AR} = 0. \tag{5.124}$$

Riverbed deformation equation:

$$\rho_s' \frac{\partial z_b}{\partial t} = \alpha\omega(C_s - C_*). \quad (5.125)$$

Suspended load nonequilibrium sediment transport equation is

$$\frac{\partial(QC_s)}{\partial x} + B\frac{\partial(hC_s)}{\partial t} = -\alpha B\omega(C_s - C_*), \quad (5.126)$$

where

Q is the water discharge (m^3/s)
B is the average sectional width (m)
h is the average water depth (m)
z_b is the elevation at the bottom of cross section (m)
g is the acceleration of gravity (m/s^2)
C is the Chezy coefficient (m$^{1/2}$/s)
R is the hydraulic radius (m)
A is the area of cross section (m^2)
ρ_s' is the sediment dry density (kg/m^3)
ω is the sediment particle fall velocity (m/s)
C_s is the sediment concentration (kg/m^3)
C_* sediment transport capacity (we can choose the appropriate sediment-carrying capacity formula to calculate; kg/m^3)
α is the coefficient of saturation recovery (dimensionless)
x is the distance (m), and
t is the time (s)

Using the above mathematical model for the calculation of ΔH_i, the maximum deposition thickness can be determined to avoid siltation affecting normal operation of water transportation.

In summary, eroding–silting-equilibrium channel optimum design mathematical model can be summarized as:

$$\begin{aligned}
&\text{Slove} \quad \boldsymbol{X} = [b \quad h]^T \\
&\text{make} \quad \Phi(\boldsymbol{X}) \to \min \\
&\text{subject to } G_1(\boldsymbol{X}) = U_H - \frac{Q}{bh + mh^2} > 0 \\
&G_2(\boldsymbol{X}) = \frac{Q}{bh + mh^2} - U_K > 0 \\
&\text{or} \quad g(\boldsymbol{X}) = \varepsilon_H - \Delta H_i \geq 0, (i = 1, 2, \ldots, n)
\end{aligned} \quad (5.127)$$

Equation (5.127) can be used to solve inequality-constrained nonlinear minimization problems. Using inner point penalty function, constraint conditions $G_i(\boldsymbol{X}) > 0$

or $g(\boldsymbol{X}) \geq 0$ in conjunction with the objective function in Eq. (5.127) can be transformed to the following unconstrained minimization problem:

$$\min F(\boldsymbol{X}, M_K) = \Phi(\boldsymbol{X}) + M_K \sum_{i=1}^{2} \frac{1}{G_i(\boldsymbol{X})}, \tag{5.128}$$

or

$$\min F(\boldsymbol{X}, M_{\mathrm{K}}) = \Phi(\boldsymbol{X}) + M_{\mathrm{K}} \frac{1}{g(\boldsymbol{X})}. \tag{5.129}$$

The iterative method can be used to obtain the minimum value of the augmented objective function $F(\boldsymbol{X}, M_K)$.

(2) Optimum results and discussion

Jinghui channel, Luohui channel, and Weihui channel in Shaanxi Province were designed based on nonsilting and noneroding balance principle. In order to avoid siltation, when the sediment concentration exceeded 15 % by weight or 165kg/m^3, the gate would be closed and the diversion was stoped. During flood seasons, the average sediment concentration of the three channels is more than 15 %. With the irrigation district area gradually expanded, supply and demand of water conflicts become increasingly prominent. In the 1960s, Luohui channel carried out hyper-concentration flow diversion trial, with sediment concentration increasing to 25 % or 296 kg/m^3. In the Luohui irrigation district, some channels have no limit on sediment concentration. In the process of the hyper-concentration flow diversion, silting may take place without interrupting normal water transportation. Silting and eroding are basically in balance within a year. These successful experiences are widely applied to Jinghui channel and Weihui channel. After several years of operation of hyper-concentrated flow diversion, the original regime channels had been continuously adjusted by themselves to reach the eroding–silting-equilibrium condition [57, 58].

The West Branch and Middle Branch Luohui channel irrigation district are used as examples to verify the validity of eroding–silting-equilibrium channel optimum design model. Channel bottom width and water depth of these channels were calculated using Eq. (5.128) to maintain eroding–silting-equilibrium within a year.

Tables 5.7 and 5.8 summarize the field data and computed results, respectively. It can be seen that calculated results from the optimum design method are in good agreement with the field data, but the results of Sha's design method are not good. Therefore, the optimum design method based on the minimum energy dissipation rate principle can be used in the design of the eroding–silting-equilibrium channels with confidence.

Table 5.8 also shows that the section of eroding–silting-equilibrium channel is narrow and deep. This has been confirmed by field data of Jinghui Channel, Luohui Channel, and Weihui Channel in Shaanxi Province.

Table 5.7 Field data of middle branch and west branch of luohui channel irrigation district

Channel	Water discharge (m^3/s)	Slope (10^{-4})	Width of channel bottom (m)	Depth of water (m)	Bank slope	Suspended sand median diameter (mm)	Roughness	C_{max} (kg/m^3)
Middle branch channel	10	4.0~5.0	1.6~4.8	1.75~2.0	1.0	0.0353~0.0420	0.025	296
West branch channel	5	4.0~6.7	1.5~2.0	1.58~1.64	1.0	0.0353~0.0420	0.025	296

Table 5.8 Calculated results of middle branch and West Branch of Luohui channel irrigation district

Channel	Optimum design method			Sha's design method		
	Slope (10^{-4})	Width of channel bottom(m)	Depth of water (m)	Slope (10^{-4})	Width of channel bottom(m)	Depth of water(m)
Middle branch channel	5.15	1.92	2.34	37.1	8.81	0.63
West branch channel	6.14	1.47	1.73	43.1	6.18	0.50

6.3 Interim Summary

1. Stable and strong transverse circulation is the key to guarantee normal operation of diversion structure. The principle of minimum energy dissipation rate can be used for the optimum design of stable diversion bend structure.
2. Minimum energy dissipation rate of flow can be used as the objective function, and the noneroding and nonsilting velocity and eroding–silting-equilibrium are the constraint conditions for the optimum design of stable channels. This method can avoid design mistakes due to the limitations of the empirical hydraulic geometry relationships, and enhance the quality of stable channel design.

7 Summary and Conclusions

1. Minimum entropy production principle is one of the basic principles of nonequilibrium thermodynamics applicable to all open systems. Minimum energy dissipation rate principle is based on the minimum entropy production principle. Minimum entropy production principle is equivalent to the minimum energy dissipation rate principle. Minimum energy dissipation rate principle states that, when an open system is at a steady nonequilibrium state, the energy dissipation

rate is at the minimum value subject to the constraints applied to the system. If the system deviates from the steady nonequilibrium state, it will go through self-adjustments to minimize its energy dissipation rate subject to the constraints applied to the system. Minimum energy dissipation rate principle or minimum entropy production principle can be considered as a stability criterion of evolution for open system in the linear range. The evolution process of the open system is complex. Not every moment of the evolution process of the open system is at a steady state. The system may experience a series of nonequilibrium state until it reestablishes a new steady state.

2. RNG $k-\varepsilon$ turbulence model and GMO model are applied to simulate fluid motion in a straight rectangular flume. The results show that the energy dissipation rate reaches a constant minimum value when the flow is steady. The energy dissipation rate of unsteady flow may increase or decrease during the evolution process before a minimum value is reached. When the external constraints of the system vary greatly, the flow may deviate from the original state through unsteady flow transition process to achieve a steady flow state. The new steady state is compatible with the new external constraints. Because each steady flow is subject to different external constraints, the minimum energy dissipation rate value of each steady is different.
3. The relative equilibrium of a river is a dynamic equilibrium. Therefore, the energy dissipation rate fluctuates around its average when a river is at a relative equilibrium state. By calculation and analysis of unit stream powers of 21 years of field data for six reaches of three river patterns, unit stream power of the wandering reach is the highest. The meandering reach has the lowest unit stream power. Unit stream power of the transitional reach is between that of the meandering reach and wandering reach. The stability of a river decreases with increasing unit stream power.
4. Self-adjustments of alluvial rivers may reach a relative equilibrium state through a long-term interaction between flow and alluvial river boundaries. Hydraulic geometry relationships based on minimum energy dissipation rate principle are theoretically sound. Theoretically derived channel geometry is in good agreement with field data.
5. The principle of minimum energy dissipation rate and optimization techniques can be used for stable channel design. This theoretically sound approach can be used for engineering design to avoid design mistakes from using empirical methods.

References

1. Yang, C. T. (1971). Potential energy and stream morphology. *Water Resources Research, 7*(2), 311–322.
2. Yang, C. T. (1971). Formation of riffles and pools. Water Resources Research, 7(6), 1567–1574.
3. Yang, C. T. (1972). Unit stream power and sediment transport. Journal of the Hydraulics Division, 98(HY10), 1805–1826.

4. Yang, C. T. (1973). Incipient motion and sediment transport. *Journal of the Hydraulics Division, 99*(HY10), 1679–1704.
5. Yang, C. T., & Stall, J. B. (1976). Applicability of unit stream power equation. *Journal of the Hydraulics Division, 102*(HY5), 559–568.
6. Yang, C. T. (1976). Minimum unit stream power and fluvial hydraulics. *Journal of the Hydraulics Division, 102*(HY7), 919–934.
7. Yang, C. T., & Song, C. C. S. (1979). Theory of minimum rate of energy dissipation. *Journal of the Hydraulics Division, 105*(HY7), 769–784.
8. Song, C. C. S., & Yang, C. T. (1979). Velocity profiles and minimum stream power. *Journal of the Hydraulics Division, 105*(HY8), 981–998.
9. Yang, C. T., & Song, C. C. S. (1979). Dynamic adjustments of alluvial channels. In D. D. Rhodes & G. P. Williams (Eds.), *Adjustments of the fluvial systems* (pp. 55–67). Dubuque: Kendall/Hunt Publishing Company.
10. Song, C. C. S., & Yang, C. T. (1980). Minimum stream power: Theory. *Journal of the Hydraulics Division, 106*(HY9), 1477–1487.
11. Yang, C. T., Song, C. C. S., & Woldenberg, M. J. (1981). Hydraulic geometry and minimum rate of energy dissipation. *Water Resources Research, 17*(4), 1014–1018.
12. Song, C. C. S., & Yang, C. T. (1982). Minimum energy and energy dissipation rate. *Journal of the Hydraulics Division, 108*(HY5), 690–706.
13. Yang, C. T., & Molinas, A. (1982). Sediment transport and unit stream power function. *Journal of the Hydraulics Division, 108*(HY6), 774–793.
14. Yang, C. T. (1984). Unit stream power equation for gravel. *Journal of Hydraulic Engineering, 110*(HY12), 1783–1797.
15. Molinas, A., & Yang, C. T. (1985). Generalized water surface profile computation. *Journal of Hydraulic Engineering, 111*(HY3), 381–397.
16. Yang, C. T., & Song, C. C. S. (1986). Theory of minimum energy and energy dissipation rate. In *Encyclopedia of fluid mechanics* (Vol. 1, Chapter ll, pp. 353–399) Houston: Gulf Publishing Company.
17. Yang, C. T., & Kong, X. (1991). Energy dissipation rate and sediment transport. *Journal of Hydraulic Research, 29*(4), 457–474.
18. Yang, C. T. (1994). Variational theories in hydrodynamics and hydraulics. *Journal of Hydraulic Engineering, 120*(6), 737–756.
19. Chang, H. H., & Hill, J. C. (1977). Minimum stream power for rivers and deltas. *Journal of the Hydraulics Division, 103*(HY12), 1375–1389.
20. Chang, H. H. (1979a). Geometry of river in regime. *Journal of the Hydraulics Division, 105*(HY6), 691–706.
21. Chang, H. H. (1979b). Minimum stream power and river channel patterns. *Journal of Hydrology, 41*(3/4), 303–327.
22. Chang, H. H. (1980). Stable alluvial canal design. *Journal of the Hydraulics Division, 106*(5), 873–891.
23. Chang, H. H. (1983). Energy expenditure in curved open channels. *Journal of Hydraulic Engineering, 109*(HY7), 1012–1022.
24. Chang, H. H. (1984). Analysis of river meanders. *Journal of Hydraulic Engineering, 110*(HY1), 37–50.
25. Prigogine, I. (1977). *Self-organization in non-equilibrium systems*. New York: *Wiley-Interscience*.
26. Wisniewski, S., Staniszewski, B., & Szymanik. R. (1988). *Thermodynamics of nonequilibrium processes*. [In Chinese.] (trans: J. Chen, K. Yin, H. Li). Beijing: Higher Education Press.
27. Li, R. S. (1986). *Non-equilibrium thermodynamics and dissipative structures*. [In Chinese.] Beijing: Tsinghua University Press.
28. Gong, M. Z. (1998). *Thermodynamics*. [In Chinese.] Wu Han: Wuhan University Press.
29. Xu, G. B., & Lian, J. J. (2003). Theories of the minimum rate of energy dissipation and the minimum entropy production of flow (I). [In Chinese.] *Journal of Hydraulic Engineering*, (5), 35–40.

30. Xu, G. B., & Lian, J. J. (2003). Theories of the minimum rate of energy dissipation and the minimum entropy production of flow (II). [In Chinese.] *Journal of Hydraulic Engineering,* (6), 43–47.
31. Wu, W. Y. (2004). *Fluid mechanics*. [In Chinese.] Beijing: University Press.
32. Chang, M., & Xu, G. B. (2013). Numerical simulation of fluid motion in flume based on theory of minimum rate of energy dissipation. [In Chinese.] *Journal of Sediment Research,* (2), 67–71.
33. Qian, N., Zhang, R., & Zhou, Z. B. (1987). *Fluvial process*. [In Chinese.] Beijing: Science Press.
34. Xu, G. B., & Lian, J. J. (2008). Principle of the minimum rate of energy dissipation for fluid based on the theory of thermodynamics. [In Chinese.] *Advances in Science and Technology of Water Resources, 28*(5), 16–20.
35. Xu, G. B., & Yang, C. T. (2012). Analysis of river bed changes based on the theories of minimum entropy production dissipative structure and chaos. [In Chinese.] *Journal of Hydraulic Engineering, 43*(8), 948–956.
36. Leopold, L. B., & Langbein, W. B. (1962). The concept of entropy in landscape evolution. U.S. Geological Survey Professional Paper 500–A, pp. 1–20.
37. Xu, G. B., & Lian, J. J. (2004). Changes of the entropy, the entropy production and the rate of energy dissipation in river adjustment. [In Chinese.] *Advances in Water Science, 15*(1), 1–5.
38. Xie, J. H. (1997). *Fluvial process and regulation* (2 nd version). [In Chinese.] Beijing: Water Resources and Electric Power Press.
39. Xu, G. B. (2011). *River engineering*. [In Chinese.] Beijing: China Science and Technology Press.
40. Huang, H., He, J. X., & Wang, X. Z., et al. (2008). Research on influencing factors and criterion of channel patterns in alluvial rivers. [In Chinese.] *Journal of Xinjiang Agricultural University, 31*(6), 76–79.
41. Xu, G. B., & Zhao, L. N. (2013). Analysis of fluvial process based on information entropy. [In Chinese.] *Journal of Tianjin University, 46*(4), 347–353.
42. Xu, G. B., & Zhao, L. N. (2013). Analysis of channel pattern changes in the lower Yellow River based on the rate of energy dissipation. [In Chinese.] *Journal of Hydraulic Engineering, 44*(5), 622–626.
43. Hu, Y. S., Zhang, H. W., & Liu, G. Z., et al. (1998). *River training in the wandering reach in the lower Yellow River*. [In Chinese.] Zhengzhou: Yellow River Conservancy Press.
44. Zhang, R. J. (1998). *River sediment dynamics* (2 nd Version). [In Chinese.] Beijing: China WaterPower Press.
45. Xu, G. B. (1993). Calculation of the outlet sluice width in low dam diversion works. [In Chinese.] *Journal of Sediment Research,* (4), 65–71.
46. Xu, G. B. (1992). Optimized design of bend in bend-type water diversion headworks. [In Chinese.] *Journal of Sediment Research,* (3), 65–69.
47. Song, Z. Z., Xu, X. T., & Zhang, S. J. (1989). *Headworks* (2 nd Version). [In Chinese.] Beijing: Water Resources and Electric Power Press.
48. Yan, X. D., Liu, X. D., & Li, G. Q. (1990). *Sediment control of low-head diversion work*. [In Chinese.] Beijing: Water Resources and Electric Power Press.
49. Zhang, K. Q. (1982). The judgment for circulation strength and using spiral stream for sediment ejection. [In Chinese.] *Journal of Xinjiang Water Conservancy Science and Technology,* (4), 35–43.
50. Xie, Z. G. (1982). Bend-type diversion headworks experience in Xin Jiang. [In Chinese.] *Journal of Sediment Research,* (3), 84–88.
51. Quarteroni, A., Sacco, R., & Saleri, F. (2006) *Numerical mathematics. Foreign famous math book series* (Photocopy Edition) 5. Beijing: Science Press.
52. Xi, S. L., & Zhao, F. Z. (1983). *Optimization methods*. [In Chinese.] Shanghai: Shanghai Science and Technology Press.
53. Xu, G. B. (1993). An optimum design method of the regime channel. [In Chinese.] *Journal of Hydrodynamics, 8*(B12), 567–570.

54. Xu, G. B. (1996). An optimum design method for stable canals. [In Chinese.] *Journal of Hydraulic Engineering,* (7), 61–66.
55. Sha, Y. Q. (1959). A method of sedimentation balance and stable channel design. [In Chinese.] *Journal of Hydraulic Engineering,* (4), 23–42.
56. Northwest Institute of Hydraulic Research. (1959). The canal sediment and channel design. [In Chinese.] Shaanxi People's Publishing House, Xi'an, China.
57. Hyper-concentrated sediment muddy water irrigation experience summary of Yin Luo channel. (1987). Selected from Book One of the First Series of Yellow River Sediment Research Report, Yellow River Sediment Research Work Coordination Group, Zhengzhou, China, pp. 139–157 [In Chinese.].
58. Hyper-concentrated sediment diversion experimental group in Shaanxi Province. (1976). Hyper-concentrated sediment diversion preliminary summary in Jing River, Luo River and Baojixia Irrigation Districts of Wei River, Selected from the Third Series of Yellow River Sediment Research Report, Yellow River Sediment Research Work Coordination Group, Xi'an, China, pp. 107–137 [In Chinese.].
59. Sha, Y. H. (1965). *Introduction of sediment kinematics*. [In Chinese.] Beijing: China Industry Press.
60. Qian, N., & Wan, Z. H. (1983). *Mechanics of sediment transport*. [In Chinese.] Beijing: Science Press.

Chapter 6
Hydraulic Modeling Development and Application in Water Resources Engineering

Francisco J.M. Simões

Contents

Abstract The use of modeling has become widespread in water resources engineering and science to study rivers, lakes, estuaries, and coastal regions. For example, computer models are commonly used to forecast anthropogenic effects on the environment, and to help provide advanced mitigation measures against catastrophic events such as natural and dam-break floods. Linking hydraulic models to vegetation and habitat models has expanded their use in multidisciplinary applications to the riparian corridor. Implementation of these models in software packages on personal desktop computers has made them accessible to the general engineering community, and their use has been popularized by the need of minimal training due to intuitive graphical user interface front ends. Models are, however, complex and nontrivial, to the extent that even common terminology is sometimes ambiguous and often applied incorrectly. In fact, many efforts are currently under way in order

F. J.M. Simões (✉)
US Geological Survey Geomorphology and Sediment Transport Laboratory, 4620 Technology Drive, Suite 400, Golden, CO 80403, USA

C. T. Yang, L. K. Wang (eds.), *Advances in Water Resources Engineering*,
Handbook of Environmental Engineering, Volume 14, DOI 10.1007/978-3-319-11023-3_6

to standardize terminology and offer guidelines for good practice, but none has yet reached unanimous acceptance. This chapter provides a view of the elements involved in modeling surface flows for the application in environmental water resources engineering. It presents the concepts and steps necessary for rational model development and use by starting with the exploration of the ideas involved in defining a model. Tangible form of those ideas is provided by the development of a mathematical and corresponding numerical hydraulic model, which is given with a substantial amount of detail. The issues of model deployment in a practical and productive work environment are also addressed. The chapter ends by presenting a few model applications highlighting the need for good quality control in model validation.

Keywords Water resources · Hydraulics · Modeling · Shallow-water equations · Numerical model · Graphical user interface

Nomenclature

$\mathbf{A}$	Jacobian matrix (–)
C	Chezy's roughness coefficient ($L^{1/2}/T$)
C_D	Wind drag coefficient (dimensionless)
C_f	Skin friction coefficient (dimensionless)
D_i	Diffusion coefficient in the ith Cartesian direction ($i=1, 2$) (L^2/T)
$\mathbf{E}$	Matrix containing the inviscid terms (–)
f	Coriolis parameter (1/T)
Fr	Froude number (dimensionless)
F_x, F_y	Components of the body forces in the Cartesian directions, same as F_i ($i=1, 2, 3$) (L^2/T^2)
g	Acceleration due to gravity (L/T^2)
h	Water depth (L)
h_{dry}, h_{wet}	Threshold for drying and wetting of computational cells (L)
h_{eff}	Effective water depth (L)
l	Length of control volume edge (L)
n	Manning's roughness coefficient ($T/L^{1/3}$)
n_x, n_y	Components of the unit normal vector to control volume edges (dimensionless)
p	Pressure ($M/(LT^2)$)
q_x, q_y	Unit discharge in the Cartesian directions (L^2/T)
$\vec{r}$	Position vector (L)
$R, \overline{R}$	Residual and its average, respectively (–)
$\mathbf{R}$	Matrix containing the viscous terms (–)
$\mathbf{S}$	Matrix containing source/sink terms (–)
$S0_i$	Component of the bed gradient in the ith Cartesian direction ($i=1, 2$) (dimensionless)
t	Time (T)
u, v, w	Components of the velocity vector in Cartesian coordinates, same as u_i ($i=1, 2, 3$) (L/T)

U, V	Components of the depth-averaged velocity vector in Cartesian coordinates (L/T)
u_w, v_w	Components of the wind vector in Cartesian coordinates (L/T)
u_*	Magnitude of the shear velocity (L/T)
x, y, z	Orthogonal Cartesian coordinate directions (L)
z_b	Elevation of channel bed above datum (L)
α_{ij}, β_i	Coefficients of the SSPRK scheme (dimensionless)
δ_e	A very small number (L/T)
δ_{ij}	Kronecker delta (dimensionless)
δ_w	Parameter used in dry computational cells (L)
Δt	Time step size (T)
ε_t	Parameter in eddy viscosity formula (dimensionless)
Φ	Limiter in the MUSCL reconstruction (dimensionless)
η	Water-surface elevation above datum (L)
λ_i	Eigenvalues of the Jacobian matrix (–)
ν	Kinematic molecular viscosity (L^2/T)
ν_t	Turbulent eddy viscosity (L^2/T)
θ	CFL coefficient (dimensionless)
θ_c	Flume contraction angle (dimensionless)
θ_s	Angle of the shock (dimensionless)
ρ, ρ_0	Density of water (M/L^3)
ρ_a	Density of air (M/L^3)
Ω_i	Area of control volume i (L^2)
τ_{bx}, τ_{by}	Components of the bed shear stress vector in the Cartesian directions ($M/(LT^2)$)
τ_{ij}	Radiation stresses (L^2/T^2)
τ_{wx}, τ_{wy}	Components of the wind stress at the free surface ($M/(LT^2)$)

1 Introduction

The study of natural river changes and the interference of man in natural water bodies is a difficult but important activity, as increasing and shifting populations place more demands on the natural sources of freshwater, and the impacts of a potential climate change bring new unknowns and renewed urgencies. Although the basic mechanical principles for these studies are well established, the complexity of the fluvial system involves not only the river corridor proper but also the surrounding geographic complex of hillslopes and floodplains, and their respective physiography (e.g., soil composition and properties, vegetation cover, etc.). To understand the fluvial system as a whole, therefore, one must understand not only how the individual parts of the system behave in isolation but also how they interact with each other. These interactions and feedbacks, which may be deterministic or random, can be very complex and extend over wide scales in time and space. Table 6.1 presents

Table 6.1 Synopsis of the variables involved in modeling fluvial systems, showing the complex dependencies between the different forcing phenomena and concomitant subsystem adjustments

	Scale	Variables	Dependencies
Independent variables	Global	Climate	
		Geology	
	Basin	Physiography	
		Vegetation	Climate, geology
		Soil type	Climate, geology
		Land use	Vegetation, soil type, biophysical agents
	Channel	Valley slope	Basin physiography
		Flow discharge	Basin physiography, climate
		Sediment load	Basin physiography, climate, vegetation, basin soil type, land use, bank material
		Bank material composition	Geology, vegetation, soil, land use
Dependent variables	Basin	Meander wavelength and sinuosity	Bank material, discharge, sediment load, channel width and depth
	Channel and subchannel	Bed slope	Valley slope, flow discharge, sediment load, sediment transport rate, bed material size
		Width and depth	Flow discharge, flow velocity, sediment load, bed material size, bank material composition
		Flow velocity	Discharge, channel width and depth, frictional resistance
		Frictional resistance	Bedform geometry, channel width and depth, bed material size
		Bedform geometry	Sediment transport rate, channel width and depth
		Sediment transport rate	Sediment load, flow velocity and turbulence, bed material size, channel slope
		Bed material size	Sediment load, sediment transport rate

a synoptic view of scales, variables, and processes involved in fluvial hydrology, which helps to paint a view of the dynamic dependencies of the morphology of natural river channels. Modeling is one of the tools available to study the behavior of environmental systems, and has become a crucial vehicle to represent and understand—and even communicate—the effects of the interactions of real-world variables in the fluvial corridor.

What is a model? A common definition is that a model is an abstraction of reality [1]. In other words, it is the methodical organization of data and knowledge—as well as assumptions—about the specific system of interest. Modeling is, therefore, a creative process that allows a systematic analysis of a problem. It can be as simple as a schematic diagram sketched in the back of an envelope, or as complex as those used in large supercomputers for weather forecasting, containing millions of lines of code and using detailed data at the planetary scale. Indeed, there are many types of models, of which computer models are a subset. Programmable computers offer a powerful tool to implement physically and mathematically sophisticated models, which nowadays are used routinely in aeronautics, acoustics, medicine, astrophysics, etc.

In hydrology and fluvial hydraulics, the use of computer models has increased significantly in the past couple of decades, due to the availability of affordable computer power, and due to advancements in large-scale data collection techniques, such as remote sensing using laser scanning. The availability of data to build, calibrate, and test a model has also become less of a limiting factor in the application and development of hydrological models. These factors have led to the proliferation of numerous software packages that are increasingly used by the nonmodeling community (i.e., by those that use models, but that do not develop them), which has brought an increased potential for misuse and misunderstanding in the capabilities of the models and in the interpretation of their results. Models are, after all, always limited by the scales, purposes, and assumptions used in their development and, consequently, always have limited application. This chapter illustrates the difference between modeling and software development by following the steps necessary to go from conceptual modeling all the way to computer software implementation. The next section presents a synoptic view of the steps involved in the development of a hydraulic model, which is followed by the detailed description of the development of a model to simulate free surface flows in rivers, reservoirs, and estuaries. Finally, this chapter concludes with the examples of application of the model to laboratory and to real-world problems.

2 Concepts and Development

The terms *model, numerical model,* and *modeling software* are often used interchangeably to mean the same thing. This, however, constitutes an imprecise, even incorrect use of the terms. To understand the difference between these terms, we must look at the whole discipline of modeling. A model is a representation of a portion of the physical world, i.e., it is an abstraction of it. This is the top level in modeling: It is the most general use of the term *model* which, in this sense, means the knowledge we have of the physical phenomena of interest, as is represented by the laws of physics and by data. Oftentimes, there are gaps in the knowledge of the system, caused by lack of data or by incomplete understanding of its physical nature, or both. These gaps must be sated with consistent simplifications, assumptions, and hypothesis. In fluvial hydraulics, a model is often used to represent a segment of a river, a lake, or an estuary, and the basic laws of Newtonian fluid dynamics are employed, which describe the behavior of fluids, flow turbulence, sediment transport mechanisms, mixing processes, etc. The data consist of boundary conditions, such as bathymetry, water discharges, sediment particle-size distributions, vegetation types, sizes and locations, etc. Common assumptions used include flow resistance laws, sediment transport capacity relations, linear dependency of turbulent diffusion on bed shear, etc.

The next level of modeling involves the conceptualization of the physical phenomena into its mathematical formulation, which results in the mathematical model. Stepping from the physical phenomena to the mathematical model involves a simplification, whereby the complexity of natural environments is reduced to a limited set of relations between the dependent and the independent variables[1]. This is accomplished by a process of successive elimination, where the relative importance of the different phenomena is compared and those with the smallest influence are eliminated—or alternatively by starting with an oversimplified view and adding those with the largest influence. Examples of simplifying approximations are steady versus unsteady and one- versus two- versus three-dimensional (3D) formulations; simplifying descriptions of turbulence; etc. This process depends on the modeling objectives, on our knowledge of the physical system, on the availability and quality of data, and even on the perception of the modeler(s). The mathematical model is, therefore, composed by a number of mathematical formulae (governing equations) that must be solved under particular sets of initial and/or boundary conditions.

In fluvial hydraulics, one usually arrives to the setup of a boundary-value problem whose governing equations contain partial differential equations with nonlinear terms. These are based on the 3D Navier–Stokes equations—for example, see [2] for the derivation and presentation of the equations—or in a simplification thereof. Useful and frequent simplifications of the governing equations are the assumption of hydrostatic pressure and the reduction of the order of the governing equations, whereby the 3D flow description is simplified to two-dimensional (2D) or one-dimensional (1D) descriptions by the use of averaging operations. Some models, however, do not use the flow equations. One alternative is provided by artificial neural network techniques, a numerical modeling technique that attempts to reproduce relations between sets of input and output data by learning from observing experimental and/or prototype tests—see [3], for example. Another alternative originated in artificial intelligence is the use of genetic programming, which is another machine-learning technique based on a set of instructions and a fitness function to measure how well a task is performed by a computer. In Ref. [4], this technique is applied to predict flow discharge in compound channels.

The processes employed to solve the governing equations result in the numerical model, which is the next level of modeling. This step is necessary because computers can only perform the most basic arithmetic operations, therefore, partial differential equations must be transformed into the simple algebraic equations that can be programmed in a computer. Furthermore, computers can only execute computations between finite numbers with a limited amount of correct digits and within a certain range. This operation has two important consequences: First, it transforms continuous equations into discrete ones and second, it transforms infinitesimal operators into finite ones. Consequently, instead of solving the original differential equations, computers only solve approximations thereof. The numerical model is, therefore, a set of algebraic equations whose solutions (hopefully) mimic those of the original differential equations.

[1] The independent variables are the inputs and the forcing quantities, such as boundary conditions and spatial coordinates; the dependent variables are the outputs and effects, such as flow velocity and water surface elevation.

The numerical model can be very simple—the direct solution of the Manning's equation on the back of an envelope, for example—or very complex. The latter often arises from the need to solve partial differential equations that have no known closed-form analytic solutions, and which require complex numerical techniques and algorithms to solve. Finite differences, finite volumes, and finite elements are some of the methodologies available and commonly used in numerical modeling, and detailed description of these can be found in basic computational fluid dynamics (CFD) textbooks, such as in Ref. [5]. Because these techniques and algorithms are approximate and do not provide the exact solution to the equations, the quality of the approximation depends on several factors, such as order of discretization, approaches to deal with nonlinear terms, discretization refinement (more about this later), choice of numerical parameters, boundary conditions, and even the experience of the modeler.

The next level of modeling involves the implementation of the numerical model in a computer, and it is called the computer model. The computer model is the result of coding the numerical model in a computer language and is, therefore, software. Often, the terms *numerical model* and *computer model* are used interchangeably, but that is not correct: The same numerical model can be coded differently by different programmers and, therefore, yield different results. This may be due to the order in which algebraic operations are implemented, resulting in different truncation errors, or it may be due to the use of numerical constants of different resolutions (say, by hard-coding $\pi = 1.1415962$ instead of using a function such as $4\ \mathrm{tg}^{-1}\ 1$). Compilers may use different libraries and/or different optimization options. In fact, the mathematical associative property does not always hold for the Institute of Electrical and Electronics Engineers (IEEE) floating-point representation of numbers by a computer: $(a+b)+c$ may not be equal to $a+(b+c)$. For example, $2^{-63}+1-1$ is equal to 2^{-63}, obviously. In a computer using double-precision floating-point numbers, however, there are two possible answers: One is $(2^{-63}+1)+(-1) \approx 1+(-1)=0$, because $(2^{-63}+1)$ rounds to 1; the other, through a modification in the order of operations, is $2^{-63}+(1+(-1)) \approx 2^{-63}+0=2^{-63}$. Both the results are correct from the computer point of view and, in such a case, the outcome depends not only on how the lines of code are written by the programmer but also on how the particular compiler used performs code optimization [6]. Different operating systems and different hardware configurations may also be the cause for numerical differences among software. And there are coding errors, something that can easily to creep into modern computer models containing thousands of lines of code.

The different steps in modeling are summarized in Fig. 6.1, where the boxes to the left represent the levels of models described above and the ones to the right represent some of the factors influencing the outcome when going from one level to the next. It is easy to see that the different levels, or types, of models have different objectives. For example, a physical model may be built to gain knowledge about a system, while computer software is implemented to perform calculations—i.e., to provide a service. In fact, the final step in modeling involves the use of the computer model (software) by an operator. Software operators, or users, often are distinct from code developers and have different objectives: While code developers are interested in implementing a computer model of some conceptualized reality,

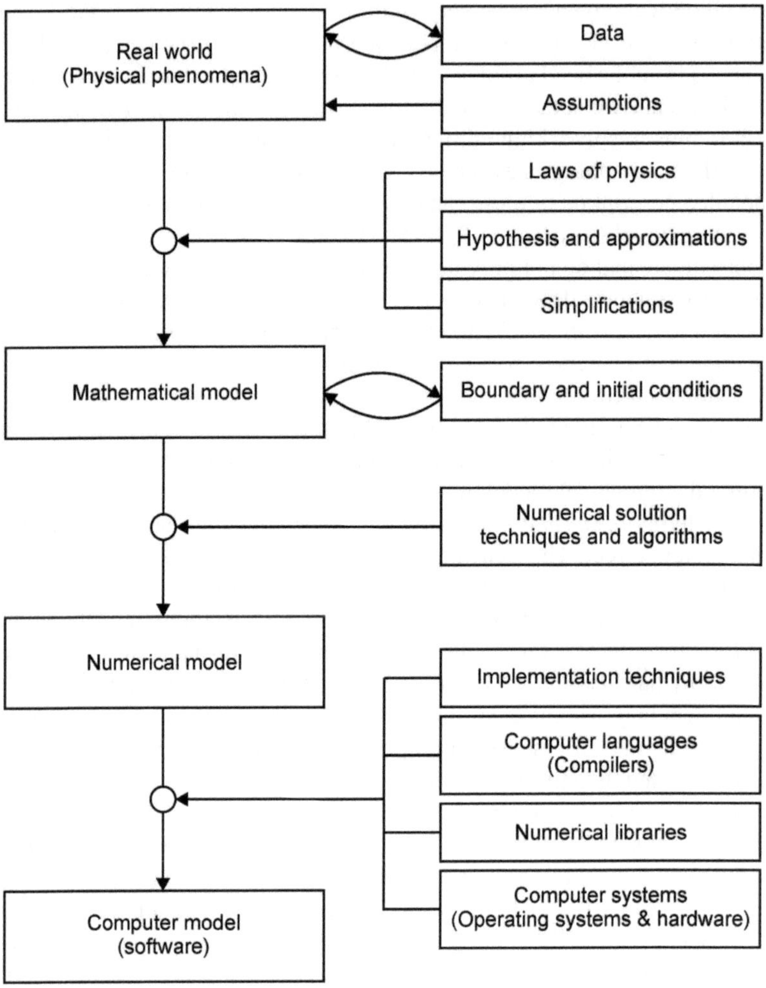

Fig. 6.1 The different modeling levels and the factors contributing to each level change

software users are usually seeking answers to management problems. This raises the issues of communication between these two groups of researchers, including documentation, terminology, best-practice guidelines, and quality assurance.

Detailed and clear model documentation should present the elements of the conceptual model and its purpose, and must use a terminology that is unambiguous and understood by all. Current practices vary widely, but efforts exist to unify and standardize terminology and procedure. Detailed terminology is proposed by [7], who also review a number of modeling guidelines and present key views of the scientific community. A comprehensive set of guidelines for model development, model evaluation, and model application is given by the US Environmental Protection Agency's Council for Regulatory Environmental Modeling [8]. Nonetheless, it is still difficult to take advantage of the structural complexity of a model, and many approximations and simplifications are introduced in this stage. Model calibration is usually achieved using time series; therefore, finding the best set of model parameters is limited by available data. An incomplete knowledge of the physical system is also common, such as imprecise bathymetry, unknown water and/or sediment discharges, friction factors, etc. Finally, introducing additional human operators in the system also raises the issues of training (i.e., the results of the simulation project depend on operator skill) and increase the likelihood of human error.

3 Developing a Surface-Water Model for Environmental Flows

One of the most important considerations when developing (and applying) a model is model purpose. In scientific and engineering applications to human and natural environmental systems and interactions, the most important modeling purposes are exploratory analysis and prognostication: While prognostication refers to the ability of a model to predict the systems' response to specific external forcing factors, exploratory analysis is concerned with mapping the space of possible development trajectories of a system, usually with the purpose to explore the possibility of unexpected behaviors and thresholds triggering abrupt change. Closely replicating physical phenomena is, therefore, the major goal when developing a hydraulic model for environmental flows.

CFD methods have been in use for over 90 years, with the first application having been developed for numerical weather prediction by L. Richardson in 1922 and published in a book titled "Weather Prediction by Numerical Process." Since then, CFD has been expanded by mathematicians and engineers for the application to flow problems in the area of industrial engineering. As the numerical techniques became more efficient and the computational power (digital computers) increasingly more affordable, CFD has expanded to many other areas of fluid dynamics. Of interest here is the realm of environmental fluid mechanics, which constitutes the study of naturally occurring fluid flows on the planet, such as air, water, and debris flows. In environmental hydraulics, we are mainly concerned with the flow of surface water within a very narrow range of temperatures and pressures. The fundamental laws and equations of fluid motion under these circumstances have long been known—conservation of mass and of momentum—and will be used here as the basis for a real-world model for computing the flow in natural bodies of water.

3.1 The Mathematical Model

Flow phenomena in natural rivers are 3D, especially those at or near meander bends, local expansions and contractions, or at hydraulic structures. Turbulence is an essentially 3D phenomenon, and 3D models are particularly useful for the

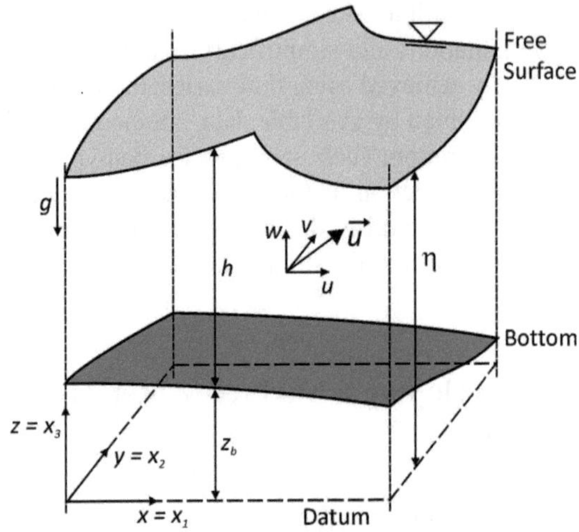

Fig. 6.2 Coordinate system used and the definition of some variables. Note that $u=u_1$, $v=u_2$, and $w=u_3$

simulation of turbulent heat and mass transport. These models are usually based on the Reynolds-averaged form of the Navier–Stokes equations, known since the nineteenth century, using additional equations of varied degree of complexity for the turbulence closure. Their derivation can be found in many basic textbooks on fluid dynamics, therefore, they will only be presented here without further consideration. The Navier–Stokes equations represent the statement of Newton's second law for fluids, i.e., the conservation of momentum, and in the Cartesian coordinate system and for incompressible fluids they can be written as

$$\frac{\partial u_i}{\partial t}+\frac{\partial u_i u_j}{\partial x_j}=\frac{F_i}{\rho}-\frac{1}{\rho}\frac{\partial p}{\partial x_i}+\frac{\partial}{\partial x_j}\left(\nu\frac{\partial u_i}{\partial x_j}-\overline{u_i' u_j'}\right) \tag{6.1}$$

where i is the Cartesian directions ($i=1$ for x, $i=2$ for y, and $i=3$ for z) as in Fig. 6.2, u_i is the Cartesian component of the velocity along the x_i direction ($i=1, 2, 3$), ρ is the fluid density, p is the pressure, F_i is the component of the body forces per unit volume in the ith direction, ν is the kinematic molecular viscosity, $-\rho\overline{u_i' u_j'}$ are the turbulent stresses; and the indexed summation convention is used. Equation (6.1) constitutes a system of equations, one for each coordinate direction, i.e., for $i=1, 2,$ and 3. The body forces include gravitational, buoyancy, and Coriolis forces, or any other body forces that may be present (such as magnetic forces in magnetohydrodynamic fluids).

Conservation of mass is expressed by the continuity equation for incompressible fluids:

$$\frac{\partial u_i}{\partial x_i}=0 \tag{6.2}$$

Table 6.2 Typical scales for the application of surface hydraulics models. The number of nodes provides a measure of the numerical burden associated with discretizing the governing equations

Dimension	Number of nodes	Spatial scale (m)	Time scale	Number of equations
1D	~ 10^2	10^3–10^5	Up to decades	2
2D	10^3–10^5	10^3	Days	3 (+2 for turbulence)
3D	10^4–10^6	10^1–10^3	Minutes to hours	4 (+8 for turbulence)

The turbulence terms ($-\rho\overline{u_i'u_j'}$) result from averaging the original Navier–Stokes equations using the Reynolds decomposition (see [9] for a more detailed explanation of the technique) and require additional closure equations. One of the commonly used closure techniques is given by the k–ε model, but there are many other alternative choices. The reader is directed to the turbulence modeling monograph [10] for further details about this subject.

When shallow-water flows are nearly horizontal, the 3D effects are not essential, and in the long-wave approximation (i.e., waves with long characteristic lengths compared to the water depth, such as tidal waves in seas and flood waves in rivers), considerable simplifications can be made to the governing equations and efficient mathematical models can be built, where depth-averaged quantities replace their fully 3D counterparts. This process is called order reduction and results in models that are useful in many practical applications, where 3D detail is not needed or when the extent of the problem is too large for the available computing power. On the other hand, 2D modeling constitutes a step-up from the much simpler 1D approach, albeit at a cost of significant additional data requirements, but the recent advancements in low-cost data acquisition techniques with high resolution and accuracy, such as light detection and ranging (LIDAR) and multi-beam echo sounding, has facilitated the fulfillment of those requirements. Table 6.2 shows the scales of applicability of the different types of modeling in view of the commonly used current-day computing facilities. In Table 6.2, the number of partial differential equations of the mathematical model is given, with additional differential equations for turbulence modeling in parenthesis.

As the moniker implies, 2D depth-averaged modeling results from averaging the 3D flow Eq. (6.1) and (6.2) along the vertical direction, and replacing the point velocities and other dependent variables with their depth-averaged counterparts. For example, for the velocity components we have:

$$U = \frac{1}{h}\int_{z_b}^{\eta} u dz \quad \text{and} \quad V = \frac{1}{h}\int_{z_b}^{\eta} v dz \tag{6.3}$$

where U and V are the depth-averaged components of the velocity in the x and y directions, respectively (see Fig. 6.2 for definition of the remaining variables). The continuity equation is the most straightforward to integrate and results in

$$\frac{\partial \eta}{\partial t} + \frac{\partial hU}{\partial x} + \frac{\partial hV}{\partial y} = 0 \tag{6.4}$$

The momentum equations can be averaged in the same way, but this time nonlinear terms appear—the full mathematical derivation of the equations is outside of the scope of this text, but the interested reader is directed to [11] for the complete details. Including the Coriolis and pressure terms, whose integration is trivial, the depth-averaged momentum equations become

$$\begin{aligned}\frac{\partial(hU)}{\partial t}+\frac{\partial(hU^2)}{\partial x}+\frac{\partial(hUV)}{\partial y}\\ =F_x+fhV-gh\frac{\partial\eta}{\partial x}-\frac{gh^2}{2\rho_0}\frac{\partial\rho}{\partial x}-\frac{\tau_{bx}}{\rho_0}+\frac{\partial(h\tau_{xx})}{\partial x}+\frac{\partial(h\tau_{xy})}{\partial y}\end{aligned} \tag{6.5}$$

$$\begin{aligned}\frac{\partial(hV)}{\partial t}+\frac{\partial(hUV)}{\partial x}+\frac{\partial(hV^2)}{\partial y}\\ =F_y-fhU-gh\frac{\partial\eta}{\partial y}-\frac{gh^2}{2\rho_0}\frac{\partial\rho}{\partial y}-\frac{\tau_{by}}{\rho_0}+\frac{\partial(h\tau_{xy})}{\partial x}+\frac{\partial(h\tau_{yy})}{\partial y}\end{aligned} \tag{6.6}$$

where f is the Coriolis parameter ($=2\Omega\sin\Phi$, with Ω=angular rate of earth's revolution), Φ is the geographic latitude, F_i is the driving forces ($i=x, y$), ρ_0 is the density of a reference state, and τ_{bi} is the bottom stresses ($i=x, y$).

The above equations are sometimes called the shallow-water equations or the depth-averaged Navier–Stokes equations. The cross-stresses τ_{ij} include viscous friction, turbulent friction, and the nonlinear terms resulting from the vertical averaging process, which are usually called the radiation stresses:

$$\tau_{ij}=\frac{1}{h}\int_{z_b}^{\eta}\left[\nu\left(\frac{\partial u_i}{\partial x_j}+\frac{\partial u_j}{\partial x_i}\right)-\overline{u_i'u_j'}+(u_i-U_i)(u_j-U_j)\right]dz \tag{6.7}$$

In most natural bodies of water, the molecular viscosity terms can be safely neglected in comparison with the turbulence terms. The radiation stresses are often neglected, but they represent important physical phenomena. For example, in bends they are at least partly responsible for shifting the high velocity part of the flow profile from the inner bank at the upstream region to the outer bank at the downstream region of the bend—see [12]. In general, however, the terms of Eq. (6.7) are collapsed in the form of diffusion coefficients and written as

$$\tau_{ij}=D_i\left(\frac{\partial U_i}{\partial x_j}+\frac{\partial U_j}{\partial x_i}-\frac{1}{2}\delta_{ij}\frac{\partial U_k}{\partial x_k}\right) \tag{6.8}$$

where δ_{ij} is the Kronecker delta (=1 if $i=k$, 0 otherwise) and D_i is the diffusion coefficient in the ith direction (in general, $D_1=D_2=D_H$). In turbulent flow, the diffusion coefficients can be prescribed or computed from any of the many existing turbulence models (see [10] for more details), and the bottom shear stresses are assumed to have the same direction of the depth-mean velocity and proportional to the square of its magnitude:

$$\frac{\tau_{bx}}{\rho_0} = C_f u\sqrt{u^2+v^2} \quad \text{and} \quad \frac{\tau_{by}}{\rho_0} = C_f v\sqrt{u^2+v^2} \tag{6.9}$$

where C_f is the standard skin friction coefficient ($C_f \approx 0.003$). Note that Eq. (6.9) can also be written in terms of the Manning's roughness coefficient, n, or in terms of Chézy's roughness coefficient, C:

$$\frac{\tau_{bx}}{\rho_0} = \frac{gn^2}{h^{1/3}} u\sqrt{u^2+v^2} = \frac{g}{C^2} u\sqrt{u^2+v^2} \quad \text{and} \quad \frac{\tau_{by}}{\rho_0} = \frac{gn^2}{h^{1/3}} v\sqrt{u^2+v^2} = \frac{g}{C^2} v\sqrt{u^2+v^2} \tag{6.10}$$

The driving forces remaining in Eq. (6.5) and (6.6) include such effects as atmospheric pressure gradients, wind stresses, density gradients, and tidal stresses. The shallow-water equations can be written in many possible forms. Those forms may include different terms than the ones considered above (corresponding to other physical effects), or may be written in terms of curvilinear coordinates, for example. Many other aspects that are of interest, but that are outside the scope of this chapter, are described with much greater detail in Ref. [13].

3.2 *The Numerical Model*

Analytical solutions for the shallow-water equations, Eq. (6.4, 6.5, and 6.6), are not known except for a very few simplified cases, therefore, application to the complex topology usually encountered in natural topography requires the application of numerical mathematics solution methods. When used in CFD, these methods transform the problem of finding the solution of a system of continuous partial differential equations into that of finding the solution of a much larger set of algebraic equations that can be solved in a computer. The algebraic equations are only approximations of the original differential equations that result from discretizing the continuous space into a finite set of points using approximation theory. There are many different methods to accomplish the discretization, and different choices result in distinct sets of algebraic equations.

The choice of a discretization technique is determined by many different factors, some of which may be subjective. For example, a mathematician may favor finite elements using weighted residual methods, while a physicist may prefer to use control volumes. Residual methods are based on minimizing some sort of error, or residual, of the governing equations and are mathematically very solid and well established, but their derivation is more laborious and loses physical meaning very early in the discretization stage, making them more difficult to implement and to debug the code. The control volume formulation, on the other hand, employs the same thought process generally used to derive the governing equations (i.e., vanishing control volumes) and the resulting algebraic equations maintain terms with physical significance all the way to the programming level, making the entire process easier to grasp and to program correctly without code bugs.

In surface hydraulics, there are other considerations to take into account. For example, in lakes and many rivers, the flow field is generally smooth and without abrupt changes; in fast rivers, however, the flow field changes rapidly and may have discontinuities, such as in areas with hydraulic jumps and regime change. Particularly in catastrophic flooding, such as that caused by dam- and dike-break events, abrupt- and fast-moving wave fronts are produced over dry and irregular beds in the downstream plains and valleys. The numerical techniques needed to compute accurately this latter type of flows are usually more complex than those needed for slow flows because they must yield low numerical dissipation, produce wiggle-free solutions at the discontinuity areas, and be robust enough to survive the regions of vanishing water depth. The numerical model presented in this section was designed having these features as the primary concern.

The control volume formulation is used here to discretize the governing equations. To simplify the mathematics, Eqs. (6.4, 6.5, and 6.6) are first simplified (uniform and constant density, no Coriolis terms) and recast in the conservative form:

$$\frac{\partial \mathbf{U}}{\partial t}+\frac{\partial \mathbf{F(U)}}{\partial x}+\frac{\partial \mathbf{G(U)}}{\partial y}-\left(\frac{\partial \mathbf{P(U)}}{\partial x}+\frac{\partial \mathbf{Q(U)}}{\partial y}\right)=\mathbf{S(U)} \tag{6.11}$$

where **U** is the vector containing the conservative variables, **F** and **G** are the inviscid fluxes, **P** and **Q** are the viscous fluxes, and **S** is the vector containing the forcing (source) terms which are defined as

$$\mathbf{U}=\begin{bmatrix} h \\ hu \\ hv \end{bmatrix},\quad \mathbf{F}=\begin{bmatrix} hu \\ hu^2+gh^2/2 \\ huv \end{bmatrix},\quad \mathbf{G}=\begin{bmatrix} hu \\ huv \\ hv^2+gh^2/2 \end{bmatrix},$$

$$\mathbf{P}=\begin{bmatrix} 0 \\ \nu_t h\dfrac{\partial u}{\partial x} \\ \nu_t h\dfrac{\partial v}{\partial x} \end{bmatrix},\quad \mathbf{Q}=\begin{bmatrix} 0 \\ \nu_t h\dfrac{\partial u}{\partial y} \\ \nu_t h\dfrac{\partial v}{\partial y} \end{bmatrix},\quad \mathbf{S}=\begin{bmatrix} 0 \\ gh\left(S_{0x}-S_{fx}\right)+\dfrac{\tau_{wx}}{\rho_0} \\ gh\left(S_{0y}-S_{fy}\right)+\dfrac{\tau_{wy}}{\rho_0} \end{bmatrix}$$

Here, lowercase letters are now used to define the components of the depth-averaged velocity, v_t is the eddy viscosity coefficient (D_i in Eq. (6.8)), S_0 ($S_{0i} = -\partial z_b / \partial x_i$) is the bed slope, and S_f $\left(ghS_{fi} = c_d u_i \sqrt{u_1^2 + u_2^2}\right)$ is the bottom friction. τ_{wx} and τ_{wy} are the wind stresses at the free surface in the x- and y-directions and are estimated from

$$\tau_{wx} = \rho_a C_D u_w \sqrt{u_w^2 + v_w^2} \quad \text{and} \quad \tau_{wy} = \rho_a C_D v_w \sqrt{u_w^2 + v_w^2}$$

where ρ_a is the density of air, u_w and v_w are the components of the wind speed at a certain prescribed height above the free surface (usually 10 m), and C_D is a dimensionless drag coefficient. The wind forcing terms may be neglected for the cases where the effects of bottom slope and friction are dominant, such as in rivers and laboratory channels, but are usually kept for lakes and reservoirs where wind-driven circulation is important.

For shallow flows, the effective eddy viscosity can be estimated well by the turbulent boundary layer theory, yielding a depth-mean value of $v_t = \varepsilon_t h u_*$, where u_* is the shear velocity $(= \sqrt{\tau_b / \rho})$ and ε_t is a parameter with a theoretical value of 0.068. Usually, $\varepsilon_t \in 0.1, 10$, but in practice, it can vary by several orders of magnitude outside this range—see [14]—and must be calibrated for each situation. Alternatively, a constant value of the turbulent eddy viscosity ε_t can also be prescribed.

The system's conservative formulation as written above, which remains valid across discontinuities in the flow variables such as hydraulic jumps, provides an ideal basis for the integral formulation over a control volume Ω used by traditional control volume methods and is adopted here. Defining $\mathbf{E} = (\mathbf{F}, \mathbf{G})^T$ and $\mathbf{R} = (\mathbf{P}, \mathbf{Q})^T$, Eq. (6.11) is now written in integral form as

$$\frac{\partial}{\partial t}\int_{\Omega} \mathbf{U} d\Omega + \oint_{\partial\Omega} (\mathbf{E}\cdot\mathbf{n}) ds - v_t \oint_{\partial\Omega} (\mathbf{R}\cdot\mathbf{n}) ds = \int_{\Omega} \mathbf{S} d\Omega \tag{6.12}$$

where **n** is the outward-pointing unit vector normal to the control volume boundary ∂Ω, and after applying Gauss' theorem to the flux integral. Equation (6.12) is the form used to discretize the shallow-water equations in the following manner:

$$\frac{\partial \Omega_i \mathbf{U}_i}{\partial t} = \sum_{k=1}^{n_C} \mathbf{E}_{ik} \Delta l_{ik} - v_t \sum_{k=1}^{n_C} \mathbf{R}_{ik} \Delta l_{ik} + \mathbf{S}_i \Omega_i \tag{6.13}$$

where $\mathbf{U}_i$ are the average values of the conserved variables over cell i, $\mathbf{E}_{ik}$ and $\mathbf{R}_{ik}$ are the inviscid and the viscous fluxes through edge k, $\mathbf{S}_i$ contains the source terms, Ω_i is the cell's area, and n_C is the number of sides of the polygonal control volume. The scheme adopted here discretizes the domain in nonoverlapping triangles ($n_C = 3$) and the values of the conserved variables are located at the geometric center of the control volumes. The linear reconstruction operators used, which must satisfy the cell averaging requirements that guarantee cell-wise discrete conservation properties, exploit the fact that the cell average is also a pointwise value of any

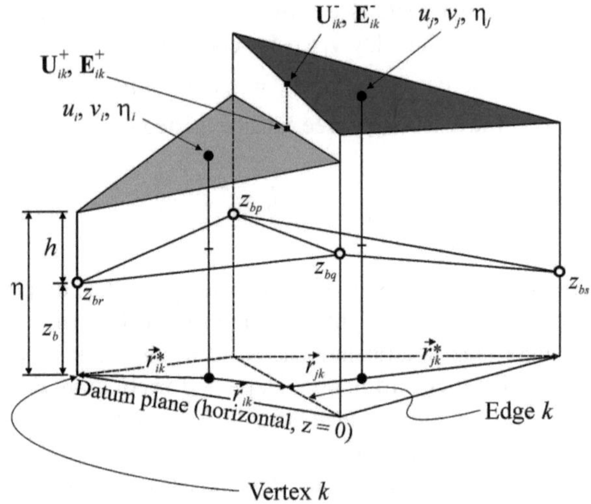

Fig. 6.3 General definition of the control volume geometry and location of the conserved variables. The dependent variables are defined at each triangle's centroid. $\vec{r}_{ik}$ is a vector that points from the centroid of triangle *i* to the midpoint of edge *k*, and $\vec{r}^*_{ik}$ is a similar vector that points to vertex *k*

permitted linear reconstruction evaluated at the centroid of the triangle (this is only true for fully wet triangles). The bed elevation above datum, however, is defined at the triangle vertices. Each triangle defines a control volume, which is used to solve Eq. (6.12). The general configuration of the discretization apparatus is shown in Fig. 6.3, where η is the free surface elevation ($=h+z_b$).

Following the principles of Godunov-type methods, the inviscid fluxes $\mathbf{E}_{ik}$ are numerical fluxes arising from a local Riemann problem at each cell edge. There are several approximate Riemann solvers that were specifically proposed for the shallow-water equations. In this work, $\mathbf{E}_{ik}$ are computed using Roe's [15] flux function at the cell edges:

$$\mathbf{E}_{ik} = \frac{1}{2}\left[(\mathbf{E}^+_{ik} + \mathbf{E}^-_{ik}) - \Gamma(\mathbf{U}^+_{ik} - \mathbf{U}^-_{ik})\right] \tag{6.14}$$

where the "+" quantities are reconstructed at the midpoint of the edge *k* using data from control volume *i* and the "–" quantities from control volume *j* (see Fig. 6.3). There are many forms of the upwinding factor $\mathbf{\Gamma}$, and an exhaustive exposition can be found in [16]. The approach of [17] is used here, which is based on the 1D Riemann problem in the direction normal to the cell edge, and which is included below for completeness. In this approach, $\mathbf{\Gamma}=|\mathbf{A}|=\mathbf{R}|\mathbf{\Lambda}|\mathbf{L}$, where $\mathbf{A}$ is the Jacobian of the flux evaluated at Roe's average state:

$$\mathbf{A} = \frac{\partial(\mathbf{E}\cdot\mathbf{n})}{\partial\mathbf{U}} = \begin{bmatrix} 0 & n_x & n_y \\ (c^2-u^2)n_x - uvn_y & 2un_x + vn_y & un_y \\ (c^2-v^2)n_y - uvn_x & vn_x & un_x + 2vn_y \end{bmatrix} \tag{6.15}$$

The quantities **R**, **L**, and **Λ** are the right and left eigenvectors and eigenvalues of **A**, respectively. The eigenvalues are given by

$$\lambda_1 = un_x + vn_y, \quad \lambda_2 = \lambda_1 - cn, \quad \lambda_3 = \lambda_1 + cn \tag{6.16}$$

where n_x and n_y are the components of the edge's normal vector **n**, n is its magnitude (=1 for a unit normal), and $c=(gh)^{1/2}$ is the wave celerity. The right and left eigenvector matrices are given by

$$\mathbf{R} = \begin{bmatrix} 0 & 1 & 1 \\ n_y & n_y - ucn_x / n & u + cn_x / n \\ -n_x & v - cn_y / n & v + cn_y / n \end{bmatrix},$$
$$\mathbf{L} = \begin{bmatrix} -\left(un_y - vn_x\right)/n & ny / n^2 & -n_x / n^2 \\ \left(un_x + vn_y\right)/2cn + 1/2 & -n_x / 2cn & -n_y / 2cn \\ -\left(un_x + vn_y\right)/2cn + 1/2 & n_x / 2cn & n_y / 2cn \end{bmatrix} \tag{6.17}$$

and Roe's average state is defined as

$$u = \frac{u^+\sqrt{h^+} + u^-\sqrt{h^-}}{\sqrt{h^+} + \sqrt{h^-}}, \quad v = \frac{v^+\sqrt{h^+} + v^-\sqrt{h^-}}{\sqrt{h^+} + \sqrt{h^-}}, \quad c = \sqrt{\frac{1}{2}g(h^+ + h^-)} \tag{6.18}$$

where all the quantities are evaluated at the edge midpoint.

Following [18], matrix **Λ** is evaluated as

$$|\Lambda| = \begin{vmatrix} f(\lambda_1) & & \\ & f(\lambda_2) & \\ & & f(\lambda_3) \end{vmatrix} \tag{6.19}$$

where $f(\lambda)$ is a function of the eigenvalues that incorporates the entropy fix, and is defined as

$$f(\lambda) = \begin{cases} |\lambda|, & |\lambda| \geq \delta_e \\ \dfrac{\lambda^2 + \delta_e^2}{2\delta_e}, & |\lambda| < \delta_e \end{cases} \tag{6.20}$$

where δ_e is a very small number. This prevents any of the eigenvalues to vanish, which causes the dissipation for that component to vanish also at that location and may lead to numerical instability. It also eliminates expansion shocks and makes Roe's flux function differentiable. The above method is particularly appropriate for discontinuous flows, where sharp gradients are important and must be accurately calculated, such as across hydraulic jumps, wet–dry fronts, and in dam-break flows.

Second-order accuracy is achieved using a piecewise linear model for the cell variables with the usual modified upwind scheme for conservation laws (MUSCL) reconstruction of [19], with limiting to enforce monotonicity near sharp gradients and discontinuities of the dependent variables:

$$V_{ik} = V_i + \Phi_i \nabla V_i \cdot \vec{r}_{ik} \tag{6.21}$$

where V_i is the variable (u, v, h) defined at the center of the control volume i, V_{ik} is the same quantity at the midpoint of the edge k (computed from the side of element i), ∇V_i is the gradient of V (piecewise linear) over the control volume, and $\vec{r}_{ik}$ is a position vector located at the centroid of the control volume and pointing to the midpoint of edge k. The flux limiter Φ_i has the objective of preventing the formation of local extrema at the flux integration points. Barth and Jespersen [20] were the first to propose a truly multidimensional limiter for unstructured grids. This limiter has been used by many, but it introduces nondifferentiability to the computation of the reconstructed function and, consequently, to the fluxes, impacting adversely the convergence properties of the solver. Venkatakrishnan [21] resolved this issue by introducing a modification to Barth and Jaspersen's limiter that makes it continuously differentiable. This limiter is applied to triangular cell-centered grids in the following manner:

1. Find the largest negative $(\delta V_{min} = \min(V_i - V_j))$ and positive $(\delta V_{max} = \max(V_i - V_j))$ difference between the solution at the centroid of the triangle, V_i, and that of all its neighbors that share an edge with it, V_j ($j = 1, 2, 3$).
2. Compute the unconstrained reconstruction value at the midpoint of each edge, $V_{ik}^* = V_i + \nabla V_i \cdot \vec{r}_{ik}$.
3. For each edge, compute the maximum allowed value of Φ_{ik}:

$$\Phi_{ik} = \begin{cases} \dfrac{1}{\delta'}\left[\dfrac{\left(\Delta V_{max}^2 + \varepsilon^2\right)\delta' + 2\delta'^2 \Delta V_{max}}{\Delta V_{max}^2 + 2\delta'^2 + \delta' \Delta V_{max} + \varepsilon^2}\right] \text{if}\{\delta > 0 \\ \dfrac{1}{\delta'}\left[\dfrac{\left(\Delta V_{min}^2 + \varepsilon^2\right)\delta' + 2\delta'^2 \Delta V_{min}}{\Delta V_{min}^2 + 2\delta'^2 + \delta' \Delta V_{min} + \varepsilon^2}\right] \text{if } \delta' < 0 \end{cases}$$

4. Choose $\Phi_i = \min(\Phi_{ik})$.

In step 3, $\Delta V_{max} = \delta V_{max} - V_i$, $\Delta V_{min} = \delta V_{min} - V_i$, δ' is given by $\delta' = sign(V_{ik}^* - V_i)(|V_{ik}^* - V_i| + \delta)$, and δ is a very small number to avoid division by zero. ε is a parameter that controls over- and undershoots and depends on the estimate of a length scale Δx, $\varepsilon = (K\Delta x)^3$. In the present work, K is set to 0.075 (found by numerical experimentation) and Δx is a local mesh length scale set to the average length of the edges of the element.

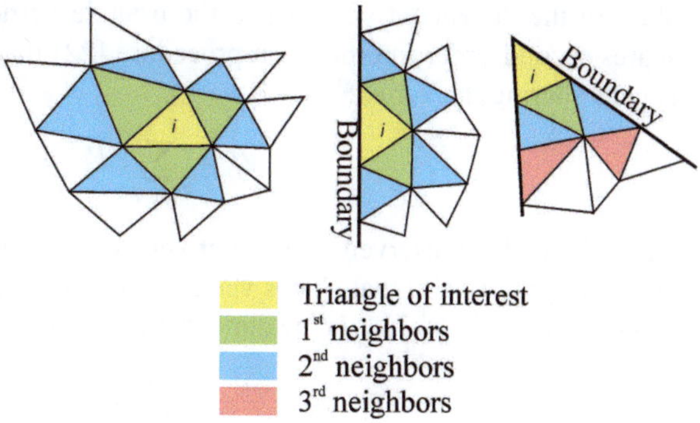

Fig. 6.4 Depiction of first, second, and third neighbors to a computational cell (cell *i*, in *yellow*) for different geometries. The *colored* area shows the stencil used in each case and the empty cells, which are third neighbors or higher, do not contribute to the computational cell

Computation of the gradients ∇V_i is done using a least squares technique which requires the solution of a two-by-two system for each cell *i* of the computational domain:

$$\begin{cases} aV_x + bV_y = d \\ bV_x + cV_y = e \end{cases} \tag{6.22}$$

where V_x and V_y are the Cartesian components of ∇V_i and

$$a = \sum_{j=1}^{n_T} w_j^2 (x_j - x_i)^2, \quad b = \sum_{j=1}^{n_T} w_j^2 (x_j - x_i)(y_j - y_i)$$

$$c = \sum_{j=1}^{n_T} w_j^2 (y_j - y_i)^2, \quad d = \sum_{j=1}^{n_T} w_j^2 (V_j - V_i)(x_j - x_i)$$

$$e = \sum_{j=1}^{n_T} w_j^2 (V_j - V_i)(y_j - y_i)$$

where n_T is the number of triangles belonging to the computational molecule. All of the above terms depend solely on the grid geometry; therefore, they can be precomputed and stored for later use by the iterative solution cycle. The system is solved directly using Cramer's rule and the inverse distance weighting is used $(w_j = ((x_j - x_i)^2 + (y_j - y_i)^2)^{-1/2})$.

The least-squares procedure thus defined requires a minimum of three control volumes, including *i*, to form a determined system. It is desirable to exceed that number by building stencils by successively adding first, second, and third neighbors until a desired size is reached. In particular, the stencils near boundaries may require farther neighbors. The additional data contained in the expanded stencil allows filtering out noise at the expense of little additional computational cost. Some of the typical stencils used are shown in Fig. 6.4.

The viscous fluxes $\mathbf{R}_{ik}$ are discretized using a central differencing approach. The viscous flux terms are computed at the edge midpoints, where *u*, *v*, and their gradients must be calculated. The velocity components at the edge midpoints are simply $V_k = (V_{k1} + V_{k2})/2$, where V_k is the velocity of interest (*u* or *v*) at the midpoint of edge *k*, and V_{k1} and V_{k2} are its values at the edge's extremities (the triangle vertices). The values of the dependent variables at the triangle vertices are obtained using least squares second-order interpolation procedure [22] that uses the information of all the cells sharing the vertex:

$$V_l^V = \sum_{Neighb} c_i V_i \tag{6.23}$$

where V_l^V is the conserved variable at vertex l, V_i denotes the conserved variable at the centroid of cell i that shares vertex l, and the sum is carried over all the cells sharing vertex $\left(x_l^V, y_l^V\right)$. c_i is the dimensionless coefficient that is defined by

$$c_i = \frac{w_i}{\sum\limits_{Neighb} w_i} \tag{6.24}$$

with

$$w_i = 1 + \lambda_x\left(x_i - x_l^V\right) + \lambda_y\left(y_i - y_l^V\right)$$
$$\lambda_x = \left(I_{xy}R_y - I_{yy}R_x\right) / \left(I_{xx}I_{yy} - I_{xy}^2\right), \quad \lambda_y = \left(I_{xy}R_x - I_{xx}R_y\right) / \left(I_{xx}I_{yy} - I_{xy}^2\right)$$
$$R_x = \sum_{Neighb}\left(x_i - x_l^V\right), \quad R_y = \sum_{Neighb}\left(y_i - y_l^V\right)$$
$$I_{xx} = \sum_{Neighb}\left(x_i - x_l^V\right)^2, \; I_{yy} = \sum_{Neighb}\left(y_i - y_l^V\right)^2, \quad I_{xy} = \sum_{Neighb}\left(x_i - x_l^V\right)\left(y_i - y_l^V\right)$$

All the quantities used in the interpolation can be computed in a preprocessing stage and stored for later use, making the method extremely fast and efficient. Because all the limiting is applied to the edge midpoints via MUSCL reconstruction (as discussed in previous paragraphs), it may appear that extrema could occur at the vertices and that some sort of limiting constraint would be required when applying the above interpolation technique. In [23], it is shown that this is not so and that, in fact, edge midpoint limiting is sufficient to ensure positivity of the cell averaged data without imposing special requirements to the quality of the computational grid.

The computation of the flow gradients at the edge midpoints follows the diamond method of Coirier [24] adapted for triangular unstructured grids. In this approach, the computational molecule associated with each edge is formed by the edge's end points and the centers of the control volumes that share that edge, as shown in Fig. 6.5. The method uses linear interpolation to determine the value of the gradient of the dependent variables over each of the shaded triangles, in a manner that is identical to the technique used in finite element interpolation (i.e., shape functions). Then, the gradient at the edge midpoint is given by the area-averaged value of the same gradients over each of the triangles (shaded areas in Fig. 6.5).

Using the notation of Fig. 6.5, the gradient of the dependent variables at the midpoint of edge k is given for interior edges as

$$\begin{aligned}
\frac{\partial V_k}{\partial x} &= \frac{1}{2(A^+ + A^-)}\left[(y_4 - y_3)(V_1 - V_2) + (y_1 - y_2)(V_3 - V_4)\right] \\
\frac{\partial V_k}{\partial y} &= \frac{1}{2(A^+ + A^-)}\left[(x_3 - x_4)(V_1 - V_2) + (x_2 - x_1)(V_3 - V_4)\right]
\end{aligned} \tag{6.25}$$

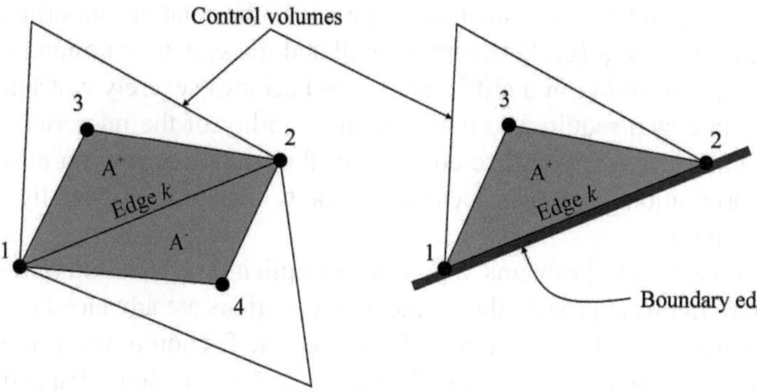

Fig. 6.5 Computational molecule used in the calculation of the viscous fluxes

and for boundary edges as

$$\begin{aligned}\frac{\partial V_k}{\partial x} &= \frac{1}{2A^+}\left[(y_2 - y_3)V_1 + (y_3 - y_1)V_2 + (y_1 - y_2)V_3\right]\\ \frac{\partial V_k}{\partial y} &= \frac{1}{2A^+}\left[(x_3 - x_2)V_1 + (x_1 - x_3)V_2 + (x_2 - x_1)V_3\right]\end{aligned} \tag{6.26}$$

where A^+ and A^- are the areas of the shaded triangles (which are equal to one third of the area of the respective control volumes). Note that points 3 and 4 are the control volume centroids, therefore the dependent variables there are a direct result of the solution of the governing equations. However, the vertex values of the same variables (points 1 and 2) are a result of the least squares second-order interpolation described in the preceding paragraphs. Consequently, the discretization of the viscous terms is also second-order accurate.

Typical environmental flows are subject to multiple forcing factors, such as bottom friction, bed slope, wind forcing, Coriolis forces, and tidal forces. It is, therefore, important to include accurate representations of these forcing terms—the source terms of Eq. (6.12)—in the conceptual and numerical model. In this section, only bed slope and bottom friction are considered, but the other terms are discretized using similar, straightforward, centered difference schemes.

Bed friction is represented by Eq. (6.9). For most circumstances, a simple treatment of these terms is sufficient. Using a standard cell-centered approach that uses only information at the cell centroids results in

$$\int_\Omega ghS_{fx}d\Omega = C_f u_i\sqrt{u_i^2 + v_i^2}\,\Omega_i \quad \text{and} \quad \int_\Omega ghS_{fy}d\Omega = C_f v_i\sqrt{u_i^2 + v_i^2}\,\Omega_i \tag{6.27}$$

where the subscript i indicates quantities evaluated at the centroid of control volume i. The main numerical problem associated with this type of formulation arises for very shallow depths, such as near advancing and receding wet–dry fronts. In very shallow water, a dimensional analysis shows that the relative importance of the convective terms in Eq. (6.11) becomes small and the system is no longer convection dominated, resulting in a stiff system that becomes severely restricted by the very small time step required to maintain the stability of the numerical solution. In this circumstance, explicit discretization of the friction terms can also produce numerical oscillations and even localized velocity inversion, especially when the roughness is high.

To circumvent these problems, a pointwise implicit discretization of the friction terms is used. In this approach, the momentum equations are advanced explicitly in time (from time step n to time step $n+1$) without the friction terms, producing the intermediate unit discharges q'_{xi} and q'_{yi} in the x- and y-directions, respectively (the unit discharges are defined as $q_{xi} = h_i u_i$ and $q_{yi} = h_i v_i$, and the subscript i denotes the control volume number). Then, the final unit discharges are updated using

$$\begin{aligned} q_{xi}^{n+1} &= \frac{q'_{xi}}{1+g\Delta t(S'_{fxi}/u'_i)} \\ q_{yi}^{n+1} &= \frac{q'_{yi}}{1+g\Delta t(S'_{fxi}/v'_i)} \end{aligned} \tag{6.28}$$

where the superscript $n+1$ denotes the time step. All the primed variables refer to quantities computed from the frictionless updated values of u and v in the time step.

Discretization of the bed slope term within the context of Godunov-type solvers has been a topic of much research in the past decade. This stems from the fact that, when using cell-centered control volumes with a MUSCL reconstruction technique, it is necessary to properly discretize the source term in order to satisfy the C-property, i.e., to ensure the balance between the flux gradient and the slope term in order to guarantee hydrostatic balance in still water conditions—e.g., see [25]. To circumvent this issue, the bed elevation, z_b, is defined at the control volume vertices instead of at its centers, as shown in Fig. 6.3. Using this approach, there is no discontinuity in the bed elevation across control volumes and the Riemann problem retains its self-similarity solution, therefore a simple centered differencing approach can be used, as indicated in [16]. Bed elevation is piecewise linear within the control volumes and its slope can be calculated using the standard interpolation techniques used in finite element methods.

Finally, the system of governing equations must be integrated in time. The choice here is to advance the solution explicitly in time using nonlinear strong stability preserving Runge–Kutta (SSPRK) schemes, also known as total variation diminishing (TVD) Runge–Kutta schemes [26]. This is done by first rewriting the governing equations as a coupled system of ordinary differential equations:

$$\Omega_i \frac{\partial q_i}{\partial t} = R_i(u, v; t), \quad i = 1, 2, 3 \tag{6.29}$$

Table 6.3 Values of the SSPRK coefficients for the schemes used. For $m=1$, the method reduces to the traditional forward Euler method

Order (m)	i ($i=1,\ldots,m$)	α_{ij} ($j=0,\ldots,i-1$)	β_j ($i=1,\ldots,m$)
1	1	1	1
2	1	1	1
	2	$\frac{1}{2}, \frac{1}{2}$	$\frac{1}{2}$
3	1	1	1
	2	$\frac{3}{4}, \frac{1}{4}$	$\frac{1}{4}$
	3	$\frac{1}{3}, 0, \frac{2}{3}$	$\frac{2}{3}$

SSPRK Strong Stability Preserving Runge–Kutta

where R_i is called the residual. Here, a simplified form of the SSPRK schemes is used, in which an m-stage SSPRK method for Eq. (6.29) is written in the form

$$\begin{cases} u^{(0)} = u^n \\ u^{(i)} = \sum_{j=0}^{i-1} \alpha_{ij} u^{(j)} + \beta_i \dfrac{\Delta t}{\Omega_i} R(u^{(i-1)}), \quad \alpha_{ij} \geq 0, \quad i = 1, \ldots, m \\ u^{n+1} = u^{(m)} \end{cases} \tag{6.30}$$

where Δt is the time step size, the superscript n denotes the time level, and the parenthetic superscripts denote the Runge–Kutta level. The coefficients α_{ij} and β_i are chosen to meet desired criteria. Second- ($m=2$) and third-order ($m=3$) optimal SSPRK methods, in the sense of the Courant–Friedrichs–Lewy (CFL) coefficient θ, are used—see [26]. The values of the coefficients are given in Table 6.3. For time-dependent cases, Δt is prescribed or it is computed from

$$\Delta t = \theta \text{Min} \left\{ \frac{l_k}{\lambda_k^*}, \frac{l_k^2}{v_k} \right\}_{k=1,\ldots,N_{we}} \tag{6.31}$$

where l_k is the length of edge k, λ_k^* is the highest eigenvalue at the edge's midpoint, v_k is the effective eddy viscosity at the same location, and N_{we} is the number of wet edges over the entire computational domain. In the latter case, θ must be prescribed. For inviscid computations, the terms containing v_k in Eq. (6.31) are dropped and only the terms containing λ_k^* are considered. Note that the presence of source terms places additional restrictions on the maximum admissible time step that preserves stability. These source terms are problem dependent; therefore, the CFL condition must be considered a general guideline and the maximum time step must be determined through numerical experimentation, by prescribing either θ or Δt.

Although for time-dependent calculations the solution must be advanced in time in a physically consistent and accurate manner, in steady-state calculations it does not, because only the converged solution needs to retain physical meaning and be consistent with Eq. (6.11). This observation can be used to devise techniques to speed convergence. For example, the Runge–Kutta scheme could use coefficients that emphasize convergence rather than accuracy. For example, two convergence acceleration techniques can be employed: local time stepping and implicit residual smoothing. Local time stepping advances the solution at each cell using a time step close to the stability limit for that cell:

$$\Delta t_i = \theta \text{Min} \left\{ \frac{l_k}{\lambda_k^*}, \frac{l_k^2}{\nu_k} \right\}_{k=1,2,3} \tag{6.32}$$

where the subscript i denotes the cell index and the subscript k denotes the edges that belong to that cell. This technique has the objective of speeding up the transport of information within the computational domain and results in an increase in the convergence rate of explicit schemes by a factor close to two—see [27].

The implicit residual smoothing suggested in [28] is another technique used to increase the maximum allowed time step. This is accomplished by enlarging the support of the scheme by averaging the residuals R_i with their neighbors by means of a smoothing operator:

$$\overline{R}_i = R_i + \kappa \sum_j (\overline{R}_j - \overline{R}_i) \tag{6.33}$$

where κ is a constant and the summation index j spans over all the neighboring elements that share an edge with element i. Using a suitable choice for κ (1/2 in this work), the resulting system is strongly diagonally dominant and can be easily solved using Jacobi iteration:

$$\overline{R}_i^{(m)} = \left(R_i + \kappa \sum_j \overline{R}_j^{(m-1)} \right) \left(1 + \kappa \sum_j 1 \right)^{-1} \tag{6.34}$$

In practice, the system does not need to be solved exactly and two or three iterations are sufficient to produce an adequately accurate approximation of $\overline{R}_i$ at the cell centers. Residual averaging is performed at every step of the Runge–Kutta marching procedure.

3.3 Numerical Implementation Considerations

Wetting and drying occur not only during the propagation of floods but also at the edges of any body of water. Thus, the dry–wet front constitutes not only a propagation problem but also a static boundary condition problem, because it defines the shoreline. It is not easy to include these effects in a straightforward manner in a numerical code and most researchers resort to different degrees of approximation. The difficulties associated with this problem were presented with detail by Horritt in [29], which also included a review of some of the approaches used by different researchers. Another review of several methods can be found in [30]. Here, a compromise between accuracy and computational efficiency is sought. The overall treatment of wet–dry lines impacts directly the quality of the solution: Mass conservation, unphysical solution behavior, and numerical stability are the major aspects of concern.

Advancing (wetting) fronts, such as those encountered during catastrophic flood events, are treated directly by the Riemann solver in Eqs. (6.15, 6.16, 6.17, and 6.18) and the numerical scheme is robust enough to treat the sharp gradients in these areas without difficulty. However, the standard Riemann solution does not apply to the dry bed problem and can result in the wrong prediction of the front velocity. A fix presented in [16] consists in wetting the dry bed of the receiving cell by some very small value δ_w. For $\delta_w > 0$, the propagating speed is not the same as for the true dry bed situation, but in the limit $\delta_w \rightarrow 0$ the two speeds coincide. In practice, due to the robustness of the numerical methods used, a value close to the machine zero can effectively be used for δ_w (currently, 10^{-12} is used).

Drying fronts are treated similarly, but they pose the additional problem that, during a drying time step, negative water depths may be reached. Mass conservation requires that the time step should be restricted to the value that corresponds to the time that takes the cell to dry out, i.e., to reach $h_i = 0$. If a control volume reaches negative depth, additional mass is introduced in the system at the receiving control volume(s), and the excess volume is equal to the "negative" volume that corresponds to the negative depth. In practice, it is too complex and time consuming to keep track of the required time step size for each control volume, which also could result in significant computational costs if too many restrictions were to be imposed on the maximum time step allowed. It is desirable to run unsteady phenomena with the largest time step possible, therefore, a less restrictive approach was chosen. After each time step, all the control volumes with negative depths are treated in the following manner: The water depth is set to zero and the amount of fluid corresponding to the "negative" volume is removed from the downstream-most wet control volume adjacent to it. If that control volume does not have enough water, the remaining volume is removed from its own downstream-most adjacent control volume (the adjacent control volume that has the lowest centroid bed elevation), and so on until all excess mass is accounted for. The process continues for every control volume with negative depth until the total mass balance is restored. A list of downstream-most elements is constructed for each control volume as part of a preprocessing stage and its computational cost is negligible.

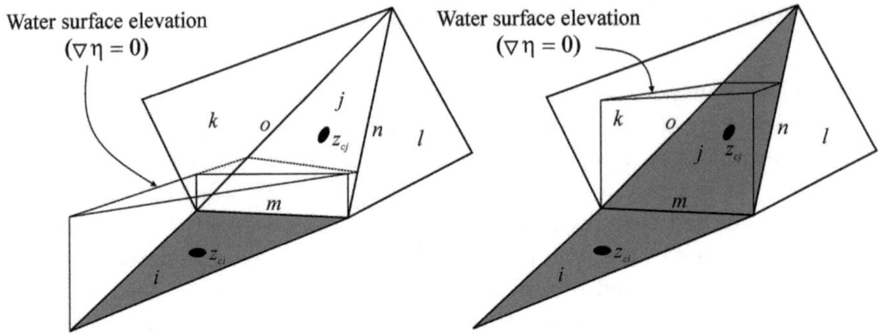

Fig. 6.6. Shoreline definition sketch. *Gray triangles* are wet control volumes. Control volumes are denoted by the letters *i, j, k,* and *l,* edges by *m, n,* and *o*. The *black dots* show the locations of the centroids of triangles *i* and *j*. On the *right*, the water-surface elevation in control volume *i* is not shown to improve clarity

The shoreline treatment is different from the two preceding cases. A shoreline is defined when all the surrounding dry triangles of a partially or fully wet control volume *i* have a mean bed elevation higher than the stage at the centroid of triangle *i*. Under this circumstance, the shoreline is defined at the control volume edges and it is treated identically to a solid wall (described later in this section): The flow velocity is set to zero and no mass can cross it. Additionally, the gradient of the water-surface elevation is set to zero in those control volumes.

Two typically occurring types of shoreline are shown in Fig. 6.6, where z_c denotes the bed elevation at the control volumes' centroid. At left, element *i* is fully wet, but the water-surface elevation is below z_{cj}, therefore, the shoreline is located at edge *m*. The gradient of the water-surface elevation in *i* is set to zero. Element *j* will become wet when the water-surface elevation in triangle *i* becomes higher than z_{cj}. On the other hand, at the right of Fig. 6.6, both elements *i* and *j* are wet. Assuming that control volumes *l* and *k* are dry, the shoreline is now located at edges *n* and *o*. The water surface is flat at control volume *j*, but not at *i* (assuming that all control volumes surrounding *i* are wet).

The decision whether a control volume is wet or dry is given by a threshold value of the water depth, h_{dry}: If $h_i > h_{dry}$ then control volume *i* is wet, otherwise it is dry. Hysteresis is introduced to prevent the triggering of instabilities, such as what might result from control volumes that become alternatively dry and wet over multiple successive time steps. This is done in the following manner: If a control volume *i* is wet, it becomes dry when h_i falls below h_{dry}; if *i* is dry, it becomes wet when h_i rises above h_{wet}. A hysteresis factor is used: $h_{wet} = \varphi h_{dry}$, where φ is set to a positive number greater than one. Note that the momentum equations are not solved for dry triangles, but the continuity equation is, which allows the continual computation of h even when $h < h_{dry}$, while simultaneously ensuring mass conservation. The threshold h_{dry} is a user set parameter, which is usually taken in the range 10^{-4}–10^{-8}.

Partially dry cells have further corrections to their area and water depth. When applying the time marching procedure to Eq. (6.13), it is assumed that $\partial\Omega_i / \partial t = 0$

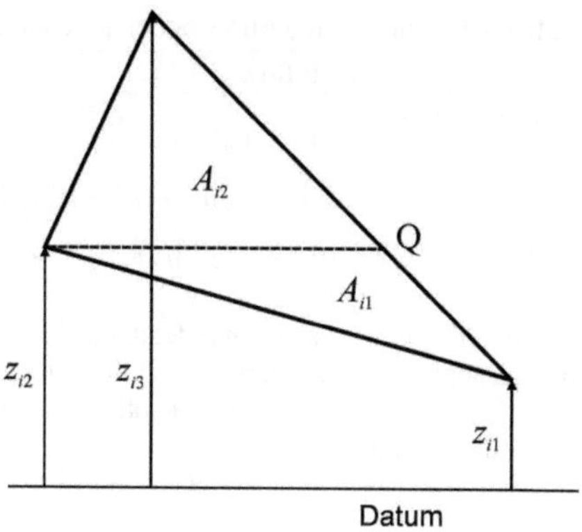

Fig. 6.7 Subdivision of a computational cell into two subtriangles for wetted area computations in partially dry cells. Note that $Q = (x_Q, y_Q, z_Q) = (x_Q, y_Q, z_{i2})$

in Eq. (6.29). In drying or wetting cells this is not true. In order to keep the computational complexity and cost low, instead of considering the time-dependent area term, the total area is simply replaced by the wetted area of the element. This area can be calculated very quickly and efficiently using the following procedure: Sort all the triangle vertices according to their vertical coordinate and renumber them appropriately, such that $z_3 \geq z_2 \geq z_1$ (see Fig. 6.7); divide the computational element in two triangles using a line connecting vertex z_2 with the opposing edge; the coordinates of point z_Q can be easily computed using the parametric equations of a line in three dimensions, namely

$$\begin{aligned} z_Q &= z_2 = z_1 + p(z_3 - z_1) \\ x_Q &= x_1 + p(x_3 - x_1) \\ y_Q &= y_1 + p(y_3 - y_1) \end{aligned}$$

where p is a parameter between 0 and 1, given by $p=(z_2 - z_1)/(z_3 - z_1)$. Now, compute the areas of the triangles A_{i1} and A_{i2}; the wetted area Ω_i^{w} can then be computed by linear interpolation:

$$\Omega_i^{w} = \begin{cases} \dfrac{\eta_i - z_{i1}}{z_{i2} - z_{i1}} A_{i1} & \text{if} \quad \eta_i < z_2 \\ \dfrac{\eta_i - z_{i2}}{z_{i3} - z_{i2}} A_{i2} + A_{i1} & \text{if} \quad \eta_i \geq z_2 \end{cases}$$

where η_i is the water-surface elevation at the element. Similarly, in a partially dry cell, the water depth is not represented well by the water depth at its centroid. An effective water depth is used instead, which is computed in the following way:

$$h_i^{eff} = \begin{cases} \dfrac{1}{3} h_{i1} & \text{if}\, \eta_i < z_{i2} \\ \dfrac{1}{3}\left(h_{i1} + h_{i2}\right) & \text{if}\, \eta_i \geq z_{i2} \end{cases}, \tag{6.35}$$

Table 6.4 Inflow and outflow boundary conditions

	Inflow	Outflow
Subcritical flow	$u^- = u_B;\ \ v^- = 0;$ $h^- = (\sqrt{h^+} - (u^- - u^+)/2\sqrt{g})^2$	$h^- = h_B;\ \ v^- = v^+;$ $u^- = u^+ + 2\sqrt{g}(\sqrt{h^+} - \sqrt{h^-})$
Supercritical flow	$h^- = h_B;\ \ u^- = u_B;\ \ v^- = 0$	$h^- = h^+;\ \ u^- = u^+;\ \ v^- = v^+$

u^-, v^-, h^-, velocity and water depth outside the computational domain, where the boundary conditions apply. u is the component of the velocity normal to the edge and v is parallel to the edge
u^+, v^+, h^+, velocity and water depth inside the computational domain, obtained by solving the governing equations
u_B, h_B, prescribed boundary conditions for the (normal) velocity and the water depth

where h_{i1}, h_{i2}, and h_{i3} are the water depths at vertices z_{i1}, z_{i2}, and z_{i3}, respectively, computed by the center-to-vertex interpolation procedure of Eq. (6.23) and (6.24). All the relevant computational parameters (z_{i1}, z_{i2}, and z_{i3}, A_{i1}, and A_{i2}) can be sorted and/or computed in a preprocessing stage and stored for later use. Note also that, using this approach, h_i^{eff} will approach h_i (the water depth when all three vertices are wet) in a smooth and continuously differentiable manner, which is a desirable property to avoid numerical instability and speed convergence. The same can be said about Ω_i^w and Ω_i.

There are several limitations of the approach described above. For example, all the wet and dry computations are done at the cell edges, i.e., at the level of the cell flux computations, following an edge-based data structure described later in this section. This results in the approximation that a wet–dry boundary is constrained to cell edges and, therefore, it is dependent on the computational mesh. Balzano [30] has shown that this is a feature common to many methods and is remediated by using mesh refinement in those areas. Flooding of dry areas can never happen faster than one grid size per time step, which may pose restrictions to the time step size, due to the explicit nature of the time marching schemes used. Finally, on partially wet edges, the computation of mass and momentum fluxes is still only an approximation, albeit one that preserves mass and momentum. However, the approximation error is not severe because, in order to be considered wet, an edge must always be dry for less than half its length.

All boundary conditions are applied at the edges of the model grid, consistently with the edge-oriented data structure used (see the paragraphs below). They use Riemann invariants and are applied in a similar manner as used by [31], i.e., the boundary fluxes are also computed by solving a Riemann problem between the interior states and the "ghost" states outside the computational domain. These "ghost" states are introduced in order to compute the boundary fluxes in a similar and consistent way to the interior fluxes. The inflow and outflow boundary conditions are summarized in Table 6.4. Note that the value of the velocity boundary condition appearing in Table 6.4 is the normal inflow velocity u_B, but in practice the discharge is the desired quantity to enforce. This is accomplished by a fixed-point iteration that computes the values of h^- and u^- successively until the desired value of the discharge is achieved. Usually two or three iterations are sufficient to match the prescribed discharge with less than 1 % of error. Additionally, it was found that convergence to steady state was improved if the stage was also enforced at the center of the control volumes at subcritical outflow boundaries. This was implemented using a relaxation parameter, instead of simply clamping the water-surface elevation at those triangles. Using the values in Table 6.4, all the fluxes can be computed at the inflow and outflow edges as if they were internal edges.

At solid walls and shorelines, a no-slip condition is used. The velocity is set to zero there and the inviscid fluxes are calculated directly, i.e.,

$$\mathbf{E}_{ik}\Delta l_{ik} = \left[0;\ \frac{1}{2}g(h^-)^2 n_{xk};\ \frac{1}{2}g(h^-)^2 n_{yk}\right]^T \Delta l_{ik}$$

where Δl_{ik} is the length of the wall edge, and n_{xk} and n_{yk} are the components of the normal unit vector pointing towards the outside of the control volume. In the simplest formulation the water depth is set as $h^- = h^+$, but the Riemann invariant can also be used: $h^- = \overline{\vec{u}^+ \cdot \hat{n}_k + 2\sqrt{gh^+}}^{\,2} / 4g$. This approach does not require the solution of the Riemann solver at this edge type, therefore, it is expeditious without losing accuracy.

For inviscid flow computations, free-slip wall boundary conditions are employed, which is done by setting a "ghost" state outside the computational cell with $h^- = h^+$, $u^- = 0$, and $v^- = v^+$, and solving the Riemann problem.

The appeal of unstructured grids resides in the flexibility of discretization of physical domains with complex geometry, and in the ease with which local mesh refinement and adaptation can be implemented. However, compared with traditional structured grids, the application of unstructured grids comes at an increased computational cost. This is due to the natural "lack of structure" suggested by the moniker, in which grid nodes are arbitrarily located over the computational domain, lacking distinct grid directions and structure. Grid nodes are combined into polygons (control volumes), but the number of polygons sharing a particular node usually varies from node to node. Connectivity tables must be established to provide the geometric links between computational nodes, edges, and control volumes necessary to carry out the algebraic integrations resulting from the discretization of the governing equations. Unfortunately, traditional computing machines perform much better if the data in their memory banks are accessed in an orderly, sequential manner. Accessing and storing data in a random way, such as what is usually required in unstructured grid solvers, result in an increased penalty that can be significant [32].

In a preprocessing stage, a reverse Cuthill–McKee reordering of the triangulation of the computational domain is performed [33]. The reordering is applied to the vertices of the triangulation. Then, reordering of the triangles' and edges' numbering is carried out following the numbering of the vertices: Starting at vertex #1, all the triangles (and edges) that share it are numbered sequentially; proceed to vertex #2 and continue numbering the triangles (and edges) that remain unnumbered; in ascending order of vertex number, and sequentially, continue until all the triangles (and edges) have been renumbered. This results in a numbering sequence that substantially reduces the overhead due to indirect memory addressing, as compared to the numbering that results from the triangulation software (i.e., from the automatic grid generators). All the data structures and look-up tables containing the linkages among nodes, edges, and control volumes are built using the newly renumbered computational grid.

Additionally, the solution methods were implemented in a computer program using an edge-based data structure. The use of edge-based data structures leads to codes with reduced computer processing unit (CPU) and memory access overhead when compared to codes that use a more traditional element-based structure. Different types of edges are classified and distributed to different, independent computational loops (dry edges, fully wet edges, partially wet edges, solid boundary edges, inflow and outflow edges, advancing front edges, receding front edges, and bank shoreline edges). The main solution cycles over the edges and the residuals are summed by scattering (antisymmetrically) the fluxes to the control volumes sharing the edge. The use of different cycles for different types of edges allows elimination of data dependencies, which results in highly optimized code in vector–parallel computers. It also allows to fully eliminate conditional statements inside those cycles, therefore, paving the way for an efficient implementation in programmable graphics hardware architectures [34].

4 Deploying the Model

Model deployment is a process with the objective to bridge the gap between model development and model application to real physical systems. This is a broad subject where controversy can easily arise, but here the discussion is limited to only a few of the relevant factors in model dissemination. The most important factors are transparency, time, modularity, and quality. Model transparency addresses the issues involved in transmitting to the user all of the relevant information concerning the model's purpose, formulations, and assumptions or, in other words, model documentation. Model documentation must inform the user about range of applicability and allow critical peer evaluation of the model and its applications. Unfortunately, a unified and definitive modeling terminology does not exist, which can make communication between model developer and user ambiguous and even difficult; therefore, particular care must be taken when producing model documentation. Documentation should be complete and include the mathematical formulation of the model and the assumptions on which it is based; the model's parametrizations and suggested parameter values and ranges, as appropriate; and the operating instructions of the model implementation in its computer software package.

The term "time" refers to the time it takes to complete a modeling project, and the purpose is to deliver a complete study within a reasonably short period of time—say, in months rather than in years. Identifying and automating the most repetitive and time consuming tasks is, therefore, of paramount importance. Data preparation has traditionally been one of the most time-consuming tasks in computer modeling, now of increased importance due to the use of digital terrain models (DTMs). In fact, the use of DTMs has become standard in many geomorphological applications, becoming a tool of choice for the extensive data requirements demanded by a complex model, such as the one described in the previous section. Increasingly available is high-resolution topographic data derived from airborne remote sensing, such as interferometric synthetic aperture radar [35], aerial digital photogrammetry [36], LiDAR [37], and terrestrial laser scanning surveys [38]. With resolutions of 1 m and higher available for even very large areas, a survey may easily contain many million points. It is a challenging task to handle such large amounts of data, while at the same time merging sets from different sources (perhaps in different coordinate systems and storage formats) while keeping a tight quality control.

Another time-consuming task is model calibration. Model calibration is necessary because some quantities cannot be directly measured in the field. An example is channel bed roughness: An indirect measurement of bed friction is instead obtained by measuring the slope of the water free-surface elevation for specific known water discharges. In this case, the calibration process has the objective of finding the correct parameters to use for bed roughness, consisting in a trial-and-error cycle in which the model operator guesses successive values for the roughness parameters until the modeled water surface matches the measured slopes. Other examples of quantities that may need calibration are turbulence parameters and diffusion coefficients. As the number of parameters needing calibration increases, the longer and more tedious the calibration exercise becomes. To streamline the process, it is important to provide quick and easy ways to enter model parameters (many of which are spatially varying) and to analyze computation results.

"Modularity" consists in the ability to use and reuse previously developed work and, possibly, to use and link to the work of others. This requires the definition and use of data transmission and storage standards. These standards should be machine independent, provide durability (the standards should not be temporary and fall from use in a short amount of time), and be widely employed, preferably across multiple disciplines (which also contributes to durability). Another advantage of using standards is the potential disambiguation in data interpretation and the facilitation of quality control. Modularity, therefore, also contributes to reducing model development time and cost, as well as providing a means for project archival.

Finally "quality" refers not only to the data and model but also to the fact that there is the need to produce results of proven quality that can be replicated by others. This objective is better met in a framework that can be used by many and, therefore, that can be applied to a wide range of different models and, preferably, in multidisciplinary projects.

The above factors must be addressed through complex integrated modeling frameworks. One such framework developed specifically for environmental flow modeling is given by the iRIC Project (http://i-ric.org/en/), which provides means to integrate diverse models within the same graphical user interface (GUI) using the same data formats and protocols. The iRIC framework provides operational facilities that are model independent, such as data input and output (multiple formats

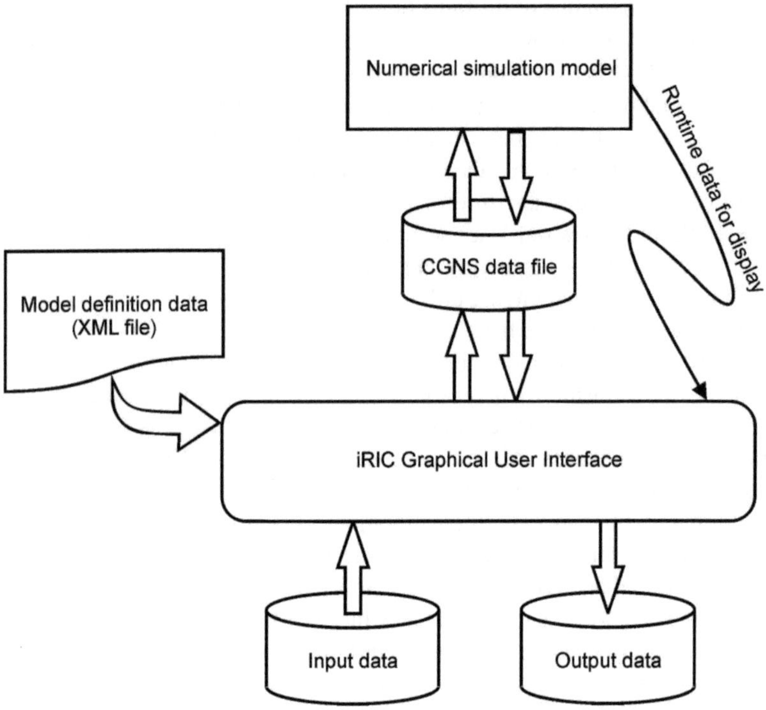

Fig. 6.8 Schematic outline of the integration of a numerical model in the iRIC graphical modeling framework

are supported), automatic grid generation, interactive visualization and editing of model input and output, ability to work with ancillary data sets for model calibration, and device-independent plotting. This functionality frees the numerical model from all of these concerns by separating the roles of model developer from those of GUI programmer, with consequential benefits to both.

A schematic view of how the iRIC software works is given in Fig. 6.8. A GUI is used as the user's front-end. The GUI communicates with the model through a device-independent file using a format that has become a standard in many applications of CFD (CGNS, see http://cgns.sourceforge.net/). Runtime information can also be displayed in a console window. The model parameter definitions needed to customize the GUI to the desired model are coded in a flat file in XML format (http://www.w3.org/XML/). The GUI can read data in a multitude of formats commonly used in hydraulics and other DTM applications. Entire GUI setups, including input data, parameter definitions, and model simulation output, can be saved in individual data files for later use, and for transmission and archival. Besides data reutilization, the GUI also provides simpler and shorter training requirements for users desiring to use different models within the same framework.

As an illustrating example, let us consider one of principal modeling preprocessing tasks (i.e., operations needed to be accomplished before the model can be run):

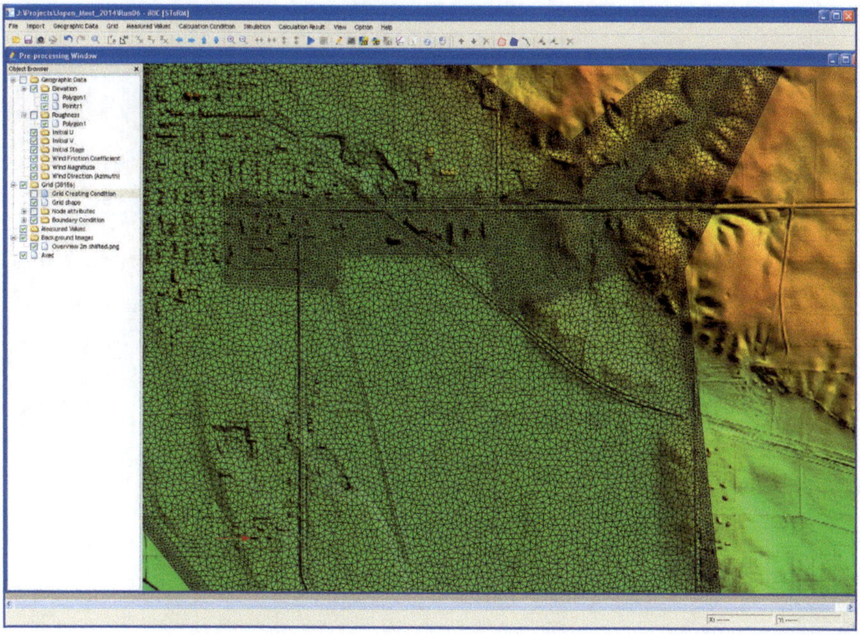

Fig. 6.9 iRIC GUI showing an automatically generated unstructured, triangular computational grid. *GUI* graphical user interface

preparing the computational grid. For the model discussed in the previous sections, this grid is formed by a lattice of nonoverlapping triangles (the control volumes) that must contain the relevant bathymetric information, and the bathymetry is provided by a DTM. The present example concerns a flood inundation study of the LaMoure Rott Municipal Airport in the city of LaMoure, ND, with the purpose of designing flood mitigation measures. The DTM consists of a set of more than 3.7 million points with 2 m of spatial resolution. The iRIC GUI allows automating the process by a few simple and quick clicks of the mouse: Start by importing the data into the GUI; next, define the area of interest where the computational mesh will be created; then enter the desired mean triangle size and click on the automatic mesh generator icon. The result is a quality grid that is automatically mapped to the terrain using the information contained in the DTM, and that is displayed on the computer screen as an overlay to the DTM. The immediate visual feedback permits the user to judge mesh quality and to make adjustments and improvements as desired. Figure 6.9 shows the result of the operation for a grid of 38,156 points (74,565 triangles) zoomed into the airport area. Note the finer grid nested within the coarser main grid, allowing for the more accurate definition of the topography in areas where higher accuracy is needed.

An example of data post-processing (i.e., graphical visualization of model results) is shown in Fig. 6.10, where the water depth contour levels are plotted over topography in the same region of Fig. 6.9. Multiple types of spatial plots (contour

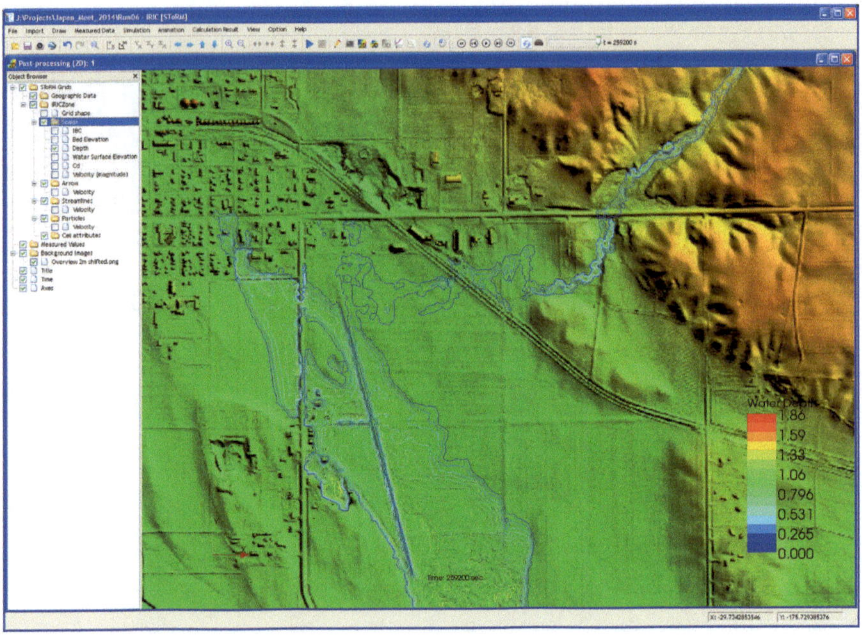

Fig. 6.10 Interactive display of computational modeling simulation results. Shown are the contour levels of water depth, colored using the color coding shown in the legend located in the lower right corner of the display

lines, color maps, vectors, etc.) allow for rapid, but detailed interactive analysis of model results and suggest areas where improvements can be made, reducing simulation production times and improving quality control. They also substantially decrease the production times of project deliverables, such as documentation and visual presentation aids, and provide a useful tool to communicate technical information to those in the management- and policy-making disciplines.

5 Model Validation Applications

Unfortunately, a model quality assurance methodology has not been established by the hydraulic modeling community. Some efforts have been made, from which terms such as model *verification*, model *validation*, and model *confirmation* have surfaced. Although universally accepted terms do not exist, Refsgaard and Henriksen [7], in large part commenting and citing the work of Oreskes and others [39], provide the following definitions:

> Verify is "an assertion or establishment of truth." To verify a model therefore means to demonstrate its truth. According to the authors "verification is only possible in closed systems in which all the components of the system is established independently and are known

> to be correct. In its application to models of natural systems, the term verification is highly misleading. It suggests a demonstration of proof that is simply not accessible." […] mathematical components are subject to verification, because they are part of closed systems, but numerical models in application cannot be verified because of uncertainty of input parameters, scaling problems and uncertainty in observations.
> The term validation is weaker than the term verification. Thus validation does not necessarily denote an establishment of truth, but rather "the establishment of legitimacy, typically given in terms of contracts, arguments and method." […] "the term valid may be useful for assertions about a generic model code but is clearly misleading if used to refer to actual model results in any particular realization."
> The term confirmation is weaker than the terms verification and validation. It is used with regard to a theory, when it is found that the theory is in agreement with empirical observations. […] such agreement does not prove that the theory is true, it only confirms it.

Applying these concepts to an instance of computer software is not trivial, but some ideas may be retained. For example, "model validation" refers to whether the model has the intended accuracy when applied to processes consistent with its designated range of application; "code verification" indicates if the model code is a true representation of the conceptual model that was used for its design; and "model confirmation" concerns the indication whether the scientific theories and hypothesis used by the conceptual model provide an adequate level of agreement to the intended real-world applications. In other words, validation checks if the right equations are being solved, while verification checks if the equations are being solved right.

Validation of a hydraulic model can be accomplished in several ways. One way is to compare the results produced by the model with the results obtained by another model applied to the same problem. This is a valid approach if the latter model has been previously verified and validated for that type of phenomenon and that range of physical parameters. A second approach is to compare model results with those obtained in a laboratory experiment. The physical and design conditions can be carefully controlled and the flow variables accurately measured in a laboratory setting, thus providing reliable data for model validation in which a scaled-down version of the real world is employed. Finally, model results can be compared with those measured *in loco*. This latter approach is the most costly and difficult, but is also the one that may produce the most desirable outcomes, because it provides data collected in the physical ambient for which the model was designed. It is not without problems: Data collection is difficult and expensive, error-free data do not exist, and real-world environments are open systems, exacerbating errors due to parameter calibration and shortcomings in model structure. This section presents examples of all three approaches to model validation.

5.1 Supercritical Flow

Supercritical flows do not often appear in natural channels, but are not uncommon in engineering works, such as spillways and flow-measuring structures. In particular, supercritical flow may occur in channels with nozzle-like constrictions, such as the Parshall flume, where steady oblique shocks are observed to develop under

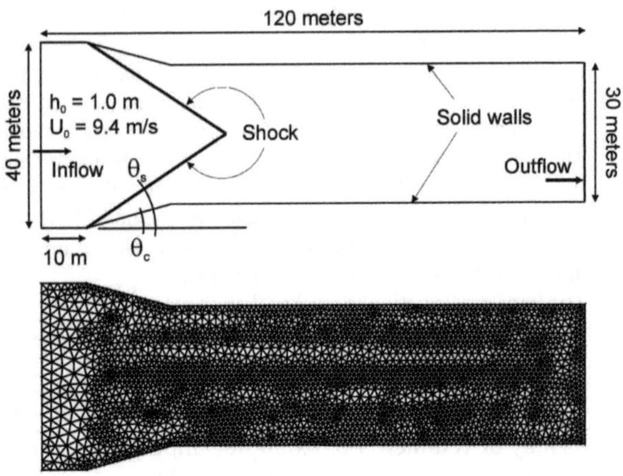

Fig. 6.11 Channel dimensions (*top*) and coarse mesh setup (*bottom*) for the symmetric contracting channel used. Flow is from *left* to *right*

appropriate conditions with supercritical flow. The jump ratio across the shock and its angle can be derived from the shock relations for the shallow-water equations and are given by

$$\frac{h_1}{h_0} = \frac{\tan\theta_s}{\tan(\theta_s - \theta_c)} \quad \text{and} \quad \sin\theta_s = \sqrt{\frac{h_1}{2h_0 Fr^2}\left(1 + \frac{h_1}{h_0}\right)}$$

where h_0 and h_1 are the depths before and after the shock, *Fr* is the Froude number of the approaching flow, θ_c is the flume's contraction angle (see Fig. 6.11) and θ_s is the angle of the shock (measured the same way as θ_c). This type of configuration is common in the literature and the case of [40] was chosen for this application of the model.

The setup used consists of a rectangular, straight 120-m long channel with an approaching width of 40 m and a symmetric constriction with $\theta_c = 15°$, reducing the width of the channel to 30 m, as depicted in Fig. 6.11. Uniform supercritical flow is prescribed at the inflow boundary with $Fr = 3.0$ and $h_0 = 1.0$ m. With these conditions, the theoretical values for the shock are $h_1 = 1.9459$ m and $\theta_s = 34.3560°$.

The computations were carried with two different computational grids: a coarse grid with 2997 nodes (5704 triangles, as depicted at the bottom of Fig. 6.11) and a finer grid with 15,895 node points (31,248 triangles, not shown). At the inflow boundary, the velocity and the water depth are prescribed ($U = U_0 = 9.3960$ m/s, $V = 0$ m/s, and $h_0 = 1.0$ m). At the outlet, no conditions are needed because the flow is supercritical. The initial conditions are the same as the inflow boundary conditions and a free-slip treatment is used at the solid walls. The inviscid flow equations are solved, without bottom friction or slope, until steady state is reached. The computational results are shown in Fig. 6.12.

The solution presents well-defined steady oblique shocks, starting at the wall points where the constriction begins and propagating downstream. The predicted

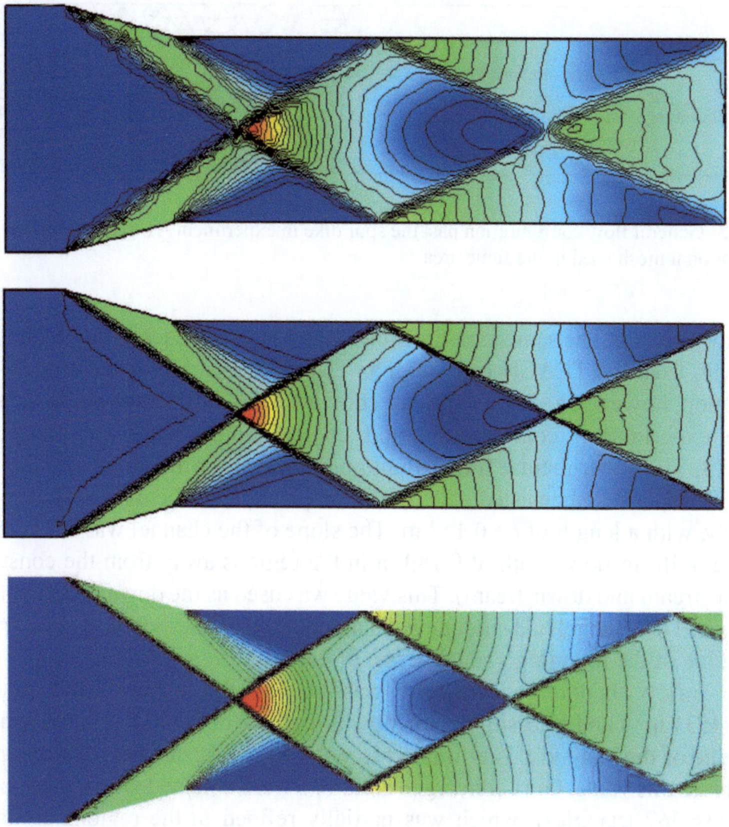

Fig. 6.12 From *top* to *bottom*: computed solution using the *coarse grid*; computed solution using the *fine grid*; reference solution of [40]. Colors represent water depth, *dark blue* is shallow water (h=1.0 m) and *red* is deep water (h=3.1 m).

angle of the shock and the water depth downstream from it are $\theta_s = 34.3487 \pm 0.10°$ and $h_1 = 1.9430 \pm 0.05$ m, which match very well the theoretical values presented above. For comparison, the solution obtained by [40] in a mesh with 23,349 computational cells is also presented.

5.2 *Recirculation Past a Spur Dike*

Groynes, spur dikes, and other structures are river-training structures used to divert the flow and protect the river banks from erosion. They are also used to improve navigability and to maintain river alignment. Rajaratnam and Nwachukwu [41] presented detailed measurements of flow velocity and bed shear stress in several experimental configurations. Their experiment A1 was chosen as a means to provide a comparison between model predictions and physical data obtained in a controlled

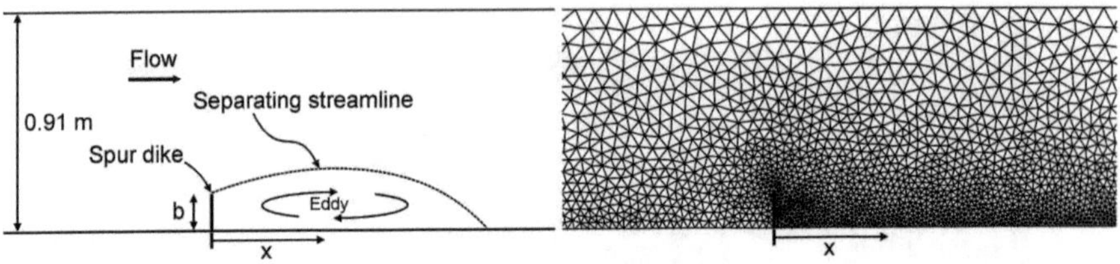

Fig. 6.13 General flow configuration past the spur dike in experiment A1 of [41] and detail of the computational mesh used in the same area

laboratory setting. The same experiment was used by Tingsanchali and Maheswaran [42], who used the TEACH code with an elaborate k–ε turbulence model. In this application, however, a simple constant eddy viscosity turbulence model is used.

The experimental configuration and computational setup are presented in Fig. 6.13. The experimental tilting flume used was 37 m in length and 0.91 m in width. The spur dike consisted of a 3-mm thin plate placed across the flow, perpendicularly, with a length of $b = 0.152$ m. The slope of the channel was set in order to obtain a uniform flow depth of 0.189 m in the regions away from the constriction (both upstream and downstream). This value was used as the downstream boundary condition. The channel had smooth walls and bottom, corresponding to an approximate value of the Manning's roughness n of 0.010.

The inflow boundary condition was set to the experimental discharge $Q = 0.04531\ m^3/s$ and a no-slip treatment was used for the solid walls and spur dike. The value of the eddy viscosity was set to $\nu_t = 0.001\ m^2/s$, but there was no attempt to calibrate this value. Grid-converged solutions were obtained with a mesh of 7693 nodes (14,467 triangles), which was partially refined in the regions around and downstream from the plate (see Fig. 6.13). A detailed view of the computed eddy streamline pattern near the plate is shown in Fig. 6.14, superimposed on the contour lines of the water depth.

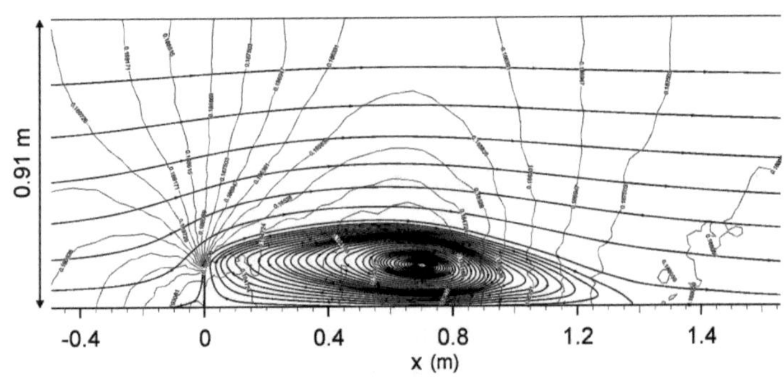

Fig. 6.14 Streamlines of the eddy formed downstream from the spur dike. The contour lines of the water depth are also shown

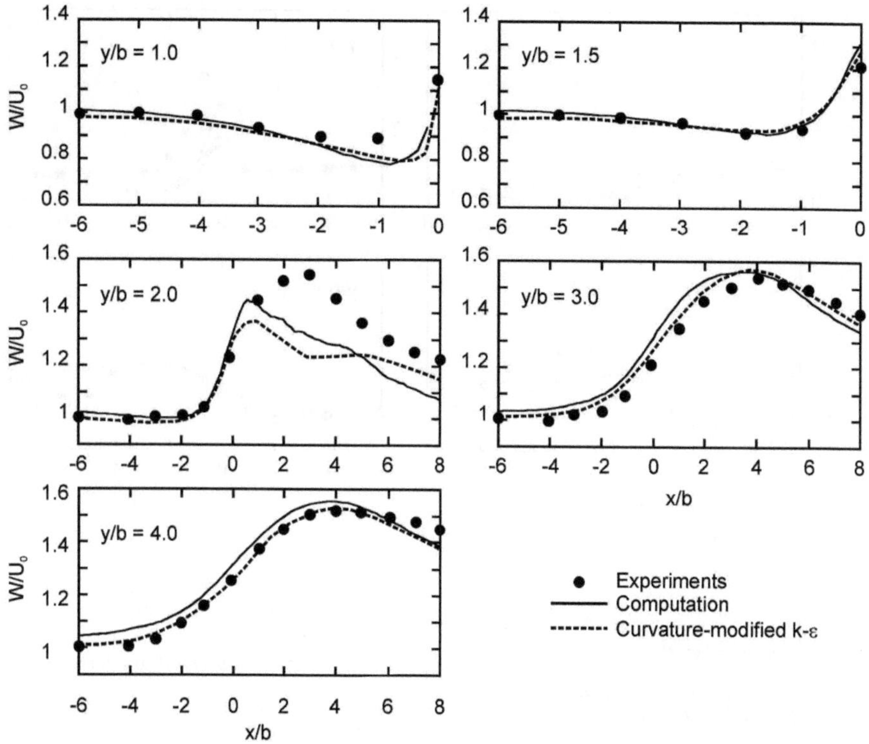

Fig. 6.15 Comparison between the computed velocity profiles (*solid line*), the calculated values of [42] (*dashed line,* only the results of the enhanced turbulence model are shown), and the experimental results (*solid circles*) for experiment A1 of [41]

The reattachment point of the eddy is located at $x/b = 9.175$. The experimental value is of approximately 12.5, resulting in an underprediction of 26.6 %. This result compares well with that of [42], which underpredicts the eddy size by 40 % using the standard *k–ε* model. Only after a streamline curvature correction factor was applied to the turbulence model did these authors improve their prediction, but their result still underestimated eddy length by 20 %. The maximum width of the eddy is $y/b = 1.824$, which is 9 % less than the experimental value of approximately 2.

Comparisons between the computed and experimental total velocities ($W = \sqrt{u^2 + v^2}$) are shown in Fig. 6.15 for several sections measured along lines of constant y. The results shown are normalized with the approaching velocity, U_0, and the plate length b. Unfortunately, there are no measurements inside the eddy region, but the results show an overall good agreement comparable to the results obtained by [42]. The poorest results are in the region $1 < x/b < 8$ at the transect $y/b = 2.0$. This is a region near the top of the eddy. The poorer predictions there do not seem to affect substantially the computed velocities in the same region of the next transect (at $y/b = 3.0$).

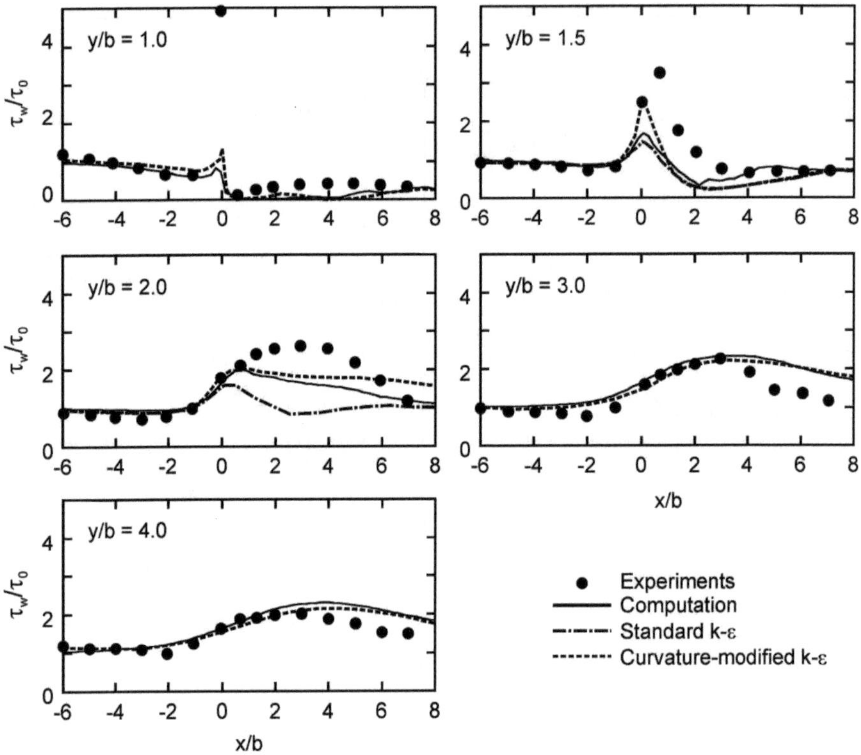

Fig. 6.16 Comparison between the computed bed shear stress profiles (*solid line*), the calculations of [42] (*dashed lines*), and the experimental results (*solid circles*) for experiment A1 of [41]

The presentation of the bed shear stresses is important because it indicates the quality of how flow resistance is modeled. Its correct computation is also important for sediment transport predictions. The computational values of the total bed shear stress, $\tau_w = \rho g n^2 W^2 / h^{1/3}$, are shown in Fig. 6.16 normalized with the approaching bed shear stress τ_0. Once again, the overall computation results agree very well with the measurements, except in the vortex regions downstream from the spur dike, where the 3D effects are significant, such as in the region $1<x<6$ for the transect located at $y/b=2.0$. Nonetheless, these results compare favorably to those in [42], being substantially better than the results obtained by the standard k–ε model, and marginally better than the modified k–ε model in some regions.

It is interesting to note that the k–ε turbulence model does not seem to produce considerably better results than the algebraic turbulence approach used in this work, in spite of the significant increase in computational cost of the former model, which requires the solution of two additional partial differential equations. It is not clear why this is so. It may be due to a breakdown of the assumptions made in deriving the 2D k–ε model, which may be valid for 3D modeling but may lose generality in shallow 2D flows. Or it just may happen that bed friction (or shear velocity) is all one has to consider in order to obtain an accurate enough evaluation of the effects of turbulence dissipation on the main flow quantities.

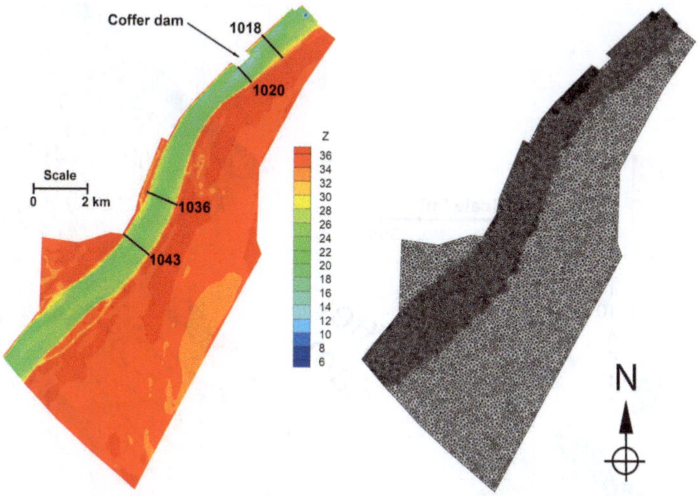

Fig. 6.17 Bathymetry (*left*) and computational mesh (*right*) used in the numerical computations. At left, the measurement transects used for verification are shown. For consistency and easy reference, the number designation of those cross sections was kept identical to the designation assigned in the data collection program. The colorization shows the bed elevation above an arbitrary datum (*Z*, in meters)

5.3 Application to a Large River

The application of hydraulic models to real-world applications is of practical importance and constitutes the final measure of the robustness and usefulness of the techniques used. The challenges posed by the scales and geometries normally encountered in environmental flows provide the best test bed for any numerical model and exercises the full range of the algorithms implemented in the computer code. For this application test case, data collected in a 15-km reach of the Ohio River near the US Army Corps of Engineering Olmsted Locks and Dam project, at river mile 964.4 are used. The data were presented in [43] and were collected using acoustic Doppler current profilers with a differentially corrected global positioning system with the purpose to provide data that could be used to calibrate and validate numerical models. The data consist of detailed bathymetry, water-surface elevations, and velocity profiles collected with known river discharges. For this study, the data corresponding to the discharge of approximately 9900 m^3/s are used.

The overall bathymetry is shown in Fig. 6.17. At the time of the data collection, a coffer dam was in place, located at the upstream end of the reach. The flow is from northeast to southwest. A computational mesh containing 10,288 nodes (20,101 triangles) was developed based on the bathymetry, using larger triangles in the flood plains, with refinements near the upstream and downstream ends of the coffer dam, and shown in the right-hand side of Fig. 6.17. The downstream water-surface elevation supplied in [43] was used for the downstream boundary, and the known discharge was prescribed at the upstream boundary.

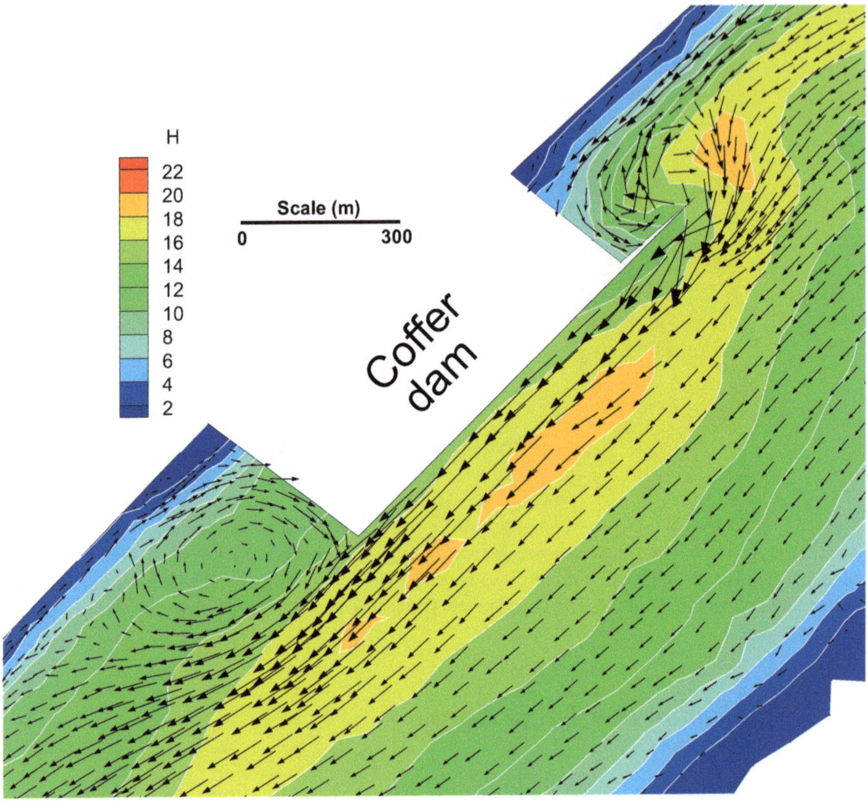

Fig. 6.18 Detailed view of the flow solution near the coffer dam. The colors indicate water depth (*H*, in meters)

The values of the bed roughness were found by calibration. The value of the water-surface elevation at the downstream corner of the coffer dam is known. Several values of the Manning's roughness *n* were tried until a good match was found. The final value chosen was $n = 0.020$, which was prescribed uniformly throughout the reach. The value of the turbulent eddy viscosity was prescribed uniformly throughout the computational domain to $\nu_t = 0.001\ \text{m}^2/\text{s}$.

The results of the computation are shown in Figs. 6.18 and 6.19. Figure 6.18 shows the known areas of recirculation around the coffer dam. The downstream eddy is smaller than the observed one by nearly 40%. The underprediction of the eddy sizes was expected, as shown in the previous validation exercise, but not by such a large amount. The simple turbulence model used has shown to provide solutions that are more accurate than this; therefore, the magnitude of this discrepancy may be attributed instead to an incorrect localized roughness value, or to the inflow boundary condition being too close to the region of interest, or both. Unfortunately, the lack of field data on the nature of the bed roughness in that region precludes a more elaborate analysis. Nonetheless, it is worthwhile to point out that the simulations carried out in [43] used different values of *n* for the banks and for the main channel.

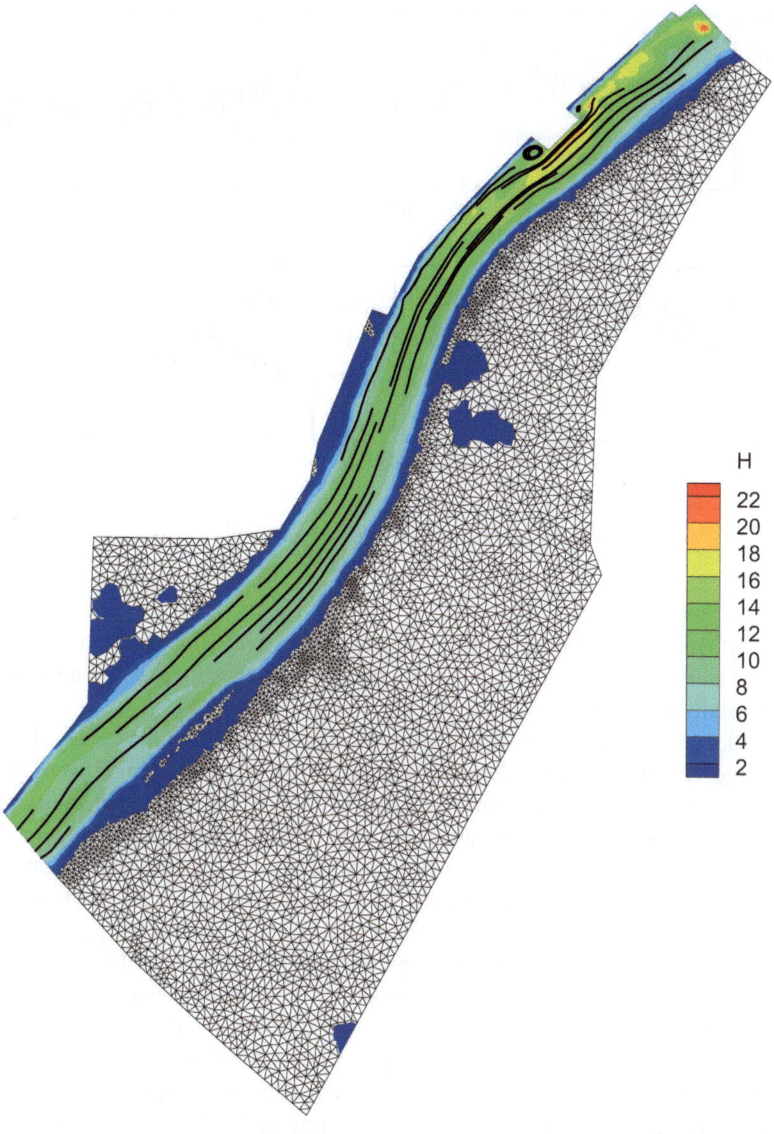

Fig. 6.19 Streamlines of the solution showing a smooth flow without oscillations. The wetted domain is shown colorized by water depth (*H*, in meters). The dry triangles are shown in black and white

The wetted areas resulting from the computation are shown in Fig. 6.19. Some of the wet areas in the flood plains resulted from the initial conditions, which were set with a high initial water-surface elevation. This resulted in some of the adjacent low-lying areas to be initially wet. During the course of the computation, water flows to the river leaving wetted ponds behind. These ponds are disconnected from

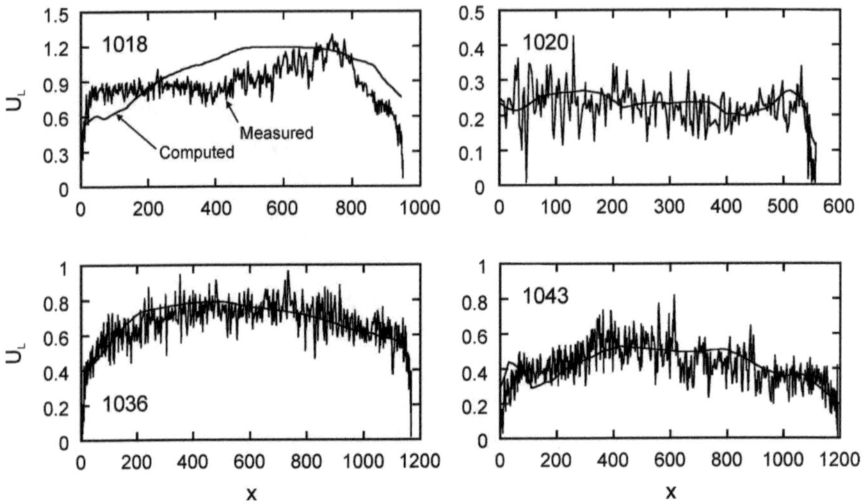

Fig. 6.20 Comparison between computed and measured longitudinal velocity profiles at selected locations

the river's main stem. These effects could have been avoided by using a much lower initial water-surface elevation, but convergence to steady state would have taken a substantially longer time.

The location of the shoreline shows a good agreement with the survey data. The measured velocity profiles taken at the transects shown in Fig. 6.17 were used to provide comparison with the corresponding profiles of the numerical solution. The results are shown in Fig. 6.20, which displays the longitudinal component of the velocities, U_L. Overall, there is good agreement between measurements and computation, with the computed velocities falling well within the band of the measurement variability. The exception resides at transect 1018, where larger differences between measurements and computation are present. This is attributed to the way the upstream boundary conditions are enforced: The total discharge is enforced by generating a velocity profile with uniform unit discharge and whose velocities are normal to the boundary line. The velocity components of the profile thus generated may be (and most likely are) substantially different from the field conditions in that region of the flow, adversely impacting the predicted velocities at transect 1018, which is located less than three channel widths away from the inflow boundary. Effects from boundaries may be seen up to 10 channel widths away from the boundary and it is suggested that, when possible, computational boundaries be placed at a large distance from critical areas. In the present situation, unavailability of bathymetric data and the fact that the modeled area is downstream from a river bend, precluded extending the modeling reach in the upstream direction in a meaningful manner. However, the flow recovers rather quickly and a good agreement between the computed and the measured longitudinal velocities seems to have already been established near the coffer dam (transect 1020).

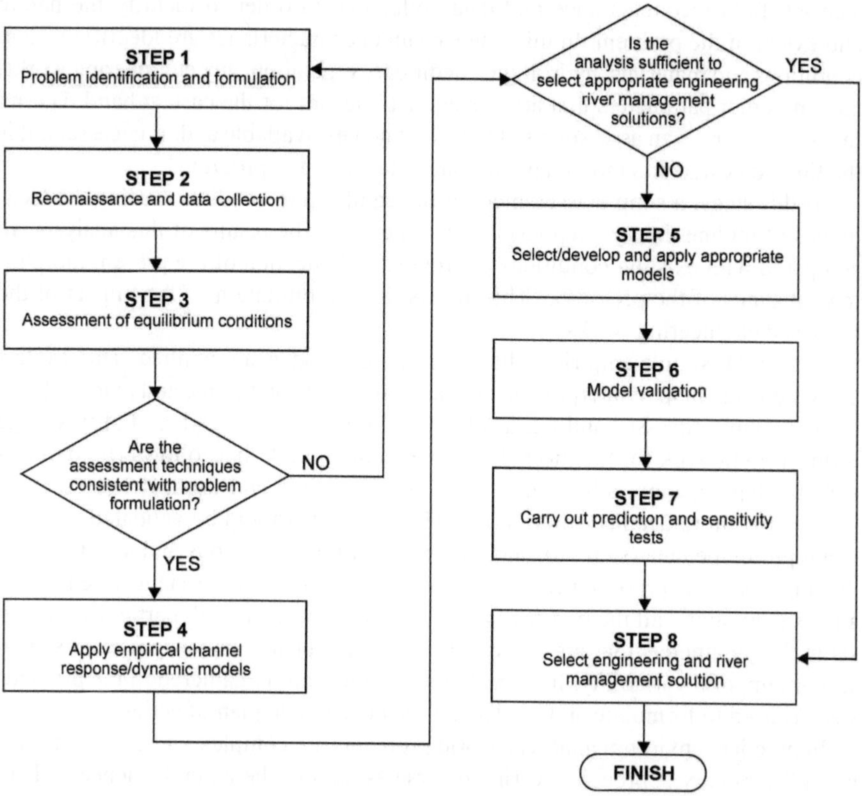

Fig. 6.21 Procedure for the identification, analysis, and modeling of river width adjustment problems, after [44], with modifications

6 Concluding Remarks

From the expositions presented in the previous sections, it has been made clear that the task of modeling environmental flows in surface hydraulics is not a simple one—certainly not as simple as just running a software program. Like in many other areas of application, the complexity and variety of geomorphic factors involved in natural phenomena require a careful and methodical approach to their modeling, with large support of field observations and prototype data. To deal with these issues, an eight-step approach is proposed in [44]. In spite of being specific to bank evolution modeling, the basic principles of these guidelines are applicable to other areas of interest in hydraulic modeling. A schematic view of the entire process is presented in Fig. 6.21.

Step #1, problem identification and formulation involves determining the factors at play, which may be associated with river engineering factors or social activities, and may relate to existing conditions or future activities. It establishes the level of analysis required and defines the appropriate level of response. It is followed by step #2, field reconnaissance and data collection, in order to identify the nature and extent of the problem. In this stage, channel characteristics are identified, bank conditions and materials, hydrologic conditions, vegetation, any engineering works, and any other parameters that are considered relevant for the case at hand. During this stage, there is an assessment of the existing data available and, if necessary, it is the time to design and mount an adequate data gathering program.

At this stage, a simple assessment of the equilibrium conditions using the techniques of regime theory (step #3) is recommended. The results of this analysis are compared with existing conditions to provide an indication of the present morphological status of the prototype. This allows the determination of the impact of the proposed engineering works.

In step #4, simple empirical channel response models are applied. This exploratory step may help to interpret existing and proposed conditions, and in identifying the dominant processes and trends at play. The information gathered at this stage forms a framework for the more detailed modeling work that follows (if appropriate). In step #5, more advanced models are developed and used, if necessary, to provide the more detailed information. These models should be validated with existing prototype data (step #6), and applied to current conditions and also to assess the impacts of the proposed engineering works (step #7). At this stage, a sensitivity analysis involving all the pertinent parameters is important, with particular emphasis in the parameters that are difficult to determine or that have significant spatial and/or temporal variability. In step #8, all the information gathered in the previous steps is used to formulate and implement the appropriate plan of action.

In practice, environmental real-world problems are complex and often require a multidisciplinary approach covering diverse aspects of the natural sciences. There is no clear-cut methodology that fits all practical cases, but the above analysis offers a rational and systematic approach that can be useful in dealing with large and small modeling projects. In the end, however, practical experience and good engineering judgment remain the fundamental tools for good decision making.

Acknowledgments The author is grateful to Chad Wagner, USGS, Louisville, KY, for kindly providing the bathymetry and experimental measurements used for the Olmsted Lock and Dam case, and for his assistance in data preparation and interpretation. The model developed and presented in this work (System for Transport and River Modeling, SToRM) was supported by the National Research Program of the US Geological Survey. The SToRM model and corresponding documentation are available for free download from the iRIC Project Web pages at http://i-ric.org/en/.

Glossary

CFD Short for computational fluid dynamics, it is a discipline that uses numerical methods and algorithms to solve fluid mechanics problems with computers.

CGNS A standard for the storage and retrieval of digital data produced in CFD applications. It stands for CFD general notation system.

Computational cell Each individual point, or volume, of the lattice (computational grid) transforming the continuous real-world domains into its discrete counterpart, suitable for numerical evaluation and implementation on digital computers.

Digital terrain model Also called a digital elevation model (DEM), it is a 3D digital representation of a terrain's surface.

Eddy viscosity A method to model the transfer of momentum caused by turbulent eddies that is mathematically similar to momentum transfer due to molecular diffusion, and that consists in replacing the fluid viscosity ν by an effective turbulent viscosity, ν_t, also called the eddy viscosity.

Froude number *(Fr)* A dimensionless quantity defined as the ratio of a characteristic velocity to a gravitational wave velocity.

Godunov scheme A conservative finite-volume numerical scheme used in the solution of partial differential equations, which solves exact or approximate Riemann problems at inter-cell boundaries.

Graphical user interface Type of computer-user interface that allows the user to interact with a computer program using pointing hardware devices, graphical icons, and other visual indicators.

Hydrology The study of flow of water over the Earth's surface

k–ε model A turbulence model based on solving two differential transport equations, one for the turbulent kinetic energy k and the other for the rate of turbulent dissipation ε.

Model A idealized representation of a system.

MUSCL Short for modified upwind scheme for conservation laws, it is a method to describe (reconstruct) the states of the variables in a computational cell based on the cell-averaged states (and their gradients) obtained in the previous time step.

Navier–Stokes equations Partial differential equations arising from Newton's second law (conservation of momentum) that describe the motion of fluids.

Runge–Kutta methods A family of implicit and explicit iterative methods used in temporal discretization for the approximation of solutions of ordinary differential equations.

Supercritical flow A flow whose velocity is larger than the wave velocity, therefore with $Fr > 1$.

TVD Total variation diminishing (TVD) is a property of certain discretization schemes used to solve hyperbolic partial differential equations that do not increase the total variation of the solution from one time step to the next.

References

1. Kirby, M. J. (1987). Models in physical geography. In M. Clark et al. (Eds.), *Horizons in physical geography.* (pp. 47–61). Totowa: Barnes and Noble.
2. Bird, R., Stewart, W., & Lightfoot, E. (1960). *Transport phenomena*. New York: Wiley.

3. Formentin, S., & Zanuttigh, B. (2013). *Prediction of wave transmission through a new artificial neural network developed for wave reflection*. Proceedings of the 7th International Conference on Coastal Dynamics, Arcachon, France, 24–28 June 2013, pp. 627–638.
4. Azamathulla, H., & Zahiri, A. (2012). Flow discharge prediction in compound channels using linear genetic programming. *Journal of Hydrology, 454–455*, 203–207.
5. Biswas, G. (2013). *Computational fluid dynamics*. Pangbourne: Alpha Science.
6. Rosenquist, T., & Story, S. (2012). Using the Intel Math Kernel Library (Intel MKL) and Intel compilers to obtain run-to-run numerical reproducible results. *The Parallel Universe*, no. 11, September 2012, pp. 26–28 Intel Corporation.
7. Refsgaard, J., & Henriksen, H. (2004). Modeling guidelines—terminology and guiding principles. *Advances in Water Resources, 27,* 71–82.
8. EPA. (2009). *Guidance on the development, evaluation, and application of environmental models*. U.S. Environmental Protection Agency, Council for Regulatory Environmental Modeling, EPA/100/K-09/003, March 2009.
9. Tennekes, H., & Lumley, J. (1972). *A first course in turbulence*. Cambridge: MIT Press.
10. Rodi, W. (1993). *Turbulence models and their application in hydraulics*. IAHR monograph. Roterdam: Balkema.
11. Pritchard, D. W. (1971). *Hydrodynamic models*. Estuarine models: An assessment (pp. 33). Tracor, Inc, for the Water Quality Office of the Environmental Protection Agency.
12. Shimizu, Y., Yamaguchi, H., & Itakura, T. (1991). Three-dimensional computation of flow and bed deformation. *Journal of Hydraulic Engineering, 116*(9), 1090–1108.
13. Vreugdenhil, C. (1994). *Numerical methods for shallow-water flow*. Boston: Kluwer.
14. Fischer, H. B., List, E. J., Koh, R. C. Y., Imberger, J., & Brooks, N. H. (1979). *Mixing in inland and coastal waters*. San Diego: Academic.
15. Roe, P. L. (1981). Approximate Riemann solvers, parameter vectors, and difference schemes. *Journal of Computational Physics, 43,* 357–372.
16. Toro, E. F. (2001). *Shock-capturing methods for free-surface shallow flows*. Chichester: Wiley.
17. Alcrudo, F., & Garcia-Navarro, P. (1993). A high-resolution Godunov-type scheme in finite volumes in 2D shallow water equations. *International Journal for Numerical Methods in Fluids, 16,* 489–505.
18. Harten, A., & Hyman, J. M. (1983). Self adjusting grid methods for one-dimensional hyperbolic conservation laws. *Journal of Computational Physics, 50,* 235–269.
19. van Leer, B. (1979). Towards the ultimate conservative difference scheme, V. A second order sequel to Godunov's method. *Journal of Computational Physics, 32,* 101–136.
20. Barth, T. J., & Jespersen, D. C. (1989). *The design and application of upwind schemes on unstructured meshes*. AIAA paper AIAA-89-0366.
21. Venkatakrishnan, V. (1995). Convergence to steady state solutions of the Euler equations on unstructured grids with limiters. *Journal of Computational Physics, 118,* 120–130.
22. Rausch, R., Batina, J., & Yang, H. (1991). *Spatial adaptation procedures on unstructured meshes for accurate unsteady aerodynamic flow computation*. AIAA Paper 91-1106.
23. Batten, P., Lambert, C., & Causon, D. M. (1996). Positively conservative high-resolution convective schemes for unstructured elements. *International Journal for Numerical Methods in Engineering, 39,* 1821–1838.
24. Coirier, W. J. (1994). *An adaptively-refined, Cartesian, cell-based scheme for the Euler and Navier-Stokes equations*, Ph.D. dissertation, Dept. of Aerospace Engineering, Univeristy of Michigan, MI.
25. Bradford, S. F., & Sanders, B. F. (2002). Finite-volume method for shallow water flooding of arbitrary topography. *Journal of Hydraulic Engineering. ASCE, 128*(3), 289–298.
26. Gottlieb, S., Shu, C.-W., & Tadmor, E. (2001). Strong stability-preserving high-order time discretization methods. *SIAM Review, 43*(1), 89–112.
27. Swanson, R. C., & Turkel, E. (1997). *Multistage schemes with multigrid for Euler and Navier-Stokes equations, components and analysis*. NASA technical paper 3631, Langley Research Center, Hampton, VA.

28. Jameson, A., & Mavriplis, D. (1986). Finite volume solution of the two-dimensional Euler equations on a regular triangular mesh. *AIAA Journal, 24*(4), 611–618.
29. Horritt, M. S. (2002). Evaluating wetting and drying algorithms for finite element models of shallow water flow. *International Journal for Numerical Methods in Engineering, 55,* 835–851.
30. Balzano, A. (1998). Evaluation of methods for numerical simulation of wetting and drying of shallow water flow models. *Coastal Engineering, 34,* 83–107.
31. Anastasiou, K., & Chan, C. T. (1997). Solution of the 2D shallow water equations using the finite volume method on unstructured triangular meshes. *International Journal for Numerical Methods in Fluids, 24,* 1225–1245.
32. Löhner, R., & Galle, M. (2002). Minimization of indirect addressing for edge-based field solvers. *Communications in Numerical Methods in Engineering, 18,* 335–343.
33. George, A., & Liu, J. (1981). *Computer solution of large sparse positive definite systems.* Computational mathematics. Englewood Cliffs: Prentice-Hall.
34. Dokken, T., Hagen, T. R., & Hjelmervik, J. M. (2007). An introduction to general purpose computing on programmable graphics hardware. In G. Hasle, et al. (Eds.), *Geometric modeling, numerical simulation, and optimization: Industrial mathematics at SINTEF* (pp. 123–161). Springer.
35. Tarpanelli, A., Brocca, L., Melone, F., & Moramarco, T. (2013). Hydraulic modelling calibration in small rivers by using coarse resolution synthetic aperture radar imagery. *Hydrological Processes, 27,* 1321–1330.
36. Gallegos, H. A., Schubert, J. E., & Sanders, B. F. (2009). Two-dimensional, high-resolution modeling of urban dam-break flooding: A case study of Baldwin Hills, California. *Advances in Water Resources, 32,* 1323–1335.
37. Hunter, N. M., Bates, P. D., Neelz, S., Pender, G., Villanueva, I., Wright, N. G., Liang, D., Falconer, R. A., Lin, B., Waller, S., Crossley, A. J., & Mason, D. C. (2008). Benchmarking 2D hydraulic models for urban flooding. *Water Management, 161,* 13–30.
38. Williams, R. D., Brasington, J., Vericat, D., & Hicks, D. M. (2014). Hyperscale terrain modelling of braided rivers: Fusing mobile terrestrial laser scanning and optical bathymetric mapping. *Earth Surface Processes and Landforms, 39,* 167–183.
39. Oreskes, N., Shrader-Frechette, K., & Belitz, K. (1994). Verification, validation and confirmation of numerical models in the earth sciences. *Science, 263*(5147), 641–646.
40. Cueto-Felgueroso, L., Colominas, I., Fe, J., Navarrina, F., & Casteleiro, M. (2006). High order finite volume schemes on unstructured grids using moving least squares reconstruction. Application to shallow water dynamics. *International Journal for Numerical Methods in Engineering, 65*(3), 295–331.
41. Rajaratnman, N., & Nwachukwu, B. A. (1983). Flow near groin-like structures. *Journal of Hydraulic Engineering. ASCE, 109*(3), 463–480.
42. Tingsanchali, T., & Maheswaran, S. (1990). 2-D depth-averaged flow near groyne. *Journal of Hydraulic Engineering. ASCE, 166*(1), 71–86.
43. Wagner, C. R., & Muller, D. S. (2002). Use of velocity data to calibrate and validate two-dimensional hydrodynamic models. Proceedings of the Second Federal Interagency Hydrologic Modeling Conference, Las Vegas, NV, July 29–August 1, 2002.
44. ASCE (1998). River width adjustment. II: Modeling. *Journal of Hydraulic Engineering. ASCE, 124*(9), 903–917.

Chapter 7
Geophysical Methods for the Assessment of Earthen Dams

Craig J. Hickey, Mathias J. M. Römkens, Robert R. Wells and Leti Wodajo

Contents

C. J. Hickey (✉) · L. Wodajo
National Center for Physical Acoustics, University of Mississippi,
P. O. Box 1848, University, MS 38677–1848, USA
e-mail: chickey@olemiss.edu

L. Wodajo
e-mail: ltwodajo@go.olemiss.edu

M. J. M. Römkens · R. R. Wells
USDA ARS National Sedimentation Laboratory, P. O. Box 1157, Oxford, MS, USA
e-mail: Matt.Romkens@ars.usda.gov

R. R. Wells
e-mail: Robert.Wells@ars.usda.gov

C. T. Yang, L. K. Wang (eds.), *Advances in Water Resources Engineering*,
Handbook of Environmental Engineering, Volume 14, DOI 10.1007/978-3-319-11023-3_7

Abstract Dams and levees are an integral part of the fluvial system in watersheds. The structural integrity of this infrastructure is of concern to the nation and to those directly impacted should failure occur. There are some 88,000 dams and 110,000 miles of levees in the USA. Many of those are earthen embankments and structures subject to failure by seepage and overtopping especially under extreme conditions of rainfall, runoff from contributing source areas, and snowmelt. They require routine inspection and the availability of technologies to assess their stability and safety conditions. This chapter discusses in a comprehensive manner the various geophysical and geotechnical techniques, and related technologies that are capable of rapidly assessing the integrity and stability of dams and levees. This chapter also discusses the underlying principles of these techniques. Finally, it presents case studies in which these techniques were used.

Keywords Dam failure · Dam stability · Geophysical techniques · Geotechnical methods · Levee failure · Levee stability

Nomenclature

α	attenuation constant
a	tortuosity
A	cross-sectional area (m^2)
ACWI	Advisory Committee on Water Information
AM	distance between electrode A to M
AN	distance between electrode A to N
β	phase factor
B	equivalent conductance of counterions as a function of pore-fluid conductivity
BM	distance between electrode B to M
BN	distance between electrode B to N
c	speed of electrical field propagation
C_c	clay content
CEC	cation-exchange capacity
δ	skin depth
E	Young's modulus
E_x	X-component of electrical field
E_{x0}	X-component of electrical field at $z=0$
ε	electrical permittivity
ε_0	permittivity of free space
ε_r	dielectric constant
f	frequency
F	formation factor
F^*	formation factor for Shaley Sand
G	shear modulus

G_d	dry shear modulus
G_s	saturated shear modulus
h	layer thickness
H	magnetic field intensity
H_p	primary magnetic field
H_s	secondary magnetic field
I	measured current
K	geometrical factor representing electrode configuration
K_b	bulk modulus of soil mineral grain
K_d	bulk modulus of dry rock
K_f	bulk modulus of fluid
K_g	bulk modulus of gas
K_l	bulk modulus of liquid
K_{sat}	bulk modulus of saturated rock
κ	propagation constant
L	length (m)
λ	Lamé constant
λ^e_{Na}	maximum equivalent ionic conductance of sodium exchange ions
m	cementation exponent
MASW	multichannel analysis of surface waves
μ	magnetic permeability
μ_0	permeability of free space
n	saturation exponent
N_B	induction number
NCLS	National Committee on Levee Safety
NID	National Inventory of Dams
NLD	National Levee Database
NRCS	Natural Resources Conservation Service
ν	Poisson's ratio
P_c	capillary pressure
P_f	primary field
P_t	threshold pressure
Q_v	concentration of sodium exchange cations associated with clay
R	electrical resistance (ohms)
RESUS-II	Reservoir Sedimentation Information System
RESSED	Reservoir Sedimentation Database
R_f	resultant field
ρ	electrical resistivity (Ωm)
ρ_b	bulk electrical resistivity
ρ_w	pore-fluid resistivity
SASW	spectral analysis of surface waves
SCS	Soil Conservation Service
S_f	secondary field
S_{irr}	irreducible water saturation
SOS	Federal Interagency Subcommittee on Sedimentation

S_w	degree of water saturation
σ	electrical conductivity
σ_e	conductivity due to clay fraction
σ_o	bulk conductivity of fully saturated soil
σ_w	conductivity of pore-fluid
USACE	US Army Corps of Engineers
USCS	Unified Soil Classification System
USDI-BR	US Department of Interior-Bureau of Reclamation
USGS	US Geological Survey
ΔV	potential voltage difference (volts)
v_1	velocity of top layer
v_2	velocity of half space
v_R	velocity of Rayleigh waves
v_p	velocity of P-waves
v_s	velocity of S-waves
V_T	total volume
V_V	volume of voids (pore space)
ϕ	porosity
x_c	crossover distance
χ_e	electrical susceptibility
χ_m	magnetic susceptibility
ω	time-harmonic frequency

1 Introduction

During the past 150 years, management of watersheds has undergone changes in functions, missions, and scope. Many of them have man-made lakes and reservoirs surrounded in full or part by water barriers. These water bodies serve our needs for water supply, flood control, power generation, water-based recreational activities, and as sedimentation basins. They are a national resource. However, they need to be maintained if they are to continue to serve us or, worse, to prevent them from becoming hazards to the general public welfare or even life. Most of these water bodies have water-retaining earthen structures and embankments.

Both dams and levees are water-retarding structures and often, especially the smaller ones, consist of earthen embankments. Dams are constructions across a stream to stop or control flows. Failure is usually due to overtopping during extreme weather events or by seepage and piping because of prolonged or permanent subsurface water pressures. Levees are retaining structures, parallel to the flow, designed to contain or guide the flow, usually for short periods. Failure is by overtopping during extreme events or by seepage and subsequent collapse. In general, the failure mechanisms are about the same, though the functions of these structures are somewhat different. In this section, we will discuss briefly the dam and levee conditions in the USA and the methods that are under development to determine the degree of instability of the structures.

The early dams were built in the Northeast part of the USA and in the eastern part of the Midwest, where the Europeans first settled, and converted swift-flowing streams or parts thereof to derive mechanical power for electricity generation and powering lumber, and grain mills. As the interior of the USA opened up, dams and reservoirs were built for similar purposes in the Midwest and elsewhere. A major construction period was from the 1930s–1970s when the Soil Conservation Service (SCS) built some 12,000 dams as part of its mission to mitigate watershed erosion, to develop sedimentation basins for trapping sediment, and to control floods. At the same time, these structures would provide water reservoirs for the irrigation of agricultural land in states like Kansas, Oklahoma, and Texas, and water supply for the public and commercial use in many other locations. The dams built were usually earthen structures. In the southeast, the Tennessee Valley Authority built dams for hydroelectric power generation which was of great significance in opening up and improving the living standard of that part of the USA. Most of these dams were concrete structures and also served as sedimentation basins and flood control.

2 Reservoir Sedimentation

The economic life for the SCS earthen structures was planned for 50 years. During this period, and depending on the location and local agricultural management practices in the upstream part of the watershed, the reservoirs behind these dams have received a significant amount of sediment. Decisions must be made as to what to do with these structures and lakes, whether to rehabilitate or to decommission them, and what to do with the sediment. Compounding the problem is the fact that one does not know the sedimentation status of most of the reservoirs. Practically no sedimentation surveys have been made during the past 30 years. Figure 7.1 shows the surveys that were made and recorded in Reservoir Sedimentation Information System (RESUS-II), which included data sets from small farm ponds to large reservoirs such as Lake Meade.

The original surveys were recorded on SCS Form 34 and compiled by the SCS. In recent years, the Federal Interagency Subcommittee on Sedimentation (SOS) of the Advisory Committee on Water Information (ACWI) that resides under the US Geological Survey (USGS) has taken custody of this data set through the cooperative role of some of its members, most notably the Natural Resources Conservation Service (NRCS), the USGS, the US Army Corps of Engineers (USACE), and US Department of the Interior-Bureau of Reclamation (USDI-BR). They have distributed this information among the members of the SOS through Sedimentation Bulletin Number 5. Figure 7.2 shows Form 34 for Grenada Lake reservoir in northern Mississippi. Some 461 surveys have been included in the Miscellaneous Publication No. 1362 through 1950. Form 34 remained the primary document for surveys through the mid-1980s, and additional surveys became part of RESUS-II.

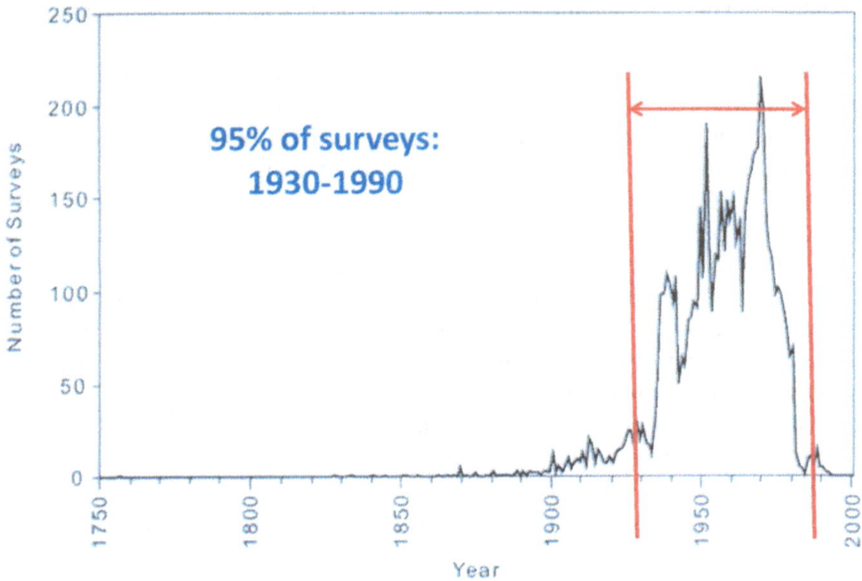

Fig. 7.1 Surveys recorded in the reservoir sedimentation information system, Reservoir Sedimentation Information System (*RESUS-II*), between 1930 and 2000

Radar observations, global positioning system (GPS) technology, and digital data recordings since the 1990–1993 time period have been included in the RESUS-II database. In 2009, this expanded database became the electronic Reservoir Sedimentation database (RESSED). Currently, the RESSED database contains records of about 1824, mostly larger, reservoirs from 6618 bathymetric surveys in the lower USA. The purpose of the surveys is to determine changes in the water storage capacity due to sediment deposition. Figure 7.3 shows the relationship between the number of reservoirs and the RESSED reservoir capacities by acre-feet classes. Note that more than half of the reservoirs have a capacity of less than 10,000 acre-feet. There are also some reservoirs with a very large capacity of more than 1,000,000 acre-feet. Two of those are located in Mississippi: Sardis Lake and Grenada Lake.

A loss in capacity due to sedimentation can be determined from repeat surveys. Since only incidental surveys have been made during the past 30 years, it is difficult, if not impossible, to obtain an accurate estimate of the capacity loss of the reservoirs. Nevertheless, it is estimated that about 436 reservoirs out of a total of 1365 have lost 10–30 % of their capacity based on surveys of 30–40 years ago, and based on extrapolations that 39 % of these 1365 have lost 60 % of their storage capacity. While the USGS does not think that these losses are that high (personal communication), losses of anywhere near this number is of serious concern from the standpoint of increasing the risk of overtopping during extreme rainfall events.

RESERVIOR SEDIMENT DATA SUMMARY	Grenada Lake NAME OF RESERVIOR	15-32 DATA SHEET NO.

DAM	1. OWNER U.S. Army Corp of Engineers	2. STREAM Yalobusha River	3. STATE Mississippi
	4. SEC 4 TWP 22N RANGE 5E	5. NEAREST TOWN Grenada	6. COUNTY Grenada
	7. STREAM BED ELEVEATION 160	8. TOP OF DAM ELEVATION 256	9. SPILLWAY CREST ELEV 231

	10. STORAGE ALLOCATION	11. ELEVATION TOP OF POOL	12. ORIGINAL SURFACE AREA ACRES	13. ORIGINAL CAPACITYACRE-FEET	14. GROSS STORAGE ACRE-FEET	15. DATE STORAGE BEGIN
RESERVIOR	a. MULTIPLE USE					
	b. FLOOD CONTROL	231	64,640	1,251,710	1,337,400	30 JUN 1953
	c. POWER					
	d. WATER SUPPLY					16. DATE NORMAL OPER BEGIN
	e. IRRIGATION					
	f. CONSERVATION	193	9,810	85,690	85,690	JAN 1954
	g. SEDIMENT					
	h. INACTIVE					

WATERSHED	17. LENGTH OF RESERVIOR 44 1/MILES		AV. WIDTH OF RESERVIOR 1.31 MILES	
	18. TOTAL DRAINAGE AREA 1,320 SQ. MI.		22. MEAN ANNUAL PRECIPITATION 54.03 INCHES	
	19. NET SEDIMENT CONTRIBUTING AREA 1,219 SQ. MI.		23. MEAN ANNUAL RUNOFF 19.60 (11.83) INCHES	
	20. LENGTH 70 MILES	AV. WIDTH 19 MILES	24. MEAN ANN RUNOFF 1,379,800 (11.83) AC -FT	
	21. MAX. ELEV. 350	MIN. ELEV. 160	25.	CLIMATIC CLASSIFICATION Humid

SURVEY DATA

26. DATE OF SURVEY	27. PERiod YEARS	28. ACCL. YEARS	29. TYPE OF SURVEY	30. NO. OF /CONTOURED INT.	31. SURFACE AREA ACRES	32. CAPACITY ACRE-FEET	33. C/W RATIO AC-FT PER SQ. MI.
30 JUNE 53			Contour, D	5-Foot	64,640	1,337,400	1013
2/	0	0	Range, D	80			
2/53-4/55 3/	11.83	11.83	Range, d	21	64,640	1,320,020 5	1000
MAY 1965 4/							

26. DATE OF SURVEY	34. PERIOD ANNUAL PRECIPITATION	35. PERIOD WATER INFLOW AC.-FT.			36. WATER INFL. TO DATE AC-FT	
		a. MEAN ANNUAL	b. MAX. ANNUAL	c. PERIOD TOTAL	a. MEAN ANNUAL	b. TOTAL TO DATE
May 1965	17,377	1469	1.205	17,377	1469	1.205

26. DATE OF SURVEY	37. PERIOD SEDIMENT DEPOSITS AC-FT			38. TOTAL SED. DEPOSITS TO DATE AC-FT		
	a. PERIOD TOTAL	b. AV. ANNUAL	c. PER SQ. FT. YR.	a. PER TOTAL	b.AV. ANNUAL	c. PER SQ. FT. YR.
May 1965	17,377	1469	1.205	17,377	1469	1.205

26. DATE OF SURVEY	39. AV. DRY WGT. LBS. PER. CU. FT	40. SED. DPT. TONS PER SQ. MI.-YR.		41. STORAGE LOSS PCT.		42. SED. INFLOW PPM.	
		a. PERIOD	b. TOTAL TO DATE	a. AV. ANN.	b. TOT. TO DATE	a. PERIOD	b. TOT. TO DATE
May 1965	60 *Assumed	1575	1575	.110	1.30	1021	1021

Fig. 7.2 Sedimentation survey results on Form 34 for Grenada Lake reservoir in northern Mississippi

3 Dam and Levee Assessment

According to the National Inventory of Dams (NID; [1]), there are about 85,000 dams distributed across the USA (Fig. 7.4) of which about 74,000, 87 % of the total number, are earthen dams. Furthermore, data from the NID indicates, that the construction of about 29,000 of these dams was completed before the year 1960. As mentioned earlier, the economic life for the SCS earthen structures was planned for 50 years, therefore, many of these structures have exceeded their economic lifetime.

The seriousness of the situation is evident when one further reviews the data from the NID, which indicates that 14.6 % of a total of 80,443 dams are classified as high-hazard dams (classification I), and 16.9 % as significant hazard dams. The "hazardous" status relates to the expected impact of a failure, rather than the quality of the structure and the probability of a failure. High-hazard dam failure would probably result in loss of life and major property damage, while significant hazard dam failure could possibly cause some loss of life and property damage.

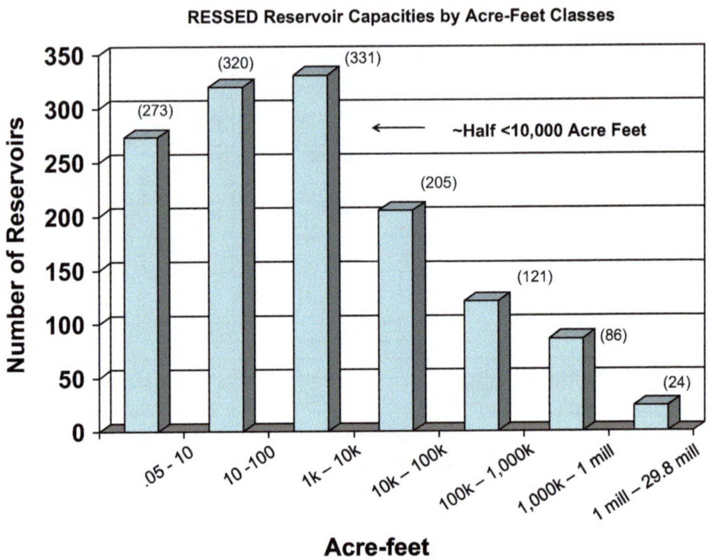

Fig. 7.3 The number of reservoirs and the Reservoir Sedimentation Database (*RESSED*) reservoir capacities by acre-feet classes

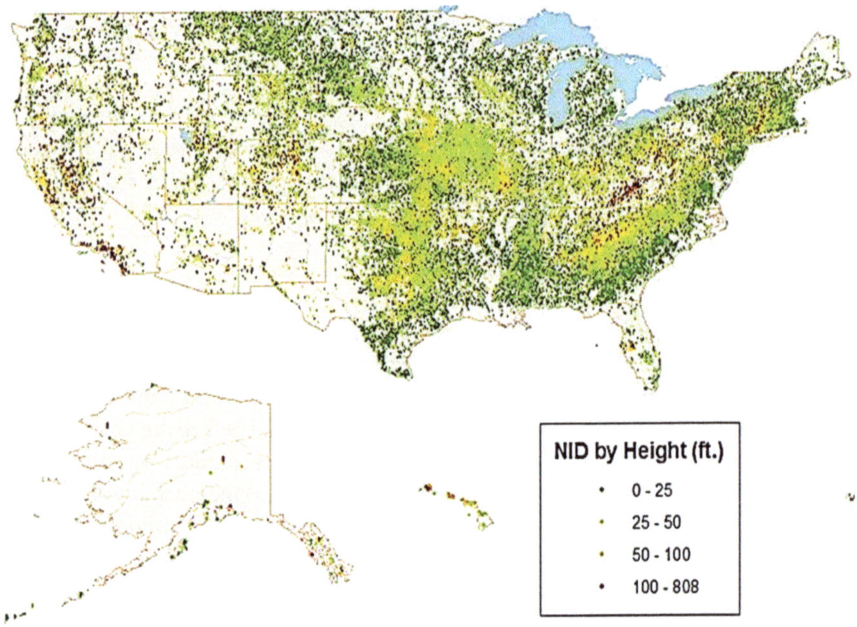

Fig. 7.4 Dam distributions in the USA by height (after NID 2009)

Levees are somewhat different than dams and are constructed along riverbanks to protect the landside of the levee from flooding when the water level on the waterside of the levee rises. These structures can be constructed from earthen materials as in the case of earthen dams or they can be made from concrete. A typical levee made from earthen materials is constructed by piling soil along the riverbanks creating an embankment with flat tops and sloping sides towards the water- and the landside. Many of the older levees have been constructed by raising their height over time. This approach for constructing a levee of suitable height can result in an inherently unstable structure. According to the National Committee on Levee Safety (NCLS) [2], there are an estimated 122,000 miles (>196,000 km) of levees currently in use in the USA. Comprehensive information about levees in the USA can be found in the National Levee Database (NLD), developed by the USACE [2].

Dams and levees have significantly different geometrical shapes and subtle differences that make their assessment procedures somewhat different. Dams are shorter structures with abutments to the native landscape, are under permanent hydraulic loads, and usually have pipes and conduits passing through the structures. Levees are much longer structures traversing different geologies with few pipes passing through the levee. Levees are dry (no hydraulic load) for most of the time and therefore critical signs of seepage are absent during many assessments.

4 Types of Dam and Levee Failures

Dam failures can be arranged into four groups: overtopping, foundation failure, structural failure, and other unforeseen failures. Data obtained from the Department of Ecology of the State of Washington [3], displayed in Fig. 7.5, show overtopping

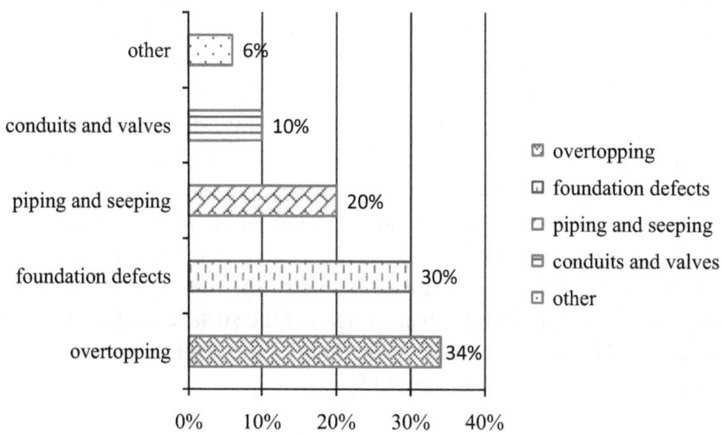

Fig. 7.5 Failures in dams. (After Department of Ecology, The State of Washington, 2007)

accounts for 34 % of dam failures nationally in the USA. Overtopping occurs when the water level on the upstream side of the dam exceeds the height of the dam and flows over the downstream side of the dam. This flow of water causes erosion of the back of the dam and can lead to a total failure. This can be caused by inadequate spillway design, debris blockage of spillway, or settlement of the dam crest. The second cause of failure accounting for 30 % of all failures nationally is the foundation defects. Foundation defects are caused by differential settlement, sliding and slope instability, high uplift pressures, and uncontrolled foundation seepage and piping. Piping and seepage is the third most frequent cause of failure, accounting for 20 % of all failures. These problems originate as seepage and erosion along hydraulic structures (outlets, conduits, or spillways), leakage through animal burrows, and cracks in dams. Piping of embankment material into conduits through joints or cracks accounts for 10 % of all dam failures, whereas the remaining 6 % is caused by undetermined factors.

The magnitude of recorded damages from earthen dams ranges from complete catastrophic failure, resulting in large property damage and loss of life, to relatively minor deterioration which may or may not necessitate remedial work. Even though catastrophic failures are not that frequent, 1090 dam safety incidents including 125 failures were reported between the years 1999 and 2004 [4]. The various forms of failures, causes, and preventive measures are listed in Table 7.1. Erosion failures, slope failure, and sliding can usually be identified during visual inspection; however, the causes of downstream slope failure requires subsurface measurements. Failures due to piping which are associated with concentrated seepage along conduit and joints, or within localized areas of the dam, are also not easily detected by visual inspection.

Levee failures are classified according to three different mechanisms: structural causes, hydraulic causes, and causes involving surface degradation [5]. Structural causes of levee failures are mainly dependent on the strength of the soil. One example of structural failure is an impact on the body of the levee during occasions of heavy storms and winds where physical objects collide with the levee. Other forms of structural failures include tree damage, slope failures, and sliding. An example of a catastrophic structural failure in levees was the 17th Street breach during Hurricane Katrina where 200 ft. (61 ft.) of the levee failed due to sliding [6].

Failure of levees due to hydraulic causes is mainly dependent on groundwater flow and pressure. This form of failure includes seepage, piping, liquefaction, and internal erosion and piping. Internal erosion and piping is associated with water passing through the levee itself and causing internal erosion or piping by washing out smaller or finer soil particles of the levee body which can lead to complete collapse [7]. Surface degradation of levees includes different types of surface erosion caused by surface water flowing over or against the surface of the levee. The most common causes of surface degradation includes: overtopping, overtopping and jetting, surface erosion, and wave impact [5].

Table 7.1 Typical failures, causes, and preventative measures in earth dams

Failure	Characteristics	Causes	Preventive measures
Overtopping	Flood water flows over embankment washing out the dam	Inadequate spillway capacity	Design the spillway with adequate capacity or provide an emergency spillway
		Clogging of spillway with debris	Install trash racks where possible or periodically remove debris
		Insufficient freeboard due to settlement or erosion of embankment	Regrade crest to design elevation
Water erosion	Upstream face eroded	Inadequate riprap or lack of bedding gravel, unprotected slope too steep	Place well-graded riprap and filters or flatten slope
Toe erosion	Erosion of embankment toe near spillway or riparian outlet	Spillway or outlet located too close to dam	Discharge water away from the embankment and provide protection
Surface erosion	Surface runoff eroding downstream face of dam	Poor surface drainage and lack of adequate grass cover on the downstream slope	Provide drains or ensure adequate grass cover on the downstream face. Regrade crest towards upstream slope
Piping	Progressive internal erosion of material usually starting from downstream side of dam or foundation and progressing upstream, eventually leading to a breach	Concentrated seepage	Install toe drains or filters
		Seepage along conduits	Grout along conduit to fill voids or replace conduit. Install filter at downstream end of conduit
		Leaking conduit joints	Seal joints
Foundation failure	Sliding of one or both slopes with heaving of the toe in direction of movement	Soft or weak foundation	Construct toe berms, flatten slope by adding material
		Excess water pressure in foundation	Provide drains and filters
Upstream slope failure	Slide in upstream slope	Slope too steep	Flatten slope, construct berm
		Rapid reservoir drawdown	Avoid rapid lowering of reservoir or flatten slope
Downstream slope failure	Slide in downstream slope	Slope too steep	Flatten slope, construct toe berm
		Saturation of slope by seepage	Provide proper drainage by installing filters

5 Inspections

The purpose of a dam or levee inspection program is to identify problems and/or unsafe conditions. Inspection is an integral part of a proper maintenance program for dams and levees. According to the Federal Emergency Management Agency [8], high-hazard dams should be fully inspected at least every 2 years. Traditionally, a full and in-depth inspection of a dam would include: (1) visual inspection, (2) formal inspection, and (3) a safety review.

A visual inspection is the simplest form of inspection. The frequency of visual inspections can be based on the dam's hazard classification. A more hazardous dam should have more frequent visual inspections, whereas a reduced frequency of inspection would suffice for a dam in a low-hazard classification.

The major purpose of performing a visual dam inspection is to gather and record all possible information about the current condition of the dam; a good record collection method should be implemented and planned before going to the dam site. A thorough dam investigation should include:

- Sketch and measurement of all the deficiencies observed.
- Description with words and photographs of all deficiencies observed.
- A clearly defined location of all deficiencies with reference to a standard reference point.

A proper dam inspection should cover all the possible areas of failure. This is best accomplished if the time of inspection is planned in a manner which accommodates the problem locations and possible occurrence times. The inspection should be planned to cover all parts of the dam; the crest, upstream and downstream embankments, the abutments, the spillway, the reservoir banks, and the area below the dam. Most states utilize a standard checklist type procedure as shown in Fig. 7.6. A typical checklist document used in the State of Mississippi is *Guidelines for Inspection of Existing Dams* and is available at [9].

A formal inspection is performed by the individual who is responsible for the safety of the dam. This can be either the owner of the dam or a qualified appropriate representative. This is an in-depth investigation of the dam and is usually done twice a year for high-hazard dams and once a year for low-hazard dams. Observations from this inspection should be recorded in a formal inspection report and should be kept by the dam owner for future references. A dam safety review involves collecting all available previous and current records of the dam. This includes all visual inspection, formal inspections, and laboratory tests on the dam soils. It also includes an in-depth investigation of the structural stability of the dam starting from the design assumptions. This type of investigation is usually performed when the dam is classified as high hazard. As mentioned earlier, the Federal Emergency Management Agency [8], recommends that high-hazard dams be fully inspected at least every 2 years.

Levees are somewhat simpler than dams in that they have minimal configurations for allowing water to reach the protected side of the levee. However, levee sections are part of a greater levee system. Levees and the levee system should be

DAM INSPECTION CHECKLIST			Date		
			Time		
NAME OF DAM					
FILE NUMBER	COUNTY	CLASS			
WEATHER & SITE CONDITIONS					
INSPECTORS					
OTHERS					
	EMBANKMENT – DIKE - LEVEE		ACTION		
CHECK AREA AS INSPECTED	CHECK/CIRCLE CONDITION NOTED	OBSERVATIONS	REPAIR	MONITOR	INVESTIGATE
U/S SLOPE	vegetation/riprap				
	beaching/slides/cracks				
	undermining/erosion				
CREST	ruts/erosion				
	cracks/settlement				
	poor alignment				
D/S SLOPE	vegetation/erosion				
	rodent burrows				
	sloughs/slides/cracks				
	seepage/wetness				
GROINS	vegetation/riprap				
	erosion				
	seepage/wetness				
ABUT-MENTS	vegetation/erosion				
	sloughs/slides/cracks				
	seepage/wetness				
TOE	cracks/slumps				
	embankment drains				
	seepage/wetness				
GENERAL COMMENTS, SKETCHES, & FIELD MEASUREMENTS					

Fig. 7.6 Typical checklist for visual inspection of embankment dams and levees

inspected for conditions that may compromise its performance. The Risk Management Yearbook [10] outlines what should be inspected. A levee inspection should include the following:

- Height of the levee
- Condition of the waterside and landside surface
- Conditions around pipes and gate structures
- Pipes and culverts other openings
- Internal drainage and pumping systems
- Water accumulation

Considering levees are much longer consisting of hundreds of kilometers of structures traversing different geologies, an efficient and cost-effective approach for inspection must be conducted in stages. The first stage is a screening stage and includes visual inspections, topographic measurements (light and radar, LIDAR) as well as some rapid electromagnetic (EM) geophysical measurements. It provides information on the entire levee and indicates areas that may weaken the levee during a flood event. The second stage focuses on smaller target locations and involves geotechnical measurements as well as higher resolution geophysical methods.

6 Geotechnical Methods

Geotechnical information is needed to refine the overall picture of the dam or levee, to define specific features that may require specific attention as they relate to design and construction, and for collecting samples to obtain specific soil engineering parameters. In situ geotechnical testing provides local physical parameters of the levee body and foundation that are required for further geotechnical analysis in order to assess the safety of the dam. Comprehensive manuals have been published by the USDI-BR [11] and the USACE [12].

The need for additional data may be relatively minor for existing dams, depending upon the adequacy of existing data and construction documentation. However, due to the historical approach for levee construction, especially smaller levees, the need for additional information may be much greater. Furthermore, if questions about a specific project cannot be properly addressed by the use of existing data, then additional site-specific field investigations are required. The evaluation of a specific structure will require information about the foundation, abutments, and structures themselves. This detailed information may only be obtained by drilling, sampling, and testing that is concentrated on the specific site areas or problems. Such investigations, when needed, should be planned to provide the engineer with information and data to answer questions on specific dam safety problems and to perform dam safety analyses.

The general types of geotechnical field explorations used to investigate sites are listed in Table 7.2. Of the different types of explorations, borings are the most practical and accurate method of obtaining subsurface information. In most instances, boreholes should be located along the center axis of the dam and at other locations to sample the foundation materials, phreatic surface, core, abutments, drains, and filter materials. In cases involving specific site problems, historical records should be used to help define locations for additional boring and sampling.

Laboratory tests are conducted on both undisturbed and disturbed soil samples. Disturbed samples are usually acquired using soil augers and split-spoon samples during boring operations and hand samples during excavations. The primary method for obtaining undisturbed soil samples is by the Shelby tube sampler. For rock sampling, core borings are the most common method. Size of the rock core borings can range in diameter, but the most common size used for exploratory work is 3 in. (75 mm). This size drilling usually produces good core recovery. Large diameter holes and special drilling equipment and methods may be justified in some types of materials if better recovery and/or identification, sampling, and testing of material are required.

The location and number of samples tested are determined by the subsurface geology and the anticipated type of failure. The sampling should allow for information about the materials in both the vertical as well as lateral directions. As the project develops, the sampling and testing requirements should be reviewed and adjusted. At a minimum, all distinct soil units should be sampled and tested. Initial soil classification can usually be conducted on disturbed samples. Soils should be classified according to the Unified Soil Classification System (USCS) which requires

Table 7.2 Geotechnical methods for dam and levee diagnosis

Method	Sampling guidelines	Purpose
Destructive drilling	Dam: 1 per 60 m down to 50% of dam height into the foundation	Obtain undisturbed soil cores (1 per 2 m of borehole)
	Levee crest: 1 per 2 km down into foundation	Estimation of depth to bedrock Samples for visual inspection
	Levee landside: 1 per 2 km down 5–10 m into foundation	
Standard penetration tests	Dam: 1 per 60 m down to 50% of dam height into the foundation	Position of interfaces between layers
	Levee: based on average size of breaches (every 200 m)	Qualitative estimates of mechanical properties
Mechanical shovel	For depths less than 5 m	Visual inspection of vertical section Disturbed soil samples
Permeability tests	Performed in boreholes as needed in critical zones	Identify high permeability zones in the foundation, abutments, and embankments
Field vane shear test	Performed in boreholes as needed in critical zones	Determine shear strength

For other methods, see Section 5–22 of US Army Corps of Engineers, EM-1110-1-1804

measures of plasticity, dilatancy, particle size, and other engineering properties. However, many engineering properties require testing on undisturbed samples. Procedures to evaluate properties of soils are described in Schroder [13] and Terzaghi et al. [14]. Soil samples recovered from the field are returned to the laboratory for testing. Table 7.3 lists some of the common geotechnical laboratory tests.

7 Geophysical Methods

Geophysical investigations are an indirect method of obtaining generalized subsurface geologic information by using special instruments to make certain physical measurements [15, 16, 17, 18, 19, 20, 21]. Geophysical investigations of a dam may provide unique and valuable subsurface information to assist with the evaluation of the integrity of an earthen dam. They have the potential to provide the precursory information about the onset of piping, seepage, and anomalous pore pressures before the actual failures occur. The cost of geophysical explorations is generally low compared with the cost of core borings or test pits, and considerable savings may often be affected by judicious use of this exploration method in conjunction with other methods. However, their reliability is only as good as their confirmation by conventional means of exploration.

Table 7.3 Geotechnical laboratory investigations

Geotechnical laboratory testing	Testing frequency	Purpose
Water content	For every sample	Indirect indication of porosity
Grain-size analysis including hydrometer analysis	2 per distinct soil unit	Used to estimate soil strength and permeability
Liquid and plastic limit	2 per distinct soil unit	Soil classification of cohesive material
Slaking tests	In moisture-sensitive clays	Effect of wetting and drying on strength of clays
Pocket penetrometer and torvane	On chunks and undisturbed samples of cohesive materials	Used as guide for shear testing
Direct shear test	1 per distinct unit	Measures shear strength of sands
1D consolidation	1 per distinct unit	Strength and deformability of material
Unconsolidated undrained triaxial test	As need in foundation material	Undrained shear strength of foundation materials
Consolidated undrained triaxial with pore pressure test	1 per distinct unit in foundation and dam or levee body	Drained shear strength of cohesive material
X-ray tests	1 per distinct unit	To determine clay mineralogy

For other methods see Chap. 7 of US Army Corps of Engineers, EM-1110-1-1804

Geophysical methods can be applied during the initial screening phase to isolate locations and define specific anomalous zones within the structure. The use of geophysical surveys for this purpose is more common in investigations of levees as there is a need for surveying long distances. The geophysical methods applied for screening must provide good quality information over long distances in short period of time. The data must cover the entire body of the structure down into its foundation. The methods best suited for screening are EM methods, which will be further discussed in Sect. 3. Localized geophysical explorations complement core drilling, test pits or other direct methods of subsurface exploration, and can provide a rapid evaluation of certain geologic conditions during a detailed investigation.

Near-surface geophysical techniques have been used for geotechnical and environmental problems, and several handbooks describing their use has been published [22, 23, 24, 25]. A recent handbook has been published on agricultural applications [26]. Geophysical methods for investigating dams and levees include: acoustic/seismic, EM and resistivity, gravity, optical sensing, and radar [22, 27, 28, 29]. These techniques are sensitive to the distribution of the bulk "geophysical" properties (elasticity, electrical resistivity, dielectric constant, etc.) in the subsurface that are in turn related to more "basic" properties (bulk density, water content, porosity, mineralogy, etc.). Table 7.4 summarizes the geophysical techniques and some engineering applications. With the exception of magnetic methods, all the geophysical techniques may have application or limited application to the investigation of hydraulically active structures such as dams and levees. A more detailed description of the methods more commonly used on dams and levees is presented below.

Table 7.4 Geophysical techniques and engineering applications

Geophysical method	Physical property	Engineering application
Seismic refraction	Shear modulus, bulk modulus, bulk density	Depth to bedrock, material strength, permafrost, fracture and seepage detection, location of voids
Seismic reflection	Shear modulus, bulk modulus, bulk density, acoustic impedance	Depth to bedrock, material strength, permafrost, fracture and seepage detection, location of voids
Seismic surface wave (MASW)	Shear modulus, bulk density	Depth to bedrock, material strength, permafrost, fracture and seepage detection, location of voids
Self-potential (SP)	Streaming potential	Seepage detection
Resistivity	Electrical conductivity	Seepage, depth to bedrock, location of voids, permafrost, metal detection
Induced potential (IP)	Chargeability	Seepage, clay content
Electromagnetic	Electrical conductivity, magnetic susceptibility	Metal detection, seepage and fracture detection, permafrost
Ground-penetrating radar (GPR)	Dielectric constant	Depth to bedrock, location of voids, seepage detection, permafrost
Gravity	Bulk density	Depth to bedrock, location of voids, geological structures
Magnetics	Magnetic susceptibility, metal content	Metal detection

7.1 Electrical Methods

Electrical methods utilize direct currents (DCs) or low-frequency alternating currents to investigate the electrical properties of the subsurface. The electrical methods are further subdivided:

1. Self-potential (SP) methods measure the natural current flow in the earth. These electrical potentials are commonly associated with electrochemical reactions in the subsurface or the flow of fluids through a porous medium. For dam and levee assessment, SPs associated with flow through and below dam and levees are important.
2. Resistivity methods measure the spatial distribution of soil resistivity in the subsurface. For dam and levee assessment, the electrolytic conduction associated with the pore-fluids and the surface conduction associated with clay mineralogy are important.
3. Induced potential or complex resistivity methods measure the capacitive nature of the subsurface. Membrane polarization occurs in clays because ion-exchange processes effectively create different mobilities for cations and anions within the pores. This effect can be exploited in some cases to distinguish clay and moisture, both of which produce a low resistivity.

The resistivity method is the most widely used geophysical method for the assessment of dams and levees. In the resistivity method, the source is an artificially generated electric current introduced into the ground using electrodes. The potential differences are measured at the surface and the pattern of potential differences provides information on the distribution of subsurface electrical resistivity.

The physical property being measured is the resistivity or the reciprocal of electrical resistivity referred to as the electrical conductivity. The resistivity of most native metals and graphite are due to the passage of electrons within the solid lattice, and most earth minerals are insulators. Rocks are an assemblage of minerals and pores, the resistivity of rocks is controlled by the passage of ions in the pore water. Factors that affect the resistivity of soil–water mixtures include ionic concentration, porosity, surface conduction, tortuosity, and connectivity of phases. The ability for resistivity measurements to detect anomalous zones within dams and levees is controlled by the contrast in resistivity of the zones with respect to the background. The factors affecting the resistivity of geologic materials are presented below.

7.1.1 Electrical Properties

Resistivity is a material property defining how strongly a material opposes the flow of electrical current. Mathematically, the resistivity of a given cube of soil can be defined as

$$\rho = R\left(\frac{A}{L}\right), \tag{7.1}$$

where the resistivity ρ (Ωm) of a given material is equal to the resistance R (ohms) of the cube of material times the ratio of cross-sectional area A (m^2) to the length L (m) of the material. The electrical resistance (R) of the material is described using Ohm's law:

$$R = \frac{\Delta V}{I}, \tag{7.2}$$

where ΔV (volts) is the potential difference across the cube of material and I (amperes) is the electrical current injected. Figure 7.7 describes the parameters used in determining the resistivity of a simple block of material. If the injected current and the geometrical parameters (A and L) are known, and the potential difference is measured across the body of material, then the resistivity can easily be calculated.

The inverse of electrical resistivity is the electrical conductivity (σ), with units of Siemens per meter or mho per meter, and is also commonly used to describe the electrical properties of soils.

Electrical resistivity measurements are greatly affected by different geologic materials that are present within a dam. Therefore, an understanding of these geologic materials and their resistivity will aid in interpreting results. Figure 7.8 is modified

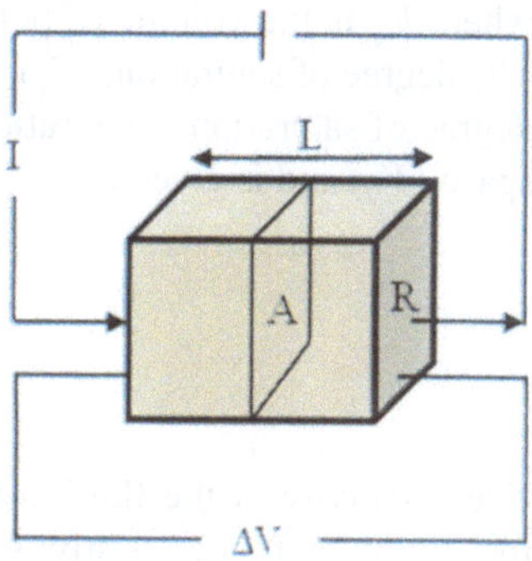

Fig. 7.7 Measuring the resistivity of a block of soil

from Palacky [30] and shows various resistivity values for different geological materials. Several of these materials have overlapping resistivity values, so some knowledge of the subsurface is needed when trying to distinguish one material from the next. Also, each material has a broad resistivity/conductivity range, which is due to several factors.

Factors contributing to the range of resistivity of a given material include: porosity (ϕ), degree of saturation (S_w), pore-fluid resistivity (ρ_w), and clay content (C_c). For additional factors affecting the resistivity of geological materials, see Freidman [31]. Porosity, (ϕ), is a measure of the void spaces in a material and is expressed by

$$\phi = \frac{V_V}{V_T}, \tag{7.3}$$

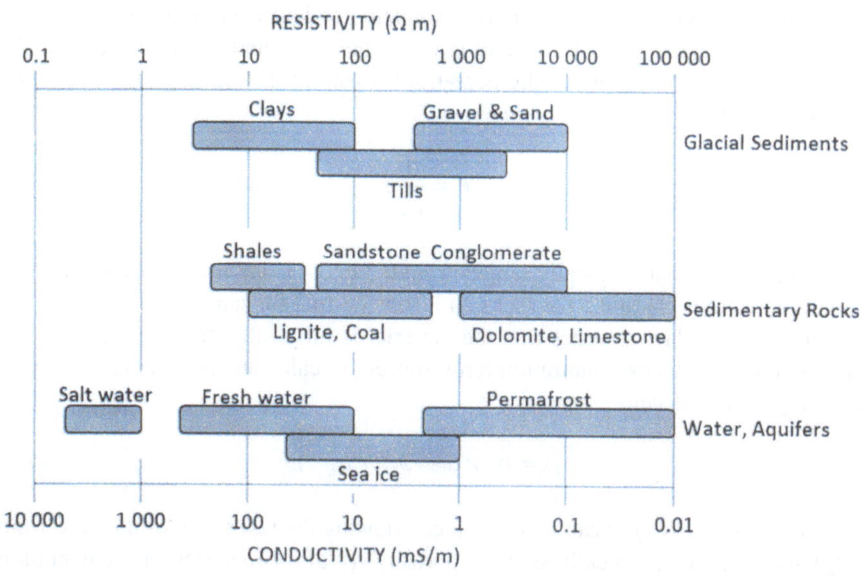

Fig. 7.8 Resistivity of various geological materials. (modified from Ref. [30])

where V_V is the volume of pore space and V_T is the total volume of the material. The degree of saturation, (S_w), is a measurement of the water content in a soil. The degree of saturation is the ratio of volume of water (V_w)to volume of void or pore space (V_V) and is given by

$$S_W = \frac{V_w}{V_v}. \tag{7.4}$$

The resistivity of the fluid in the pore space of a soil is known as the pore-fluid resistivity (ρ_w). The resistivity of the pore-fluid can be easily measured on an extracted fluid sample using a handheld resistivity meter. The clay fraction or clay content is the percentage of clay within the soil. There is some ambiguity in the definition of clay. Clay can be defined as materials having a grain size less than 2 μm or as a group of hydrous aluminum phyllosilicates, minerals which include kaolinite, montmorillonite-smectite, illite, and chlorite. The clay fraction defined by grain size is determined from a sieve analysis. An X-ray diffraction test determines the clay fraction based upon the mineralogy of the material and not the particle size. Since clay minerals carry an electric charge, the presence of this material will affect resistivity measurements.

Archie's first law is an empirical formula relating the resistivity to the porosity of a fully saturated rock (i.e., clean sand or coarse-grained material). Archie's first law is expressed as

$$\rho_b = \rho_w \times \phi^{-m} \tag{7.5}$$

The bulk resistivity (ρ_b) of a material is calculated knowing the resistivity of the pore-fluid (ρ_w), porosity of the sand (ϕ), and the cementation exponent (m). The ratio of the bulk resistivity to the pore-fluid resistivity is known as the formation factor F, where

$$F = \frac{\rho_b}{\rho_w}. \tag{7.6}$$

For sand, the porosity typically ranges from 0.3 to 0.45, and the cementation exponent ranges from 1.3 to 2.5 for rocks and from 1.8 to 2 for sands.

Archie's first law assumes that the material is fully saturated, $S_w = 1$. Archie's second law introduces a saturation term in order to calculate the bulk resistivity of partially saturated sand:

$$\rho_b = \rho_w \times a \times (\phi^{-m} \times S_w^{\ -n}) \tag{7.7}$$

The bulk resistivity (ρ_b) can be calculated knowing the resistivity of the pore-fluid (ρ_w), tortuosity (a), typically set to 1, porosity (ϕ) of the material, the cementation exponent (m), degree of saturation (S_w), and the saturation exponent (n), an empirical coefficient that depends on the pore-fluid, but is typically set to 2 when the pore-fluid of interest is water.

Archie's first and second law hold only for materials where clay is not present (i.e., clean sands). For clean sands, it is assumed that the electric current will flow through the pore-fluid; therefore, the measured bulk resistivity is directly related to the pore-fluid resistivity, the porosity of the material, and the degree of saturation. When clay is present, the path of the current is not solely through the pores but also along the surface of the clay material. Therefore, the measured bulk resistivity is now dependent on the clay content as well as the type of clay in the soil. The bulk conductivity (note we changed from resistivity to conductivity) for soils containing clay can be calculated using the Waxman–Smits formula:

$$\sigma_O = \frac{1}{F^*}(\sigma_W + \sigma_e). \tag{7.8}$$

The bulk conductivity of a fully saturated soil sample σ_o (mho cm^{-1}) is related to the formation factor of shaley sand, F^*, the conductivity of the pore-fluid, σ_w, and the conductivity due to the presence of the clay fraction, (σ_e). More specifically, σ_e is the conductance of clay counterions and has units of mho per centimeter. According to Waxman and Smits [32], F^* can be approximated using Archie's first law (combining Eqs. 7.5 and 7.6):

$$F^* = \phi^{-m}. \tag{7.9}$$

Waxman and Smits [32] constructed plots of σ_o versus σ_w based on laboratory measurements. This study illustrated that at low pore-fluid conductivity ($\sigma_w < 0.06$ mho cm^{-1}), the conductivity of the fully saturated sample increases exponentially as a function of the conductivity of the pore-fluid. When the conductivity of the pore-fluid is greater than 0.06 mho cm^{-1}, the conductivity of the fully saturated sample followed a linear trend. Ultimately, the σ_e term is affected by the pore-fluid conductivity. When σ_w is sufficiently large that the bulk conductivity is linear, then

$$\sigma_e = \frac{\lambda_{Na}^e}{1000} \times Q_V, \tag{7.10}$$

where λ_{Na}^e is the maximum equivalent ionic conductance of the sodium exchange ions with units of cm^2 equiv^{-1} ohm^{-1}. From the experimental data, λ_{Na}^e is determined to be equal to 38.3 cm^2 equiv^{-1} ohm^{-1}. Confidence levels of 10 and 90 % for λ_{Na}^e are 36.9 and 39.6 cm^2 equiv^{-1} ohm^{-1}, respectively [32]. Q_v is the concentration of sodium exchange cations associated with the clay and has units of equiv L^{-1} and can be expressed by

$$Q_V = \frac{\text{CEC}(1-\phi)\rho_0}{\phi}, \tag{7.11}$$

Table 7.5 Cation-exchange capacity (CEC) of common clays

Clay type	CEC value (meq/g)	Median CEC (meq/g)
Montmorillonite	0.8–1.5	1.15
Illite	0.1–0.4	0.25
Chlorite	0–0.1	0.05
Kaolinite	0.03–0.06	0.045

where CEC is the cation-exchange capacity, ϕ is the porosity of the clay-associated water, and ρ_0 is the mineral grain density [33].

The cation-exchange capacity, CEC, is determined from the type of clay present in the sample. Table 7.5 lists four major types of clay and the associated CEC value. Adjustments to the CEC value need to be made for instances when the clay fraction present is a result of a mixture of multiple clay types. For this case, knowing the percentage of each clay type within the mixture, an effective CEC value can be determined using:

$$\text{CEC}_{\text{mixture}} = \sum_{i=1}^{n} \frac{\text{Mass}_{\text{Clay Type}}}{\text{total Mass}_{\text{Clay}}} \times \text{CEC}_{\text{Clay Type}}. \tag{7.12}$$

For lower pore-fluid conductivity, where the increase in sand conductivity is exponential with increasing pore-fluid conductivity,

$$\sigma_e = BQ_V. \tag{7.13}$$

Equation 7.13 introduces a new term B which represents the equivalent conductance of the counterions as a function of pore-fluid conductivity (σ_w). B is expressed using the empirical formula

$$B = \left[1 - 0.6e^{\left(\frac{-\sigma_W}{0.013}\right)}\right] \times 0.046. \tag{7.14}$$

The threshold at which the sand conductivity transitions from an exponential to a linear dependence on pore-fluid is about 0.06 mhos cm^{-1}. A conductivity of this value represents a pore-fluid of salt water. For applications of dam integrity investigation, the pore-fluid present in the dam should have conductivity much less than 0.06 mhos cm^{-1}. Therefore, when using the Waxman–Smits equation to predict conductivities for the purpose of dam evaluation, Eq. 7.13 should be used.

7.1.2 Surveying Procedure

Electrical surveys are conducted using injection of electric currents from point source electrodes. Common electrical resistivity field methods use four electrodes

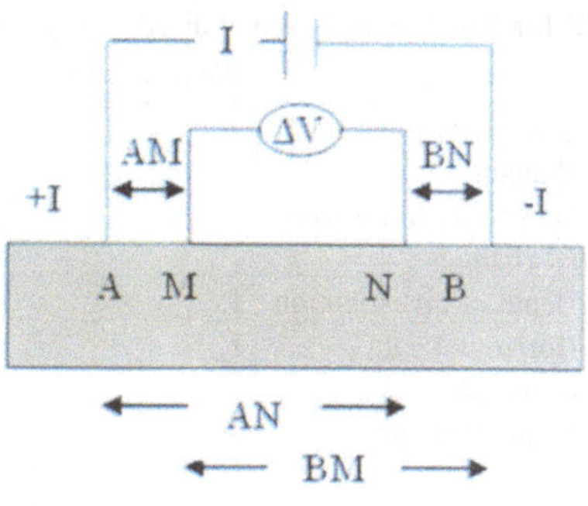

Fig. 7.9 Four-electrode configuration

to perform these surveys. In this configuration, shown in Fig. 7.9, two electrodes are injecting the current (generally labeled A and B), and the other two are the electrodes that measure the potential difference (generally labeled M and N).

The resulting resistivity of the subsurface is given by

$$\rho = \frac{2\pi \Delta V}{I} \cdot \left(\frac{1}{\frac{1}{\mathrm{AM}} - \frac{1}{\mathrm{BM}} - \frac{1}{\mathrm{AN}} + \frac{1}{\mathrm{BN}}} \right). \tag{7.15}$$

The distances between the electrodes are represented by AM (distance from electrode A and M), BM (distance from electrode B and M), AN (distance from electrode A and N), and BN (distance from electrode B and N). Equation 7.15 is often written as:

$$\rho = \frac{\Delta V}{I} \times K, \tag{7.16}$$

where ΔV is the applied voltage difference, I the measured current, and K is the geometrical factor that represents the configuration of electrodes A, B, M, and N. If the subsurface were entirely homogeneous, the measured resistivity would be independent of electrode configuration and spacing. Since the ground is not homogeneous, but actually heterogeneous, the measured resistivity is known as an "apparent resistivity."

Three common electrode configurations used in resistivity surveys are: the Wenner array, the Schlumberger array, and the dipole–dipole array. An understanding of the specific features being investigated, site noise, and equipment being used must be considered when choosing a specific array type. Each configuration will produce results that vary in resolution, sensitivity, and depth of investigation [34]. Table 7.6 is a modified table from Samouëlian et al. [34], which shows each electrode configuration and its corresponding strengths and weaknesses. The numbers range from low to high where the lower numbers represent poor sensitivity and the higher numbers represent higher sensitivity.

Table 7.6 Advantages and disadvantages of common electrode configurations

	Wenner Array	Schlumberger array	Dipole–dipole array
Sensitivity to vertical changes	4	2	1
Sensitivity to horizontal changes	1	2	4
Depth of investigation	1	2	3
Horizontal data coverage	1	2	3
Signal strength	4	3	1

Early electrical prospecting techniques were classified as vertical electrical sounding (VES) and constant separation traversing (CST). VES is a technique in which resistivity measurements are taken with increasing spacing between electrodes. An example of an application of this method is to locate the depth of the water table. The deeper soil investigation is associated with the increase in electrode spacing and provides information about the one-dimensional variation of resistivity with depth [34]. The Wenner electrode configuration is generally used when VES is performed. For simplified interpretations, the ground is assumed to consist of several horizontal layers [34].

CST, also known as electrical profiling, is another technique used to map out the variation of resistivity of the ground. For this method, the electrode spacing remains constant, and the electrode array is moved along in a straight line until the end of the survey area is reached. Since the electrode spacing is constant, the CST technique will map out lateral resistivity variations in the subsurface up to a constant depth. This surveying method is used in the levee screening phase. The dipole–dipole electrode configuration is commonly used when CST is performed [35].

The development of multichannel resistivity meters and electronic switches enables the use of many electrode configurations. This results in a measurement that is equivalent to many VESs. In electrical resistivity tomography (ERT), the forward problem uses the finite-element method to compute the electric potential response of the earth due to a given input electric current. The inverse algorithm iteratively finds the best distribution of subsurface resistivity that best fits the observed data. Daily et al. [36] presents a concise overview of ERT.

A photograph of an ERT field setup on a scaled earthen embankment is shown in Fig. 7.10. A line of 56 stainless steel electrodes are inserted into the crest of the dam at equal intervals and connected to the yellow electrode cables by stainless steel springs which are then connected to a multiplexer box, and consequently to the resistivity meter. The resistivity meter and multiplexer box is located at the center of the electrode spread for this survey. Two 12-V batteries power the system.

A command file containing the type of electrode configuration, electrode spacing, and other acquisition parameters must be created using the administrator software and uploaded to the resistivity meter. A quality-control measure is usually performed before a survey is conducted known as a contact resistance test. This test evaluates the electrical contact between the stainless steel electrodes and the ground.

Fig. 7.10 Photograph of an electrical resistivity tomography (*ERT*) field setup

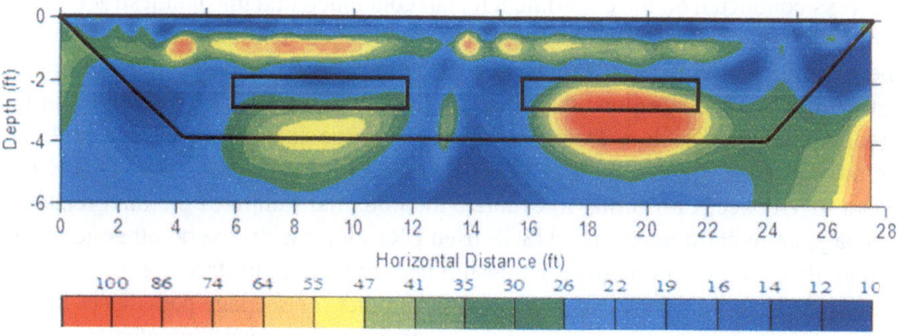

Fig. 7.11 Electrical resistivity tomogram for survey conducted on a scaled embankment dam with compromised zones

Poor contact (high resistance values) can produce noisy data and water is usually poured in the area surrounding the electrode to enhance electrical contact. Once the contact resistance is checked, the command file is executed and the measurements are obtained. Depending on the number of electrodes and the electrode configuration the measurement may take from several minutes to several hours to complete.

The resulting tomogram for an ERT survey using a dipole–dipole configuration with an electrode spacing equal to 0.15 m on a scaled embankment dam is shown in Fig. 7.11. The tomogram represents a slice through the embankment with the X-axis being the distance across the dam crest and the Y-axis being the depth in the dam. The intensity values for high resistivity are assigned colors of red and orange and low resistivity values have colors of blue. Lower resistivity values indicate areas of higher moisture and clay content while dryer area are indicated by higher resistivity.

A schematic of the dam is drawn on top of each tomogram to help illustrate the size and placement of known compromised zones. The results show that both the sand zone on the right and the dry clay zone on the left are imaged as zones of high resistivity in the tomogram and appear in their appropriate locations. Higher levels of moisture (low resistivity) are detected around both the left and right abutments. Also, a high-resistivity band is evident at a depth of 30 cm in the dam. This might be related to a variation in moisture of the clay material used during dam construction.

The long acquisition times for ERT surveying are an issue with wide spread use of this method for dam and levee assessment. Survey methods using capacitive-coupled electrodes have been proposed and tested on levees suitable for vehicles [37, 38]. Advances in such surveying methods may allow for expedient acquisition and wider spread use of ERT.

7.1.3 Case Studies

Case [39] examined the feasibility of ERT to map and monitor internal compromised zones within earthen embankment dams. Time-lapse ERT surveys were conducted on scaled embankment dam over a period of a year. The results from ERT surveys conducted on the experimental dam concluded that this nondestructive geophysical method is effective in identifying weak zones in an earthen dam that would be susceptible to seepage. Al-Fares [40] conducted ERT surveys to determine the cause of water leakage along the Afamia B Dam in Syria. ERT results helped outline geologic features such as fractures which were associated with seepage problems in the dam [40]. Bedrosian et al. [41] performed ERT surveys on the Martis Creek Dam in Truckee, California, to evaluate the potential failure of the dam related to seepage or an earthquake. The USGS used ERT along with several other geophysical methods to determine the relationship between geologic structure, seepage patterns, and reservoir depth. Its study concluded that the seepage paths were located along the interface between sedimentary deposits and the overlying glacial outwash [41]. Minsley et al. [42] performed electrical resistivity investigations at the Hidden Dam in Raymond, California, with the goal of identifying seepage paths associated with changes in the subsurface geology. Low-resistivity anomalies located on the right side of the dam were associated with groundwater seepage through a sediment channel. Weller et al. [43] performed several resistivity surveys on a series of dikes located in North Vietnam along the Red River. Additional case studies on the use of ERT for detecting compromised zones within embankments have been discussed in the Advance Geosciences, Inc. website [44].

7.2 *Seismic Methods*

Seismic methods for subsurface investigation use the fact that, elastic waves in soil and rock travel with different propagation velocities. The procedure for seismic investigations involves creating elastic waves using a seismic energy source, e.g., explosives, weight drops, sledge hammers, and measuring the arrival time of the elastic wave at a number of other locations using electromechanical transducers called geophones. The time taken for the elastic waves to travel from the seismic source to the receivers depends on the path travelled and the velocity in the material along the path. Seismic waves provide indirect information about the subsurface through its propagation velocity. Therefore, understanding the factors affecting seismic velocities is vital for extracting relevant information from seismic survey results.

Seismic methods are classified as active when some form of known, controlled energy source is used. Common active seismic techniques include reflection, refraction, surface wave, and borehole methods. Surface wave and refraction techniques are most commonly used in the assessment of dams and levees.

Passive seismic methods listen to seismic signals produced by fluid flow and mechanical failures within the embankments. A clear advantage of the passive techniques is they do not require active sources, only passive receivers. More recently, interferometric techniques incorporate aspects of both passive and active methods. These techniques use background seismic "noise," such as seismic vibrations created by automobile traffic, as the energy source for imaging the subsurface.

7.2.1 Seismic Theory and Elastic Properties

The simplest model for the propagation of seismic waves in soils and rocks is based on the theory of elasticity. The elastic properties of the subsurface materials are governed by two elastic moduli, which define the linear relationship between the stress and strain of the soil. The bulk modulus (K_b) of a soil body represents its resistance to a change in volume due to a normal compression. The shear modulus (G) of a soil is defined as the resistance of the soil to shear stresses.

Seismic waves are vibrations that travel through the earth carrying the energy released during natural events such as earthquakes or artificial sources such as explosives or impacts on the ground. Seismic waves are divided into two groups known as body waves and surface (interface) waves. Seismic body waves travel through the interior of an elastic body and always travel faster than surface waves. Body waves are divided into two wave types known as compressional or P-waves and shear or S-waves. The velocity for a P-wave in terms of the elastic moduli and density is

$$v_p = \left[\frac{K_b + \frac{4}{3}G}{\rho}\right]^{\frac{1}{2}}. \tag{7.17}$$

Shear waves propagate by a pure shear strain and the S-wave velocity is given by

$$v_s = \left[\frac{G}{\rho}\right]^{\frac{1}{2}}. \tag{7.18}$$

Table 7.7 Typical values of P-wave and S-wave velocities

Material	P-wave velocity (m/s)	S-wave velocity (m/s)
Air	332	
Water	1400–1500	
Petroleum	1300–1400	
Steel	6100	3500
Concrete	3600	2000
Granite	5500–5900	2800–3000
Basalt	6400	3200
Sandstone	1400–4300	700–2800
Limestone	5900–6100	2800–3000
Sand (unsaturated)	200–1000	80–400
Sand (saturated)	800–2200	320–880
Clay	1000–2500	400–1000
Glacial till (saturated)	1500–2500	600–1000

S-waves can be polarized because the particles oscillate along a defined line perpendicular to the direction of wave propagation. In general, there can be two planes of polarization and therefore two shear waves. These are sometimes referred to as horizontally and vertically polarize shear waves. S-waves can only propagate in materials that have shear strength and therefore do not propagate through fluids. P-waves travel faster than S-waves. The ratio $\frac{v_p}{v_s}$ is independent of density and can be written in terms of Poisson's ratio (v) of the material as:

$$\frac{v_p}{v_s} = \left[\frac{2(1-v)}{(1-2v)}\right]^{\frac{1}{2}}. \tag{7.19}$$

P-wave and S-wave velocities for different materials are listed in Table 7.7. The large range of velocities for geo-materials such as soils and rocks suggests that these are mechanically complex and the elastic model for these materials is a very simplified representation.

Surface waves at the surface of the earth (assumed to be a free surface) are divided into Rayleigh waves and Love waves. Love waves are horizontally polarized shear waves (SH-waves) guided by an elastic layer. Love waves travel slower than P- or S-waves, but faster than Rayleigh waves. These waves are observed only when there is a low-velocity layer overlying a high-velocity layer/sublayers. Few studies have attempted to use Love waves for dam and levee assessments.

The propagation of Raleigh waves is by a combination of parallel and perpendicular particle movements to the direction of propagation. In other words, they are composed of compressional and shear wave particle movements. The equation for the propagation velocity of Rayleigh waves (V_R) is given by Achenbach [45], in terms of the P- and S-wave velocities of the material, as:

$$\left[2-\left(\frac{V_R}{v_S}\right)^2\right]^2 - 4\left[1-\left(\frac{V_R}{v_P}\right)^2\right]^{1/2}\left[2-\left(\frac{V_R}{v_S}\right)^2\right]^{1/2} = 0. \tag{7.20}$$

The propagation velocity of Rayleigh waves is slightly less than that of S-waves.

In practice, Rayleigh waves travelling at the surface of the Earth are observed to be dispersive; their waveform is undergoing progressive change during propagation as a result of the different frequency components traveling at different velocities. This dispersion is directly attributed to velocity variation with depth and is the bases for spectral analysis of surface waves (SASW) method and multichannel analysis of surface waves (MASW) methods for dam and levee assessment. The use of surface waves to measure the shear-wave velocity distribution of the subsurface was first introduced in the SASW method and is based on a two-receiver acquisition and processing scheme, but fails to account for unfavorable waves such as higher modes of surface waves, body waves, and ambient waves [46]. In the early 1990s, a new method utilizing multiple channels was developed known as MASW. This method uses an acquisition similar to traditional seismic exploration acquisitions, where multiple receivers (geophones) are used along a straight survey line.

There are numerous factors that can affect seismic velocities through soils and rocks [33, 47]. Some of them are: lithological properties (grain sizes, grain shape, grain type, grain size distribution, amount of compaction, amount of consolidation, and cementation), physical properties (porosity, permeability, density, degree of saturation, pressure, and temperature), and elastic properties of the constituents (shear modulus (G), bulk modulus (K_b), Young modulus (E), Poisson's ratio (v), and Lamé constant (λ)).

A more complicated model for a soil to account for the effects of the pore-fluid is to assume it is a fully saturated porous material. Considering a porous soil mass where the interconnected pores of the soil are filled with water, the presence of water in the pores acts to stiffen the soil and causes the flow of pore water by diffusion between regions of higher and lower pore pressure. According to Biot's deformation model, the total stress of a soil mass consists of both the effective stress given by the strain of the solid structure and the pore pressure associated with the water in the pores.

Gassmann [48] formulated equations, which can be used to predict changes in seismic velocity for fully saturated rocks in the low-frequency regime. The equations, which can be used for estimating the effect of full saturation on the bulk modulus of a porous material is given by

$$K_{sat} = K_d + \Delta K_d, \tag{7.21}$$

where

$$\Delta K_d = \frac{K_o(1 - K_d/K_o)^2}{(1 - K_d/K_o) - \phi[1 - K_o/K_f]} \tag{7.22}$$

and

$$G_s = G_d. \tag{7.23}$$

K_o is the bulk modulus of the soil mineral grain, K_f is the bulk modulus of the fluid, K_d is the bulk modulus of the dry rock, and K_{sat} is the bulk modulus of the saturated rock. G_s is the saturated shear modulus and G_d is the dry shear modulus of the soil.

The presence of water in the pore space increases the bulk modulus of the rock; this overshadows the effect of increased density resulting in an increased P-wave speed. The shear modulus is not affected by the water in the pores, so the S-wave velocity decreases due to the increase in density.

Gassmann's equations indicate that the bulk modulus of a saturated soil depends on the porosity of the soil mass. However, pore spaces of a soil mass can be either fully saturated or partially saturated. The bulk modulus of the pore-fluid that is a liquid–gas mixture (partially saturated) is related to the moduli of the fluid components as:

$$\frac{1}{K_f} = \frac{S_w}{K_l} + \frac{1 - S_w}{K_g}, \tag{7.24}$$

where K_f is the effective bulk modulus of the liquid–gas mixture, S_w is the degree of saturation, K_l is the bulk modulus of the liquid, and K_g is the bulk modulus of the gas. Equation 7.24 does not account for capillary pressure. Brooks and Corey [49] formulated a relationship relating the degree of saturation to the capillary pressure as shown below [50]:

$$S_w = S_{irr} + (1 - S_{irr})\left(\frac{P_t}{P_c}\right)^{\lambda} \tag{7.25}$$

where S_{irr} is the irreducible water saturation, P_t is the threshold pressure, P_c is the capillary pressure, and λ is a constant for a given lithology. The threshold pressure is defined as the measure of gas pressure required to initiate the displacement of water and is a property of the soil or rock. The empirical formula in Eq. 7.25 provides insight into the complex interrelationship between factors such as degree of saturation, capillary pressure, permeability, and porosity. For embankment materials composed primarily of soils, the capillary pressure could provide a significant contribution to the grain-contact stiffness and thereby affect the seismic velocity.

7.2.2 Surveying Procedure

The most common form of shallow seismic surveying records signals from a surface source into a line of equally spaced geophones along the crest of the dam or levee to produce a cross section of the subsurface. This produces a good balance among acquisition speed, processing ease, interpretation convenience, and anomaly characterization. The resulting cross section appears less ambiguous and more easily interpreted than other methods.

Seismic refraction surveying has been widely used for near-surface and high-resolution investigations. Data acquisition for seismic refraction surveying, shown

Fig. 7.12 Photograph of a seismic refraction survey on the crest of a dam. A sledgehammer seismic source and a line of 24 geophone receivers are shown

in Fig. 7.12, requires placing a line of multiple geophones on the ground surface and creating seismic waves using an impact source at a shot point location. The seismic energy at the shot point will travel directly through the upper layer and arrive at the geophones as a direct arrival, or it may travel down to deeper layers, refract back to the surface, and arrive at the geophones as a refracted wave. The geophones record these energies, which are stored as a time-dependent waveform on a seismograph referred to as a seismic trace. The record containing the information from all the geophones in the spread is referred to as a seismogram.

Seismic refraction requires quite simple processing; the first arrival time is the relevant information required from the field seismograms. The first arrival time is the time it takes for the first seismic energy to travel from the source to a geophone. These first arrival times are picked for all geophones of the spread and are used to determine the velocity of seismic waves in the subsurface. Figure 7.13 is an example of a shot gather from a P-wave refraction survey. The *red line* on the shot gather indicates the location of the first arrival picks. The first arrival times from the field data must be further analyzed using a traditional approach, which is subject to restrictive assumptions or more advance tomography inversion methods.

The traditional method of refraction surveying utilizes the time–distance $(t - x)$ graph to determine the seismic velocity of the subsurface and depth to the different layers. Methods of analysis can be found in many introductory geophysics texts [15]. In overview, an offset (x) is defined as the distance between a seismic source and a geophone. For receivers (geophones) at a very short offset, the first arrivals of

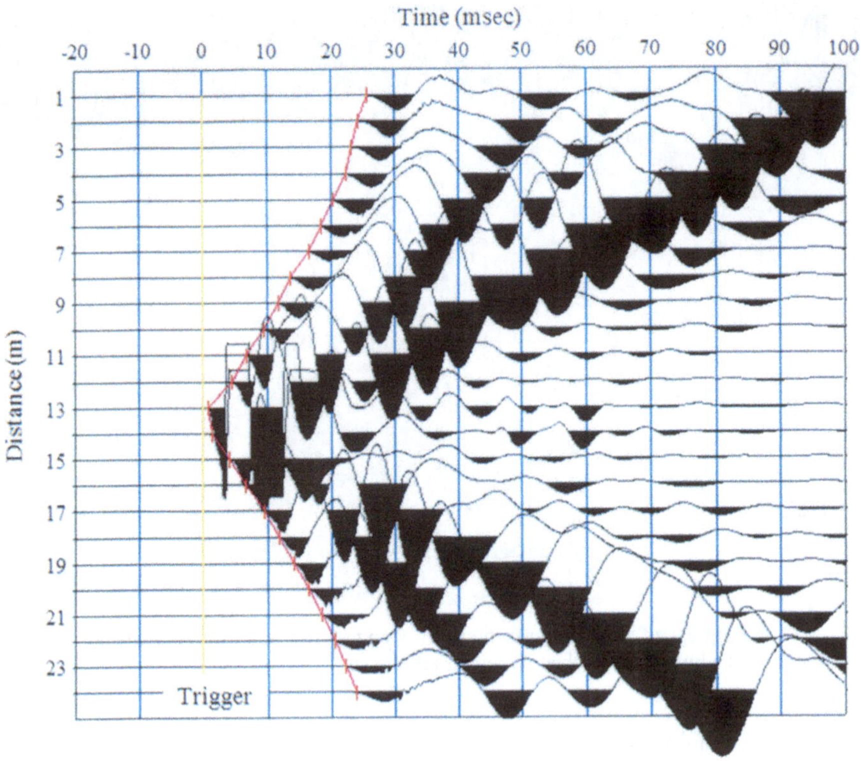

Fig. 7.13 A shot gather from a P-wave refraction survey. The *red line* on the shot gather indicates the location of the first arrival picks

seismic energy are always the direct waves. Beyond an offset known as the crossover distance, (x_c), refracted rays from the lower layer arrive first at the geophones:

$$x_c = 2h\left(\frac{v_2 + v_1}{v_2 - v_1}\right)^{\frac{1}{2}}, \tag{7.26}$$

where v_1 is velocity of the top layer, v_2 is velocity of the halfspace, and h is the thickness of the layer. For a horizontal layer case, the velocity of the waves in the first layer can be obtained by the inverse of the slope of the arrival times of the direct waves. Similarly, the inverse of the slope after the crossover distance is used to obtain the velocity in the second layer. The crossover distance or intercept time is used to calculate the thickness of the layer. More advance methods of processing include:

- Plus–minus method
- Generalized reciprocal method
- Ray-tracing methods
- Wavefront methods

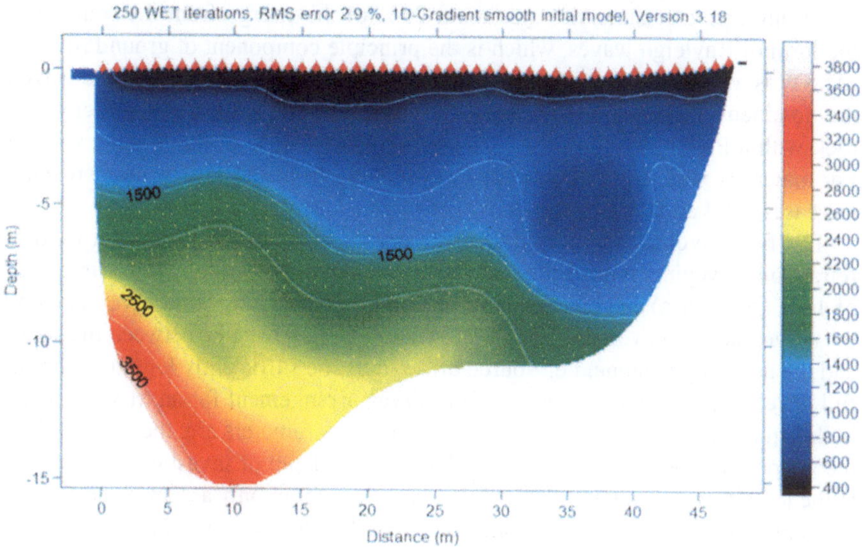

Fig. 7.14 An example of a P-wave velocity tomogram for a 48 geophone survey line. The low-velocity (*blue*) anomaly on the right suggest weak or porous zone at a depth of about 5 m

The traditional approach for interpreting refraction data uses the $t-x$ graph to determine the depth and velocity of different layers subject to somewhat restrictive assumptions. More advanced methods that have less restrictive assumptions are currently being developed. One very prominent method is known as seismic tomography. Seismic tomography is based on travel-time anomalies observed for many ray paths. The travel times associated with different source–receiver combinations is inverted for a "best-fit" velocity distribution of the subsurface. Images obtained from seismic tomography are referred as seismic tomograms.

To build tomograms, multiple shot records must be obtained with multiple geophones. All these shot records are then loaded into processing software to obtain the first arrival picks. A 2D velocity tomogram is a station location (distance) versus depth image showing the velocity distribution in the subsurface. Figure 7.14 shows an example of a P-wave velocity tomogram for a 48 geophones survey line. The locations of the geophones are indicated by the red triangles in the figure. The velocity tomogram is plotted using color scales depending on the value of the velocity obtained for each grid after processing the first arrival times. The low-velocity (*blue*) anomaly on the right of the tomogram suggests a weak or porous zone at a depth of about 5 m.

Tomograms are ideal for investigating the interior of dams and levees because they provide a good spatial image. This is a great advantage over boring programs where data are collected at specific locations of interests and interpolated for the areas in-between boreholes.

During active seismic surveys, about 70 % of the energy from the source is in the form of Rayleigh waves, which is the principle component of ground roll [51]. MASW is a seismic surveying method based on the knowledge that the S-wave is the dominant influence on the Rayleigh wave. The Rayleigh wave velocity for a rock with a Poisson's ratio of 0.25 is approximately 92 % of the S-wave velocity. For materials with higher Poisson's ratio ranging from 0.4 to 0.5, the percentage increases to 94 and 95.5 %, respectively.

The field layout for active MASW surveying is similar to the layout used for refraction surveying. It utilizes a linear array of geophones (usually lower frequency of 4-Hz geophones) and the geophone spread is "rolled along" the survey line until the survey area is covered. For an MASW survey, the distance between the source and the nearest geophone, i.e., source offset, must be sufficiently long to insure the full development of surface waves. The survey arrangement for an MASW survey is shown in Fig. 7.15. The three important survey parameters are: the source offset, the geophone spacing, and the spread length. The depth of investigation is related to the longest wavelength of the surface wave. It is commonly assumed the depth of investigation is from one half to one quarter of the longest wavelength. In practice, the spread length must be equal to the maximum wavelength and the source offset must be equal to one half of the spread length. The geophone spacing is usually determined by the spread length and the number of channels in the data acquisition system. As an example, consider a required depth of investigation of 50 m. Assuming that the half wavelength criteria would require the analysis of Rayleigh waves with a wavelength of 100 m. Measuring these Rayleigh waves requires a spread length of 100 m and an offset of at least 50 m. If the recording system has 48 channels, then the geophone spacing would be 100 m/47 or approximately 2 m.

Data processing with commercial software is quite straight forward. Each shot record represents the ground vibration as a function of time for each geophone (or space). Data from multiple shots are sorted into common midpoint (CMP) gathers [15]. This data are then transformed to the frequency–wave number ($f-k$) domain using Fourier transforms on the time and space domain. The $f-k$ diagram is then used to develop a wave velocity versus frequency, or dispersion curve, for the Rayleigh wave. This dispersive property of the Rayleigh wave is utilized to investigate the change in embankment material as a function of depth. The final step is the inversion of the Rayleigh wave dispersion curve to obtain an S-wave velocity profile (one-dimensional S-wave velocity function, v_s vs. depth). This S-wave profile represents an average over the spread length and is plotted at the CMP of the spread. A two-dimensional cross section of the S-wave velocity is generated by interpolation between different profiles [52].

Although the MASW survey can provide a two-dimensional map of the shear-wave velocity distribution faster than S-wave refraction survey, there are some limitations on the method. One of the limitations is that MASW velocity maps are obtained by interpolation of multiple one-dimensional S-wave velocity profiles located at the CMP. This interpolation reduces the resolution of the S-wave velocity map compared to the S-wave refraction velocity tomograms.

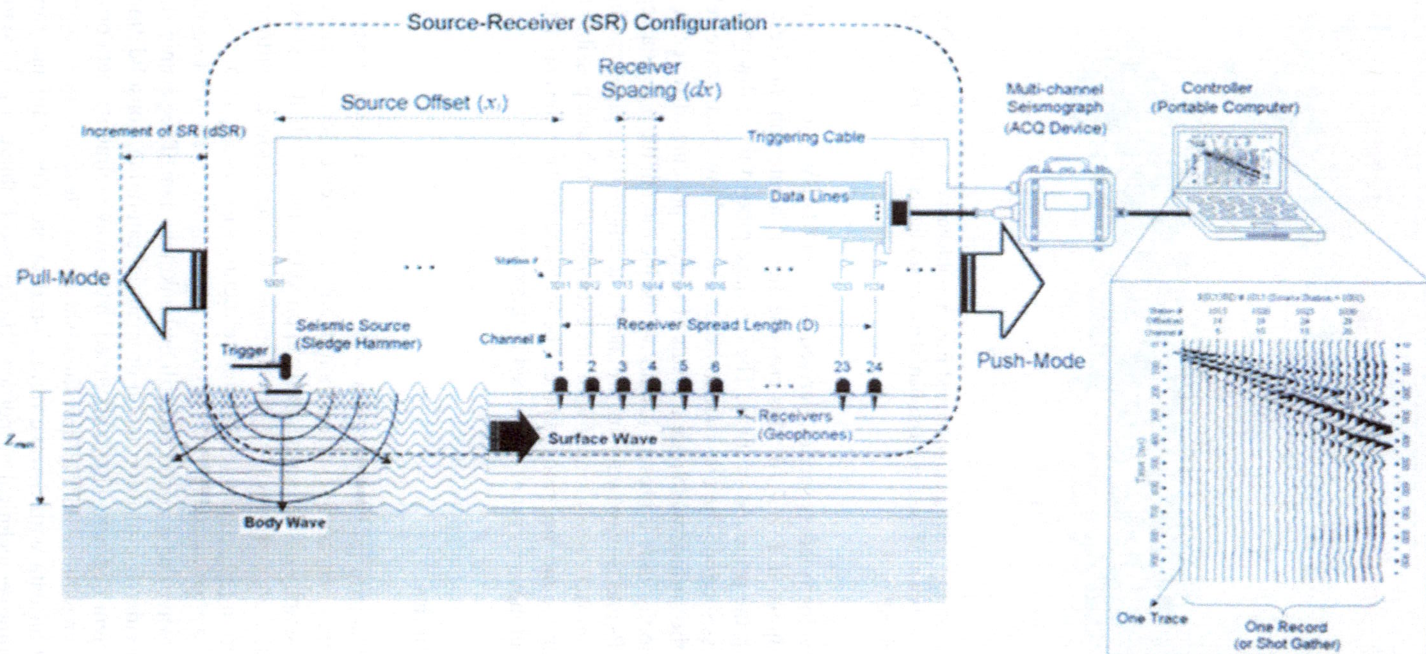

Fig. 7.15 Multichannel analysis of surface waves (MASW) survey arrangement and data acquisition. Important parameters are source offset, geophone spacing, and spread length. (http://www.masw.com)

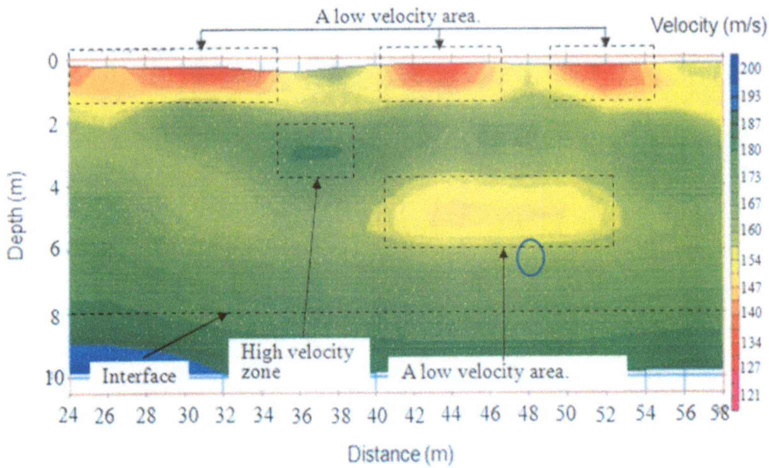

Fig. 7.16 S-wave velocity cross section derived from a multichannel analysis of surface waves (*MASW*) survey on a dam in Mississippi. This part of the survey line is located over the subsurface pipe of the principle spillway

An S-wave velocity cross section derived from an MASW survey on a dam in Mississippi is shown in Fig. 7.16. This cross section is located on the crest of the dam in line with the principle spillway. The drainage pipe of the spillway is annotated on the cross section. The cross section shows several low shear-wave velocity zones close to the surface which are a result of surface ponding of water. A wide, low S-wave velocity zone is present just above the location of the pipe at a depth of 5 m. Both low P- and S-wave velocity around the location of the pipe is an indication of lower compaction associated with the excavated trench and possible issues with compaction around the drainage pipe. The interface between the native ground and embankment material is not delineated in this cross section.

7.2.3 Case Studies

Osazuwa and Chinedu [53] conducted a seismic refraction survey to image high-permeability zones beneath an earthen dam in Nigeria. They observed that seismic refraction tomography can be successfully applied to view the shallow subsurface high-permeability zones responsible for seepage of water below an earthen dam. Powers and Burton [54] conducted seismic refraction surveys as part of a series of geophysical investigation using DC resistivity and seismic refraction tomography (P-wave and S-wave) on a zoned earthfill embankment dam. The seismic surveys on the dam provided useful information on material saturation, consolidation, and depth to competent bedrock. They concluded that S-wave refraction data provided results consistent with traditional downhole shear wave tests and that combining P-wave refraction data with S-wave refraction data provides additional capabilities for evaluating geotechnical properties. Rucker and Holmquist [55] conducted surface seismic surveys for locating and tracing earth fissures and other significant discontinuities in cemented unsaturated soils and earthen structures. They concluded that seismic refraction is a cost-effective and noninvasive way to detect anomalies in unsaturated soils in arid and semiarid environments. Kilty et al. [56] conducted a P-wave seismic refraction survey using controlled exploding sources at Horse Mesa Dam. They concluded that seismic refraction surveys can be used to reduce the number of drillings required to provide the location of bedrock. Ivanov et al. [57] used seismic methods on five levees in southern Texas to determine compressional and shear velocity (v_s) distribution within the body of the levees and examined the relationship to existing core taken from the levee and airborne EM data. Inazaki and Sakamoto [37] conducted a number of geophysical surveys including MASW, resistivity, and multifrequency EM method, for the geotechnical characterization and safety assessment of a levee. Ivanov et al. [58] conducted multiple seismic surveys for the evaluation of Ball Mountain Dam in Vermont. The survey included MASW, refraction tomography, and vertical seismic profiling. Wodajo [59] studied three seismic survey techniques; seismic refraction, MASW, and shear wave surveys for the purpose of dam integrity assessment. Multiple techniques were applied to an earthen embankment dam in Mississippi, and one in Alabama. Inferences were made on how seismic tomography images can be used to detect compromised zones (i.e., seepage, piping, etc.) within earthen embankment dams.

7.3 Electromagnetic Methods

EM methods were first developed during the 1920s in Scandinavia, the USA, and Canada where the detection of conductive base-metal deposits were facilitated by their large contrast with resistive host rock [15]. Geophysical EM methods involve the propagation of continuous wave or transient EM fields in or over the surface of the earth. Electromagnetic methods involve both natural and artificial sources. Methods based on natural sources are most often referred to as magnetotelluric methods. Telluric currents occur in response to ionosphere tidal effects and lightning. Magnetotelluric methods can be used to investigate greater depths than other EM methods. Very low frequency (VLF) measures the EM fields generated by military communication transmitters. An advantage of VLF is that the source transmitter does not have to be transported into the field by the survey crew. The depth of investigation ranges from less than 30 m in high-conductivity terrains but is typically 35–60 m in saturated overburden.

The more common EM methods using artificial sources can be subdivided into frequency domain (electromagnetic interference EMI) methods and time domain methods. The frequency domain methods measure the subsurface electrical conductivities by low-frequency sinusoidal EM induction. The time domain (or transient) electromagnetic methods (TDEM) measure the change in response as a function of time after a source is abruptly turned off. Geoelectric sounding interprets the decaying magnetic field as a function of time with respect to the subsurface resistivity as a function of depth. Common EM methods used in engineering are listed in Table 7.8.

Table 7.8 Commonly used electromagnetic (EM) methods in environmental and engineering surveys

Methods	Source characteristics	Frequency	Parameters measured	Principal application
VLF	Far-field source, radio transmitter	Single	Tilt angle	Mapping structure
AMT	Natural EM source	Broadband	*E*, *B*	Resistivity sounding
CSAMT	Intermediate-field, electric bipole	Broadband	*E*, *B*, *Ǿ*	Resistivity sounding
HLEM (slingram)	Near-field source, mobile loop	Variable	*B*	Mapping, shallow sounding
GCM (ground-conductivity meter)	Near-field source, mobile loop	Single	*B*	Mapping, shallow sounding
TDEM/TEM	Near-field source, mobile loop, x(T, R) variable	Time domain	d*B*/d*t*	Mapping, shallow sounding Sounding
	Central loop, x(T, R) fixed		d*B*/d*t*	Sounding

VLF very low frequency, *AMT* audio-frequency magneto tellurics, *CSAMT* controlled source audio-frequency magneto tellurics, *HLEM* holomorphic embedding load-flow method, *TDEM* time domain electromagnetic method, *TEM* transient electromagnetic method

7.3.1 Electromagnetic Theory and Properties

The theory governing EM behavior is very well developed and the governing system of equations is known as Maxwell's equations. There are numerous physics textbooks on this subject at all levels of mathematical complexity [60, 61, 62]. The description for a time-harmonic electric (E) and magnetic field intensity (H) in a uniform medium is given by

$$\nabla\vec{\mathrm{E}} + \kappa^2\vec{\mathrm{E}} = 0 \tag{7.27}$$

and

$$\nabla\vec{\mathrm{H}} + \kappa^2\vec{\mathrm{H}} = 0 \tag{7.28}$$

The propagation constant is given by

$$\kappa^2 = \omega^2\mu\varepsilon - i\omega\mu\sigma, \tag{7.29}$$

where ω is the time-harmonic frequency and μ, ε, σ are characteristic material properties. The electrical conductivity (σ) is the inverse of the resistivity that was discussed earlier and relates the current to the electrical field. A list of electrical resistivity of typical earth materials are presented in Table 7.9. The magnetic

Table 7.9 Resistivity of earth materials

Crystalline rocks	Range of ρ
Granite	10^2–10^6
Diorite	10^4–10^5
Gabbro	10^3–10^6
Andesite	10^2–10^4
Basalt	10–10^7
Peridotite	10^2–10^3
Schist	10–10^4
Gneiss	10^4–10^6
Slate	10^2–10^7
Marble	10^2–10^8
Quartzite	10–10^8
Sedimentary rocks	Range of ρ
Shale	10–10^3
Sandstone	1–10^8
Limestone	50–10^7
Dolomite	10^2–10^4
Unconsolidated sediment	Range of ρ
Sand	10–10^3
Clay	10–10^2
Marl	10–10^2
Groundwater	Range of ρ
Portable well water	0.1–10^3
Brackish water	0.3–1
Seawater	0.2
Supersaline brine	0.05–0.2

Table 7.10 Relative magnetic permeability for minerals

Mineral	Relative permeability
Magnetite	5
Pyrrhotite	2.55
Titanomagnetite	1.55
Hematite	1.05

permeability (μ) describes a materials' ability to become magnetized and is often related to the magnetic susceptibility (χ_m) by

$$\mu = \mu_0 (1 + \chi_m), \tag{7.30}$$

with $\mu_0 = 4\pi \times 10^{-7}$ being the permeability of free space. Most minerals have a small magnetic susceptibility and therefore a relative permeability close to one except those listed in Table 7.10. The values in Table 7.10 are normalized by the permeability of free space. The electrical permittivity of the material (ε) is related to the electrical susceptibility (χ_e) as:

Table 7.11 Dielectric constants of rocks and minerals

Rock, mineral	Dielectric constants
Calcite	7.8–8.5
Rock salt	5.6
Anthracite	5.6–6.3
Gypsum	5–11.5
Granite (dry)	4.8–18.9
Gabbro	8.5–40
Diorite	6.0
Serpentine	6.6
Gneiss	8.5
Sandstone (dry to moist)	2.9–105
Packed sand (dry to moist)	2.9–105
Soil (dry to moist)	3.9–29.4
Basalt	12
Clays (dry to moist)	7–43
Petroleum	2.07–2.14
Water (20 °C)	80.4
Ice	3–4.3

$$\varepsilon = \varepsilon_0 (1 + \chi_e) \tag{7.31}$$

with the permittivity of free space $\varepsilon_0 = 8.85 \times 10^{-12}$ $C^2\ N^{-1}\ m^{-2}$. The permittivity is more often referenced in terms of the dielectric constant (ε_r):

$$\varepsilon_r \equiv \frac{\varepsilon}{\varepsilon_0}. \tag{7.32}$$

Water has a large dielectric constant value of 81. Dielectric constants for different rocks and minerals are listed in Table 7.11.

Although not very useful for geophysical methods on dams and levees, an interesting limit of Eq. 7.29 is that of free space, where

$$\begin{aligned} \mu &\Rightarrow \mu_0 = 4\pi \times 10^{-7}, \\ \varepsilon &\Rightarrow \varepsilon_0 = 8.85 \times 10^{-12}, \\ \sigma &= 0, \end{aligned} \tag{7.33}$$

so that the speed of propagation of the field is given by

$$c \equiv \frac{\omega}{\kappa} = \frac{1}{\sqrt{\mu_0 \varepsilon_0}} = \frac{1}{\sqrt{(4\pi \times 10^{-7})(8.85 \times 10^{-12})}}, \tag{7.34}$$

which equates to

$$c = 2.9986 \times 10^8\ ms^{-1} \tag{7.35}$$

or the speed of light in a vacuum.

At low frequencies, usually for $f < 10^5$ Hz,

$$\mu\varepsilon\omega^2 \ll i\omega\mu\sigma \tag{7.36}$$

and

$$\kappa^2 \approx i\omega\mu\sigma. \tag{7.37}$$

This is known as the inductive regime so the "propagation" depends on the magnetic permeability and electrical conductivity. As stated above, the magnetic permeability does not change for most materials and therefore measurements in the inductive regimes are used to measure the conductivity of the ground.

In the high-frequency limit, $f \gg 10$ MHz and low conductivity,

$$\pi\varepsilon\omega^2 \gg \omega\mu\sigma \tag{7.38}$$

the propagation constant is:

$$\kappa^2 \approx \omega^2\mu\varepsilon \tag{7.39}$$

In this regime, the propagation constant is real valued and depends only on the magnetic permeability and electric permittivity or dielectric constant. Airborne radar systems for aviation and law enforcement fall into this regime. Ground-penetrating radar (GPR) systems are based upon propagating waves, but in most cases the conductivity must be accounted for and attenuation must be included.

In general, when attenuation must be accounted for, the electric and magnetic fields are described by a complex propagation constant. This complex propagation constant, implies that the field decays as it propagates in space. For example, the $x-$component of electric field can be written as:

$$E_x = E_{x0}e^{-i\kappa z}, \tag{7.40}$$

where the field E_{x0} is referenced at $z = 0$.

Writing in terms of a complex propagation constant,

$$E_x = E_{x0}e^{i(\beta+i\alpha)z}, \tag{7.41}$$

where β is the phase factor and α is the attenuation factor.

For the propagation constant derived earlier,

$$\beta = \omega\left[\frac{\mu\varepsilon}{2}\left\{\left(1+\frac{\sigma^2}{\varepsilon^2\omega^2}\right)^{1/2}+1\right\}\right]^{1/2} \tag{7.42}$$

and

$$\alpha = \omega\left[\frac{\mu\varepsilon}{2}\left\{\left(1+\frac{\sigma^2}{\varepsilon^2\omega^2}\right)^{1/2}-1\right\}\right]^{1/2}. \tag{7.43}$$

These expressions are used to calculate the velocity of propagation and attenuation of a GPR signal as it propagates through the ground.

The depth of penetration of an EM signal is usually defined as the skin depth. The skin depth is the distance (depth) at which the amplitude of the field has decayed by $1/e$ of its original amplitude,

$$\delta = \frac{1}{\alpha}. \tag{7.44}$$

In the case where conduction currents dominate over displacement currents:

$$\alpha = \beta = \left(\frac{\omega\mu\sigma}{2}\right)^{1/2}, \tag{7.45}$$

then the skin depth is:

$$\delta = \left(\frac{2}{\omega\mu\sigma}\right)^{1/2}. \tag{7.46}$$

This shows that the depth of penetration decreases with increasing frequency and conductivity.

7.3.2 Surveying Procedure

Most EM methods applied to dam and levee assessment are those methods which involve the use of coils for the source and receiver. The ability to conduct surveys without contacting the ground, makes these methods appropriate for surveying large lateral distances associated with the screening phase or obtaining plan view data. This data are usually an average response over a range in depth of the ground. EM surveys have also been conducted using airplanes and helicopters to cover large sections of levees [63, 64].

The basic concept of EM methods using two coils is shown in Fig. 7.17. A primary field (P_f) from the source coil spreads out above and below the ground. In the presence of a conducting body, the primary magnetic field produces eddy currents in the conducting body. These eddy currents produce a secondary field (S_f) which is added to the primary field. The resultant field (R_f) produces a current in the receiver coil which is the measured response. Different data are extracted from the measured

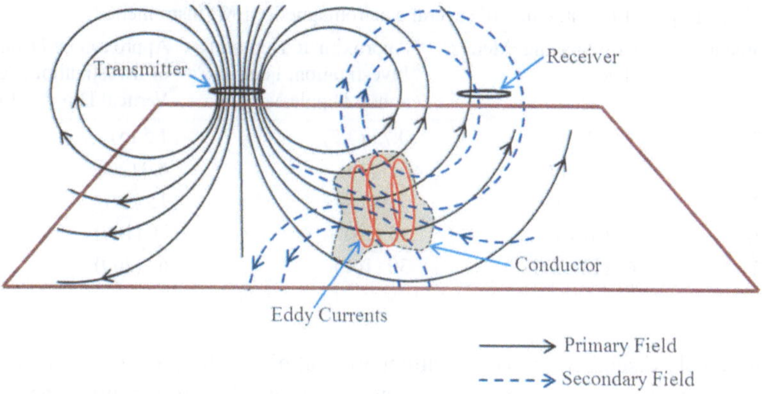

Fig. 7.17 Electromagnetic induction

resultant field and include: the tilt angle, phase difference, amplitude ratio, and time decay. The decay time is usually measured in TDEM methods. Although these time domain methods use coils, the transmitter coil is usually quite large and not easily moved for surveying purposes.

The frequency domain or EMI methods have been more widely used for dam and levee screening and assessment. The instrumentation can have fixed coils and some methods allow for different source–receiver coil spacing. Figure 7.18 shows a picture of an EM-31 fixed coil instrument and an EM-34 which allows for the coil spacing to be changed [65]. When selecting an EM instrument for surveying there are usually three factors to consider:

- Frequency of operation
- Coil spacing
- Coil orientation

Fig. 7.18 Electromagnetic (EM) induction instrumentation using **a** EM-31 and **b** EM-34. (http://www.geonics.com)

Table 7.12 Depth of investigation of several electromagnetic (EM) instruments

Instrument	Coil Spacing, Meters (Feet)	Approximate Depth of Investigation, Feet, Horizontal Dipole Mode	Approximate Depth of Investigation, Feet, Vertical Dipole Mode
EM-38	1 (3.28)	0.75 (2.5)	1.5 (5)
EM-31	3.7 (12.14)	3.0 (10)	6.0 (20)
EM-34	10 (32.81)	7.6 (25)	15 (50)
EM-34	20 (65.61)	15 (50)	30 (100)
EM-34	40 (131.22)	30 (100)	60 (200)

As previously discussed above, the attenuation at higher frequency limits the depth of interrogation. The usual rule is that the maximum detection depth is about half the transmitter–receiver coil spacing. However, it is obvious that the frequency and coil separation must be related for the proper operation of the instrument and ease in data interpretation. The induction number is defined as the ratio of the coil spacing (a) to the skin depth (δ):

$$N_B = \frac{a}{\delta} = a\sqrt{\frac{\mu_0 \omega \sigma}{2}}. \tag{7.47}$$

For small induction numbers, the ratio of the secondary to primary magnetic field is given by

$$\frac{H_S}{H_P} = i\frac{\mu_0 \omega \sigma a^2}{4}, \tag{7.48}$$

such that the ground conductivity can be determined from the quadrature decomposition of the signal.

Coil orientation also influences the relative contributions to the secondary magnetic field from all materials below a given depth. A measurement with a horizontal loop configuration is referred to as a vertical dipole and coils oriented in a vertical fashion are referred to as a horizontal dipole. Effective exploration depth is about 0.25–0.75 the intercoil spacing for the horizontal dipole and 0.5–1.5 intercoil spacing for the vertical dipole. The depths of investigation of several EM instruments are listed in Table 7.12.

The EM-34 instrument has been used in the horizontal loop EM method (also known as the Slingham method) as well as in a vertical loop configuration for dam and levee screening [66, 67]. The coil configuration for the Slingham configuration is shown in Fig. 7.19. The transmitter and receiver coils are about 1 m in diameter and are usually carried horizontally with a 30–100-m separation. Coils are connected by a cable which carries a reference signal to compensate for the primary field so that the system output is due to only the secondary field. A decomposer splits the secondary field into real and imaginary components which are used to determine the conductivity and magnetic susceptibility of the ground.

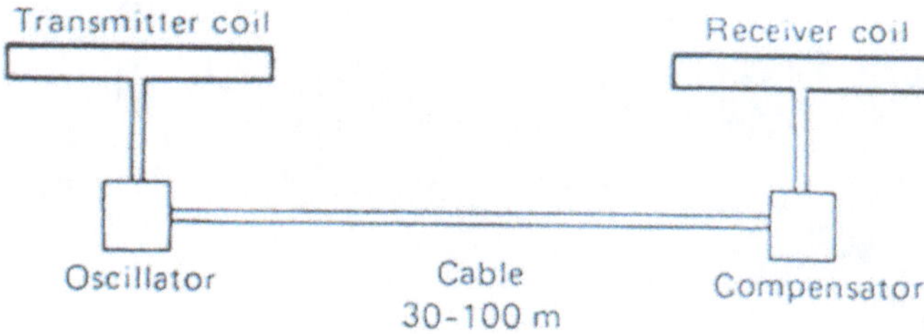

Fig. 7.19 Holomorphic embedding load-flow method (*HLEM*) or Slingham method of surveying

Traverses are made parallel to the levee or dam and the readings are plotted at the midpoint of the coils. Brackett [67] investigated two Mississippi levee location showing signs of anomalous seepage and the onset of piping. Electromagnetic surveys were conducted to measure the lateral variation in conductivity at the sites in order to delineate the geology controlling the flow of water from the river side to the landside. Figure 7.20 shows a map of conductivity on the landside of a Mississippi levee using an EM-34 in a horizontal dipole configuration. The results show that the material on the landside field of the levee is more conductive than the levee apron itself. This high-conductivity material is due to thick clay overburden. The measured apparent conductivity of the material is believed to be highest where the clay is the thickest. The two existing sand boil locations coincide with locations of low conductivity suggesting the clay overburden is thin or nonexistent, thereby allowing water seepage to the surface.

7.3.3 Case Studies

EM methods are a rapid and cost-effective approach for obtaining spatial information at a site of interest. Many investigations collect EM data to constrain lateral variations in geology and map dam and levees structures. Other geophysical methods are usually employed for assisting with understanding features as a function of depth. Dunbar et al. [63] have developed a procedure called LevCat using airborne EM measurements to assess the condition of the foundation of levees. Amine et al. [68] used a helicopter-based EM system to collect data in Sacramento, California, to obtain insight into the layers beneath the levee. Brackett [67] conducted ERT and EM-34 surveys to constrain the geological interpretation underlying sand boil development at two different Mississippi mainline levee locations. Dalton [69] used data from an EM-38 to assist with constraining the extent of shallow slough slides along a secondary levee in Mississippi. Llopis et al. [38] use data from a towed EM-34 to assess the levee conditions in California. Bedrosian et al. [41] used time domain EMs to assist with constraining the geology at Martis Creek Dam in California. Viganotti et al. [70] use EM sounding to assess estuarine earthen flood defense embankments in Ireland. Al-Fares [40] used EM and other geophysical methods to help determine probable leakage paths through a dam and to outline

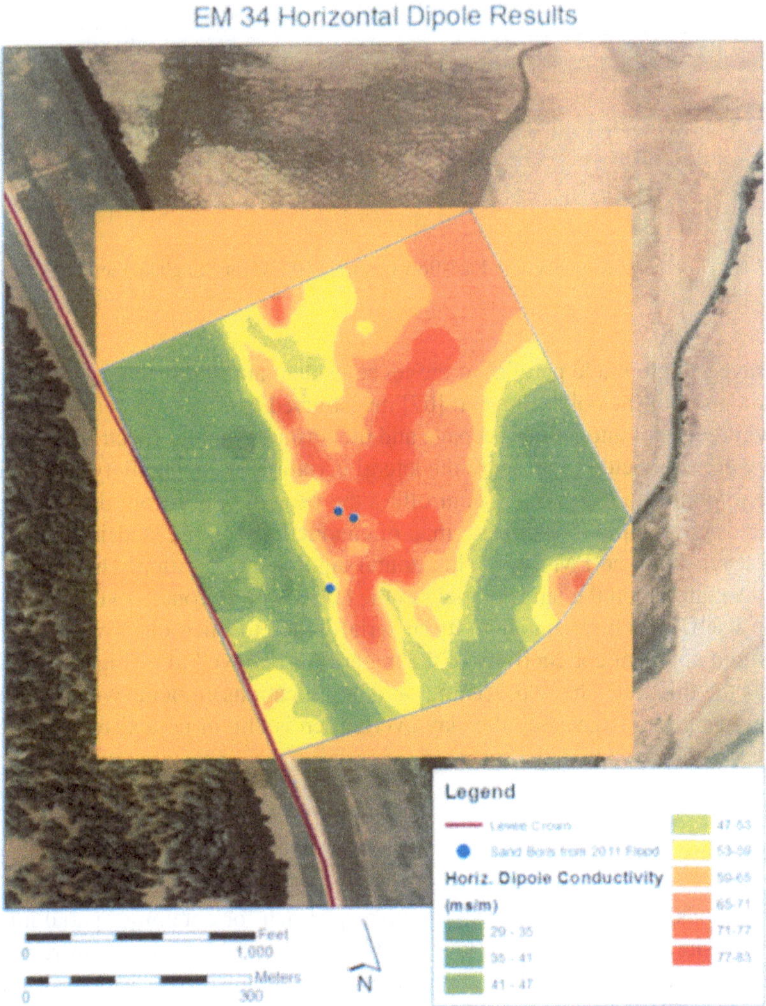

Fig. 7.20 Horizontal dipole data at a Mississippi levee location collected with an EM-34

subsurface faults and fractures. Darby [71] used an EM-31 to the groundwater flow near the dam at Alpine Lake, West Virginia.

8 Multiple Geophysical Methods (Joint Inversion)

Earthen dams and levees have spatially varying properties that can lead to false interpretations. The use of data from multiple geophysical techniques may reduce ambiguity in interpretation associated with the natural heterogeneity of dams and

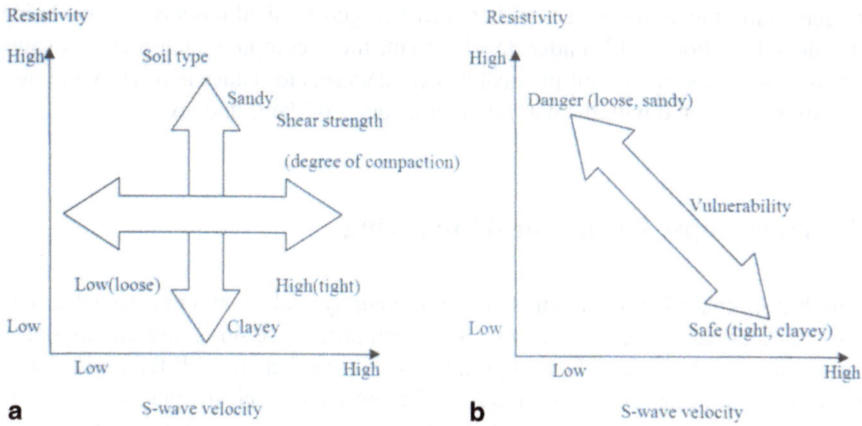

Fig. 7.21 Cross-plotting approach for relating geophysical observables to levee vulnerability. (After Hayashi and Konishi 2010)

levees. Approaches are being developed based on the premise that compromised zones within these embankments will have a *common set of geophysical signatures*. For example, electrical resistivity is strongly dependent on water content, clay fraction, and the presence of soluble salts. Seismic velocities and absorption depend on the elasticity and bulk density of the subsurface. The elasticity of soil is strongly dependent on the cohesion, degree of cementation, state of stress, and water content. Therefore, a compromised zone within a levee might be characterized by a low electrical resistivity due to increased water content and a lower seismic velocity due to the loss of cohesion and the change in state of stress. It is therefore possible to use multiple geophysical techniques to reduce the number of predicted anomalies (number of possible compromised zones to be validated) due to the heterogeneity of the surrounding material.

As an example of this approach, Hayashi and Konishi [72] integrate S-wave data obtained using MASW surveys and DC resistivity measurements for the assessment of levees in Japan using a cross-plotting approach. The cross plot shown in Fig. 7.21 relates the resistivity to a dominant grain size. High resistivity is associated with sandy soils and low resistivity is associated with clayey soils. The S-wave velocity is related to the degree of compaction and infers a shear strength of the soil. The dominant grain size and degree of compaction are used to establish general vulnerability criteria. High soil resistivity and low S-wave velocity infer a sandy, poorly compacted levee with high vulnerability. A low resistivity and high S-wave velocity infers a clayey, well-compacted levee with low vulnerability.

The practice of performing collocated DC resistivity and seismic refraction surveys in order to obtain a more accurate characterization of the interior of dams and levees is becoming more common. Quantitative approaches for "joint inversion" of these data sets are under development [73, 74, 75, 76]. Some inversion methods make use of common petrophysical characteristics to relate different geophysical properties. For example, resistivity and seismic velocity both depend upon water saturation and porosity. Inversion approaches based upon boundaries of geological targets being the common factors between two geophysical models are also being developed. Although still under development, these computational methods show promise for integrating multiple geophysical data sets to reduce ambiguity in interpretation associated with the natural heterogeneity of dams and levees.

9 Time-Lapse Geophysical Surveying

Another approach in reducing the number of geophysical anomalies associated with the effects of natural heterogenety is to perform time-lapse surveying. In time-lapse surveying, the same type of geophysical survey is carried out at different times and the change between surveys is analyzed. The success of such an approach is clearly dependent on the times at which the different surveys are conducted. Depending on the number of repeated investigations and the data analysis method involved, the application might be able to detect weak zones that would otherwise be missed by a one-time investigation. For example, surveys might be conducted during the dry season and subsequently during the wet season. Physical properties of the embankment properties will change because of temperature, rainfall, water table, and river level rise. These global property changes will produce a seasonal trend in the geophysical data. However, if compromised zones have more pronounced property changes, then this could be used as an indicator of progressive deterioration (e.g., water infiltration, seepage, internal erosion, and piping). Another approach might be surveys conducted during an imposed change in conditions such as the filling or draining of an impounded reservoir. In this case, multiple surveys might be used to determine the critical reservoir height for the onset of seepage.

One can take advantage of these seasonal differences to improve compromised zone detection by implementing temporal approaches based on repeated geophysical investigations. Such approaches have proved very effective and are now being gradually applied to dam and levee assessment and surveillance. Ideally, one should always try and repeat geophysical investigations at different times when weak zones are suspected to show the greatest contrast.

As an example, a P-wave time-lapse survey to monitor the evolution of internal erosion in a small experimental earthen embankment dam was discussed by Hickey et al. [77]. An experimental dam was constructed with a metal pipe within the dam body. The pipe was removed to initiate internal erosion by creating a passway for water on the upstream side. A series of six seismic surveys were conducted on the dam, starting from the removal of tarps from the embankment until the collapse of part of the embankment due to cavity formation from internal erosion. The timing of seismic surveys is presented in Table 7.13.

Each acquisition of field data was post processed using Rayfract™ software [78]. This method uses an inversion technique based upon a Fresnel volume approach to represent the wave propagation instead of simple rays. The first arrival time (in milliseconds, relative to time when shot was fired) that seismic energy arrives at the corresponding receiver and the source–receiver locations are used to determine the

Table 7.13 Time schedule of the series of seismic surveys

Time	Survey name	Figure	Comments
Day 1-10:00 a.m.	Baseline survey 1	Fig. 7.2a, b, c	Tarps were removed from embankment
Day 2-8:00 a.m.	Baseline survey 2	Fig. 7.2d, e, f	Embankment was allowed to dry and some surface cracking was visible
Day 2-9:00 a.m.	Reservoir loading	Fig. 7.3a, b, c	The reservoir was filled with water, and changes in the seismic veloc-ity could be due to the additional loading by the reservoir of the addition of moisture
Day 2-11:00 a.m.	Initial breaching	Fig. 7.3d, e, f	The metal pipe was pulled at approximately 9:30 a.m. and good flow of water was now evident through the embankment
Day 2-12:30 p.m.	Intermediate stage of erosion	Fig. 7.4a, b, c	The flow rate and erosion channel with the embank-ment has evolved
Day 2-3:00 p.m.	Late stage of erosion	Fig. 7.4d, e, f	Late stage of erosion, where it is believed that an air cavity has developed above the water under the geophone spread. Part of the embankment collapses at 6 p.m., so seismic acqui-sition was ended

distribution of seismic velocities in the subsurface. This inversion method generates continuous velocity versus depth profiles for all the stations. These profiles are processed with Golden Surfer Software [79] to produce velocity contour maps.

The first survey was conducted immediately after the tarps were removed from the embankment. The surface of the embankment was moist with no noticeable surface cracks. A photograph of the embankment with the geophone spread and the source is shown in Fig. 7.22a. The origin of the survey, station 0 or offset 0 m is shown in the back left-hand side of the figure. The velocity tomogram, representing P-wave velocity within the embankment, is shown in Fig. 7.22b. The tomograms have about a 1:10 vertical to horizontal aspect ratio. The velocities inverted from the refraction analysis increases with depth from about 200 ms^{-1} near the surface to 450 ms^{-1} at a depth of about 1 m, which is typical for compacted soil materials. The velocity gradient of the material and the spread length (restricted due to the length of the embankment) results in an imaging depth of about 0.8 m.

Seismic travel time tomography requires optimization of both the distribution of seismic velocities and the paths along which the wave has traveled, or ray paths.

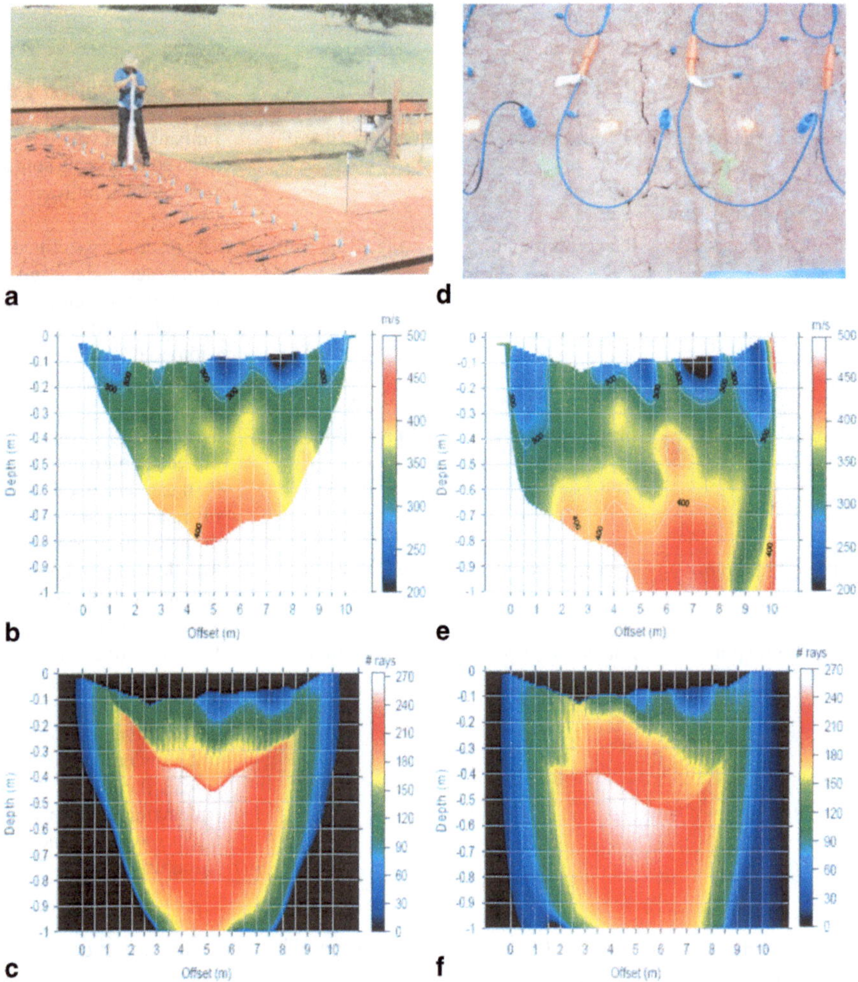

Fig. 7.22 Results of the two first seismic surveys performed on the earthen embankment. The time elapse between these two surveys is approximately 22 h

Figure 7.22c displays the distribution of ray paths corresponding to the optimum solution. The obvious use of this map is to establish the relative sampling of the subsurface. Results in areas with low ray coverage would have lower confidence than areas of high ray coverage, and Rayfract™ uses the ray coverage map to establish spatial bounds on the velocity tomogram. The ray coverage could also be a representative of the relative flux of seismic energy associated with the sum of all surface sources and can be used as an attribute in itself. For this survey, the ray coverage is highest in the middle of the survey and lower near the surface and edges, which is typical for a material with a fairly uniform vertical velocity gradient. From Fig. 7.22b, c, there is no strong evidence of the metal pipe known to be located at a depth of about 0.9 m and at an offset of about 5.4 m.

The second survey was conducted the following morning and the photograph of the embankment surface (Fig. 7.22d) shows that, the surface was subjected to drying with the formation of surface cracks. Closer inspection revealed that these surface cracks were less than a few centimeters deep. The difference in the state of the embankment relative to the first survey is the drying and a slight change in temperature. These changes could change the distribution of internal stresses. This is quite important because the seismic behavior of granular materials is highly dependent on the intergranular contact strength that is influenced by the applied forces at the contacts. The resulting velocity tomogram (Fig. 7.22e) has the same range of velocities as in survey 1. The notable differences are: the localized velocity maximum at a depth 0.4–0.5 m and offset 6–6.5 m, a low-velocity anomaly at a depth of 0.5–0.6 m and offset 5–5.5 m, and a low-velocity zone extending along the right abutment of the dam. The low-velocity anomaly is in the correct lateral position, but it is not deep enough to be representative of the metal pipe. The ray coverage (Fig. 7.22f) is not as symmetric as was calculated in the first survey.

Survey 3 was conducted approximately 1 h later after the reservoir was filled with water. Figure 7.23a shows a photograph of the embankment at the beginning of the third survey. The height of the water in the reservoir was kept constant using the overflow concrete wall seen at the back of the photograph. The metal pipe was still in the embankment during this survey, so the significant difference between this survey and survey 2 would be the additional load on the embankment from the water in the reservoir, and the contact of water against one face of the dam. Since this embankment was well compacted, changes in water content within the body of the embankment would be primarily through capillary mechanisms and not traditional seepage. The resulting velocity tomogram is shown in Fig. 7.23b. The notable feature is the enhancement of the low-velocity anomaly at a depth 0.6–0.7 m and offset 5.0–5.5 m. The ray cover shown in Fig. 7.23c shows adequate ray coverage in the region to allow good confidence in the velocity tomogram. The anomaly is still shallow compared to the true location of the metal pipe, but could easily be representative of the location of the pipe. If it is associated with the pipe, then the velocity in this region might have changed due to changes in water content facilitated by the interface between the pipe and soil, or a specific redistribution of the reservoir load enhancing the compaction and presence of the metal pipe.

Survey 4 was conducted after the metal pipe was removed and a continuous flow of water was established through the embankment. The flow of water could be seen on the down streamside of the embankment in Fig. 7.23d. This survey represents an early stage of breaching, but would represent a highly evolved stage of seepage in a typical earthen dam. The velocity tomogram (Fig. 7.23e) now has a very prominent low-velocity anomaly in the area of the erosion channel. The anomaly extends slightly deeper, but more notable is how the low-velocity anomaly now extends continuously towards the surface. Removing a rigid metal pipe and replacing it with a water-filled channel clearly has a nonlocal effect on the embankment. One could quite successfully predict the lateral position of the erosion channel using the velocity tomogram. The ray coverage shown in Fig. 7.23f reveals adequate ray coverage in the area of the anomaly.

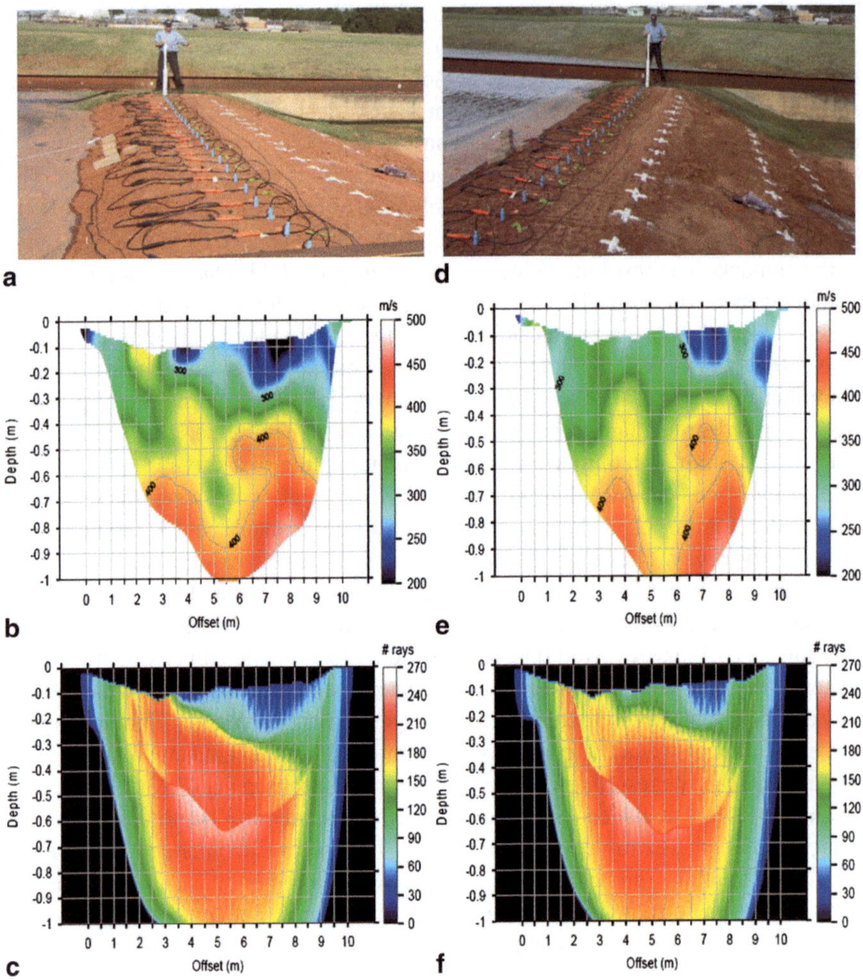

Fig. 7.23 Results associated with reservoir loading and the early stage of internal erosion

Survey 5 was conducted several hours after survey 4 and represents an intermediate stage of erosion. The flow rate of water through the embankment has increased substantially as shown in Fig. 7.24a. The velocity tomogram (Fig. 7.24b) now shows a laterally expanded low-velocity anomaly associated with the expanding water channel. The anomaly appears more diffused than was seen in survey 4. This may be due to relaxation of stresses or possibility a more uniform redistribution of moisture in the vicinity of the erosion channel. The ray coverage (Fig. 7.24c) shows adequate coverage in the vicinity of the anomaly.

Survey 6 was conducted approximately 2 h after survey 5 and represents a late stage of erosion. The erosion channel has widened and evolved to a state where an air cavity is now present over the water. From inspection, it appears that this air

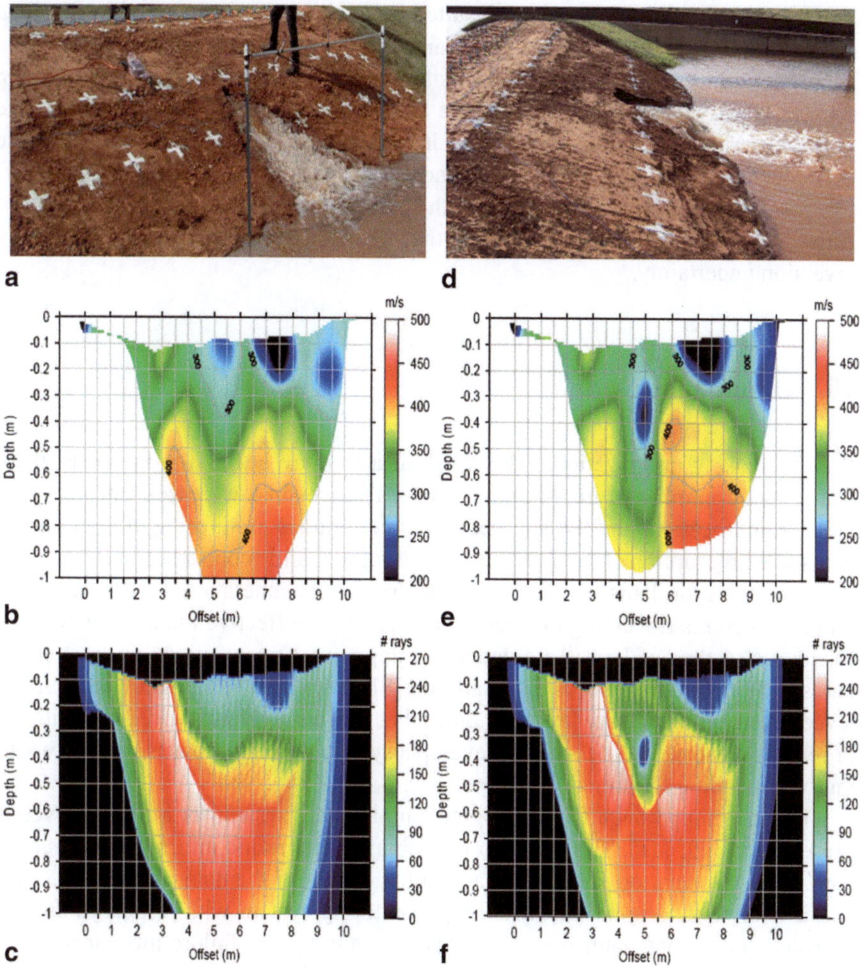

Fig. 7.24 Results at an intermediate and late stage of erosion

cavity has migrated into the embankment to such an extent as to be present under the geophone spread. The velocity tomogram shown in Fig. 7.24d is dramatically differentthan the one from survey 5. The lateral boundaries of the erosion channel appear to be better delineated by the low-velocity (say 375 m s^{-1}) contour. There is now a very low-velocity (200 m s^{-1}) anomaly located at a depth of 0.3–0.4 m and offset 4.75–5.25 m which is in the region of the overlying air cavity. The differences in seismic velocity signatures between early-stage erosion from survey 4 (Fig. 7.23e) and late-stage erosion from survey 6 (Fig. 7.24e) are very dramatic. The ray coverage (Fig. 7.24f) from this survey is very different than the previous surveys. The ray coverage shows an isolated minimum at a depth of 0.3–0.4 m and offset 4.75–5.25 m. This is believed to be associated with the presence of the air cavity.

This general methodology is well suited for analyzing most types of geophysical data, such as ERT measurements, SP, and seismic refraction measurements. The analysis might be as simple as a qualitative interpretation of differences between successive surveys or more quantitative by differencing of data to more elaborate time constrained inversions. Time-lapse inversion approaches are available in some ERT processing software. In such approaches, geophysical data sets obtained at different times are inverted jointly rather than separately, which provides imaging of zones of significant temporal variations that are more relevant and less disturbed by inversion uncertainty.

10 Dam Monitoring

With increasing urbanization and critical infrastructure in flood-prone areas, there is a growing number of high-hazard dams. Advances in low-power sensors, wireless communication, and the internet makes continuous monitoring of high-hazard dams a viable option. Dams are shorter structures and monitoring systems have been designed for dams associated with hydroelectric generation. Levees are much longer structures traversing different geologies. Cost-effective monitoring systems for long stretches of levees are under development. Organizations in the Netherlands such as Deltares [80] and Urbanflood [81] have been working towards "smart levee" systems, which are levees or dikes that are remotely monitored using in situ or remote sensors. Their goal is to develop a system that can provide an early warning of impending failure.

The general approach for developing a monitoring system for a dam or levee is presented in Fig. 7.25. The first step is to obtain background information about the dam. This information should be easily obtained from the dam assessment study and is used to determine the goals of a monitoring system with respect to the anticipated failure mechanism and hazards. The anticipated failure mechanism will dictate the type of meaurements and sensors required.

Many of the different parameters that can be measured on or in earthen embankments are detailed in manuals by the USACE [82] and the USDI-BR [83]. These include water content, water tension, hydraulic head, temperature, salinity, tilt, acceleration, and electric potentials. Numerous sensors are available such as piezometers, thermocouples, time domain reflectometers (TDR)s, fiber optic cables, microelectromechanical systems (MEMS) accelerometers, tilt meters, pH meters, etc. Table 7.14 lists various sensors that could provide information for monitoring a specific failure.

The hazard classification will influence the duration of the monitoring and the data collection frequency. For high-hazard applications, the monitoring system may be permanent with continuous monitoring. The location and access to the site will dictate what type of communication is required between the individual sensors and the base station or the base station and the data collection center. Present-day wireless communications will easily allow for the data to be transferred to a central

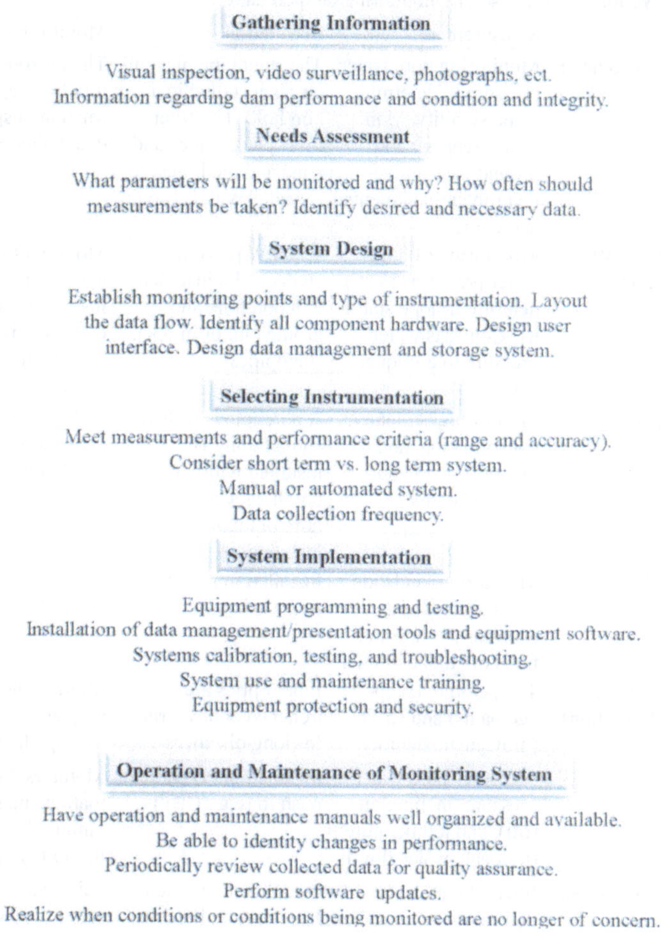

Fig. 7.25 Flow chart outlining the steps for designing and implementing a remote monitoring system

location for processing. Figure 7.26 illustrates a system being designed to monitor excessive pore pressures at the toe of a dam. The system consists of several vibrating wire piezometer installed in stand pipes, a weather station, and a reservoir water-level sensor. In order to alleviate issue with ground maintenance, the individual sensors communicate with the base station using a wireless link. The base station consists of a datalogger and a cellular modem communication for transmitting data to a central data server. At the server location, the data are archived and processed, with results forwarded to various agencies. The simplest processing for a monitoring system is based on the difference in the average values of measured parameters between two time windows. Large changes in the average value with time would be an indication of changes in the levee. More detailed monitoring might involve geotechnical stability calculations based on the data received.

Table 7.14 Various sensors used in monitoring earthen dams

Sensor	Measurement	Installation	Monitoring failure type
Stand pipe piezometer	Monitoring pore water pressure to determine slope stability. Monitoring seepage and groundwater movement in embankments and dams	The standpipe piezometer is installed in a borehole. The filter tip is placed in a sand zone and the borehole is sealed with bentonite	The piezometer monitors for seepage failure (manual inspection and data collection)
Vibrating wire (VW) piezometer and thermistor	Monitoring of pore water pressures to determine slope stability. Converts water pressure to a frequency signal	The VW piezometers can be installed in a borehole and grouted in, or in open standpipes	Monitors for seepage failure. Allows instrumentation to be configured wirelessly in order to collect data real time
Extensometers	Monitoring settlement, consolidation, and movements in abutments in dams	A borehole is drilled and a tubing is installed within the borehole. Can be horizontal or vertical installation	Deformation within the dam, slope stability, and failure
In-place inclinometer	Monitoring landslide areas above dams. Monitoring deformations of embankments	Installed in bore hole with casing	Monitors the dam for slope stability
Fiber optic (temperature and deformation)	Temperature (seepage paths) and strain (movement/stability)	Fiber optics are trenched and can run for long distances	Monitor the dam for seepage failure as well as slope failure
GeoBeads	Measures pore water pressure, inclination (tilt), and temperature (to locate water flow)	Installed in borehole or push in (e.g., CPT)	Monitors dam deformation and seepage. Failure due to slope instability and seepage
ShapeAccelArray, pore pressure	Internal dam body deformation	Installed in borehole, PVC casing, horizontal, or vertical	Failure due to slope stability as well as seepage
Settlement pins w/ automated total station (survey)	Surface deformation and displacement	The ground surface	Monitor the dam for slope failure
Soil moisture probe/sensor	Soil moisture, soil moisture percentage, and temperature	Buried in the soil at several different depths	Monitor the dam for seepage and slope failure
Weather station (environmental)	Measure barometric pressure, temperature, relative humidity, and wind speed and direction	The ground surface	Monitors dam site for environmental conditions, e.g., excessive rainfall (slope or overtopping failure)
Water level (reservoir)	Measure the water level of the lake, reservoir or the particular body of water the dam is containing	Installed to be a non-contact device or in the water (VW piezometer) and secured to some type of fixture	Monitor the dam for overtopping

PVC polyvinyl chloride

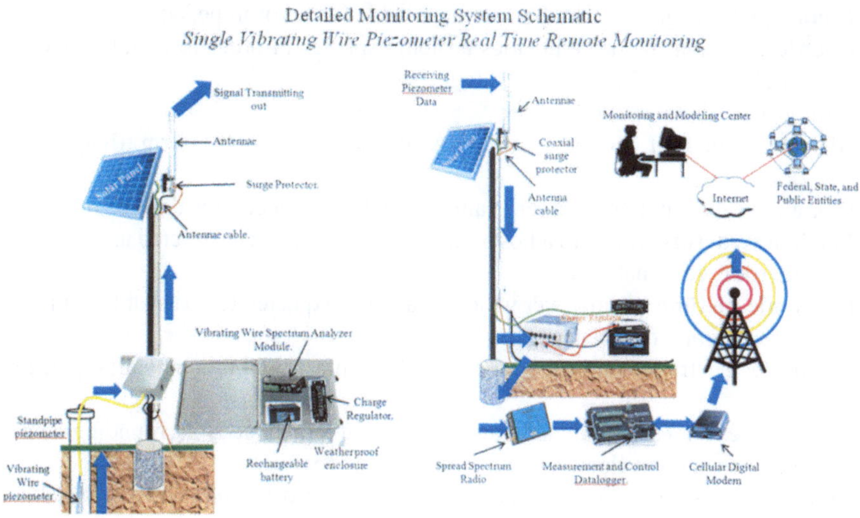

Fig. 7.26 Schematic for monitoring excessive pore pressures at a remote earthen dam site

11 Summary

Screening and rapid assessment of the integrity of the existing large number of levees and earthen dams will require advanced screening tools based on remote sensing or geophysical techniques. Subsurface geophysical imaging provides unique, valuable information regarding the integrity of dams or levees. Some methods for screening, assessment, and monitoring have been discussed. Geophysical information should be part of a systematic approach in levee assessment that is implemented based upon the evidence from visual inspection but before the use of traditional boring programs.

Glossary

Abutment connecting element or structure used for support.
Algorithm step-by-step procedure for calculations.
Anion negatively charged ion that is attracted to the anode in electrolysis.
Bulk modulus material resistance to uniform compression.
Catastrophic failure sudden and total failure of some system from which recovery is impossible.
Cation positively charged ion that is attracted to the cathode in electrolysis.
Cementation hardening and welding of sediments by the precipitation of mineral matter in the pore spaces. Affects porosity and permeability.
Conductivity measure of the ease at which an electric charge or heat can pass through a material.

Connectivity degree to which movement is facilitated or impeded.

Counterions ion that accompanies an ionic species in order to maintain electric neutrality.

Dam barrier that impounds water.

Dielectric constant ratio of permittivity of a substance to the permittivity of free space.

Domain set of input or argument values for which a function is defined.

Earthen structure refers to a host of artificial structures constructed using earthen (soil, rock, etc.) materials.

Economic lifetime period over which an asset is expected to be useable, with normal repairs and maintenance.

Electrode electrical conductor used to make contact with a nonmetallic part of a circuit.

Embankment an artificial bank raised above the immediately surrounding land to redirect or prevent flooding by a river, lake, or sea.

Erosion process by which soil and rock are removed by the action of wind and water.

Flood control—all methods used to reduce or prevent the detrimental effects of flood waters.

Formation process of orderly grouping.

Frequency number of occurrences of a repeating event per unit time.

Geophone device that converts ground movement into voltage.

Geophysics study of earth materials using quantitative physical methods.

Geotechnical study of the behavior of earth materials.

Hydrous aluminum phyllosilicates clay minerals.

Hydraulic study of the mechanical properties of liquids.

Instability the tendency to behave in an unpredictable, changeable, or erratic manner.

Interferometric techniques in which waves are superimposed in order to extract information about the waves.

Ionic conductance movement of an ion from one site to another.

Levee elongated naturally occurring ridge or artificially constructed wall that regulates water levels.

Overtopping to extend or rise over or beyond the top of a dam or levee.

Permeability measure of the ability of a porous material to allow fluids to pass through it.

Permittivity measure of the resistance that is encountered when forming an electric field in a medium.

Phreatic surface commonly called the water table. Location where the pressure head is zero.

Piping soils being washed out of an earthen structure through an unfiltered exit.

Porosity measure of the empty pore spaces in a material. Defined as the ratio between the volume of voids (i.e., empty pore space) and the total volume.

Potential energy stored in a system of forcefully interacting physical entities.

Reservoir a natural or artificial lake, storage pond, or impoundment from a dam which is used to store water.

Resistance opposition to the passage of electrical current.

Resistivity an intrinsic property of a material, measured as its resistance to current per unit length for a uniform cross section.

Saturation the ratio of the volume of a particular fluid to the total volume of the void space.

Sedimentation basin earthen or concrete structure using sedimentation to remove settleable matter and turbidity.

Seepage process of liquid leaking through a porous substance.

Seismic refraction investigates the subsurface by generating arrival time and offset distance information to determine the path and velocity of the elastic disturbance in the ground.

Seismograph instrument for measuring and recording the vibrations of earthquakes.

Shear modulus ratio of stress to strain that describes deformation that takes place when a force is applied parallel to one face of an object while the opposite face is held fixed.

Sodium exchange capacity measure of the soil's ability to hold and release sodium.

Sounding a mechanism of probing the environment by sending out some kind of stimulus.

Spillway structure used to provide controlled release of flows from a dam or levee into a downstream area.

Survey a method of collecting quantitative and qualitative information.

Tomography imaging by sections or sectioning through the use of a penetrating wave.

Tortuosity degree to which a path is curved.

Water recreation diving, fishing, swimming, surfing, etc.

Water-retarding structure designed to hold back water to prevent downstream flooding.

Water supply provision of water by public utilities, commercial organizations, community endeavors, or by individuals, usually via a system of pumps and pipes.

Wavenumber spatial frequency of a wave.

References

1. NID. (2009). National inventory of dams database. http://geo.usace.army.mil/pgis/f?p=397:1:0. Accessed 23 June 2014.
2. USACE. (2013). National levee database. http://nld.usace.army.mil/egis/f?p=471:1. Accessed 23 June 2014.
3. State of Washington Department of Ecology. (2007). http://www.ecy.wa.gov/ecyhome.html. Accessed 23 June 2014.
4. Lane, N. (2008). Aging infrastructure: Dam safety. CRS report for Congress.
5. USACE. (2000). Engineering and design: Design and construction of levees. EM 1110-9–1913.
6. Heerden, I. (2005). Preliminary report on the performance of the New Orleans levees.

7. Shamy, U. E., & Aydin, F. (2008). Multiscale modeling of flood-induced piping in river levees. Journal of Geotechnical and Geoenvironmental Engineering. *ASCE, 134*(9), 1385–1398.
8. FEMA. (2008). Dams sector security awareness guide: Levees. A guide for owners and operators (p. 16).
9. MDEQ. (2013). State of Mississippi guidelines for inspection of existing dams. http://www.deq.state.ms.us/MDEQ.nsf. Accessed 24 June 2014.
10. Skinner, P. (2006). Levees-taking the long view. In *The 2006 risk management yearbook* (p. 17). http://www.riskinstitute.org/peri/images/file/06RMYearbook_levees.pdf. Accessed 24 June 2014.
11. USBR. (1983). *Safety evaluation of existing dams. A water resources technical publication* (p. 163). US Department of the Interior, Bureau of Reclamation.
12. USACE. (2001). Geotechnical investigations. EM-1110-1-1804.
13. Schroder, W. L. (1984). *Soils in construction* (p. 330). Prentice Hall.
14. Terzaghi, K., Peck, R. B., & Mesri, G. (1996). *Soil mechanics in engineering practice* (3rd edn.). Wiley-Interscience.
15. Telford, W. M., Geldart, L. P., & Sheriff, R. E. (1998). *Applied geophysics* (2nd edn., p. 770). Cambridge University Press.
16. Sharma, P. V. (1997). *Environmental and engineering geophysics* (p. 475). Cambridge University Press.
17. Kearey, P., Brooks, M., & Hill, I. (2002). *An introduction to geophysical exploration* (p. 254). Oxford: Blackwell Publishing Company.
18. Milsom, J., & Eriksen, A. (2011). *Field geophysics* (4th edn.). West Sussex, UK: A John Wiley and Sons, Ltd.
19. Griffiths, D. H., & King, R. F. (1965). *Applied geophysics for engineers and geologists* (p. 223). Pergamon Press.
20. Reynolds, J. M. (2011). *An introduction to applied and environmental geophysics* (2nd edn. p. 696). Wiley-Blackwell.
21. Butler, D. K. (2005). *Near-surface geophysics*. Tulsa: Society of Exploration Geophysicists.
22. USACE. (2005a). Geophysical exploration for engineering and environmental investigations, EM 1110-1-1802.
23. USACE. (2005b). Using geophysics to assess the condition of small embankment dams, ERDC/GSL TR-05–17.
24. EPA. (1993). Use of airborne, surface, and borehole geophysical techniques at contaminated sites: A reference guide, EPA/625R-92/007.
25. ASCE. (1998). *Geophysical exploration for engineering and environmental investigations, technical engineering and design guides as adapted from the USACE* (No. 23). Reston: ASCE Publishing.
26. Allred, B. J., Daniels, J. J., & Ehsani, M. R. (2008). *Handbook of agricultural geophysics*. Boca Raton: CRC Press.
27. Butler, D. K., Llopis, J. I., & Deaver, C. M. (1989). Comprehensive geophysical investigation of an existing dam foundation, Geophysics. *The Leading Edge Of Exploration, Geotechnical Applications*, *5*(part 1), 10–18.
28. Butler, D. K., Llopis, J. I., Dobecki, T. L., Corwin, R. F., Wilt, M. J., & Olhoeft, G. (1990). Comprehensive geophysics investigation of an existing dam foundation; Engineering geophysics research and development, Geophysics. *The Leading Edge of Exploration, Geotechnical Applications*, *5*(part 2), 44–53.
29. Fauchard, C., & Mèriaux, P. (2004). *Méthodes géophysiques et géotechniques pour le diagnostic des digues de protection contre les crues. (Geophysical and Geotechnical Methods for the Diagnosis of Anti-flood Dykes) Guide pour la mise en oeuvre et l'interprétation.* Cemagref editions.
30. Palacky, G. J. (1987). Resistivity characteristics of geologic targets. *Electromagnetic Methods in Applied Geophysics, 1,* 53–129.

31. Friedman, S. P. (2005). Soil properties influencing apparent electrical conductivity: A review. *Journal of Computers and Electronics in Agriculture, 46*(1–3), 45–70.
32. Waxman, M. H., & Smits, L. J. M. (1968). Electrical conduction in oil-bearing sands. *Society of Petroleum Engineers Journal, 8,* 107–122.
33. Mavko, G., Mukerji, T., & Dvorkin, J. (1998). *The Rock Physics Handbook* (p. 329). Cambridge: Cambridge University Press.
34. Samouëlian, A., Cousin, I., Tabbagh, A., Bruand, A., & Richard, G. (2005). Electrical resistivity survey in soil science: A review. *Soil and Tillage Research, 83,* 173–193.
35. Loke, M. H. (2011). Tutorial: 2-D and 3-D electrical imaging surveys. http://www.geotomosoft.com/downloads.php. Accessed 24 June 2014.
36. Daily, W., Ramirez, A., Binley, A., & LeBrecque, D. (2004). Electrical resistance tomography, Society of Exploration Geophysicists. *The Leading Edge, 23*(5), 438–442.
37. Inazaki, T. & Sakamoto, T. (2005). *Geotechnical characterization of levee by integrated geophysical surveying.* Proceedings of the International Symposium on Dam Safety and Detection of Hidden Troubles of Dams and Dikes.
38. Llopis, J. L., Smith, E. C., & North, R. E. (2007). *Geophysical surveys for assessing levee foundation conditions.* Sacramento: Sacramento River Levees. ERDC/GSL TR-07-21.
39. Case, J. S. (2012). *Inspection of earthen embankment dams using time lapse electrical resistivity tomography.* M.S. Thesis, University of Mississippi, University, Misissippi.
40. Al-Fares, W. (2011). Contribution of the geophysical methods in characterizing the water leakage in Afamia B. Dam, Syria. *Journal of Applied Geophysics, 75,* 464–471.
41. Bedrosian, P. A., Burton, B. L., Powers, M. H., & Minsley, B. J. (2012). Geophysical investigations of geology and structure at the Mathis Creek Dam, Truckee, California. *Journal of Applied Geophysics, 77,* 7–20.
42. Minsley, B. J., Burton, L. B., Ikard, S., & Powers, M. H. (2011). Hydrogeophysical investigations at Hidden Dam, Raymond, California. *Journal of Environmental and Engineering Geophysics, 16,* 145–164.
43. Weller, A., Cahn, T., Breede, K., & Trong Vu, N. (2006). Multi-electrode measurements at Thai Binh Dikes (Vietnam). *Near Surface Geophysics, 4,* 135–143.
44. www.agiusa.com/. Accessed 23 June 2014.
45. Achenbach, J. D. (1975). *Wave propagation in elastic solids.* Amsterdam: North-Holland.
46. Park, C. B., Miller, R. D., & Xia, J. (1997. Multi-channel analysis of surface waves (MASW), Kansas Geologic Survey, Open-file Report.
47. Santamarina, J. C., Klein, K. A., & Fam, M. A. (2001). *Soils and waves* (p. 488). New York: John Wiley and Sons.
48. Gassmann, F. (1951). Über die elastizität poröser medien: Vierteljahrss-chrift der Naturforschenden Gesellschaft in Zurich, 96, 1–23. The English translation of this paper is available at http://sepwww.stanford.edu/sep/berryman/PS/gassmann.pdf. Accessed 23 June 2014.
49. Brooks, R. H., & Corey, A. H. (1964). *Hydraulic properties of porous media, Hydrology paper 3.* Fort Collins: Civil Engineering Department, Colorado State University.
50. Knight, R., Dvorkin, J., & Nur, A. (1998). Acoustic signatures of partial saturation. *Geophysics, 63*(1), 132–138.
51. Ivanov, J., Park, C. B., Miller, R. D., Xia, J. (2001). *Modal separation before dispersion curve extraction by MASW method.* Proceedings of the SAGEEP 2001, Denver, Colorado.
52. Xia, J., Miller, R. D., Park, C. B., & Ivanov, J. (2000). *Construction of 2-D vertical shear-wave velocity field by the multichannel analysis of surface wave technique.* Proceedings of the SAGEEP 2000, Arlington, Virginia.
53. Osazuwa, I. B., & Chinedu, A. D. (2008). Seismic refraction tomography imaging of high-permeability zones beneath an earthen dam, in Zaria area, Nigeria. *Journal of Applied Geophysics, 66,* 44–48.
54. Powers, M. H., & Burton, B. L. (2008). *Seismic refraction tomography in an urban environment using a vibrator source.* Proceedings of the SAGEEP 2008, Philadelphia.

55. Rucker, M. L., & Holmquist, O. C. (2006). Surface seismic methods for locating and tracing earth fissures and other significant discontinuities in cemented unsaturated soils and earthen structures. *American Society of Civil Engineers, 147,* 601–612.
56. Kilty, K. T., Norris, R. A., McLamore, W. R., Hennon, K. P., & Euge, K. (1986). Seismic refraction at Horse Mesa Dam: An application of the generalized reciprocal method. *Geophysics, 51*(2), 266–275.
57. Ivanov, J., Miller, R. D., Ballard, R. F., Dunbar, J. B., & Stefanov. J. (2004). *Interrogating levees using seismic methods in southern Texas*. SEG expanded abstracts, Denver, CO.
58. Ivanov, J., Johnson, C. D., Lane, J. W., Miller, R. D., & Clemens, D. (2009). *Near-surface evaluation of Ball Mountain Dam, Vermont, using multi-channel analysis of surface waves (MASW) and refraction tomography seismic methods on land-streamer data*. 2009 SEG Annual Meeting, Houston, TX.
59. Wodajo, L. T. (2011). *Seismic tomography for the integrity assessment of earthen dams*. M.S. Thesis, University of Mississippi, Oxford, MS.
60. Jackson, J. D. (1962). *Classical electrodynamics* (p. 641). New York: John Wiley and Sons.
61. Morse, P. M., & Feshbach, H. (1953). *Methods of theoretical physics*. McGraw-Hill Science.
62. Griffiths, D.J. 2012. *Introduction to electrodynamics* (4th edn.). Boston: Addison-Wesley.
63. Dunbar, J. B., Wakeley, L., & McKenna, J. (2007). Levee Condition Assessment Technology (LevCAT). ERDC/GSL Capability Service Bulletin, May 2007.
64. Doll, W., Norton, J., Gamey, J., Holladay, J. S., West, O., & Tatum, B. (2013). *Validation of TEM –8 airborne conductivity data with well borings from the Cairo Il levee system*. Proceedings of the SAGEEP 2013, Denver, Colorado.
65. http://www.geonics.com. Accessed 23 June 2014.
66. Llopis, J. L., & Finnegan, D. (2007). Geophysical surveys for rapid assessment of levees and levee foundation conditions. ERDC/GSL Capability Service Bulletin, May 2007.
67. Brackett, T. C. (2012). *Use of geophysical methods to map subsurface features at levee seepage locations*. M.S. Thesis, University of Mississippi, University, Mississippi.
68. Amine, D., Marlow, D., Woldringh, B., Hodges, G., & Selvamohan, S. (2010). *Levee evaluation studies in Sacramento, California: Correlating helicopter-borne EM data, borings and geology*. GeoFlorida 2010, Advances in Analysis, Monitoring and Design, ASCE, West Palm Beach, FL.
69. Dalton, L. M. (2012). *Using geophysics to characterize levee stability*. M.S. Thesis, University of Mississippi, Oxford, MS.
70. Viganotti, M., Jackson, R., Krahn, H., & Dyer, M. (2013). Geometric and frequency EMI sounding of estuarine earthen flood defense embankments in Ireland using 1D inversion models. *Journal of Applied Geophysics, 92,* 110–120.
71. Darby, T. (2003). *Terrain conductivity investigation of groundwater flow near the dam at Alpine Lake, West Virginia*. Sr. Thesis, West Virginia University, Morgantown, WV.
72. Hayashi, K., & Konishi, C. (2010). *Joint use of a surface-wave method and a resistivity method for safety assessment of levee systems*. GeoFlorida 2010, Advances in Analysis, Monitoring and Design, ASCE, West Palm Beach, Fl.
73. Gallardo, L. A., & Meju, M. A. (2004). Joint two-dimensional DC resistivity and seismic travel time inversion with cross-gradient constraints. *Journal of Geophysical Research, 109,* 1–11.
74. Gallardo, L. A., & Meju, M. A. (2011). Structure-coupled multiphysics imaging in geophysical sciences. *Reviews in Geophysics, 49*(1), 1–19.
75. Doetsch, J., Linde, N., & Binley, A. (2010). Structural joint inversion of time-lapse crosshole ERT and GPR traveltime data. *Geophysical Research Letters, 37,* L24404.
76. Karaoulis, M., Revil, A., Zhang, J., & Werkema, D. D. (2012). Time-lapse joint inversion of crosswell DC resistivity and seismic data: A numerical investigation. *Geophysics, 77*(4), D141–D157.
77. Hickey, C. J., Ekimov, A., Hanson, G. J., & Sabatier, J. M. (2009). *Time-lapse seismic measurements on a small earthen embankment during an internal erosion experiment*. Proceedings of the SAGEEP 2009. Fort Worth, TX.

78. http://www.rayfract.com/help/manual.pdf. Accessed 23 June 2014.
79. http://www.goldensoftware.com. Accessed 23 June 2014.
80. http://www.deltares.nl/en. Accessed 23 June 2014.
81. http://www.urbanflood.eu/Pages/Workshop2012.aspx. Accessed 23 June 2014.
82. USACE. (2005c). Instrumentation for Embankment Dams and Levees. EM 1110-2-1908.
83. USBR. (1987). *Embankment dam instrumentation manual. A water resources technical publication* (p. 250). US Department of the Interior, Bureau of Reclamation.

Chapter 8
Soil Erosion on Upland Areas by Rainfall and Overland Flow

Mathias J. M. Römkens, Robert R. Wells, Bin Wang, Fenli Zheng and Craig J. Hickey

Contents

M. J. M. Römkens (✉) · R. R. Wells
USDA ARS National Sedimentation Laboratory, P. O. Box 1157, Oxford, MS, USA
e-mail: Matt.Romkens@ars.usda.gov

R. R. Wells
e-mail: Robert.Wells@ars.usda.gov

B. Wang
Beijing Forestry University, Beijing, 100083, China
e-mail: wangbin1836@bjfu.edu.cn

F. Zheng
Northwest Agriculture and Forestry University, Yangling,
Shaanxi Province 712100, China
e-mail: flzh@ms.iswc.ac.cn

C. J. Hickey
National Center for Physical Acoustics, University of Mississippi,
P. O. Box 1848, University, MS 38677-1848, USA
e-mail: chickey@olemiss.edu

C. T. Yang, L. K. Wang (eds.), *Advances in Water Resources Engineering*,
Handbook of Environmental Engineering, Volume 14, DOI 10.1007/978-3-319-11023-3_8

Abstract Soil erosion in agricultural watersheds is a systemic problem that has plagued mankind ever since the practice of agriculture began some 9000 years ago. It is a worldwide problem, the severity of which varies from location to location depending on weather, soil type, topography, cropping practices, and control methods. Research to address and predict soil loss from agricultural land and in watersheds began in earnest in the 1930s following the events of the Dust Bowl. Early research primarily consisted of monitoring of soil loss from natural runoff plots and small watersheds. Gradually and over time, the focus shifted toward the development of prediction equations based on the acquired soil loss database. With computer technology, modeling watershed erosion and sedimentation processes became routine. Also, fundamental research was conducted to acquire a better understanding of the complex aspects of soil erosion and sediment transport processes and to fill in knowledge gaps in cases where data were not readily available. In recent years, most soil loss from upland areas occurs as gully erosion. This chapter presents a background of the knowledge that was systematically acquired in predicting soil erosion from upland areas and the technology that was developed and is used today. This chapter does not address all the aspects of upland soil erosion, but focuses primarily on the erodibility (K-factor) and hydrological aspects (R-factor) of the most widely used erosion prediction equations: the revised universal soil loss equation, version 2 (RUSLE2) and water erosion prediction project model (WEPP) models-based formulae. This chapter also includes a presentation of the Chinese approach of adapting gully erosion predictions according to the universal soil loss equation (USLE) format. Finally, ongoing research and technology development using light detection and ranging (LiDAR) and photogrammetry in gully erosion predictions is discussed.

Keywords Gully · Erosion models · RUSLE · Soil conservation · Soil erosion · USLE · Upland erosion · WEPP

Nomenclature

A	Average annual soil loss per unit area (Mg/ha year)
A_{wr}	Average annual soil loss per unit area (wheat and range region) (Mg/ha year)
Al_2O_3	Aluminum oxide
AGNPS	Agricultural nonpoint pollution system
APEX	Agricultural policy/environmental eXtender model
ARS	Agricultural research service
β	Correction factor
cm	Centimeter

C	Cover and management factor
Ca	Calcium
$CaCO_3$	Calcium carbonate
CREAMS	Model for chemicals, runoff, and erosion from agricultural management systems
d_b	Flow depth at brink (m)
d_d	Downstream flow depth (m)
d_u	Upstream flow depth (m)
d_t	Depth of depositional bed (m)
D_i	Interrill detachment rate (kg/m^{2}s)
D_c	Detachment capacity of clear water
D_g	Geometric mean particle diameter (m)
D_r	Soil detachment by rill flow (kg/m^{2}s)
e	Storm kinetic energy (Nm)
e_{max}	Maximum unit energy (Nm)
E	Rainfall energy (MJ/ha)
EUROSEM	European soil erosion model
Fe_2O_3	Iron oxide
G	Ephemeral gully erosion factor (China)
GLEAMS	Groundwater loading effects of agricultural management systems model
GUESS	Griffith University Erosion Sedimentation System Model
h	Vertical distance from brink to pool surface (m)
i_m	Rainstorm intensity (mm/h)
I	Rainfall intensity (mm/h)
I_{30}	30-min maximum rainfall intensity (mm/h)
K	Potassium
K	Soil erodibility factor (Mg ha h/ha MJ mm)
K_d	Soil erodibility (cm^3/Ns)
K_i	Interrill erodibility constant (cm^3/Ns)
k_0	Organic matter sub-factor
k_p	Soil profile permeability sub-factor
K_r	Rill erodibility (1/s)
k_s	Soil structure sub-factor
k_t	Soil texture sub-factor
k_{t68}	Base soil texture sub-factor
K_{wr}	Estimated soil erodibility for winter period (wheat and range region) (Mg ha h/ha MJ mm)
L	Slope length factor (m)
m	Meter
mm	Millimeter
M	Silt term (%)
Mg	Magnesium
M_h	Annual soil erosion rates with ephemeral gully erosion (China) (Mg/ha year)

M_n	Annual soil erosion rates without ephemeral gully erosion (China) (Mg/ha year)
M_r	Migration rate (m/s)
MUSLE	Modified universal soil loss equation
NSERL	National Soil Erosion Research Laboratory
NSL	National Sedimentation Laboratory
O_m	Percent organic matter in unit plot condition (%)
OM	Organic matter percentage (%)
p'	Rainfall amount of individual rainstorms greater than (>) 10 mm (mm)
P	Support practice factor
P_{cl}	Percent clay (%)
P_m	Soil permeability (mm/h)
P_r	Permeability class
P_{sl}	Percent silt (%)
P_{vfs}	Percent very fine sand (%)
P_{wr}	Support practice factor for contouring; assumed to be 1 for winter
PSD	Particle-size distribution
θ	Slope angle (degrees)
θ_e	Jet entry angle (degrees)
q_s	Suspended sediment flux (kg/s)
Q	Incoming flow discharge (L/min)
Q_s	Rate of sediment transport (kg/s)
ρ	Density of water (kg/m^3)
r_h	Hydraulic radius (m)
R	Rainfall and runoff factor/erosivity factor (MJ mm/ha h year)
R_{eq}	Equivalent rainfall and runoff factor/erosivity factor (MJ mm/ha h year)
R_s	Sub-factor for runoff effect
RTK	Real-time kinetic
RUSLE	Revised universal soil loss equation
RUSLE2	Revised universal soil loss equation, version 2
s	Hydraulic gradient (m/m)
S	Slope-steepness factor
S_D	Scour depth (m)
S_f	Slope factor
S_s	Soil structure code
SAR	Sodium adsorption ratio
$(\mathrm{SLR})_{\mathrm{wr}}$	Soil loss ratio for rilling in the winter period (wheat and range region)
SWAT	Soil-water management tool
T_c	Maximum transport capacity (kg/m s)
τ	Hydraulic shear of flowing water (kg/m^2)
τ_c	Critical shear stress (kg/m^2)
USDA	US Department of Agriculture
USLE	Universal soil loss equation
wr	Wheat and range region
WEPP	Water erosion prediction project model

1 Introduction

Soil erosion is a worldwide concern for the sustainable management of agricultural land. Rates of soil erosion from agricultural areas are often one to two orders of magnitude higher than the combined rates of soil formation and losses from nonagricultural areas [1–3]. According to the US EPA [4, 5], sedimentation is one of the leading causes of water quality impairment of surface water bodies. Reports have estimated the annual costs of soil erosion in the USA to be between US$ 30 and 44 billion [6–8]. This degradation of soil can devastate crop yields and cause famine and starvation [9].

In agricultural watersheds, soil erosion is a systemic problem that has existed to various degrees ever since man became engaged in agricultural practices some 9000 years ago. Because of man's poor land-use practices, severe soil erosion problems developed, landscape degraded, livelihood on the land diminished, and civilizations disappeared. There is ample historic evidence of the disastrous consequences of severe soil erosion, as may be concluded from observations in and around the Mediterranean Basin, the Chinese Loess Plateau, in the countries of eastern Africa, and many other places in the world.

In the USA, serious wind erosion problems developed in the Southern Plains in the 1930s due to a severe drought coupled with strong winds over plowed-up sods. In the Southeast and Central part of the USA, serious erosion problems by water had developed due to severe and intense rainstorms on poorly covered sloping land. As a result and for the first time in recorded history, major soil conservation and soil erosion control research programs were initiated. Natural runoff plots and research watersheds were established to monitor soil erosion from land in various agricultural practices to determine their erosion control effectiveness, to clean up silted-up stream systems and channels, to prevent or reduce flooding hazards, to support and encourage reforestation programs, and to put in place mechanical erosion control practices such as contouring, strip cropping, and terraces. With these programs, applied and fundamental research was conducted to develop and improve monitoring and measurement techniques, to gather databases, and to develop and refine analytical capabilities for better conservation techniques. As time progressed, process-oriented models were developed in the latter part of the twentieth-century for erosion prediction, first the universal soil loss equation (USLE) for application on upland areas to preserve the soil productivity and subsequently models that included the stream system such as agricultural non-point pollution system (AGNPS) and soil-water management tool(SWAT). In recent years, there is an increasing concern about the stability of earthen dams built to protect landscapes from flooding and to provide water for irrigation and for domestic and industrial use. This chapter describes the more recent aspects of these newly developed technologies and programs for upland areas and gully erosion research. Acoustic/seismic technology is being developed to evaluate the stability of earthen dams and levees in regard to soil structural changes indicative of potential failures. This chapter is also designed to familiarize the Engineering Handbook Readership with the complex details of soil erosion problems and processes, and conservation technologies that agriculture has been faced with for many years. While the primary purpose of this chapter is to serve as a guide to new technologies to erosion and sedimentation prediction and control, a discussion of the background and the physics of these problems is essential in order to understand the shortcomings and to properly evaluate the technology and techniques used.

1.1 Upland Erosion Research: A Brief Historic Perspective

The early efforts of soil erosion control research in watersheds consisted almost exclusively of: (1) measurements of runoff and soil loss from fields and natural runoff plots established at various locations in the USA on different soils and for different locations' characteristic cropping practices, (2) monitoring and stream gauging of sediment in streams, draining unstable watersheds, and (3) implementing appropriate measures (channelization, placing control structures) to stabilize these streams. The focus of upland erosion research was local on preserving the productivity of the land and soil. There was a little national coordination. A National Runoff and Soil Loss Data Center was established in 1954 at Purdue University in West Lafayette, Indiana, which offered the opportunity to obtain factor-based soil loss relationships that ultimately culminated into the development of the USLE and its updated versions. Much of the erosion prediction and management technology used today is based on the data collected during the years following the Dust Bowl [10, 11]. The development of erosion prediction models and their underlying factor relationships was greatly aided by the use of field rainfall simulators [12, 13] that could generate an erosive power similar to that of a severe natural rainstorm of similar intensity. Many agronomic conditions relative to crop type, crop stage, soil type, and tillage management and surface condition could be tested under controlled rainfall regimes that were otherwise difficult to evaluate. Also, the search for process-based relationship involving rainfall and runoff and parameter determinations for various soil types, soil surface, and profile conditions was facilitated. Since the 1980s, the development and use of physically based computational erosion models [14] greatly speeded up the erosion assessments for conditions, locations, and climate, that had never been evaluated.

1.2 Erosion Predictive Relationships

The large soil erosion database collected in the USA enabled the development of the USLE and other specialized models such as the modified universal soil loss equation (MUSLE), model for chemicals, runoff, and erosion from agricultural management systems (CREAMS), groundwater loading effects of agricultural management systems model (GLEAMS), erosion productivity impact calculator (EPIC), agricultural policy/environmental extender model (APEX), and revised universal soil loss equation (RUSLE) with the universal soil loss equation (USLE) as the parent relationship. Process-based erosion prediction technology such as water erosion prediction project (WEPP), which is foremost a research model, was developed in the 1980s parallel with the scientifically improved versions of the USLE [15, 16]. Other process-based models, such as GUESS and EUROSEM, are not discussed in this chapter given their limited field applicability.

1.2.1 The USLE–RUSLE Erosion Prediction Models

The USLE was steadily improved and eventually became RUSLE [16], and it most recently developed the scientifically enhanced and improved relationship RUSLE2 [15]. The structure of the USLE and RUSLE equations is identical, except that the underlying factor relationships have been improved as more data became available. The USLE–RUSLE models are multiplicative models of factors, each of which represents a distinct and independent aspect of soil erosion conditions. It is commonly formulated as:

$$A = RKSLCP \tag{8.1}$$

where A is the computed average annual soil loss per unit area, expressed in the units selected for K and R. In the USA, A is usually expressed in tons per acre per year, but can also be given in terms of metric tons per hectare per year, as is done in most countries [17]. R is the rainfall and runoff factor, or erosivity factor, and represents the erosive power of rainfall, runoff, and seepage on soil loss. K is the soil erodibility factor representing the soil loss rate per erosivity index unit as measured on a unit plot defined as 22.13 m (72.6 ft) long of uniform 9% slope steepness with tilled fallow management. L represents the slope length effect and is defined as the ratio of soil loss from the field slope length to that from a 22.13-m (72.6 ft) unit plot length under otherwise identical conditions. S, the slope-steepness factor, is the ratio of soil loss from the field slope gradient to that from a 9% slope under otherwise identical conditions. R, K, S, and L are the intrinsic factors for a given location and cannot be modified. C is the cover and management factor and is defined as the ratio of soil loss from an area with specified cover and management practice to that from an identical area in clean soil tilled up-and-down slope. By selecting appropriate agronomic practices, one can control soil loss to predetermined levels. The P factor is the support practice factor, which represents the ratio of soil loss with an erosion control practice, such as contouring, strip cropping, or terracing, to that with straight row farming soil-manipulating implements up and down slope [11]. Large datasets were collected over the years for different cropping practices and soil manipulations [18, 19]. These datasets have been tabulated and are readily available in the USLE and RUSLE computer programs. RUSLE2 is the most advanced and scientifically up-to-date field-oriented soil erosion prediction and conservation management tool. It is used worldwide and can address erosion on agricultural fields, grassland, forestland, and rangeland, as well as on construction sites. In using the USLE and RUSLE models or to conduct experiments on natural runoff or rainulator plots, one must be aware of the standard or unit plot requirements. Also, because of the many time-variant factors and properties, the ULSE relationship should not be used for soil loss predictions of individual storms.

Several observations are pertinent regarding type, manner, and frequency of tillage of erosion research plots for K-factor determinations. The recommended preparation is as follows[1]: "Plow to normal depth and smooth immediately by

[1] Administrative communication from D. D. Smith to runoff plot managers (January 1, 1961): "Instructions for establishment and maintenance of cultivated fallow plots..

disking and cultivating two or more times, except for areas where wind erosion during the winter poses a serious hazard. In the latter case, disking or cultivating should be delayed until spring. Plowing shall be each year at the time continuous row crop plots are plowed. Cultivation shall be at new crop planting, routine, cultivating times and when necessary to eliminate serious crust formations. Chemical weed control may be used if cultivation does not control weed growth. Plowing and cultivation should be up and down slope and should not be on too wet soil." It is evident from this description that, in early investigations, in spite of multitudes of factors with all its gradations, detailed thought was given to the standards of plot preparation in order to guarantee as much as possible the consistency in methodology and reliability in the data obtained [20].

Soil conservation efforts have primarily focused on the *C*-factor regarding crop type, crop stage, cropping systems, residue management, tillage type, and cover crops to control soil loss. The more basic research aspects have focused on the *R*- and *K*-factors. These two factors are linked, in that the *R*-factor represents the exogenic force, while the *K*-factor represents the response, in terms of soil loss, to the exogenic force. This article restricts itself to discussing the *R*-factor and *K*-factor evaluations. For detailed information about the *S*, *L*, and *C*-factor, the reader is referred to Agriculture Handbook 703 [16] and the computerized RUSLE2 model [15]. These two sources provide numerical values in tables and figures for a large array of agronomic and topographic conditions.

1.2.1.1 *R*-Factor

The *R*-factor includes, in principle, all exogenic forces that contribute to the removal of soil from a given area. These forces may consist of raindrop impact during rainfall, external shear from runoff, and internal shear due to hydraulic forces by seepage or exit gradients. Of these, rainfall and runoff are normally considered to be the principal driving forces. Film flow and drop impact in rainfall are the primary factors on interrill areas and scour by runoff in rills. Rills are usually described as small incisions on sloping land in which flow concentrates. Seepage effects are rarely considered and are not directly or separately identified as a contributing factor in the current soil loss equations. However, they are indirectly accounted for in soil loss measurement by increases in the detached soil through the *K*-factor.

The main erosivity parameters affecting soil loss is the amount of rainfall or rainfall energy (*E*) and the rainfall intensity (*I*). How these two parameters interrelate has been the subject of much research in earlier days [21]. Figure 8.1 shows the relationships between rainfall energy and intensity obtained from 315 raindrop samples collected at Holly Springs, MS and Washington, D.C. While there appears to be a good correlation in the very-low to low-intensity range up to about 12 mm/h, for higher intensities the data show a large amount of scatter. The product of the energy and an intensity term, the EI_{30}, was best correlated to soil loss in the upper Midwest of the USA, where most of the early erosion research was done. *E* represents the total amount of rainfall energy in a storm and I_{30} is the maximum

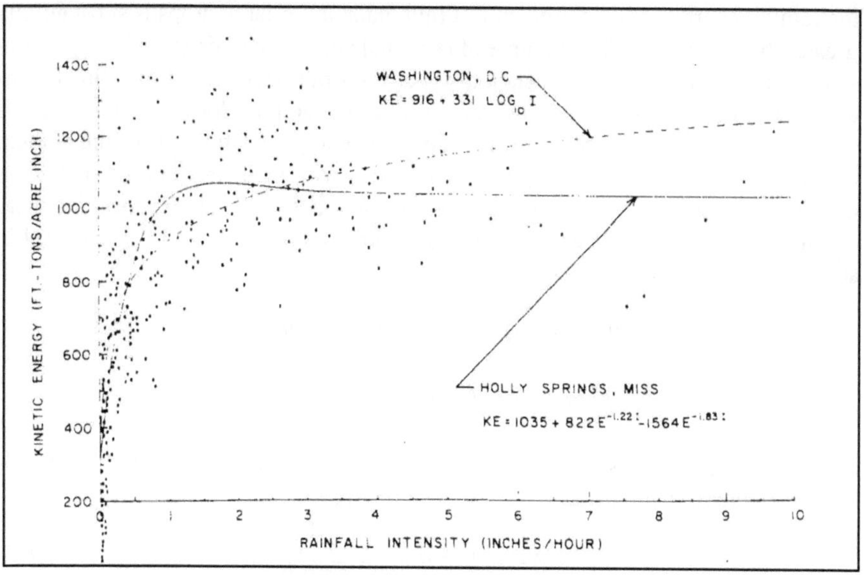

Fig. 8.1 Relationship of kinetic energy and rainfall intensity computed from 315 raindrop samples collected at Holly Springs, Mississippi, compared with that derived from raindrop samples collected at Washington, D.C., and extrapolated intensities of 4 in./h (10.2 cm/h). Note: The coordinate parameters are given in the US customary units. For the corresponding SI metric units, one must multiply the abscissa coordinate by 25 to yield mm/h and the ordinate coordinate by 2.625×10^{-4} to yield MJ/ha/mm

rainfall intensity during any consecutive 30 min of the rainstorm [10, 11]. E is expressed as the amount of raindrop energy which, by itself, is not a good indicator of the erosive potential of a storm. This may readily be concluded considering that the raindrop energy in a low-intensity storm may equal that of a high-intensity storm of shorter duration, though the erosive power of the latter storm as measured by soil loss may be appreciably higher. The intensity modifies the erosivity of the storm. Rainfall energy is determined by the raindrop size distribution and raindrop impact velocity which, in turn, is determined by raindrop size, the downdraft and updraft of the air in which the drops are suspended, and the associated drop-size formation dynamics. Therefore, the rain energy for a given intensity may be highly variable for comparable rainfall intensities. Also, raindrops may break up during their fall trajectory, further complicating erosivity estimations.

Other practical considerations in calculating the rainfall erosivity at a given location are the finding in the USA that rainfall of less than 1.13 cm and separated from other rainstorms by more than 6 h were not included in the calculation unless 0.613 cm fell within a 15-min period [11]. This reflects rain absorption by the soil and surface storage. Details about erosivity calculations and the subsequent development of iso-erodent maps have been published by Wischmeier and Smith [11] and Renard et al. [16], and more recently in the updated version of the RUSLE2 [15]. Since the drop-size distribution of rain varies appreciably from storm to storm, the rainfall energy for a volume unit of rain made up of large drops is substantially greater than for an identical volume of rain made up of small drops. Also, the drop-size distribution may vary substantially between rainstorms of similar intensities due to a host of factors, such as temperature, geographic location, dynamics of formation, wind effects, etc. Thus, the rainstorm energy is a function of the amount of rain and all the storm's component intensities with their inherent differences in drop-size distribution. Nevertheless, a relationship between the rainstorm intensity (i_m) and the storm kinetic energy (e) was determined by Laws and Parsons [22] in still air, which in metric units is:

$$\begin{aligned} e &= 0.119 + 0.0873 \cdot \log(i_m) \quad i_m < 76 \text{ mm/hr} \\ e &= 0.283 \quad i_m > 76 \text{ mm/hr} \end{aligned} \tag{8.2}$$

where e is expressed as MJ ha^{-1} mm^{-1} and i_m in mm [28]. Brown and Foster [23] used a different relationship:

$$e = e_{max} \cdot \left[1 - a \cdot exp(-b \cdot i)\right] \tag{8.3}$$

where e_{max} is a maximum unit energy as intensity approaches infinity and a and b are the coefficients with a=0.72 and b=0.05. Based on the field data, the highest correlation of measurement was observed with the product EI_{30} [11, 16, 21].

Given the occasional use that RUSLE still enjoys, this section follows largely the RUSLE approach by Renard et al. [16] described in Handbook 703 on obtaining R-values for the continental USA. R-values are obtained from iso-erodent maps. These maps show lines representing the locations with equal rainfall erosivities. Erosivity values for locations between iso-erodent lines are obtained by linear interpolation. For the eastern part of the USA, individual R-values were computed from 22-year station rainfall records by computing the EI-value for each storm. Iso-erodents were then obtained using rainfall-frequency data from the US Weather Bureau 1958 and topographic maps cf. [16]. Because of relatively fewer number and insufficient long-term recording rain gauge records in the western part of the USA, iso-erodents there could not accurately be established. Iso-erodents for that area were obtained in 1976 by an extended estimating procedure [11].

Erosivity values in the western plains and North-Central States of the USA were approximated with the relationship R=27.38$P^{2.17}$, where p represents the 2-year frequency, 6-h rainfall amount [24]. The iso-erodents developed from these values were not sufficiently accurate to reflect the spatial variability of the mountain and valley topography of this region. Instead EI-values were calculated on a 15-min measurement interval basis and a 60-min measurement interval basis for all storms at 713 stations. Of these stations, 225 had record periods of 12 years or longer with precipitation resolution of 0.25 mm. Then, the correlations were performed yielding coefficient of the determination of better than 0.8 using the relationship $(EI)_{15}$=b $(EI)_{60}$, with b ranging from 1.08 to 3.16 for the 120 homogeneous climate zones of the contiguous USA. The climate zones represent regions in which the

EI-distribution as a percentage of the annual EI for 24 periods of 15 days each are known and can be found in tables from Agriculture Handbook 703 [16].

Agriculture Handbook 703 [16] presented the additional information in obtaining more accurate iso-erodent maps for the western plain and North Central States Region by calculating the EI_{60} for 1082 stations of which 790 had a 20-year record of length or longer. This information was in turn used to calculate the EI_{15} values based on the aforementioned regression relationship using the b-value as appropriate for each climate zone. The R-value or erosivity value was then obtained by the adjustments using R=1.0667 [25]. The R-value could now be used to develop a more accurate iso-erodent map for the western part of the USA.

The effect of the runoff component in the R-factor during the early erosion studies did not receive much attention. Soil loss was seen as a rainfall-caused phenomenon in which rainstorm parameters formed exclusively the driving force. Most of the runoff plots were concentrated in the Midwest and Southeast of the USA, where rainfall was the dominant erosive force. Separating the erosivity factor, the relative contribution of rainfall-caused soil erosion from the surface-flow erosion was of particular interest in the development of process-based erosion models. Since 1975, the accepted premise was that soil erosion consisted of interrill erosion with rainfall as the driving force and rill erosion with concentrated surface flow as the driving force [6] This issue was especially urgent with the need for more effective soil conservation methods in the dry farmed cropland of the Northwest wheat and range region of the USA, where as much as 90 % of the annual soil loss is attributed to runoff from melting snow, light rain on snow, and rain on thawing land. Attempts were made to somehow include erosion predictions from low-rainfall regions such as the Northwest and on rangeland, in the convenient and simple structure of the USLE.

A sub-factor R_s was introduced in the erosivity term of the USLE to account for the runoff effect. Based on limited information, Wischmeier and Smith [11] found that R_s was about 1.5 times the December–March precipitation amount measured as inches of water. If the precipitation during December 1–March 31 is 12 in. (30.5 cm), and the annual EI_{30} for this region is 20 (US customary units) then the average annual R would be $20+1.5\times12=38$ erosivity index units. This approach was a practical measure based on the experience and available local data. A more rational approach based on the field measurements of eroded soil volumes (A_{wr}) taken over a 10-year period was obtained from strips of soil over a 72–80-km-wide transect in eastern Washington State and northern Idaho. Similar measurements were made in southeast Idaho over a 4-year period and from north central Oregon over a 15-year period. An equivalent $R(R_{eq})$ for cropland was derived cf. [16] from measured soil loss values. The effects of soil moisture, surface residue, growing canopy, and surface roughness were accounted for in arriving at this value. This relationship is structurally identical to that of the USLE/RUSLE, and is given in Agriculture Handbook 703 [16]:

$$(R_{eq})_{wr} = \frac{A_{wr}}{K_{wr}\cdot(LS)_{wr}\cdot(SLR_{wr})\cdot P_w} \tag{8.4}$$

where the subscript "*wr*" refers to the wheat and range region, K_{wr} is the estimated soil erodibility for the winter period obtained from seasonal *K*-values for this region [16], (Chap. 3), *SL* is the topography relationship based on slope length and steepness for this landscape [16] (Chap. 2, p. 33), $(SLR)_{wr}$ represents the soil loss ratio for rilling in the winter period [16] (Chap. 2, p. 33), and P_{wr} is the supporting practice factor which is mostly for contouring and is assumed to be unity for the winter period [16] (Chap. 2, p. 34). By correlating the calculated $(R_{eq})_{wr}$ based on soil loss measured over a certain period (10–7 years) with the annual p (precipitation), the following relationship was obtained:

$$(R_{eq})_{wr} = 110.3 + 10.78 \cdot P \qquad r^2 = 0.98 \tag{8.5}$$

$(R_{eq})_{wr}$, the soil loss by rilling in the winter period, must be adjusted for (1) soil loss due to interrill erosion and (2) soil loss in the nonwinter period. Renard et al. [16] indicated that field rill erosion measurements at the Columbia Plateau Conservation Research Center near Pendleton, OR, estimated that 95 % of the soil loss was attributed to rilling. He also indicated that soil loss estimates by rilling on the nearby Pullman Conservation Field Station in the State of Washington were 75 %, while direct measurements over a 4-year period at Pullman yielded 85–90 % soil erosion due to rilling. Renard et al. [16] concluded that these estimates were very imprecise and implied that the contribution to total soil loss by rilling were high and used 90 % as the percentage soil loss by rill erosion during the winter period (October 1 to March 31).

Renard et al. [16] indicated that the soil loss in the nonwinter period was rainfall based and was about 5 % of the total annual soil loss. Based on this information, the R_{eq} for the Northwest Wheat region was given as:

$$R_{eq} = (R_{eq})_{wr} \cdot 100/90 \cdot 100/95 \tag{8.6}$$

and

$$R_{eq} = -129 + 12.61 \cdot P \quad 7.5 < P < 15.0 \tag{8.7}$$

For the lower precipitation region in the Northwest Wheat and Rangeland area with frozen soil problems, the following relationship was giving a smooth transition from R_{eq} to the non-R_{eq}:

$$R_{eq} = 1.602 \cdot exp(0.2418 \cdot P) \tag{8.8}$$

Thus, the total erosivity factor for this region and presumably for regions with similar conditions elsewhere is:

$$Total\ R = R(rainfall) + R_{eq} \tag{8.9}$$

In short, for high rainfall areas one may use the USLE/RUSLE erosivity factor based on rainfall effects, while for low rainfall areas with frozen soil conditions, approaches of deriving an erosivity term based on the method used in the northwest of the USA should be considered. It should also be evident that much more needs to be done in regard to the erodibility and erosivity relationships of frozen soils from both a physics and engineering perspective.

1.2.1.2 *K*-Factor

Soil erodibility involves fundamentally two mechanisms: detachment and transport of soil particles. Each one of these mechanism involves a different mix of erosive forces and have dependencies on a different set of intrinsic soil properties. It is generally understood that the erosive nature of rainfall which relates to the intense localized energy expenditure by impacting rain drops is primarily a detachment process that is mostly controlled by chemical and physical soil properties. Once these bonds are broken, the breakup of the raindrop into smaller droplets following impact imparts the balance of the available impact energy into a transport phase. The size of these droplets and their kinetic energy determines what size and how far the loosened soil particles will move, which in turn depends on the soil textural composition and micro-surface roughness. On the other hand, the erosive nature of surface flow relates to the ability of this flow to dislodge soil particle from its matrix by breaking particle bonds and then to carry the loosened soil particles down slope. In this process, the dominant aspect is transport. The larger the soil particles the more energy is required for transport. It is therefore natural that the erodibility factor model is made of textural properties and constituents that relate to bonding and aggregation whether chemical (organic and inorganic constituents) or physical (interlocking or compaction).

The challenging aspect of soil erodibility determinations is to be able to relate the incident of exogenic, hydrologic, and hydraulic forces to the endogenous or intrinsic soil properties. The discovery, based on field data that soil loss was linearly related to the EI_{30} enabled the separation of the contributing factors of erosivity and erodibility to soil loss [21]. Details about the development and evaluation of the *K*-factor are given in Agriculture Handbook 537 [11], Agriculture Handbook 703 [16], and by USDA-ARS [15]. Several major studies were conducted to relate experimentally determined *K*-factors to soil properties [20].The most important and consequential was conducted by Wischmeier and Mannering [27] on 55 mostly medium-textured soils in the Midwest of the USA, complemented with the laboratory determinations of an array of 24 soil properties. These properties consisted mostly of interaction terms of various particle fractions with organic matter, structure, and an aggregation index. Subsequent analysis of this dataset indicated that a correlation existed between size fractions 2 and 100 μm and the measured soil erodibility. Additional analysis of the same database led to the formulation of the now widely known standard soil erodibility nomograph [28] in which the number of soil erosion sensitive parameter could be reduced to 5: the particle-size fraction between 0.002 and 0.1 mm defined as a “new silt” term, the “new sand” fraction between 0.1 and 2 mm, the organic matter content (*OM*), a structure code (*S*), and a permeability

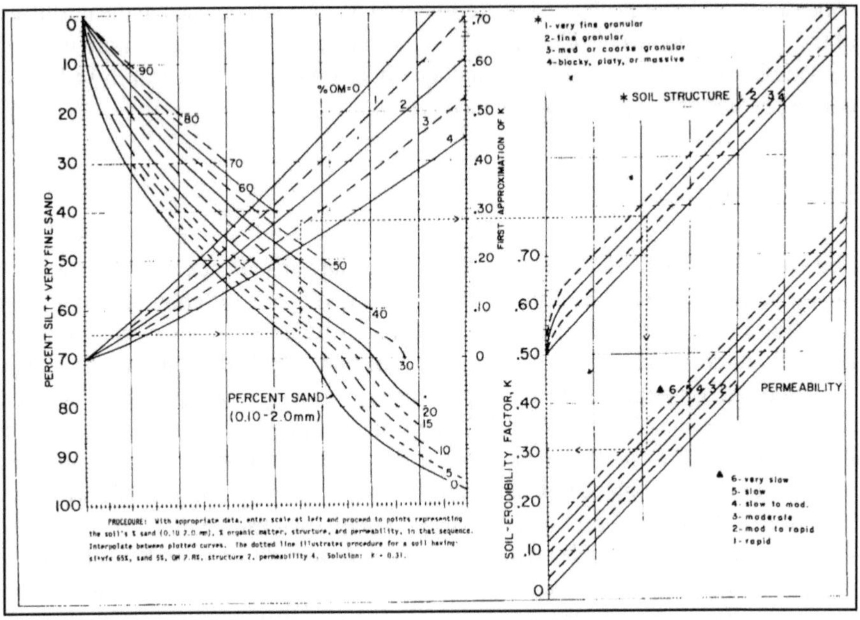

Fig. 8.2 Soil-erodibility nomograph. For conversion to SI, divide K values of this nomograph by 7.59 K as in US customary units [28].

class (P) (Fig. 8.2). The "new silt" term was reflected in the nomograph through the parameter M being equal to the product of (% new silt) × (% new silt + % new sand).

The algebraic version of the nomograph for soils with less than 70 % "new silt" is:

$$100K = 2.1 \times 10^{-4} M^{1.14} \cdot (12 - OM) + 3.25 \cdot (St - 2) + (Pm - 3) \qquad (8.10)$$

where OM is the percent organic matter. The indices St and Pm stand for structure and permeability, respectively, and are to be obtained from the USDA Soil Survey Manual [29]. The newly defined parameter M accounted for 85 % of the variation in the measured K-values for the predominantly medium-textured soil. The strength of this relationship is: (1) the highly erodible soils are mostly medium-textured soil for which this relationship was obtained, and (2) the number of parameters is small and can be obtained from standard soil and soil profile descriptions. The validity of the nomograph has been tested for other classes of soils with different mineralogy or genesis such as high-clay subsoils [30], volcanic soils in Hawaii [31], Minnesota soils [32], and found to be less applicable, though the textural properties were always the most significant ones. The dominance of a textural parameter in the nomograph expression was the impetus for additional research in which all known erodibility factors in the literature were pooled and grouped into ten texture size classes and related to the geometric mean particle diameter (D_g) of the texture classes [33]. Since the USA data were reliably obtained using a consistent methodology, they were analyzed by themselves (138 soils), but also analyzed in a larger, global dataset of 225 soils. In both instances, bell-shaped relationships were obtained between the log (D_g) and the average soil erodibility factor of the size classes [30]. These relationships (Fig. 8.2) are for the global erodibility database:

$$K = 7.594\left\{0.0034 + 0.0405 \cdot exp\left[-\frac{1}{2}\left(\frac{\log(D_g)+1.659}{0.7101}\right)^2\right]\right\} \tag{8.11}$$

where D_g = geometric mean particle diameter expressed as

$$D_g(mm) = exp\left(0.01\sum f_i ln(m_i)\right) \qquad r^2 = 0.983 \tag{8.12}$$

and for the American soil erodibility database:

$$K = 7.594\left\{0.0017 + 0.0494 \cdot exp\left[-\frac{1}{2}\left(\frac{\log(D_g)+1.675}{0.6986}\right)^2\right]\right\} \tag{8.13}$$

Both relationships indicate that fine textured soils high in clay had a low *K*-value, suggesting the greater resistance to detachment by rainfall and runoff on these more cohesive soils. Likewise, the coarse textured soils were usually higher in their sand content and, thus, had more infiltration and less runoff, leading to lower *K*-values. The medium-textured soils had the highest erodibility values, indicating the greater detachability and transportability of these soils in erosion processes. The significance of these bell-shaped relationships is the transferability of this information to other parts of the world that have different soil textural classification systems. Römkens et al. [34] related the variation in *K*-value within each size class to other soil properties, such as organic matter and clay percentage, but found only marginal improvement in the *K*-factor predictions when this factor was included. Wang et al. [35] showed that the bell-shaped relationship between the log (D_g) parameter and the soil erodibility factor also applied for the loess soils of the Chinese Loess Plateau. Also, a soil erodibility review paper was prepared by Wang et al. [36]. Successful efforts to relate soil erodibility to soil properties, even on a regional or soil class basis, would be of great benefit in soil loss predictions and for soil conservation programs.

Substantial progress has been made in soil conservation practices on upland areas. The first 60 years were devoted primarily to acquiring a huge database to support the statistically based USLE–RUSLE models of a primary conservation tool; much science has gone into the development of these models to make them useful for conditions not covered in the database. While these efforts were underway, substantial efforts were also made in developing process-based soil erosion prediction models. The most important one, certainly in the USA, is the water erosion prediction project (WEPP) Model. Process-based models are, by their nature, very complex, including many more physical parameters, each of which is usually site specific.

1.2.2 WEPP Erosion Prediction Model

WEPP is a process-based, distributed parameter, continuous one-dimensional simulation erosion prediction model based on the concept that erosion on upland areas predominantly takes place in rills and interrill areas and consists of detachment and transport components. These concepts were introduced by Ellison [37] in the 1940s and in the early 1970s by Foster and Meyer [26], respectively. To a large degree, these concepts have dominated the analytical thinking of erosion research during the past 40 years. Originally, WEPP was designed for hillslope erosion profile application but has, in more recent years, been expanded to applications of small agricultural watersheds by including channels and impoundments. The model includes a large array of processes such as erosion, sediment transport, and deposition; infiltration and other hydrologic processes; soil roughness and consolidation; the effect of plant growth, canopy and residue effects on erosion; sediment movement and runoff; and impoundments. These all-encompassing areas of concern are, by far, too many to be discussed in this chapter. Therefore, this section is limited to discussing soil erosion and transport processes on upland areas per se, and how they are related to the exogenic and endogenic factors, especially relative to soil property relationships. For related and more detailed aspects, one is referred to earlier accounts of the WEPP model discussed by Foster [38], Elliott et al. [39], Nearing [40], Flanagan and Nearing [41], Flanagan et al. [42], Flanagan et al. [43], and Alberts et al. [44]. Current WEPP model software, science and user documentation can be obtained from the National Soil Erosion Research Laboratory (NSERL) website [43].

1.2.2.1 WEPP Field Study

From an engineering perspective, the most relevant aspects are soil detachment and transport of rill and interrill processes and their relationship to soil properties. In that regard, the most significant study was the field process study on soil erodibility by Elliot et al. [39] in which 36 soils, located in 20 states, were evaluated on site for their erodibility using a standard approach. Most of the information given below is taken from this work.

The soils represented a wide range of textural, geomorphic, and chemical properties and were selected for their dominant agricultural importance of a selected region and their highly erodible nature [1]. Soil selection included commonly used standard soil survey analysis and profile descriptions. The research sites had slopes varying from 4 to 6%. In contrast to the USLE plots, the WEPP plots had been planted to the local specific site crop the year prior to the erosion studies with removal of excess residue following harvest, followed by deep tillage and light disking 3–12 months after residue removal [45].

Three types of experiments were performed: (1) 50-cm-wide and 75-cm-long uniform sloping infiltration plots, in which water entry into a typical profile is monitored; (2) 50-cm-wide and 75-cm-long sloping plots with 50 % side slopes toward a trough at the center for interrill erosion measurements; and (3) 9 m long, 46 by 50 cm wide rills for runoff measurements. In the latter experiments, a longer-slope length could be simulated with incremental flow at the top end.

All plots were located within the area covered by the rotating boom rainfall simulator [13] equipped with Veejet 80100 nozzles at a pressure yielding a 62-mm/h rainfall intensity. Most interrill plots were replicated four times and infiltration plots were replicated two times. Measurements included runoff rates, sediment rates, application rates, runoff velocities, and rill cross-sectional areas at the end of the experiment. From these data and the soil data obtained in the laboratory analysis, the rill and interrill erodibility, as well as shear stresses, were determined, and correlations were made with the laboratory measured intrinsic soil property thought to be related to soil erodibility.

1.2.2.2 WEPP Analysis Approach

For interrill erosion, the erodibility exogenic force is the rainfall intensity given by the relationship [46–49]:

$$D_i = K_i \cdot I^2 \tag{8.14}$$

where

D_i Interrill detachment rate
K_i Interrill erodibility constant
I Rainfall intensity

Based on fundamental reasoning and experimental evidence, interrill erosion is also affected by slope steepness [49–51] according to the expression:

$$D_i = K_i \cdot I^2 \cdot S_{\mathrm{f}} \tag{8.15}$$

where S_f is the slope factor given by the relationship

$$S_{\mathrm{f}} = 1.05 - 0.85 \cdot \int exp\left[-4 \cdot \sin(\theta)\right] \tag{8.16}$$

with θ as the slope angle.

For rill erosion, the erodibility exogenic force is the scour equation as formulated by Foster and Meyer [26] and presented by Elliot et al. [39]:

$$D_c = K_r \cdot (\tau - \tau_c)^b \tag{8.17}$$

where

D_c Detachment capacity of clear water (Kg M⁻ S)
K_r Rill erodibility (S M^{-1})
τ_c Critical shear stress below which there is no soil detachment (Pa)
τ Hydraulic shear of flowing water (Pa)
b Constant

and

$$\tau = \rho \cdot r_h \cdot s \tag{8.18}$$

where

ρ Density of water (N m^{-3})
r_h Hydraulic radius (m)
s Hydraulic gradient

As rill flow rate increases and exceeds the critical shear stress, the sediment load increases, but not necessarily linearly ($b \neq 1$), though it is often presented as a linear relationship [39]. Once the maximum transport capacity is reached, sediment deposition occurs. The effect of suspended sediment on soil detachment in the rills is given by the expression (26):

$$\frac{D_r(x)}{D_c} + \frac{Q_s(x)}{T_c} = 1 \tag{8.19}$$

where

D_r Soil detachment by rill flow at position x (Kg m^{-2} s^{-1})
D_c Maximum rill detachment capacity (Kg m^{-2} s^{-1})
Q_s Rate of sediment transport down rill at x (Kg m^{-2} s^{-1})
T_c Maximum transport capacity of rill flow (Kg m^{-2} s)

The soil detachment rate is related to the transport rate according to Eq. 8.19 which upon substitution of

$$D_r = \frac{dQ_s(x)}{dc} \tag{8.20}$$

yields:

$$\frac{dQ_s(x)}{dx} = D_c\left[1 - \frac{Q_s(x)}{T_c}\right] \tag{8.21}$$

If the interrill contribution, E, to the sediment load in the rill is added, then Eq. 8.21 becomes

$$\frac{dQ_s(x)}{dx} = D_c\left[1 - \frac{Q_s(x)}{T_c}\right] + \mathrm{E} \tag{8.22}$$

which, upon integration over the rill length, l, with a maximum detachment capacity, D_c, becomes

$$D_c = -\frac{T_c}{l} ln\left[1 - \frac{Q_s}{T_c} \cdot \frac{D_c}{D_c + E}\right] \tag{8.23}$$

Relationship (Eq. 8.23) allows the determination of the detachment capacities for various flow conditions, which, in turn, permit the determination of K_r and τ_c as per Eq. 8.17. Having determined these soil parameters, relationships can now be sought with intrinsic soil and soil profile properties thought to be important in rill erodibility and critical shear strength.

1.2.2.3 WEPP Soil Erodibility Coefficients

Intrinsic soil property relationships. Soils consist of thousands of physical, chemical, and mineralogical constituents, each of which may have properties that in some way affect a soil's susceptibility to erode. Many of these constituents interact, creating in the process new parameters and mechanisms of soil detachment. In general, the approach has been to seek a combination of soil textural parameters which are more indicative of transportability properties, particularly the silt (2–50 μm) and very fine sand (50–125 μm) fractions and binding or cementing agents indicative of the soil detachability (organic matter, sesquioxides, high-valent cations, etc.). The best example of this physical model approach has been the work discussed by Römkens et al. [30] in their studies on the soil erodibility factor of the USLE and RUSLE. The routine has been to use a stepwise linear regression approach of relating the measured soil erodibility parameters to an array of laboratory and profile-determined soil properties presumed to be significant in soil erosion. The highest value of the coefficient of determination was then chosen as the basis for the most significant erodibility prediction relationship.

The WEPP studies used a similar approach. Their analysis used (1) the textural parameters % clay (<2 μm), % silt (2–50 μm), % very fine sand (50–100 μm), % sand (100–2000 μm), % rock (>2000 μm), % dispersible clay, and the *M*-parameter; (2) the selected chemical properties the % F_2O_3, the % Al_2O_3, the amount of Ca, Mg, Na, K, and $CECNH_4$ as the mill-equivalent/100 gm soil, and the sodium adsorption ratio (SAR); (3) the selected soil physical properties conductivity, % organic carbon, % $CaCO_3$, coefficient of linear extensibility in cm/cm; and (4) soil water content at the 1/3 and 15 bar suction, aggregate stability and the specific surface in milligram ethylene glycol monoethyl ether (EGME) per gram of soil.

In using the stepwise linear aggression technique, they arrived at the following predictive relationship listed in Table 8.1, between interrill, rill, and critical shear stress on one hand, and selected combinations of the property identified as erosion susceptible on the other hand (Table 8.1).

It is evident that, even though appreciable values for the coefficient of determination were obtained for all but rill erodibility, the relationships were

Table 8.1 Regression equations to predict soil erodibility from measurable soil properties

Interrill erodibility (Ki) kg s m^{-4}	
For clay >35 %	$\frac{K_i}{10^6} = 2.67 - 0.115\ (Ln\ (18 - Stability)^2)$ $r^2 = 00.82$ $\left(\text{if } 17 < \text{Stab} < 19, \text{ then } \frac{K_i}{10^6} = 2.67\right)$
For clay ≤35 %	$\frac{K_i}{10^6} = -2.92 - 2.71\left(\frac{WDClay}{Clay}\right) - 0.51\ Mg + 0.10\ WDClay$ $+4.19\left(\frac{Clay}{Fe + Al}\right)^{0.16} + 1.24\ Cond$ $r^2 = 0.77$
Rill erodibility (K_r), s m^{-1}	$\frac{K_r}{10^{-3}} = 196 + 0.015\ M^{0.2}(M^{0.8} - 3500) - \frac{32.7}{CECNH_4^{0.4}}$ $+\frac{0.16}{SAR^{0.75}}\left(\frac{1}{Al} - 50\ SAR^{1.75}\right) + 7.98\frac{(OM^{1.5} + 3.5)}{OM^{0.5}}$ $r^2 = 0.10$
Critical shear (τ_c), Pa	
For clay >30 %	$\tau_c = 0.5 - 284\ MC(MC - 0.3)$ $r^2 = 0.86$
For clay 30 %	$\tau_c = -2.85 - \frac{8.87}{(VFS + 0.1)^{0.2}} - 0.16\ CaCo_3 + 3.65\ SAR$ $+3.79\ SS^{0.2} + 28.1\left(\frac{WDClay}{Clay}\right)^{0.8}\frac{1}{Sand^{0.3}}$ $r^2 = 0.79$

complex. For the WEPP soils with its broad variety of properties, a simple physical model of a combination of a textural property with a bonding agent did not seem to exist, suggesting that soil erodibility as it relates to a specific exogenic force is far more complicated than has been assumed.

1.3 Status Upland Soil Erosion Research

The need to update soil erosion prediction technology was motivated by the realization based on practical experience that the existing technology had many shortcomings due to the complex nature of predicting soil loss, inadequately based science, higher requirements for consistency given the dependence of farm programs on conservation compliance and benefits of farm support programs, etc. The development of the USLE was a major breakthrough in identifying and quantifying the major soil erosion causing factors: weather, soils, topography, and land-use and control practices. These factors were at that point in time considered to be independent. RUSLE recognized and considered a limited interdependence of some of these major factors such as an erosivity weighted values for the soil erodibility and cropping management factors. As was discussed above, in RUSLE, the important concepts of rill and interrill erosion were introduced in which processes were affected by soil properties, topography, and cover management. RUSLE2 improved further the evaluation of interaction effects between main factors and introduced many more detailed sub-factor relationships based on scientific findings that addressed various aspects of the main factors. For instance, Dabney et al. [52] used RUSLE2 to estimate time varying runoff and process-based sediment transport routines to quantify sediment transport/deposition/delivery on complex slopes. Dabney et al. [53] showed how RUSLE2 could be used to include sediment transport and deposition process-based relationships and to improve in a major way the capability of predicting soil movement on irregular and complex slopes, including slopes with impoundments, and slopes with rough surfaces, such as mechanically disturbed land. RUSLE2 can be used to evaluate the impact of climate changes on runoff and sediment yield [54]. RUSLE2 is capable of segmenting topography, soil and management, and make temporal and spatial estimates of soil movement. RUSLE2 is land use independent and can be used on cropland, pastureland, rangeland, construction sites, reclaimed mine land, landfills, mechanically disturbed land, burned land and forestland, military sites, etc. It estimates average annual rill and interrill erosion on site-specific conditions. Much of the information in this section is taken from the documentation of the RUSLE2, Version 2, Science and User's Manual [15].

For most soil erosion control and conservation practitioners, the Revised Universal Soil Loss Equation, Version 2 (RUSLE2) is the most reliable and widely used management and prediction tool that is available for estimating average annual soil losses. RUSLE2 consists of three components: (1) a science component that includes the mathematical equations to compute soil erosion and deposition based on the major erosion relevant factors: climate, soil, topography, and land use, (2) a database from which factor values can be selected in relation to site-specific conditions, and (3) a powerful computer program that handles and uses the relevant mathematical equations, databases, and a computer interface that accepts and displays inputs and computed values. Besides its strong physical base of rill and interrill erosion, its strength and usefulness has been very much enhanced with a large database collected over 80 years of research representing all kinds of conditions in which the database has been extensively used for validation purposes. This database contains information about climate, soils, cover management, flow barriers and diversions, terrace channels and impoundments, surface drainage systems, irrigation, overland flow paths, etc.

RUSLE2 has retained the basic structure of its predecessors RUSLE and the USLE. While the latter is an index-based empirical model, RUSLE and RUSLE2 are hybrid models that combine index and process-based equations [55]. RUSLE2 is scientifically enhanced over RUSLE and uses different mathematical integrations. The reader is referred to USDA-ARS-NSL [15] home page for the details on Science and the Users Model.

For the remainder part of this section, the specific approaches concerning the *R*-factor and *K*-factor evaluations of the RUSLE2 will be briefly discussed.

1.3.1 *R*-Factor

In the USLE and RUSLE, the *R*-factor was represented by the EI_{30}-parameter, which was computed for each storm-producing runoff. From these storm values, the annual and monthly EI_{30}-values could be computed and iso-erodent lines could be developed when enough point rainfall records were available, such as in the eastern part of the USA, where they could be generated from $P_{10,24}$ data 10-year-24-h precipitation amounts that correlated well with EI_{30}-values computed from locations with sufficient record in the western USA. Nevertheless, rainfall amounts and erosivity values greatly vary and a sufficient length of record is needed to obtain reliable averages. To address this issue, the erosivity density concept was introduced which is the ratio of the monthly erosivity to monthly precipitation. This index is far more stable, does not require as long a rainfall record, and can overcome to a better degree situations with missing data points. Erosivity densities vary spatially and temporarily and is higher in the southern USA than in the northern part and do not vary, based on available data, up to an elevation of 3000 m [55]. Summer erosivity density is greater than winter erosivity in the eastern part of the USA. Monthly erosivity density maps throughout the continental USA have been given in the user's reference guide of RUSLE2 [15]. Figures 8.1, 8.2, 8.3, 8.4, 8.5, 8.6 and 8.7 shows the monthly erosivity density map for January and July. The erosivity values are directly proportional to the average monthly 30-min rainfall intensity values and reflect in effect seasonal variations in rainfall intensities [56]. The ability to partition EI_{30} values to shorter time periods greatly enhances the capability of obtaining more accurate soil loss values when used in combination with other main temporal and location soil erosion factors.

1.3.1.1 K-Factor Evaluations

Not many studies of the USLE soil erodibility factor were conducted since the 1970s. As was indicated, the most commonly used tool for determining the soil erodibility factor is the nomograph. To a large degree, the determination of this factor has not changed and the standard nomograph is still used for selecting *K*-factors. This standard nomograph [28] was derived from studies [27] relating to rainulator produced soil loss data to laboratory-measured soil properties and soil profile permeability characteristics. The soils were mostly medium-textured material and thus the nomograph fits best to these type of soils. The nomograph equation [57] for soils with less than 68 % (70 % in 1996) silt plus very fine sand consist of four sub-factors representing the sum effect of four properties: texture, organic matter, soil structure, and the soil profile permeability. These are essentially the same factors used for *K*-factor estimation in the USLE [16]. In RUSLE2 [15], each of these sub-factors was given a greater definition, based on regression analysis, by relationships covering certain particle-size ranges and fitting them to the graphically drawn relationships based on judgment in HdBk 537 [11]. However, in applications involving highly disturbed lands, a modified nomograph is used. In disturbed land, mixing of soil

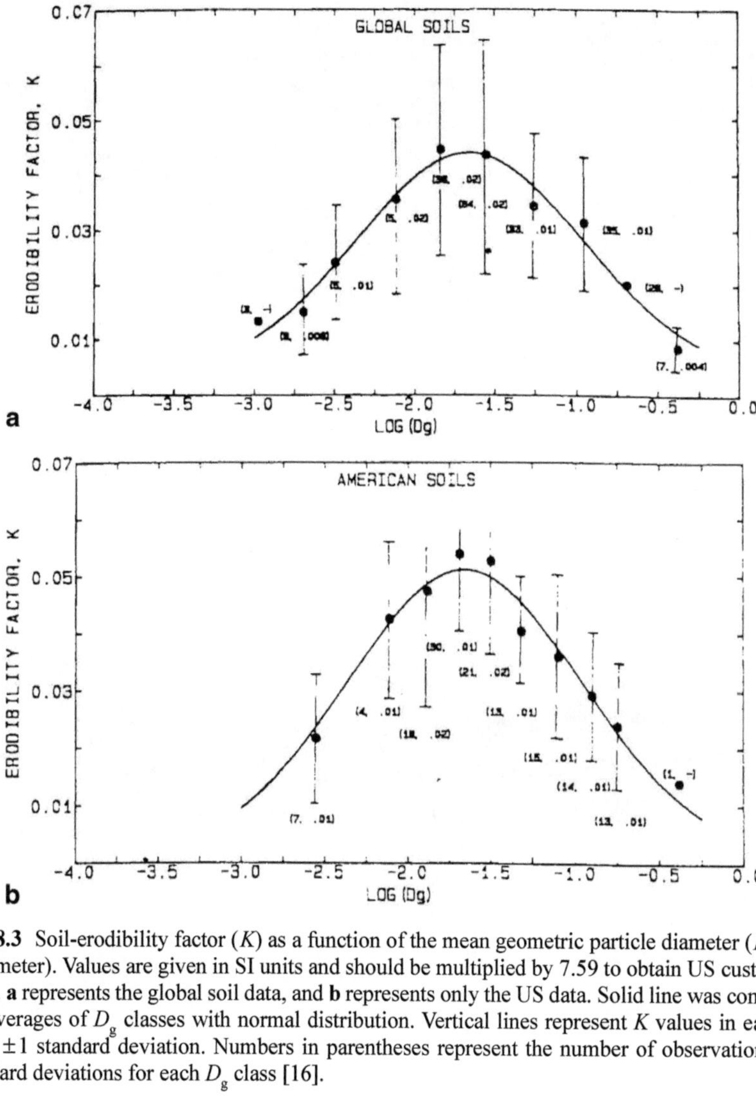

Fig. 8.3 Soil-erodibility factor (K) as a function of the mean geometric particle diameter (D_g) (in millimeter). Values are given in SI units and should be multiplied by 7.59 to obtain US customary units. **a** represents the global soil data, and **b** represents only the US data. Solid line was computed for averages of D_g classes with normal distribution. Vertical lines represent K values in each D_g class ± 1 standard deviation. Numbers in parentheses represent the number of observations and standard deviations for each D_g class [16].

material from different soil horizons with a different textural/chemical/mineralogical composition occurs, which alters the erodibility characteristics. Also, provisions have been made by the USDA- Natural Resources Conservation Service (NRCS) in their database for the user to select K-values.

The standard soil erodibility factor relationship (see Eq. 8.10) is:

$$K = \frac{(k_t \cdot k_o + k_s + k_p)}{100} \tag{8.24}$$

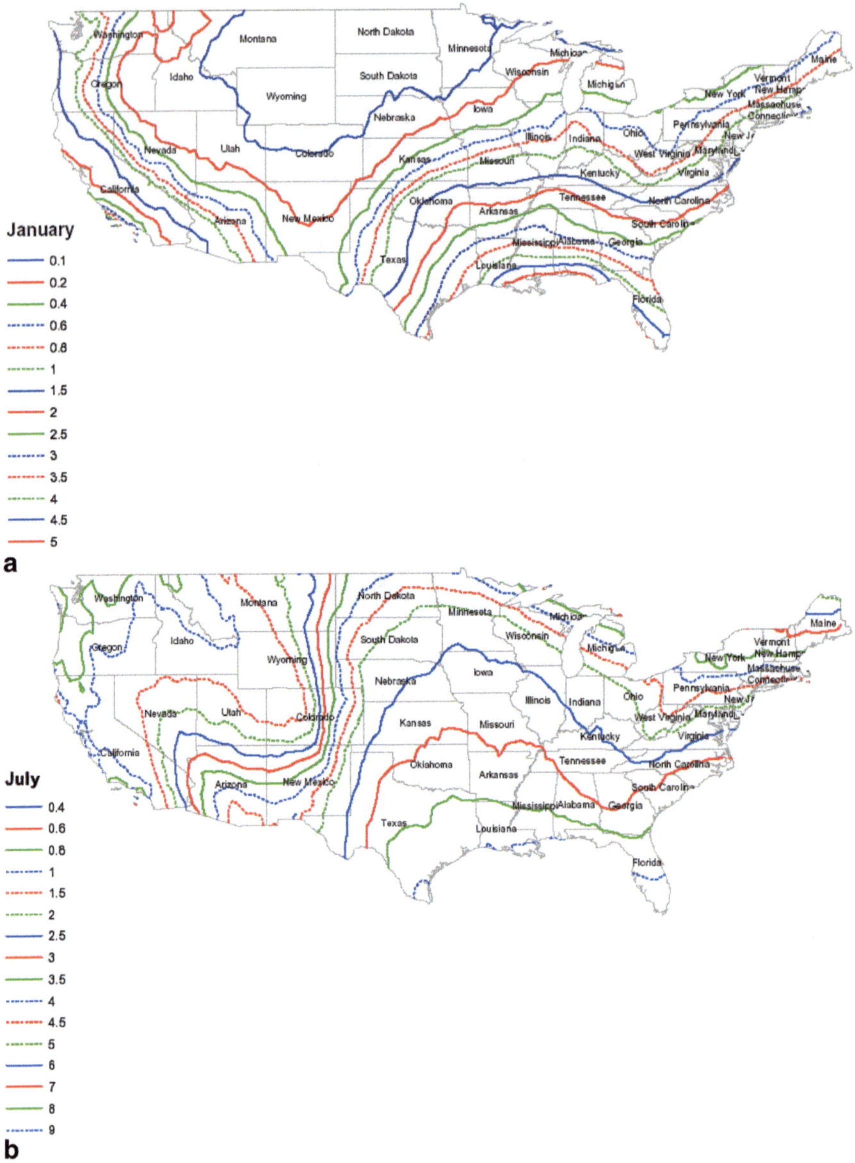

Fig. 8.4 **a** Monthly erosivity density [monthly erosivity (SI units)/monthly precip (mm)] for January. **b** Monthly erosivity density [monthly erosivity (SI units)/monthly precip (mm)] for July [16].

where K is the soil erodibility factor, k_t is the texture sub-factor, k_o is the organic matter sub-factor, k_s is the soil structure sub-factor, and k_p is the soil profile permeability sub-factor.

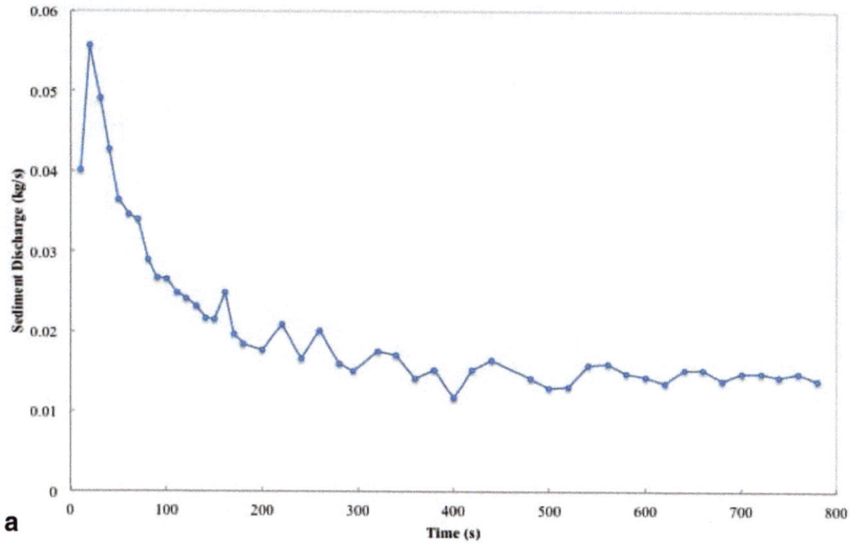

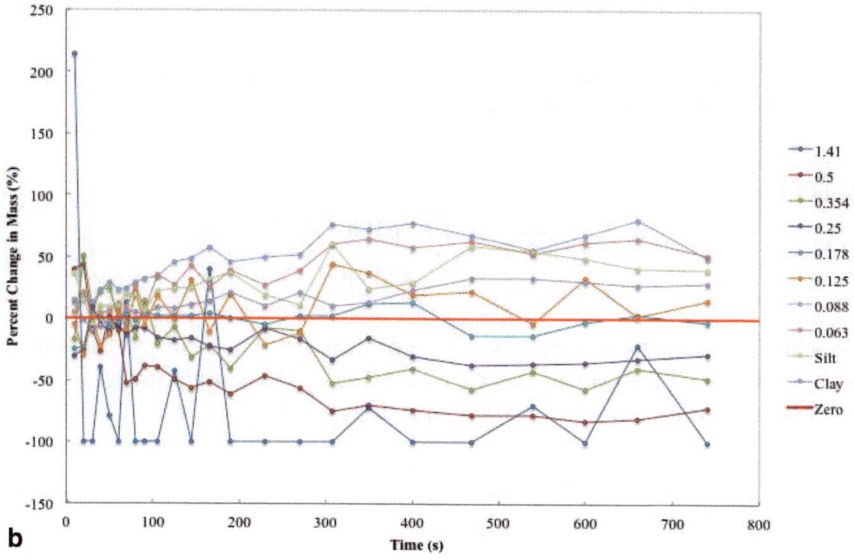

Fig. 8.5 **a** Time series of sediment discharge. **b** Percent change in PSD size class of the sediment discharge material as compared to the original soil material. PSD particle-size distribution.

1.3.1.2 The Texture Sub-Factor

The texture sub-factor is further broken down to reflect the effect of certain size classes, where:

$$k_t = k_{tb} \quad P_{sl} + P_{vfs} < 68\% \tag{8.25}$$

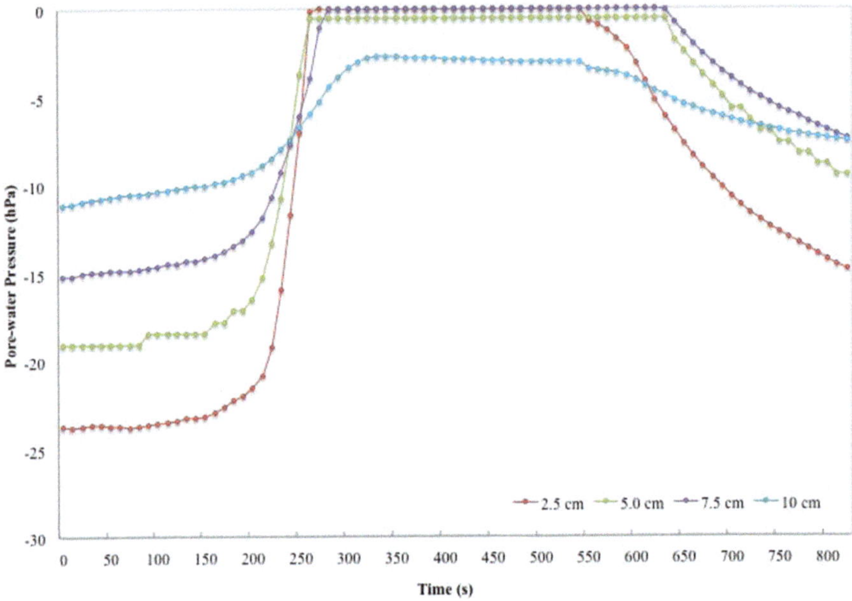

Fig. 8.6 Time series of pore-water pressure data, displayed with reference to depth below the surface, as a headcut moves past the sensors

with

$$k_{tb} = 2.1 \cdot \frac{(P_{sl} + P_{vfs})(100 - P_{cl})^{1.14}}{10{,}000} \tag{8.26}$$

P_{sl} is the percent silt (2–50 mu), P_{vfs} is the percent very fine sand (50–100 mu) in the soil, and P_{cl} is the percent clay (0–2 mu). The k_{tb} relationship was obtained by regression analysis. If $P_{sl} + P_{vfs} > 68\%$, then these relationships are:

$$k_t = k_{tb} - \left[0.67\left(k_{tb} - k_{t68}\right)^{0.82}\right] \qquad P_{sl} + P_{vfs} > 68\% \tag{8.27}$$

with

$$k_{t68} = 2.1 \cdot \frac{68 \cdot (100 - P_{cl})^{1.14}}{100} \tag{8.28}$$

where k_{t68} is the base soil texture sub-factor in the soil erodibility nomograph when $P_{sl} + P_{vfs} > 68\%$. The k_t relationship (Eq. 8.27) was obtained by fitting the different organic matter graphically drawn curves of the Wischmeier nomograph [28].

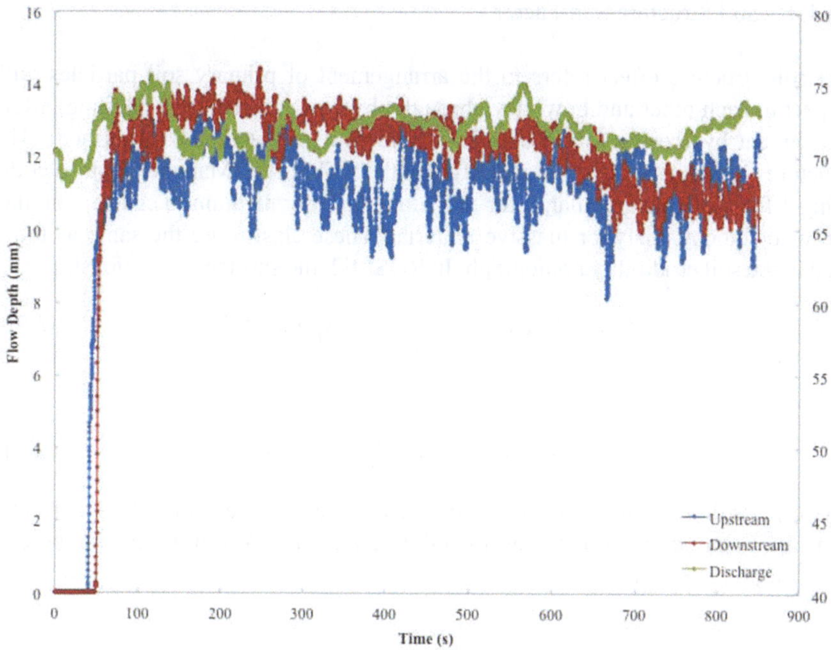

Fig. 8.7 Time series of overland flow discharge and water depth using a magnetic flow meter and ultrasonic depth sensor, respectively

1.3.1.3 The Organic Matter Sub-Factor

The organic matter sub-factor for the erodibility nomograph is given by the relationship:

$$k_o = (12 - O_m) \tag{8.29}$$

where O_m is the percent soil organic matter of the soil in unit plot condition. The soil database for the erodibility nomograph consisted of (1) data from a relatively few natural runoff plots that had been and were maintained in fallow condition for an extended period of time and in which the organic matter content was "native" to the soil and (2) data from a larger rainfall simulator plots on soils that had been in crops and were only kept in fallow condition for several months following the removal of surface residue. In these plots, the subsurface residue had not fully been converted to a "native" status. Therefore, different *K*-values could be arrived at though the significance of their differences is probably marginal.

1.3.1.4 Soil Structure Sub-Factor

The soil structure effect refers to the arrangement of primary soil particles with respect to each other and how they, through physical and chemical bonding, affect detachment by erosive forces of rainfall and flow (surface and subsurface). The structure classes used are those identified in the Soil Survey Manual [29] by a code, being 1 for very fine granular, 2 for fine granular, 3 for medium to coarse granular, and 4 for blocky, platy, or massive material. These classes are the same as those used in the soil erodibility nomograph. In RUSLE2, the sub-factor relationships are:

$$k_s = 3.25 \cdot (S_s - 2) \qquad \text{if } (k_t \cdot k_o + k_s) > 7 \tag{8.30}$$

and

$$k_t \cdot k_o + k_s = 7 \qquad \textit{if } (k_t \cdot k_o + k_s) < 7 \tag{8.31}$$

where S_s is the soil structure code as defined above. The reason for these two relationships is based on the fact that this information was taken from the nomograph which shows a bend [11].

1.3.1.5 Soil Profile Permeability Sub-Factor

The soil permeability sub-factor accounts for intrinsic differences in soil to absorb and transmit water into the soil profile in unit plot conditions. As was done for the soil structure sub-factor, the permeability sub-factor is based on a classification system of the USDA-SCS [29] which codes as follows: 1 is rapid and has low runoff potential, 2 is moderate to rapid, 3 is moderate, 4 is slow to moderate, 5 is slow, and 6 is very slow. This approach was also used in the USLE. The relationship for this sub-factor is:

$$k_p = 2.5 \cdot (P_r - 3) \tag{8.32}$$

where P_r is the permeability class as defined.

1.4 Recommendations

From a conservation standpoint, RUSLE2 is the most advanced, accurate, and science-supported soil erosion prediction model and conservation management tool [58]. It can be applied for a wide range of land uses. Its major use has been in agriculture, with the expansion of analytical approaches and database usages; it has also become useful for nonagricultural applications. Of the factor relationships, the erosivity density, known on monthly basis, is the most up-to-date erosivity relationship. The erodibility factor relationships have not changed very much since the development of the erodibility nomograph. For certain classes of soils (clay- and volcanic subsoils) the special relationships obtained in certain studies are still current [59]. The bell-shaped erodibility relationships may be very helpful for erosion predictions outside the USA. For the erosion predictive relationship for slope length, steepness, and cover-management practices factors, one is referred to the RUSLE2 documentation [15].

2 Concentrated Flow Erosion

2.1 Introduction

Soil erosion by water occurs if the combined power of rainfall energy (mechanical detachment) and overland flow energy (hydraulic detachment) exceeds the resistance of soil to detachment. The form of soil erosion may be viewed as a continuum of scales, not necessarily continuous, but in some cases evolving from one to the other, ranging from rill to gully [60]. Rill erosion occurs on sloping fields in which numerous and randomly occurring small incisions of only several centimeters deep are formed. Gully erosion is caused by concentrated overland flow or sub-surface flow of water during and immediately following heavy rains [61]. Both rill and gully erosion are driven by hydraulic processes related to shear of overland flow.

Soil erosion has been recognized as a primary cause of soil degradation. Gully erosion has been cited as a particularly nasty form of erosion and as the dominant source of sediment within watersheds [62–64]. Gullies are typically classed by size, i.e., classical, bank, or ephemeral, and often relate to whether or not the erosion feature may be crossed by farm implements [61]. However, each of these distinctive gullies is formed under similar processes of migrating or advancing headcuts. Headcuts are areas of intense, localized erosion, most commonly associated with a step-change in bed elevation [62]. The occurrence and migration of headcuts are commonly associated with significant increases in sediment yield [57, 65–70].

Soil conservationists have been increasingly emphasizing the importance and recognition of gully erosion on agricultural fields. Along with this heightened awareness, techniques to quantify the processes driving erosion and sedimentation have emerged. Within this section, the methodology, techniques, and implications for discovery will be discussed.

2.2 Gully Erosion Prediction in the Chinese Loess Plateau

Gully erosion has been the subject of research for a long time, by many scientists, from diverse scientific disciplines and different interests in soil erosion and sedimentation problems. Yet, no definitive or generally accepted prediction equation for gully erosion has been obtained. In recent years, Chinese researchers [71–73] have

Table 8.2 Comparison of soil loss data with and without ephemeral gully erosion in Loess Plateau, China

Year	Annual precipitation (mm)	Rainfall from June to september (mm)	Soil loss with EG ($t\ km^{-2}\ a^{-1}$)	Without EG ($t\ km^{-2}\ a^{-1}$)	Soil loss ratio with EG and without EG
1985	667.4	536.2	16,328.1	12,092.2	1.35
1986	390.5	290.9	4471.5	2057.6	2.17
1987	413.1	220.8	8122.3	5169.7	1.57
1988	726.4	612.0	56,471.1	40,875.4	1.38
1989	567.1	402.1	31,160.2	23,556.0	1.32
1990	581.0	365.5	6390.7	4746.5	1.35
1991	613.4	347.7	23,334.5	15,565.3	1.50
1992	604.9	478.9	1072.0	620.5	1.73
Annual	570.5	406.8	18,418.8	13,085.4	1.41

EG ephemeral gully

attempted to incorporate soil loss from ephemeral gullies into the structure of the widely known USLE. In their approach, a gully erosion factor, G, is added to the USLE as a multiplicative factor, that reflects the increase in soil loss at a given location due to ephemeral gully formation. Then, the USLE (Eq. 8.1) becomes

$$A = RKSLCPG \qquad (8.33)$$

with

$$G = 1 + \beta \qquad (8.34)$$

Where β is the correction coefficient that accounts for increased soil loss due to ephemeral gullies and is determined by a host of factors including rainfall, upslope runoff rate, slope length, slope degree, soil, and soil surface conditions.

The data for the G-factor development were obtained from soil loss measurements from two runoff plots in fallow condition, adjacent to each other; one with an ephemeral gully and one without. The plots were established at the Ansai Soil Conservation Experiment Station in Shaanxi Province (Institute of Soil and Water Conservation, Northwest A&F University, China). Rainfall and soil loss data were collected in a conventional manner from 1985 to 1992 (Table 8.2). Based upon a statistical analysis of this erosion data, the following equations were obtained:

$$M_h = 7.376\left(\sum P' I'_{30}\right)^{1.682} \qquad R^2 = 0.937 \qquad (8.35)$$

$$M_n = 2.337\left(\sum P' I'_{30}\right)^{1.849} \qquad R^2 = 0.957 \qquad (8.36)$$

where, M_h and M_n are annual erosion rates ($t\ km^{-2}\ a^{-1}$) with and without ephemeral gully erosion, respectively; P' is the rainfall amount of individual rainfall events greater than 10 mm of rain (mm); and I'_{30} is the maximum 30-min rainfall intensity (mm/min).

Dividing Eq. 8.35 by [35] yields:

$$G_o = 3.156\left(\sum P'I'_{30}\right)^{-0.167} \tag{8.37}$$

Within the Loess Plateau, a critical condition for the occurrence of ephemeral gully erosion is a slope greater than 15° [74]. The correction coefficient, β, may then be defined as:

$$\beta = \left(\frac{\theta - 15}{15}\right)\left[3.156\left(\sum P'I'_{30}\right)^{-0.167} - 1\right] \tag{8.38}$$

which leads to

$$G = 1 + \left(\frac{\theta - 15}{15}\right)\left[3.156\left(\sum P'I'_{30}\right)^{-0.167} - 1\right] \tag{8.39}$$

Equation 8.39 is valid for slope angles greater than 15°. For $\leq 15°$, $G = 1$. Also, for areas where no rainfall and erosion data are available and slopes are greater than 15°, the following equation is proposed:

$$G = 1 + 1.60\sin(\theta - 15) \tag{8.40}$$

Where farmland is composed primarily of gentle slopes less than 15°, the following equation is used [75]:

$$G = 1 + 1.20\sin^{0.5}(\theta) \tag{8.41}$$

The above relationships were obtained as adjustments to account for the soil loss contribution by gullies to the total soil loss in places with ephemeral gullies. These findings based on the Ansai data are site specific. Given the time frame of these measurements, no direct measurements of soil loss due to gullying were made with the available new technologies of remote sensing. Thus, much more data of the type collected by the Chinese study must be obtained under a variety of conditions if the *G*-factor approach is to gain universal acceptance.

2.3 Gully Erosion Measurement Methods

For some time, researchers have developed theories that have guided gully erosion studies, inspired new theories, and/or led to problem solutions. In the area of soil erosion by water, many have distinguished themselves as pre-eminent thinkers and leaders. From this framework, numerous studies were designed to address specific parameters within the erosion sequence. Variables of importance in these studies included: time; drainage area; surface/subsurface runoff sources and quantities; shapes and lengths of surface slopes; duration and frequency of rainfall intensity; grain-size distribution; soil permeability; soil structure and percent organic matter; cropping and land management history; and soil erodibility. Critical to the quantification of the processes is an accurate definition of the variable under investigation. Hence, a discussion of the measurement techniques used to evaluate the soil-water system and processes seems warranted.

A specific experimental program was designed to address, under rigidly controlled conditions and in mechanistic terms, the development of actively migrating headcuts. Headcuts are step changes in bed surface topography which are basic to ephemeral gully development. The experimental variables and constraints included: discharge, slope, step-height, soil texture, pore-water pressure and tail-water depth manipulation, sediment inflow, and soil stratification. One key result from these studies was the observation of steady-state soil erosion: after an initial period of bed adjustment, a headcut migrated upstream at a constant rate and the shape and form of the headcut and total sediment discharge exiting the flume remained nearly invariant with time during this process [62]. Moreover, the flow structure within steady-state headcut scour holes was analogous to reattached wall jets [76, 77]. Larger scour holes with higher sediment discharges were observed with higher overland flow discharge rates [62], higher bed slopes [78], larger initial step heights [79], and homogeneous strata [80]. Furthermore, the susceptibility to significant erosion losses was related to soil textural properties at comparableplied energy [81]. Increased tail-water depths downstream of the headcut led to an immediate cessation of the erosion process [82]. Varying subsurface pore-water pressure can either enhance or suppress the erosion process [70, 82]. The hydraulics and morphodynamics of the headcut erosion process are markedly affected by sand loads greater than about 30 % of the sand flux emanating from the migrating headcut [83]. The change in channel width was shown to be a function of slope and discharge [84].

The aforementioned studies relied upon sediment discharge measurements exiting a flume, bulk density measurements, particle-size distribution [85], pore-water measurement, acoustic flow-depth measurement, video recordings, total station and real time kinetic (RTK) survey, terrestrial light detection and ranging (LiDAR), and photogrammetry of morphologic change. Companion studies of soil erodibility utilizing in situ submerged jet-testing methods [86–88] were carried out to assess the properties of soils. Each of these methods will be briefly outlined below.

2.3.1 Sediment Discharge

Water and sediment exiting the soil cavity were captured in 0.5-L glass bottles at preset intervals (10–30 s) as the headcut developed and migrated upstream. Smaller time increments were used in the early stages of development and increased to larger time increments as steady conditions were achieved. Sediment concentration samples were weighed and placed in an oven at 105 °C for 24 h, then reweighed. Sediment yield (kg/s) is the mass of oven-dry soil (kg) divided by the time (s) required to fill the sample bottle; where, this time is simply the mass of water (kg) divided by the product of flow discharge (L/s) and fluid density (1 kg/cm^3). The time series, for one experiment, is shown in Fig. 8.5a. Integration under the curve gives the total soil loss as 22.7 kg, which in this case is a 20 % reduction from the original soil mass.

2.3.2 Bulk Density

Dry bulk density was determined from core samples (74-mm inside diameter, 50-mm height) driven into un-eroded soil and the depositional bed [89]. The cores were excavated, cleaned, weighed, oven-dried at 105 °C, and reweighed to obtain the mass of soil solids. The dry bulk density was calculated from the ratio of total mass of oven-dried soil solids and total volume of the core.

2.3.3 Particle-Size Distribution

Particle-size distribution (PSD) analysis is a measurement of the size distribution of dispersed primary particles in a soil sample. In the USDA classification system, soil texture refers to the relative proportions of clay, silt, and sand on a <2 mm basis. The procedure follows the method of Day [85], where <2 mm diameter soil aggregates are dispersed into discrete units by chemical, mechanical, or ultrasonic means, followed by separation or fractionation of these particles according to size limits through sieving and sedimentation [90]. A 10 g sample is dried in the oven to obtain the initial weight, dispersed with a sodium hexametaphosphate solution, and mechanically shaken. The sand fraction is removed from the suspension by wet sieving and then fractionated by dry sieving. The clay and fine silt fractions are determined using the suspension remaining from the wet sieving process. The silt/clay suspension is diluted to 1 L in a sedimentation cylinder, stirred, and 25 mL aliquots removed with a pipet at calculated, predetermined intervals based on Stokes' law [91]. The aliquots are dried at 105 °C and weighed. The amount of each size-fraction is then gravimetrically measured as a percent of the total sample weight on an oven-dry basis. Soil separates of total sand (0.05–2.0 mm), silt (0.002–0.05 mm), and clay (<0.002 mm), with subclasses of sand (coarse, medium, fine, and very fine) are determined.

Samples from the original soil material, consisting of soil cores obtained from the deposited material and sediment discharge samples, were examined following the procedures outlined above. From the PSD of these various samples, an elementary particle tracking and mass balance exercise may be performed to determine the change in specific size classes of material relative to the original soil. From one experiment, measurements of the PSD of the original soil material (58 % sand, 18 % silt, 24 % clay) and the deposited core (BD samples) material (69 % sand, 13 % silt, 18 % clay) were obtained and used in conjunction with the sediment flux samples. As the headcut develops and moves upstream, the sediment flux is enriched with fine material and the depositional bed is enriched with coarse material (Fig. 8.5b).

As noted previously, there is an overall 20 % reduction in soil mass due to this erosion feature and the dominant sizes of exported material were the finer PSD classes.

2.3.4 Pore Water

Pore-water pressure, in the vadose zone, refers to the absorption pressure of water by the surrounding soil matrix. The matric pressure is critical to understanding the stress state of the soil and, in general, at any point above the water table, the effective stress is equal to the total stress [92]. In the aforementioned experiments, the pore-water potential was measured by using tensiometers which allow the pore water of the soil to equilibrate with water inside a sealed tube and porous cup in contact with the soil. As the soil dries, water potential decreases (tension increases) and the tensiometer gauge reading increases. Conversely, as the amount of soil water increases, i.e., rainfall or irrigation, the water potential increases (tension decreases) and the tensiometer gauge reading decreases. Therefore, the water energy state may be recorded to monitor soil-water fluctuations.

In our experiments, tensiometers were installed in 2.5-cm depth increments. The wetting front passed each prescribed depth during rainfall application and was monitored throughout overland flow application. Figure 8.6 shows the time series of pore-water pressure data as a headcut moves past the tensiometer positions.

2.3.5 Flow Depth

A noncontact ultrasonic distance sensor measures flow depth in these experimental studies. Measurements to any object are obtained by reflected ultrasound. Calibration of the voltage output is essential for optimum performance and should be performed on a regular basis. The sensor is anchored above the test bed and initial bed height measurements are obtained. Then, the difference between the initial bed height and measured flow surface during flow release represents the flow depth time series for the experimental run (Fig. 8.7). The sensors may also be used to perform bed surface profiles when combined with a string potentiometer. The string potentiometer provides the distance along the measurement axis and the ultrasonic sensor provides surface measurements; therefore, a thalweg profile (y-axis) or cross section (x-axis) or both, given two string potentiometers, are possible.

2.3.6 Video Analysis

Top and side views of headcut migration and scour hole morphology were recorded on videotape using two video cameras. The video observations were transferred from tape to computer, and a morphologic analysis began by capturing individual frame recordings at prescribed intervals. Each captured frame was digitized to obtain measurements, relative to a preset datum, of typical scour hole parameters (maximum scour depth, length from brinkpoint to maximum scour depth, nappe entry angle,

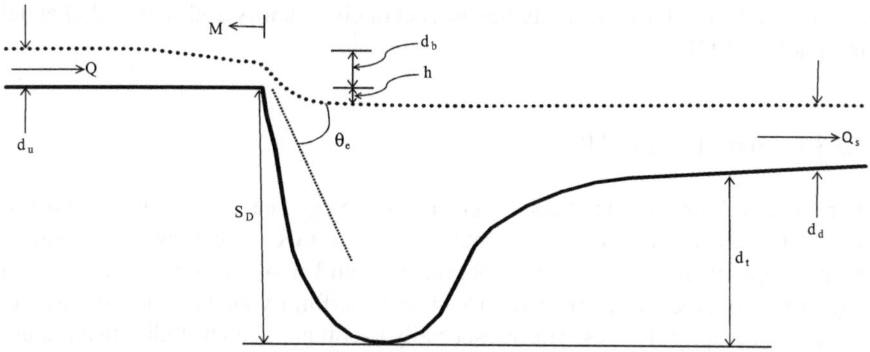

Fig. 8.8 Definition sketch of key morphologic parameters of the headcut, where M is the migration rate, Q is the incoming flow discharge, d_u is the upstream flow depth, d_b is the flow depth at the brink, d_d is the downstream flow depth, S_D is the scour depth, θ_e is the jet entry angle, d_t is the depth of depositional bed, Q_s is the sediment discharge, and h is the vertical distance from brink to pool surface

brinkpoint position, thickness of the depositional bed, upstream and downstream flow depths, and digitized dimensions of the scour hole morphology). A definition sketch of the key morphologic parameters of the headcut is shown in Fig. 8.8.

2.3.7 Total Station and RTK Survey

A total station is an electronic/optical theodolite integrated with a distance meter whereby coordinates of an unknown point relative to a known coordinate can be determined as long as a direct line-of-sight can be established between the two points. The basic concept of RTK is to have a base station receiver set on a point with known coordinates in the project site. The base station receiver sends correction data to the surveyor who is operating the survey receiver (rover). The correction data is typically sent via UHF or spread spectrum radios that are built specifically for wireless data transfer. The corrections from the base station receiver can be sent to an unlimited number of rovers.

2.3.8 LiDAR and Photogrammetry

Capturing erosion dynamics poses a challenge in monitoring efforts, process descriptions, and validation of theoretical models that calculate soil erosion losses and translocations within agricultural watersheds [93]. Typically, studies designed to understand erosion processes use digital elevation models (DEM) to formulate theories that relate topography changes and erosion [94–96]. Despite the availability of DEMs at regional and local scales (spatial resolution ranging from 1 to 30 m), these datasets often do not offer the necessary spatial resolution required in rill and gully investigations. DEMs with spatial resolution ranging between 5 mm and 5 cm are required in gully erosion studies to accurately quantify soil loss and channel morphology [97].

2.3.8.1 Terrestrial LiDAR

Recent developments in laser scanner technology have provided improved ways of studying gully erosion. Ground-based LiDAR systems provide detailed multi-temporal analysis of micro-topography. With LiDAR, it is possible to obtain digital terrain models, estimate soil roughness, perform volumes calculations, and terrain topography changes. This is especially useful in research studies that require multiple, repeat surveys over extensive periods of time to monitor erosion/deposition changes due to rainfall events and topography changes due to management and/or conservation practice assessment. These systems are capable of collecting information with a wide range of ground sampling densities as a result of operator-controlled factors such as the scan angles, average point density, and degree of overlap between scans. Higher point density can be achieved by higher sensor resolution, smaller vertical field of view angles, and multiple scans of the same ground location. Multiple scans can be used to collect data at the same geographical location resulting in increased sampling density that overcomes problems such as shadowing and limited coverage due to vegetation. For a more detailed explanation of this technology, see Momm et al. [98].

Since these datasets typically contain millions of points, displays are not often very clear and effective; however, Fig. 8.9 showing a gully near Hutchinson, KS, indicates the degree of survey clarity.

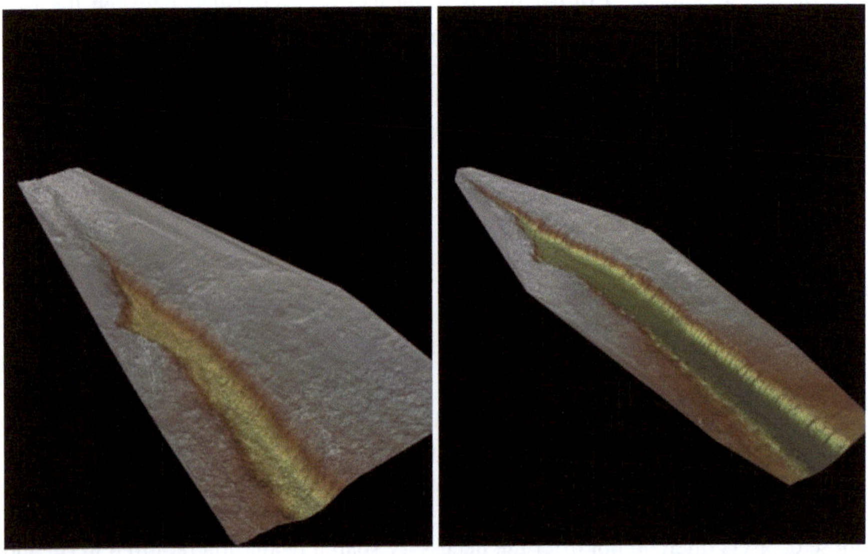

Fig. 8.9 LiDAR survey data from a gully near Hutchinson, KS, from (*left*) March 2010 and (*right*) November 2010. LiDAR light detection and ranging

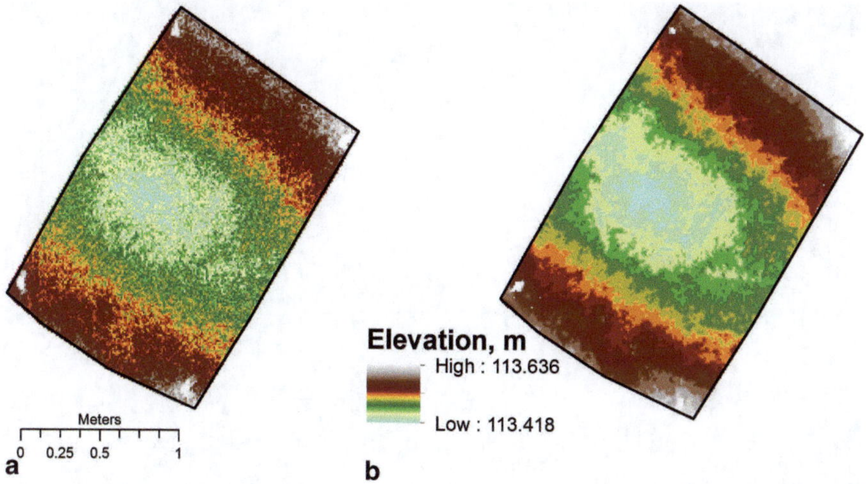

Fig. 8.10 Comparison of **a** terrestrial LiDAR and **b** photogrammetry survey data. Both are tied to real Earth coordinates using GPS survey techniques. LiDAR light detection and ranging

2.3.8.2 Photogrammetry

Overlapping photographs (stereo pair) of a certain area of interest are acquired. Control targets within the stereo pair are surveyed with an electronic total station. Bearings (horizontal and vertical angles) to the targets from multiple stations are converted to three-dimensional coordinates using intersection and triangulation methods as described by Chandler et al. [99]. Control coordinates, interior and exterior camera orientation information, and lens distortion model parameters are linked to image pairs using commercially available software. Control points are manually identified within each image pair and other geometrically similar tie points are automatically identified through the software to obtain a triangulated solution for each image pair. A single DEM is generated for each image pair and these may be combined to produce a DEM of any size. Uncertainties from all sources (surveying, control point digitizing, triangulation, and DEM interpolation) produce combined errors on the order of 1–3 mm in the horizontal and vertical dimensions, respectively [see 99, 100]. Figure 8.10 shows a comparison between data collected using terrestrial LiDAR and photogrammetry.

Gordon et al. [101] implemented these techniques in a study of rill network development and evolution, and Wells et al. [84] in a study on channel widening.

2.3.9 In Situ Submerged Jet Test

The in situ jet-test apparatus [86, 87] consists of a pump, adjustable head tank, jet submergence tank, jet nozzle, jet tube, point gage, and head tank (Fig. 8.11). The submergence tank is driven into the soil bed (40 mm), the nozzle of the jet (6.4 mm

Fig. 8.11 Photo of jet-tester device and scour hole

diameter nozzle) is set at an initial height (6–12 nozzle diameters; i.e., greater than jet core length) above the soil bed, and submerged by water supplied from the head tank. The jet velocity at the nozzle origin, jet height, and jet diameter control the initial stress at the bed. The adjustable head tank allows the user to set a desired jet velocity. Changes in maximum bed scour are measured using a point gage. The point gage is aligned with the jet nozzle so that the point gage can pass through the nozzle to the bed to measure scour depth (i.e., flow is effectively shut off). Maximum scour measurements are taken at interval sequences of 5 s, 10 s, 20 s, 30 s,1 min,5 min, 10 min, 20 min, and 30 min over a 60-min period. An important consideration in using this test is that agricultural soils require special attention in measuring scour depth. Note: The user must assure that the point gage is not pushed down into the bed and maximum scour depth assessment at each interval requires the user to feel the scour hole bottom immediately before the end of the time step, as eroded material may not be evacuated from the scour hole after dislocation. This is accomplished by feeling (finger in the hole) when the rod just touches the bottom of the scour hole.

Calculation of equilibrium scour depth [102], based upon the measured maximum scour depth series and pressure head configuration, is used to estimate the critical shear stress, τ_c, of the soil. Soil erodibility, K_d, is determined by standard error minimization for predicted time versus measured time of scour [88].

Glossary

Aggregation soil particles bound to each other by moist clay, organic matter, organic compounds, and fungal structures. Well-aggregated soils are more stable and less susceptible to erosion.

Agricultural land land suitable for agricultural production, both crops and livestock.

Contouring farming practice of tilling sloped farmland along the lines of constant elevation.

Density a physical property of all matter and is given as mass per unit volume.

Detachment process of tearing loose soil particles.

Dust Bowl period of severe dust storms (1930s) that greatly damaged the ecology and agriculture of the USA and Canadian prairies.

Ephemeral gully erosion topographically driven erosion caused by runoff concentration within a few natural waterways or swales. Typically, these features are larger than rills and may be erased/removed by tillage operations.

Erodibility susceptibility of a soil to be detached and transported. Defined as the amount of soil loss per unit exogenic force of rainfall and overland flow.

Erosion process by which soil and rock are removed by the action of wind and water.

Erosion control mechanical practice of preventing or controlling wind or water erosion in agriculture, land development, and construction.

Erosivity power of a storm or surface flow to erode soil, usually determined from storm characteristics such as rainfall intensity and energy or flow volume and flow gradient.

Exogenic external forcing

Gully erosion slope incisions by flowing water that erodes soil to form channels deeper than 30 cm. Typically, these features require extensive and expensive treatment methods to abate.

Headcut step change in bed topography. Primary position of soil detachment within rills and gullies.

Interrill erosion process of soil detachment by raindrops and transport in thin sheet flow.

Intrinsic soil properties field observable soil attributes like texture, structure, organization, color, features, and consistence.

Iso-erodent lines lines on a map that join points having the same value for erosivity.

Landscape comprises visible features of an area of land, including physical elements (i.e., landforms), living elements (i.e., land cover), human elements (i.e., land use), and transitory elements (i.e., climate).

Models schematic description of a system that accounts for known or inferred properties and may be used to study system characteristics.

Morphology scientific study of form and structure.

Nomograph two-dimensional diagram designed to allow the approximate graphical computation of a function.

Organic matter matter composed of organic compounds that has come from the remains of once-living organisms such as plants and animals and their waste products in the environment.

Permeability measure of the ability of soil to transmit fluids.

Rainfall intensity measure of the amount of rain per unit area and unit time.

Rainfall simulator tool that produces controlled parameter (intensity, duration, drop size) rainstorms over a confined soil surface. Also known as a rainulator.

Rill natural fluvial topographic feature. These channels are shallow and narrow, and form in multiples, parallel to each other.

Runoff excess water flow that occurs when the soil infiltration capacity is exceeded during a rainstorm event, melt water, or other sources of flows over the land.

Runoff plots field plots of various size (standard USLE plot size is 3.7 m wide and 18.3 m long) to monitor runoff volumes, soil loss, chemical transport, etc.

Sediment naturally occurring soil and gravel material that is broken down by processes of weathering and erosion, and is subsequently transported.

Sedimentation the tendency for particles in suspension to settle out of the fluid in which they are entrained, and come to rest.

Soil conservation set of management practices to prevent soil from being eroded.

Soil erosion natural process that occurs when soil is removed through the action of wind and/or water.

Soil production the rate of bedrock weathering into soil as a function of soil thickness.

Soil texture refers to the size and size distribution of the particles that make up the soil.

Strip cropping growing crops in an arrangement of lines in order to reduce erosion.

Sustainable management concept of keeping a system running indefinitely without depleting resources, while maintaining economic viability and providing for the needs of present and future generations.

Terrace a piece of slope plane that has been cut into a series of successively receding flat, horizontal surfaces which resemble steps, for the purpose of decreasing erosion and surface runoff.

Water quality impairment description of diminished strength or value based upon designated water use. Pollutants and sources are considered.

Unit plot standard plot condition to determine soil erodibility. Conditions for the plot are LS factor = 1 (slope = 9 % and length = 72.6 ft), plot is fallow, tillage is up and down slope, and no conservation practices are applied (CP = 1).

References

1. Montgomery, D. R. (2007). Soil erosion and agricultural sustainability. *Proceeding of the National Academy of Science, 104,* 13268–13272.
2. Nearing, M. A., Romkens, M. J. M., Norton, L. D., Stott, D. E., Rhoton, F. E., Laflen, J. M., Flanagan, D. C., Alonso, C. V., Bingner, R. L., Dabney, S. M., Doering, O. C., Huang, C. H.,

McGregor, K. C., & Simon. A. (2000). Discussion of measurements and models of soil loss rates. *Science, 290,* 1300–1301.
3. Trimble, S. W., & Crosson, P. (2000). U.S. soil erosion rates-myth and reality. *Science, 289,* 248–250.
4. U.S. Environmental Protection Agency. (1998). *National water quality inventory: 1998 report to congress* (Report EPA841-R-00-001). Washington, D.C.
5. U.S. Environmental Protection Agency. (2000). *The quality of our Nation's waters: A Summary of the National water quality inventory: 1998 report to congress* (Report EPA841-S-00-001). Washington, D.C.
6. Lal, R. (2001). Soil degradation by erosion. *Land Degradation and Development, 12,* 519–539.
7. Pimentel, D., Harvey, C., Resosudarmo, P., Sinclair, K., Kurz, D., McNair, M., Crist, S., Shpritz, L., Fitton, L., Saffouri, R., & Blair, R. (1995). Environmental and economic costs of soil erosion and conservation benefits. *Science, 267,* 1117–1123.
8. Uri, N. D., & Lewis, J. A. (1999). Agriculture and the dynamics of soil erosion in the United States. *Journal of Sustainable Agriculture, 14,* 63–82.
9. Lal, R. (2009). Soils and food sufficiency: A review. *Agronomy for Sustainable Development, 29,* 113–133.
10. Wischmeier, W. H., & Smith, D. D. (1965). *Predicting rainfall erosion losses from cropland east of the Rocky Mountains—Guide for selection of practices for soil and water conservation.* (Agriculture Handbook, 282) Washington, D.C.
11. Wischmeier, W. H., & D. D. Smith. (1978). *Predicting rainfall erosion losses: A guide to conservation planning.* (Agricultural Handbook, 537) U.S, Dep.
12. Meyer, L. D., & McCune, D. L. (1958). Rainfall simulator for runoff plots. *Agricultural Engineering, 39,* 644–648.
13. Swanson, N. A. (1965). Rotating doom rainfall simulator. *Transactions of the American Society of Agricultural Engineers, 8,* 71–72.
14. Foster, G. R., & Meyer, L. D. (1972). A closed form soil erosion equation for upland areas. In H. W. Shen (Ed.) *Sedimentation (Einstein)* (Chap. 12, pp. 1–19). Fort Collins: Colorado State University.
15. USDA-ARS (2014). Science Documentation Revised Universal Soil Loss Equation, Version 2 (RUSLE2). Washington, D.C. http://www.ars.usda.gov/sp2UserFiles/Place/64080510/RUSLE/RUSLE2 Science Doc.pdf. Accessed June 6, 2014.
16. Renard, K. G., Foster, G. R., Weesies, G. A., McCool, D. K., & Yoder, D. C. (1997). Predicting soil erosion by water: A guide to conservation planning with the revised universal soil loss equation (RUSLE). *Agriculture Handbook, 703,* 384.
17. Foster, G. R., McCool, D. K., Renard, K. G., & Moldenhauer W. C. (1981). Conversion of the universal soil loss equation to SI metric units. *Journal of Soil and Water Conservation, 36,* 355–359.
18. Wischmeier, W. H. (1960). Cropping management factor evaluations for a universal soil loss equation. *Soil Science Society of America Proceedings, 23,* 322–326.
19. Wischmeier, W. H. (1966). Relation of field plot runoff to management and physical factors. *Soil Science Society of America Proceedings, 30,* 272–277.
20. Römkens, M. J. M. (1985).The soil erodibility factor: A perspective. In S. A. El-Swaify, W. C. Moldenhauer, & A. L. Low (Eds.), *Soil erosion and conservation* (pp. 445–461). Ankeny: Soil Conservation Society.
21. Wischmeier, W. H. (1959). A rainfall erosion index for a universal soil loss equation. *Soil Science Society of America Proceedings, 23,* 246–249.
22. Laws, O. J., & Parsons, D. A. (1943). The relation of raindrop size to intensity. *Transactions of the American Geophysical Union, 24,* 452–460.
23. Brown, L. C., & Foster, G. R. (1987). Storm erosivity using idealized intensity distributions. *Transactions of the ASAE, 30,* 379–386.
24. Wischmeier, W. H. (1974). *New developments in estimating water erosion.* Proceeding of the 29th Annual Meeting of. (pp. 179–186). Madison: Soil Science Society America.

25. Weiss, L. L. (1964). Ratio of true to fixed-interval maximum rainfall. *Journal of Hydraulic Division ASCE, 90*(HY1), 77–82.
26. Foster, G. R., & Meyer, L. D. (1975). Mathematical simulation of upland erosion by fundamental erosion mechanics. In *Present and prospective technology for predicting sediment yield and sources* (pp. 190–207). ARS S 40: USDA Agricultural Research Service.
27. Wischmeier, W. H., & Mannering, J. V. (1969). Relation of soil properties to its erodibility. *Soil Science Society of America Proceedings, 33,* 131–137.
28. Wischmeier, W. H., Johnson, C. B., & Cross, B. V. (1971). A soil erodibility nomograph for farmland and construction sites. *Journal of Soil and Water Conservation, 26,* 189–193.
29. USDA-SCS. (1951). *Soil survey manual: Agriculture handbook 18.* U.S. Government Printing Office. Washington, D.C.
30. Römkens, M. J. M., Roth, C. B., & Nelson, D. W. (1977). Erodibility of selected clay subsoils in relation to physical and chemical properties. *Soil Science Society of America Journal, 41*(5), 954–960.
31. El-Swaify, S. A., & Dangler, E. W. (1976). Erodibility of selected tropical soils in relation to structural and hydrologic parameters. In G. R. Foster (Ed.), *Soil erosion: Prediction and control* (pp. 105–114). Ankeny: Soil Conservation Society of America.
32. Young, R. A., & Mutchler, C. K. (1977). Erodibility of some Minnesota soils. *Journal of Soil and Water Conservation, 32*(4), 180–182.
33. Römkens, M. J. M., Prasad, S. N., Poesen, J. W. A. (1986). Soil erodibility and properties. *Transactions of the XIII Congress of the International Society of Soil Science, 5,* 492–504.
34. Römkens, M. J. M., Poesen, J. W. A., & Wang, J. Y. (1988). Relationship between the USE soil erodibility factor and soil properties. In Rimwanichland, S. (Ed.), *Conservation for future generations* (pp. 371–385). Bangkok: Department of Land Development.
35. Wang, B., Zheng, F., & Römkens, M. J. M. (2013a). Comparison of soil erodibility factors in USLE, RUSLE2, EPIC, and Dg-models based on Chinese erodibility database. *Acta Agriculturae Scandinavica, Section B-Soil and Plant Science, 63,* 69–79.
36. Wang, B., Zheng, F., Römkens, M. J. M., & Darboux, F. (2013b). Soil erodibility for water erosion: A perspective and Chinese experiences. *Geomorphology, 187,* 1–10.
37. Ellison, W. D. (1947). Soil erosion studies, part I. *Agricultural Engineering, 28*(4), 145–146.
38. Foster. G. R. (1987). *User requirements USDA water erosion prediction project (WEPP): Draft 6.2.* Purdue University West Lafayette: USDA ARS National Soil Erosion Research Laboratory.
39. Elliot, W. J., Laflen, J. M., & Kohl, K. D. (1989). *Effect of soil properties on soil erodibility* (Paper no. 89–2150). St. Joseph, MI: American Society of Agricultural Engineers.
40. Nearing, M. A., Foster, G. R., Lane, L. J., & Finkner, S. C. (1989). A process-based soil erosion model for USDA- water erosion prediction project technology. *Transactions of the ASAE, 32,* 1587–1593.
41. Flanagan, D. C., & Nearing, M. A. (1995). USDA water erosion prediction project—hillslope and watershed model documentation (NSERL Report No. 10). West Lafayette: USDA ARS National Soil Erosion Research Laboratory.
42. Flanagan, D. C., Gilley, J. E., & Franti, T. G. (2007). Water Erosion Prediction Project (WEPP): Development history, model capabilities, and future enhancements. *Transactions of the ASABE, 50*(5), 1603–1612.
43. Flanagan, D. C., Frankenbenberger J. R., & Ascough, J. C. (2012). WEPP: Model use: Calibration, and validation. *Transactions of the ASABE, 55*(4), 1463–1477.
44. Alberts, E. E., Holzhey, C. S., West, L. T., & Nordin, J. O. (1987). Soil selection USDA Water Erosion Prediction Project (WEPP) (Paper no. 87–2542). St. Joseph, MI: American Society of Agricultural Engineers.
45. Laflen, J. M., Thomas, A. W., & Welch, R. (1987). Cropland experiments for the WEPP project (Paper no. 87–2544). St. Joseph, MI: American Society of Agricultural Engineers.
46. Foster, G. R., Meyer, L. D., & Onstad, C. A. (1977). An erosion equation derived from basic erosion principles. *Transactions of the ASAE, 19*(4), 678–682.

47. Lane, L. J., Foster G. R., & Nicks A. D. (1987). Use of fundamental erosion mechanics in erosion prediction (Paper no. 87–2540). St. Joseph, MI: American Society of Agricultural Engineers.
48. Meyer, L. D., & Harmon W. C. (1984). Susceptibility of agricultural soils to interrill erosion. *Soil Science Society of America Journal, 48,* 1152–1157.
49. Watson, D. A. & Laflen, J. M. (1986). Soil strength, slope, and rainfall intensity effects on interrill erosion. *Transactions of the ASAE, 29,* 98–102.
50. Lattanzi, A. R., Meyer, L. D., & Baumgardner, M.F. (1974). Influence of mulch rate and slope steepness on interrill erosion. *Soil Science Society of America Journal, 38,* 946–950.
51. Singer, M. J., & Blackard, J. (1982). Slope angle-interrill soil relationship for slopes up to 50 %. *Soil Science Society of America Journal, 46,* 1270–1273.
52. Dabney, S. M., Yoder, D. C., Vieira, D. A. N., & Bingner, R. L. (2011). Enhancing RUSLE to include runoff-drivenphenomena. *Hydrological Processes, 25,* 1373–1390.
53. Dabney, S. M., Vieira, D. A. N, & Yoder, D. C. (2013). Effects of topographic feedback on erosion and deposition prediction. *Transactions of the ASABE, 56*(2), 727–736.
54. Dabney, S. M., Yoder, D. C., & Veira, D. A. N. (2012b). The application of the Revised Universal Soil Loss Equation, Version 2, to evaluate the impacts of alternative climate change scenarios on runoff and sediment yield. *Journal of Soil Conservation, 67*(5), 343–353.
55. Foster, G. R., Toy, T. J., & Renard, K. G. (2003). *Comparison of the USLE, RUSLE i.06c, and RUSLE2 for application to highly disturbed land.* First interagency conference on research in the Watersheds (pp. 154–160). Washington, D.C: USDA–ARS.
56. Dabney, S. M., Yoder, D., & Renard, K. G. (2012a). Contributions of RUSLE2 to TMDL development. *EWRI,* 1–10.
57. Bryan, R. B. (1990). Knickpoint evolution in rillwash. In R. B. Bryan (Ed.), *Soil erosion-experiments and models* (pp. 111–132). Germany: Catena Supplement 17.
58. Renard, K. G., Yoder, D. C., Lightle, D. T., & Dabney, S. M. (2011). Universal Soil Loss Equation and Revised Soil Loss Equation. In *Handbook of erosion modeling* (pp. 137–167). Oxford: Blackwell.
59. Römkens, M. J. M. (2010). Erosion and sedimentation research in agricultural watersheds in the USA: From past to present and beyond. *Sediment dynamics for a changing future: Proceedings.of the ICCE Symposium* (pp. 17–26). Warshaw University: IAHS Publ. 337.
60. Grissinger, E. H. (1996). Rill and gullies erosion. In M. Agassi (Ed.), *Soil erosion, conservation, and rehabilitation.* New York: Marcel Dekker, Inc.
61. Soil Science Society of America. (2008). *Glossary of soil science terms.* Madison: Soil Science Society of America.
62. Bennett, S. J., Alonso, C. V., Prasad, S. N., & Römkens, M. J. M. (2000). Experiments on headcut growth and migration in concentrated flows typical of upland areas. *Water Resources Research, 36,* 1911–1922.
63. Casalí, J., Bennett, S. J., & Robinson K. M. (2000). Processes of ephemeral gully erosion. *International Journal of Sediment Research, 15,* 31–41.
64. Poesen, J., Nachtergaele, J., Verstraeten, G., & Valentin, C. (2003). Gully erosion and environmental change: Importance and research needs. *Catena, 50,* 91–133.
65. Helming, K., Römkens, M. J. M., & Prasad, S. N. (1998). Surface roughness related processes of runoff and soil loss: A flume study. *Soil Science Society of America Journal, 62,* 243–250.
66. Meyer, L.D., Foster, G.R., & Nikolov, S. (1975). Effect of flow rate and canopy on rill erosion. *Transactions of the ASAE, 18,* 905–911.
67. Mosley, M. P. (1974). Experimental study of rill erosion, *Transaction of the ASAE, 17,* 909–913.
68. Römkens, M. J. M., Prasad, S. N., & Gerits, J. J. P. (1995). Seal breakdown on a surface soil and subsoil by overland flow. In H. B. So, S. R. Raine, B. M. Schafer, & R. J. Loch (Eds.), *Sealing, crusting and hardsetting soils: Productivity and conservation* (pp. 139–144). Brisbane: Australian Society of Soil Science.
69. Römkens, M. J. M., Helming K., & Prasad S. N. (2000). Sediment yield-surface topography relationships for selected Mississippi soils. *International Journal of Sediment Research, 15,* 1–16.

70. Römkens, M. J. M., Helming, K., & Prasad, S. N. (2001). Soil erosion under different rainfall intensities, surface roughness, and soil water regimes. *Catena, 46,* 103–123.
71. Jiang, Z. S., & Zheng F. L. (2005). Prediction model of water erosion on hillslopes. *Journal of Sediment Research, 4,* 1–6. [In Chinese with English abstract].
72. Jing, K., Wang, W. Z., & Zheng, F. L. (2005). *Soil erosion and environment in China* (p. 359). Beijing: Science. [In Chinese].
73. Zheng, F. L., & Kang, S. Z. (1998). Erosion and sediment yield in different zones of loess slopes. *ACTA Geographica Sinica, 54*(5), 422–428. [In Chinese with English abstract].
74. Jiang, Z. S., Wang Z. Q., & Liu Z. (1996). Study on the use of GIS to estimate soil erosion in a small watershed in the Loess Hilly Region. *Research of soil and water conservation, 3*(2), 84–97. [In Chinese with English abstract].
75. Zhao, X. G., Wu, F. Q., & Liu, B. Z. (1999). *Analysis of runoff and soil loss on gentle fallow slope land in gully region of loess plateau.* Proceedings of international symposium of floods and droughts, October, 733–739. [In Chinese with English abstract].
76. Bennett, S. J., & Alonso C. V. (2005). Kinematics of flow within headcut scour holes on hillslopes. *Water Resources Research, 41,* W09418. doi:10.1029/2004WR003752.
77. Bennett, S. J., & Alonso, C. V. (2006).Turbulent flow and bed pressure within headcut scour holes due to plane reattached jets. *Journal of Hydraulic Research, 44,* 510–521.
78. Bennett, S. J. (1999). Effect of slope on the growth and migration of headcuts in rills. *Geomorphology, 30,* 273–290.
79. Bennett, S. J., & Casalí, J. (2001). Effect of initial step height on headcut development in upland concentrated flows. *Water Resource Research, 37,* 1475–1484.
80. Gordon, L. M., Bennett, S. J., Wells, R. R., & Alonso, C. V. 2007. Effect of soil stratification on the development and migration of headcuts in upland concentrated flows. *Water Resource Research, 43,* W07412. doi:10.1029/2006WR005659.
81. Wells, R. R., Alonso, C. V, & Bennett, S. J. (2009a). Morphodynamics of headcut development and soil erosion in upland concentrated flows. *Soil Science Society America Journal, 73*(2), 521–530.
82. Wells, R. R., Bennett, S. J., & Alonso, C. V. (2009b). Effect of soil texture, tailwater height, and pore-water pressure on the morphodynamics of migrating headcuts in upland concentrated flows. *Earth Surface Processes and Landforms, 34,* 1867–1877.
83. Wells, R. R., Bennett, S. J., & Alonso, C. V. (2010). Modulation of headcut soil erosion in rills due to upstream sediment loads. *Water Resource Research, 46,* W12531. doi:10.1029/2010WR009433.
84. Wells, R. R., Momm, H. G., Rigby, J. R., Bennett, S. J., Bingner, R. L., & Dabney, S. M. (2013). An empirical investigation of gully widening rates in upland concentrated flows. *Catena, 101,* 114–121.
85. Day, P. R. (1965). Particle fractionation and particle-size analysis. In C. Black (Ed.), *Method of soil analysis. Part 1. Physical and mineralogical properties, including statistics of measurement and sampling* (pp. 545–567). Madison: American Society of Agronomy.
86. Hanson, G. J. (1990). Surface erodibility of earthen channels at high stresses. Part II. Developing an *in situ* testing device. *Transactions of the ASAE, 33*(1), 132–137.
87. Hanson, G. J. (1991). Development of a jet index to characterize erosion resistance of soils in earthen spillways. *Transactions of the ASAE, 34*(5), 2015–2020.
88. Hanson, G. J., & Cook K. R. (1997). *Development of excess shear stress parameters for circular jet testing.* Proceedings of the American Society of Agricultural Engineers Annual International Meeting, Minneapolis, MN. August 10–14, 1997. St Joseph, MI: American Society of Agricultural Engineers.
89. Blake & Hartge, (1986). Bulk density. In A. Klute (Ed.), *Method of soil analysis. Part 1. Physical and mineralogical methods* (pp. 363–375). Madison: American Society of Agronomy.
90. Gee, G. W., & Bauder J. W. (1986). Particle-size analysis. In A. Klute (Ed.), *Method of soil analysis. Part 1. Physical and mineralogical methods* (pp. 383–411). Madison: American Society of Agronomy.

91. Kilmer, V. J., & Alexander, L. T. (1949). Methods of making mechanical analyses of soils. *Soil Science, 68,* 15–24.
92. Terzaghi, K. (1948). *Theoretical soil mechanics*. New York: Wiley.
93. Casali, J., Lopez, J., & Giráldez, J. (1999). Ephemeral gully erosion in southern Navarra (Spain). *Catena, 36,* 65–84.
94. Cerdan, O., Souchère, V., Lecomte, V., Couturier, A., & Le Bissonnais, Y. (2002). Incorporating soil surface crusting processes in an expert-based runoff model: Sealing and transfer by runoff and erosion related to agricultural management. *Catena, 46,* 189–205.
95. Parker, C., Thorne, C., Bingner, R., Wells, R., & Wilcox, D. (2007). *Automated mapping of potential for ephemeral gully formation in agricultural watersheds laboratory publication*. National Sedimentation Laboratory, 56.
96. Woodward, D. (1999). Method to predict cropland ephemeral gully erosion. *Catena, 37,* 393–399.
97. Schmid, T., Schack-Kirchner, H., & Hildebrand, E. (2004). A case study of terrestrial laser-scanning in erosion research: Calculation of roughness indices and volume balance at a logged forest site. *International Archives of Photogrammetry, Remote Sensing and Spatial Information Sciences, 36*(8), 114–118.
98. Momm H. G., Bingner, R. L., Wells, R. R., & Dabney, S. (2011). Methods for gully characterization in agricultural croplands using ground-based light detection and ranging. In F. Bhuiyan (Ed.), *Sediment transport—flow and morphological processes* ISBN: 978-953-307-374-3. InTech.
99. Chandler, J. H., Shiono, K., Ponnambalam, R., & Lane, S. (2001). Measuring flume surfaces for hydraulics research using a Kodak DCS460. *The Photogram Record, 17,* 39–61. doi:10.1111/0031-868X.00167.
100. Heng, B. C. P., Chandler J.H., & Armstrong A. (2010). Applying close range digital photogrammetry in soil erosion studies. *The Photogram Record, 25,* 240–265.
101. Gordon, L. M., Bennett, S. J., & Wells, R. R. (2012). Response of a soil-mantled experimental landscape to exogenic forcing. *Water Resource Research, 48,* W10514. doi:10.1029/2012WR012283.
102. Blaisdell, F. W., Anderson, C. L., & Hebaus G. G. (1981). Ultimate dimensions of local scour. *Journal of Hydraulic Division ASCE, 107*(3), 327–337.

Chapter 9
Advances in Geofluvial Modeling: Methodologies and Applications

Yong G. Lai

Contents

Y. G. Lai (✉)
Technical Service Center, U.S. Bureau of Reclamation,
Bldg. 67, P.O. Box 25007, Denver, CO 80225, USA
e-mail: ylai@usbr.gov

C. T. Yang, L. K. Wang (eds.), *Advances in Water Resources Engineering*,
Handbook of Environmental Engineering, Volume 14, DOI 10.1007/978-3-319-11023-3_9

Abstract Stream bank erosion is an important form of channel change in alluvial environments; it should be accounted for in geomorphic studies, river restoration, dam removal, and channel maintenance projects. Recently, one-dimensional (1D) and two-dimensional (2D) flow and mobile-bed numerical models are becoming useful engineering tools for predicting channel morphological responses to stream modifications. Most, however, either ignore bank erosion or implement only simple ad hoc methods. A combined modeling of vertical and lateral fluvial processes in streams, i.e., geofluvial modeling, is important yet still at its research stage. In this chapter, advances in geofluvial modeling are presented. First, a literature review is provided in the area of geofluvial modeling. Second, important geofluvial processes are discussed as they need to be incorporated into models. Third, a recently developed 2D geofluvial model, SRH-2D, is described. SRH-2D incorporates processed-based bank erosion modules into a 2D mobile-bed module; it is developed with the primary objective of advancing the geofluvial modeling toward a practical engineering tool. Bank erosion modeling consists of a uniform retreat module and mechanistic failure module; both are suitable for uniform and multilayer banks with noncohesive or cohesive materials. The coupling techniques are developed that emphasize ease of use and model robustness (stability). Fourth and final, a number of laboratory and field cases are selected to validate and verify the geofluvial model; application cases are also presented to demonstrate the practical aspects of the model. It is found that the state-of-the-art 2D geofluvial modeling is becoming practical to assist project planning, design, and evaluation. SRH-2D can predict accurately for some streams, but only qualitatively for more complex streams.

Keywords Geofluvial model · Bank erosion · 2D mobile-bed model · Coupled modeling · Fluvial processes · Sediment transport modeling

Nomenclature

A	Area (m^2)
c'	Effective cohesion (kPa)
F_s	Factor of safety, defined as the ratio between the resisting and driving forces (−)
$\lvert GD \rvert$	Horizontal distance between G and D (m)
h	Water depth (m)
h_0	Initial water depth at toe (m)
H_0	Initial bank height (m)
k	Erodibility coefficient (m s^{-1})
L_j	Length of the slice base (m)
$\vec{q}$	Scalar flux vector (ϕm/s)
r_B	Bank retreat distance (m)
$\vec{s}$	Edge length of a cell (m)
S_j	The driving force, a component of the weight of the soil block (kPa m)
S_S	Source/sink term of the scalar equation (ϕm/s)
$\vec{S}_V$	Source/sink term of the momentum equation (m^2/s^2)

t Time (s)
$\vec{V}$ Fluid flow velocity vector (m/s)
$\vec{V}_g$ Velocity vector of the moving mesh (m/s)
α Bank angle or angle of repose (–)
ϕ A scalar variable that is being transport (ϕ)
ϕ' Effective angle of internal friction (°)
ϕ^b Angle describing the increase in shear strength due to an increase in matric suction (°)
ε_L Lateral erosion rate (m s^{-1})
σ Normal stress on the shear plane at the base of the slice (kPa)
$\vec{\vec{\sigma}}$ Stress tensor (m^2/s^2)
τ Shear stress on the bank node (Pa)
τ_c Critical shear stress (Pa)
ω_V Vertical toe erosion distance predicted by the 2D model (m)
ω_L Lateral toe erosion distance computed by equation (m)
μ_a Pore-air pressure (kPa)
μ_w Pore-water pressure (kPa)

1 Introduction

Stream bank erosion is a natural geomorphic process occurring in all alluvial streams. It is an important mechanism by which a channel adjusts its size, shape, and slope to convey water and sediment supplied to it.

Bank erosion has both benefits and costly consequences. Bank erosion is important to waterway ecology as bank changes may create a variety of habitats for flora and fauna which contribute to ecological diversity [1, 2]. Bank protection is being removed by some restoration projects to let banks experience natural erosive forces to benefit the ecosystem (e.g., [3]). Bank erosion, however, may also be detrimental for many streams. It may cause road collapse and land loss and become a resource maintenance problem [4]. Accelerated bank erosion can also be a significant point source pollutant, presenting challenges to river and reservoir managers. In disturbed systems, stream bank erosion has commonly been found to contribute more than 50 % of the total load (e.g., [5–8]).

Many watershed-based erosion models have been developed; some, e.g., System Hydrologic Europeen Sediment (SHESED), Water Erosion Prediction Project (WEPP), Chemicals, Runoff, and Erosion from Agricultural Management Systems (CREAMS), Mike-11, and Sedimentation and River Hydraulics Watershed (SRH-W), included modules to predict in-stream sediment transport and/or bed changes. However, they often neglected the contribution of bank erosion to sediment load [9]. Stream-specific mobile-bed sediment transport models are also widely available such as Hydrological Engineering Center River Analysis System (HEC-RAS), Generalized Sediment Transport model for Alluvial River Simulation version 3 (GSTARS3), Sedimentation and River Hydraulics 1D (SRH-1D), Conservational Channel Evolution and Pollutant Transport System (CONCEPTS),

Center for Computational Hydroscience and Engineering 2D (CCHE-2D), and Sedimentation and River Hydraulics 2D (SRH-2D). They are more versatile, and offer more capabilities and choices in modeling the vertical streambed changes than the watershed erosion models. However, most of them did not take lateral bank erosion into consideration, except for CONCEPTS, and cannot be used to predict lateral stream changes [10].

The above discussion shows the need for geofluvial models to predict or estimate lateral stream changes since field study is both costly and limited. In this chapter, the terminology used by American Society of Civil Engineers (ASCE) (22, p. 910) is adopted: a geofluvial model refers to one that is capable of coupling in-channel "flow and sediment-routing models with bank erosion and mass wasting algorithms." A historical geomorphic study is often used for project planning purpose and it will continue to be useful. However, the method lacks the capability of predicting future changes, particularly when streams are disturbed due to human intervention. As the literature review shows below, process-based geofluvial models are needed for more accurate, predictive tools. They, however, are still at the research stage at present and few models are available for project use. The remaining of the chapter is organized as follows: A review is first provided with previous geofluvial modeling efforts; discussion is then given with regard to the important geofluvial processes that should be modeled. A recently developed 2D geofluvial model, SRH-2D, is the primary focus of the chapter. SRH-2D represents the current advancement of geofluvial modeling towards a practical engineering tool.

2 A Literature Review

Significant effort has been expended in the past on developing tools to estimate and/or predict lateral stream bank erosion. Two categories of modeling tools may be identified [11]: empirical models and process models. Empirical models attempt to predict equilibrium channel width using either regime equations developed through regression on data collected from the field or extremal hypotheses that assume that alluvial channels attain equilibrium when an "indicator variable" reaches a maximum or minimum. Empirical models are generally used for the estimation of long-term changes of channel width; they are suitable for dynamically stable channels. Process models, on the other hand, attempt to explicitly simulate the physical processes most important for quantifying stream vertical and lateral changes. They may be used to provide short- to medium-term predictions of channel changes in both stable and unstable streams.

Most empirical models rely on regime formulas developed from field observations and data. For example, equilibrium channel geometry has been correlated to basin characteristics including drainage area [12, 13], bank-full or effective discharge [14, 15], or sediment size [16, 17]. Empirical equations have also been developed to estimate bank retreat or reach-averaged meander migration rate. For example, bank retreat was regressed against bank-full discharge [18] and soil and flow depth characteristics [19]; while the reach-averaged meander migration rate was correlated to basin characteristics, channel geometry, and flow parameters [20–22]. A widely used method was proposed by Rosgen [23, 24] who related the bank erosion rate to the bank erosion hazard index (BEHI), velocity gradient, and near-bank shear stress. BEHI is based on field observations or surveys of bank geometry, stratigraphy, soil type and composition, and protection. Velocity gradient and shear stress are based on published data of typical velocity profiles in streams. A serious drawback of using the empirical relations is that the regression coefficients may vary significantly from one locality to another [22]. For example, Larsen [25] analyzed nearly 200 bends in the Mississippi River and found that the bend migration rate showed no significant correlation with the local radius of main channel curvature. He also showed that empirical or statistical methods may be applicable only to the sites studied and may be inadequate in establishing a general physical law that governs the evolution of river meander. The limitations of these methods, particularly the BEHI method, have been extensively discussed by, e.g., Simon et al. [26]. These methods are problematic in predicting the unsteady response of alluvial streams to disturbances.

Another category of empirical models uses the extremal theories to estimate the equilibrium channel form. The extremal theories assume that alluvial channels attain equilibrium when an "indicator variable" reaches the maximum or minimum [22]. Various indicator variables, such as the minimum unit stream power [27] or stream power [28], the maximum sediment transport efficiency [29], the maximum sediment transport capacity [20–32], the minimum variance [33], and the principle of least action, have been proposed [34, 35]. The width of the equilibrium channel was also determined by using the threshold concept of sediment incipient motion [36, 37], or by accounting for the lateral diffusion of downstream momentum [38–42]. Despite the theoretical attractiveness and some claims of success, extremal theories in general are used infrequently to predict the width adjustment process, which often is of greater interest than the shape of the equilibrium channels [43].

Empirical models are relatively simple to use and will continue to be useful where and when an estimate of the equilibrium channel dimensions is beneficial. They, however, are inappropriate for short- and medium-term predictions of unsteady geomorphic response of streams to disturbances.

Process-based models attempt to explicitly simulate the physical processes that are important for quantifying channel changes and thus aim to provide reliable short- to medium-term predictions of channel morphology in both stable and unstable channels. Process-based models are often used to estimate local bank retreat rates; and frequently, they are incorporated into analytical or numerical models. ASCE [44] provided a review of the geofluvial models that existed before 1997, Rinaldi and Darby [45] updated and expanded this review to include seepage modeling, Langendoen and Simon [46] provided a review focusing primarily upon the geotechnical modeling element, Motta et al. [47] provided a review of models that linearized the 2D mass and momentum equations, and recently Lai [10] provided a review of many process-based models until 2013.

In the area of bank erosion modeling, an early attempt was made independently by Hasegawa [48] and Ikeda et al. [49] who assumed that bank retreat rate was proportional to the difference between the depth-averaged near-bank velocity and cross-sectional mean velocity (termed the linear velocity defect model). The model was employed in numerous subsequent studies, primarily for meander migration modeling (e.g., [25, 50–53, 54–57]). More recently, Rüther and Olsen [58] used a three-dimensional (3D) model to simulate the formation of two laboratory meandering streams. They obtained impressive results despite omitting bank erosion processes. Other simple bank erosion models include those of Crosato [59] and Odgaard [53] who assumed that bank retreat rate was proportional to the excess near-bank water depth, and Mosselman [60] who related the combined excess shear stress and excess bank height to the retreat rate. More recent models in the category include the effort by Spruyt et al. [61]. In general, these simple models are restricted to artificial morphologies tied to idealized representations of the river planform, such as the most probable path or sine-generated curve. They are too simplistic as they fail to take into account other importance processes such as the near-bank flow hydraulics, removal or accumulation of sediment at the bank toe, bank geometry, geotechnical properties, etc.

Osman and Thorne [62] introduced probably the first bank model to explicitly consider both basal erosion and mass failure of cohesive sediments in which the bank geotechnical properties were taken into account. The ratio of driving force to resistance force on a bank, the factor of safety, was used to determine the mass failure of a bank. This study represented an important advance towards process-based retreat modeling of homogeneous cohesive banks; and the model has since been widely used with various modifications and improvements. A spreadsheet-based model was developed by Thorne and Abt [63] based on the work of Osman and Thorne [62]. The model required inputs such as total and submerged bank heights, bank slope, soil density, friction angle, cohesion, and tension crack index. Darby and Thorne [64] added a quasi-2D flow component to the model to obtain the lateral shear stress, suggested a probabilistic approach to predict the streamwise length of geotechnical failures, and proposed a dimensionless parameter to assess whether a failure block would disaggregate into smaller pieces following an impact with the bank face, bank toe, or water body. Darby and Thorne [65, 66] added pore-water pressure and hydrostatic confining force terms to the Osman–Thorne model and relaxed the restriction that the failure plane must pass through the toe of the bank. Simon et al. [67] and Rinaldi and Casagli [68] added the effects of bank pore-water pressure and the confining hydrostatic pressure to the model. Dapporto et al. [69] studied the role of river stage and pore-water pressure in triggering planar and cantilever failures. Two types of stability analyses were carried out: (a) the limit equilibrium method was used to predict the effect of pore-water pressure on bank stability and (b) a seepage analysis based on hydrographs of different return periods was used to assess the effect of river stage and pore-water pressure on bank stability. Pollen and Simon [70] added the bank vegetation effect into the model by developing a root reinforcement model. Their model was able to study the effect of different tree species on bank stability.

Most bank erosion models mentioned above prescribed an idealized geometry and greatly simplified the bank stratigraphy; bank materials were often assumed to be homogeneous. Although they only simulated planar failures, the models of Simon et al. [71] and, later, Langendoen and Simon [46] permitted the use of actual bank geometries and also accounted for multiple stratigraphic layers. Langendoen and Simon [46] coupled a geotechnical sub-model to an unsteady 1D flow and mobile-bed model CONCEPTS [72]. Their geotechnical algorithm generalizes the limit equilibrium method of Simon et al. [71] by employing vertical slices to distribute the weight of the failure block along the failure plane and enabling the automatic detection and insertion of tension cracks. They also used a search routine to identify the minimum factor of safety. Langendoen and Simon [46] and Langendoen et al. [73] presented results of a number of applications of CONCEPTS. Motta et al. [47] recently coupled the geotechnical algorithms within CONCEPTS to the linearized and nondimensionalized 2D mass and momentum equations; the model produced promising results. However, this approach is strictly valid only for the central region of mildly curved channels in which helical flow can be neglected. Motta et al. [47] also acknowledged a number of simplifying assumptions, including constant discharge and constant channel width during hydrodynamic and bed morphodynamic computations. Also, all eroded and failed bank materials were immediately transported out of the reach as suspended load, and sediment was transported in equilibrium with uniform bed material and without net aggradation or degradation along the reach. They stressed the need to couple sediment transport and bank erosion sub-models because of the destabilizing influence of bed degradation (or stabilizing influence of bed aggradation) and the protection potentially afforded by failed bank material.

In recent years, Darby, Rinaldi, and coworkers [74–77] have produced a series of papers documenting the sequential and iterative use of separate models to simulate the components of the bank retreat process. For example, Rinaldi et al. [76] used a suite of four separate models to simulate the impact of a single flood event on a bend of the Cecina River, Italy. First, they applied a commercial 2D depth-averaged flow model (Deltares Delft-3D) to predict spatiotemporal distributions of shear stress during a flood event. Second, the predictions of near-bank shear stresses were inputted into a separate fluvial erosion model and the bank face and bank toe geometry was updated. Third, this updated geometry was inputted into a commercial 2D groundwater model (GeoSlope SEEP/W) to predict patterns of pore-water pressure within the stream bank. Fourth, geotechnical stability was assessed with a commercial 2D rotational failure model (GeoSlope SLOPE/W) and, when cantilevers had been predicted to form, a shear-type cantilever model. Steps two to four were then repeated iteratively until the end of the flood event. Luppi et al. [77] extended the analysis to multiple events. These studies contributed significantly to our understanding of the interactions between fluvial erosion, pore-water pressure variations, and mass failure during flood events, but their success owed much to tedious and time-consuming manual remeshing between each time step. Furthermore, interactions between the flow model, fluvial erosion, and mass failures were only loosely accounted for; no feedback occurred between the morphology of the eroding bank and the flow. Bed topographic changes were also ignored.

Despite much progress over the past decade, a fully coupled geofluvial model that integrated the bank erosion modules into stream mobile-bed models and that is generally applicable and practical to field projects is still elusive. Existing geofluvial models suffer from one or several of the following limitations: (a) use of static, rigid meshes to simulate a moving boundary problem; (b) use of steady or quasi-steady flow models; (c) limited consideration of the secondary currents that are characteristic of natural meander bends; (d) simplifications to sediment transport and bed deformation sub-models, making them applicable only to idealized cases (e.g., equilibrium transport of single grain sizes); (e) simplistic bank retreat models lacking key physical processes (e.g., explicitly accounting for only fluvial erosion or mass failure or neither) and requiring a number of "fudge factors" to obtain realistic behaviors; and (f) inappropriate or nonexistent coupling procedures. Also, most previous efforts have been focused on the mass failure process modeling, and less on the coupling between the lateral bank processes and the vertical stream processes. In many models, hydraulic and erosion parameters in the stream are obtained in the field or estimated through separate analyses. Consequently, a continuous dynamic simulation of simultaneous lateral and vertical processes is rarely done. In recent years, mobile-bed models that predict stream vertical erosion have been advanced to a point they are being used routinely for morphological predictions of alluvial channels. Since these mobile-bed models may provide more accurate data near banks than the empirical methods, it becomes natural to develop geofluvial models capable of simultaneous modeling of bank and stream erosion over an unsteady continuous hydrograph. Such efforts have been attempted and some are reviewed next.

One of the earliest attempts to develop continuous integrated geofluvial models was reported by Mosselman [60] who incorporated the excess bank height-based bank module into a quasi-steady 2D flow model. The model simulated equilibrium sediment transport of a single grain size, but it suffered from numerical truncation when the mesh became overly skewed and/or distorted and hence required quasi-regular manual remeshing. The model was applied to a reach of the River Ohre, but with poor results. He attributed the poor performance to the formulation of the bank erosion mechanism as well as the simplicity of the 2D flow model. He recommended the improvements of the 2D flow model, such as the use of 3D numerical models. Darby et al. [78] incorporated the Darby and Thorne [64] model within that of Mosselman [60] and used an improved 2D depth-averaged model named RIPA. They, however, found that the predictive ability of the coupled model did not significantly improve the previous model. They ascribed this to systematic underprediction of transverse bed slopes by the flow model and systematic overprediction of bank stability by the geotechnical module. Nagata et al. [79] developed a geofluvial model with the governing equations cast in a moving boundary-fitted coordinate system and a new formulation of nonequilibrium sediment transport. The research model was applied to examine the morphological behaviors of experimental channels with good results. Their bank erosion module, however, was limited to noncohesive, uniform banks with the angle of repose imposed for bank retreat. In addition, the modeling neglected the difference between basal erosion (a lateral process) and bed degradation (a vertical process). The model's ability to simulate more complex streams in the field was not demonstrated. Duan et al. [80] reported a geofluvial research model that integrated a noncohesive uniform bank module with a 2D mobile-bed model. An analytical bank retreat equation was derived based on mass conservation of sediment transport near the bank and the constant angle of repose assumption. The rate of bank retreat was linked to the longitudinal gradient of sediment transport, strength of secondary flow, and erosion of sediment from the bank. The model was tested with flume cases and found that the simulated meandering wavelength and amplitude did not agree with the observations. Again, the model was only of research nature. A recent geofluvial model was developed by Motta et al. [47] who coupled the bank stability algorithms of Langendoen and Simon [46] to the linearized solutions of the 2D depth-averaged shallow water equations similar to that of Ikeda et al. [49]. This approach overcame the need for a migration coefficient to be specified but was strictly valid only for the central region of mildly curved channels in which helical flow could be neglected. The flow and sediment module is not general enough for practical applications.

A practical, and comprehensive geofluvial model was developed by Langendoen and Simon [46] who incorporated an enhanced bank erosion model of Simon et al. [71] into the 1D mobile-bed model CONCEPTS (Langendoen [72]). CONCEPTS simulated 1D unsteady flow, transport of cohesive and noncohesion sediments in suspension and on the bed selectively by size fractions, and bank erosion processes in stream corridors. The geotechnical algorithm generalized the limit equilibrium method of Simon et al. [71] by employing vertical slices to correctly distribute the weight of the failure block along the failure surface and enabling the automatic detection and insertion of tension cracks. A search routine was used to identify the minimum factor of safety. A sensitivity analysis showed that the factor of safety was most sensitive to variability in the elevation of the groundwater table, but that local variability in effective cohesion might also be important. Tension cracks could reduce the calculated factor of safety by as much as 30 %. Langendoen and Simon [46] applied the model to the bank retreat of a bendway on Goodwin Creek, MS, with good agreement between the model results and measured data. Discrepancies between the observed and modeled failure timings and dimensions were caused by unaccounted processes such as a loss of matric suction due to infiltrating rainfall or lateral variability in boundary shear stress.

Limited attempts have also reported by coupling the bank erosion module with the 3D numerical models. For example, Darby et al. [82] described their attempts to use a 3D computational fluid dynamics (CFD) approach to model the shear stress on a riverbank for bank erosion modeling in a meandering river. Field data, such as topography, hydraulics, bank erodibility, and retreat rates, were obtained for two rivers in Italy and UK, and they were used to parameterize and verify the model. Olsen [83] and Rüther and Olsen [58] also coupled the 3D CFD model with the bank erosion module. The model was applied to simulate the formation of a laboratory meandering stream, and promising results were obtained. In general, 3D CFD models may compute a more accurate bed shear stress and may be needed for cases where the secondary flow effects are too strong for the 2D models to be valid. Due to the large demand of computing power, however, 3D models are mostly limited to research. Their use for practical streams is yet to be studied.

In summary, this chapter shows that empirical models are relatively easy to use and will continue to be useful as project decision support tools. Their applicability ranges, however, need to be taken into account: They are useful mainly for a long-term estimate of an equilibrium channel form; they are inappropriate for the short- and medium-term prediction of geomorphic stream responses to disturbances or for unstable fluvial channels. Process models hold a better promise to predict the morphological changes vertically and laterally for both stable and unstable streams. Process-based bank modules, when uncoupled to any mobile-bed models in the stream, may be useful for qualitative assessments of banks locally. For general and quantitative usage, geofluvial models that couple the bank erosion modules into reliable stream mobile-bed models are needed. Such geofluvial models exist but suffer one or several of the following: (a) use of a simplified flow and/or mobile-bed stream model that is applicable only to idealized cases; (b) use of simplistic bank erosion modules that do not take into account key physical processes; and (c) limited calibration, verification, and demonstration of the numerical model for practical field applications.

In the following, advances towards the development of a practical geofluvial model, SRH-2D, are described. The development of bank erosion modules and the coupling of the model modules into the mobile-bed module of SRH-2D has been under research and development at the US Bureau of Reclamation since 2009. Some of the contents reported have been documented in a number of recent reports and papers by Lai and his collaborators [10, 84–87]. Some verification and application cases are new contributions. The primary objective of the SRH-2D geofluvial modeling development focused on the necessary procedures and modules for the coupling of bank and stream models so that a practical geofluvial model that is reliable and stable with minimal sets of calibration parameters may be available for engineers to use.

3 Geofluvial Processes

Geofluvial study concerns with the processes of simultaneous vertical streambed change and lateral bank retreat. Vertical fluvial processes in streams are well studied and have been discussed in numerous papers and textbooks (e.g., see the book chapter by Yang [88]). Only the lateral bank retreat processes are discussed below.

Many physical processes impact lateral stream changes. Two groups are considered [89]: hydraulic fluvial processes and mass failure processes. Hydraulic fluvial processes refer to the removal of bank materials by flowing water. Fluvial processes are often dominant in smaller streams and the upper reaches of larger streams. Mass failure, sometimes described as collapse or slumping, refers to bank failure resulting in a section of the bank sliding or toppling into the stream in a single event. Collapse following undermining of the bank toe and slumping as a result of saturation after flooding are common examples. Mass failure is often dominant in the lower reaches of large streams. The two groups are interrelated—the hydraulic fluvial processes usually induce bank undercutting which leads eventually to gravitational mass failure; materials from mass failure may protect banks which provide, as a minimal, a temporary halt of bank retreat.

Fluvial erosion results usually from an imbalance between the hydraulic shear stress on bank and the resistance to the shear. Where the shear stress exceeds a critical value (called critical shear stress), erosion is initiated. In general, bank shear stress increases with flow up to the bank-full discharge, while bank resistance typically decreases, particularly when the bank becomes saturated. Two hydraulic fluvial processes may be identified: basal erosion and basal cleanout. Basal erosion, or undercutting, refers to the direct removal of bank materials laterally by flowing water. In addition to shear force, basal erosion is also dependent on parameters such as bank properties (stratigraphy, size, composition, etc.), vegetation/debris, and soil characteristics (drainage characteristic, degree of saturation, etc.). Basal cleanout is a process of removing the mass-failure-produced materials that may provide protections of bank toe. Repeated cycles of basal cleanout, basal erosion, mass failure, and bank toe accumulation play an important role in controlling the retreat rate of a stream bank.

Effective strategies for combating mass failure are generally aimed at stabilizing the bank toe and restoring bank vegetation. Watson and Basher [89] described six types of mass failure mechanisms: dry granular flow, shallow slide, rotational failure, planar failure, cantilever failure, and wet earthflow.

Dry granular flow is a process in which individual sediment grains roll, slide, and bounce down a bank surface. It typically occurs on noncohesive banks near the angle of repose. When the bank angle is increased beyond the angle of repose due to basal erosion, dry granular flow occurs. Shallow slide is a process in which a layer of material moves along a plane parallel to the bank surface. This failure often occurs on banks where soils have low and varied cohesion and the bank is moderately steep. The process initiates when the angle of the bank exceeds the angle of internal friction of the bank material. Rotational failure is a deep-seated movement of material both downward and outward along a curved slip surface. After the failure, the upper surface of the slipped block is typically tilted inward toward the bank. The failure is often linked to high pore pressure within the bank and/or the formation of vertical tension cracks within the bank structure. Rotational failure commonly occurs on tall cohesive banks, and during rapid drawdown following high flow events. Planar (or slab) failure is the sliding and forward toppling of a deep-seated mass into the stream. It is also associated with high pore pressure in the bank, and often characterized by the development of vertical tension cracks at the top. Planar failure occurs often on steep, fine-grained cohesive banks. Planar failure is more common than rotational failure because it can occur within banks that are shorter in height and have less cohesive materials than are necessary for deep-seated rotational failures to occur. Cantilever failure is the collapse of an overhanging block into a stream and is often the result of significant undercutting. The failure tends to occur on banks with composite layers of fine/coarse and/or cohesive/noncohesive materials. Finally, the wet earthflow is a process where the soil of a bank flows as a highly viscous liquid. The flowing material is extremely weak and easily removed by hydraulic fluvial process, even at lower flows. Wet earthflow is the result of a significant loss of strength on a section of the bank due to saturation and increased bank weight, and is associated with strong seepage and poor drainage. It is typically caused by waterlogging due to high rainfall, snowmelt, or a rapid drawdown of water.

Piping failure may be singled out as a separate mechanism although it may be in the form of rotational, planar, or cantilever failure. Piping failure is the collapse of part of a bank due to seepage flows, not fluvial processes, causing a selective removal of bank layers. The failure is usually caused by preferential groundwater flows along inter-bedded saturated bank layers; the bank has lenses of noncohesive materials sandwiched between layers of finer cohesive materials. Flow is induced in the more permeable layers by changes in the river stage and/or groundwater seepage; it is often associated with rainfall events.

Thorne and Tovey [90], on the other hand, discussed three different stream bank failure mechanisms: shear, toppling, and tensile failures. Shear failure occurs when shear stress due to the weight of a potential failure block and positive pore-water pressure overcomes the shear strength of soil (cohesion, friction, and negative pore-water pressure) and confining pressure acting along a planar or circular slip surface. Toppling (or beam) failure occurs when a potential failure block rotates forward about a horizontal axis. Rotation occurs when the moment of the weight of the block about the axis is greater than the tensile and compressive resistive moments at the base of the block. Tensile failure occurs on undercut banks when the tensile stress due to the weight of the lower block overcomes the tensile strength of the soil. The result is a failure across a horizontal plane causing the lower part of a potential failure block to fall under gravity.

Stream bank failure may be induced in a number of different ways [91], including overdeepening (near-bank bed scour), oversteepening (undercutting), cantilever production/overhanging (undermining) caused by either fluvial or seepage erosion (Fox and Wilson [92] provided a review), overloading (wetting), reduction of cohesion (wetting or mineral leaching), reduction of friction angle (weathering), and reduction of failure plane length (tension cracking, including through the action of freeze–thaw cycles). Although mass stability is itself not directly related to fluvial forces, bank retreat rates over the medium to long term are fluvially controlled, regardless of the nature of the inherent bank materials, denudation processes, or mass failure mechanisms [91, 92]. This interaction is manifest in the concept of basal endpoint control [94, 95], of which three main states have been identified [96]:

1. Impeded removal—Bank failures, plus lateral and upstream inputs, supply material to the bank toe at a faster rate than it is removed. Basal accumulation results, decreasing the bank angle and height. The rate of supply decreases, tending towards the second state.
2. Unimpeded removal—Processes delivering and removing material to and from the bank toe are in balance. No change in basal elevation or slope angle occurs. The bank recedes by parallel retreat at a rate determined by the degree of fluvial activity at the bank toe. If the flow is not competent to erode material from the toe, the state of basal endpoint control is static and the bank retreat rate is zero.
3. Excess basal capacity—Basal scour has excess capacity over the supply from the bank failures and fluvial inputs. Basal lowering and undercutting occur, increas-

ing the bank height and angle. The rate of supply increases, tending towards the second state.

Correctly simulating the processes encapsulated within the basal endpoint control is the crux of the bank erosion modeling. Solving the problem requires five coupled elements: (1) a mass failure module that can evaluate the balance of forces or moments along the most critical potential failure surface that promotes and resists the downward motion of a material block (for cohesive banks); (2) a methodology by which failed bank materials can be appropriately distributed at the bank toe or dispersed; (3) a near-bank hydraulic fluvial model that can predict the complex 2D or 3D turbulent flow field at the spatial scale of irregular bank topography and vegetation; (4) a robust methodology that can simulate a fluvial (predominantly lateral) erosion of the bank face and bank toe; and (5) a far-field mobile-bed model that can predict sediment transport with sufficient accuracy to evaluate whether or not material eroded from the banks will be transported away from the bank. Further, there is always the need for a good coupling procedure so that variables predicted by the bank module and the mobile-bed stream model may be exchanged correctly.

4 A New Geofluvial Model

In this section, a practical geofluvial model, SRH-2D, developed at the US Bureau of Reclamation is described. Additional details may be found in a number of recent reports and papers by Lai and his collaborators [10, 84–87]. However, this chapter is the first attempt to put together the works carried out by the group, along with additional verification and application cases added. With SRH-2D, the main channel fluvial processes are solved with the 2D depth-averaged mobile-bed model [97], while the lateral bank erosion processes are solved with several bank modules described below. SRH–2D has been widely used for 2D depth-averaged modeling of vertical stream processes with its mobile-bed capability (e.g., [98–103]). More technical details may be found in a number of papers [97, 101, 103–105].

Two lateral bank erosion modules are incorporated into SRH-2D: the uniform retreat module and the mechanistic failure module. The two adopt the same basal erosion (lateral erosion) algorithm given a shear stress distribution on a wetted bank. The main difference lies in the mass failure algorithm. The uniform retreat module assumes that the bank is retreating uniformly as a whole. Mass failure process is computed by assuming that a constant bank angle (e.g., the angle of repose) is maintained, and the loss of the bank material equals to the basal erosion. The uniform retreat module is developed primarily for uniform noncohesive banks undergoing the dry granular and shallow slide processes. It, however, may also be applied to cohesive banks or banks with other mass failure processes, (a) if only gross bank retreat amount is of the interest, (b) when actual physical processes are too complex to model, or (c) if there is a lack of measured bank data. The mechanistic failure module is developed primarily for multilayer cohesive banks; the development is the result of collaboration with Andrew Simon and his team at the US Department of Agriculture-Agricultural Research Service (USDA-ARS). Key algorithms are based on the Bank Stability and Toe Erosion Model (BSTEM) as reported by Simon et al. [71, 81] and Langendoen and Simon [46]; the integration of key BSTEM algorithms into SRH-2D was documented by Lai et al. [10, 84, 86]

4.1 SRH-2D Model

The in-stream flow hydraulics and vertical fluvial processes are simulated with the SRH-2D mobile-bed module—a 2D depth-averaged hydraulic and sediment transport model. Its hydraulic modeling theory was documented by Lai [104, 105]. The model adopts the arbitrarily shaped element method of Lai et al. [106], the finite-volume discretization scheme, and an implicit integration scheme. The numerical procedure is sufficiently robust that SRH-2D can simultaneously model all flow regimes (sub-, super-, and trans-critical flows) and both steady and unsteady flows. The special wetting–drying algorithm makes the model very stable to handle the flow over dry surfaces. The mobile-bed sediment transport module adopts the methodology of Greimann et al. [99]; theories have been described in the reports and papers by Lai and his colleagues [97, 101, 103–105]. The mobile-bed module predicts vertical streambed changes by tracking multisize, nonequilibrium sediment transport for suspended, mixed, and bed loads, and for cohesive and noncohesive sediments, and on granular, erodible rock, or nonerodible beds. The effects of gravity and secondary flows on the sediment transport are accounted for by displacing the direction of the sediment transport vector from that of the local depth-averaged flow vector.

4.2 Basal Erosion

Basal erosion is the direct removal of bank materials laterally by flowing water in a stream. A number of methods may be used to compute basal erosion. In one method, lateral erosion of a bank node is computed using the vertical erosion predicted by the depth-averaged mobile-bed model over the wetted bank. This approach is similar to the control volume approach proposed by Hasegawa [107], and subsequently adopted by Nagata et al. [79], Duan et al. [80], and Chen and Duan [11]. The approach, however, works well only for special cases; it may incur large uncertainties for steep banks. Only a few mesh nodes may be used to define steep bank geometry and depth-averaged solutions near the bank ace may be inaccurate.

A different approach is adopted by the SRH-2D geofluvial model. An arbitrary number of banks can be selected for simultaneous bank retreat modeling and coupling with the main channel processes. Each bank can also be well represented by an arbitrary number of bank nodes for bank geometry definition; the bank nodes do not need to be the same as the 2D mesh. For each bank, the toe and top nodes are marked out in both the list of bank nodes and the 2D mesh that represent the bank.

The only requirement is that the two nodes match each other within the two bank representations. At the top node, zero vertical or lateral erosion is assumed. At the toe node, vertical erosion, along with other fluvial variables, are predicted by the 2D mobile-bed module, while the lateral erosion of the wetted bank face is computed using a semiempirical excess shear stress equation (e.g., [62, 78, 46]) expressed as:

$$\varepsilon_L = k\left(\frac{\tau}{\tau_c} - 1\right), \tag{9.1}$$

where

ε_L is the lateral erosion rate ($m\,s^{-1}$),
k is the erodibility coefficient ($m\,s^{-1}$),
τ is the shear stress on the bank node (Pa), and
τ_c is the critical shear stress (Pa).

This equation introduced two empirical coefficients (critical shear stress and erodibility) that need to be determined, and it has also been widely used for erosion of cohesive streambed [108, 109].

Shear stress distribution along the wetted bank is needed to use Eq. (9.1), but it is not readily available from the 2D depth-averaged mobile-bed model (3D model is needed instead). Two methods are adopted with SRH-2D: the linear distribution method and the ray–isovel method. With the linear method, the shear stress is assumed to decrease linearly from the bank toe to the intersect point of the bank and water election. The shear stress at the toe is computed using the 2D mobile-bed model while the stress is zero at the intersect point. The linear method is appropriate with the uniform retreat module since the bank has been lumped together as a single unit for lateral retreat modeling. And uncertainties induced by the linear assumption of shear stress distribution may be compensated by calibrating the erodibility coefficient. The ray–isovel method is more involved and it has been reviewed by a number of authors [110–114]. SRH-2D adopts a noniterative procedure as described by Lai et al. [10, 86]. Briefly, the shear stress distribution on a wetted bank is determined by dividing the total flow area at the bank cross-section into area segments. Each area segment corresponds to each bank and streambed node and it is affected by the roughness of the bank or the bed, as well as the local bank geometry. Once the area segment is obtained, the shear stress at each wetted bank node, say i, is computed by $\tau_i = \tau_{toe} R_i / R_{toe}$, where τ_{toe} = bank toe shear stress (Pa), R_i = the hydraulic radius of the area segment (m), and R_{toe} = bank toe hydraulic radius (m). The bank toe shear stress τ_{toe} is obtained from the 2D mobile-bed model.

4.3 Uniform Retreat Module

With the uniform retreat module, the mass failure is computed, after basal erosion, as follows: The bank profile from toe to top is assumed to be a straight line at a user-specified angle (e.g., at the angle of repose) and the bank retreat is computed such

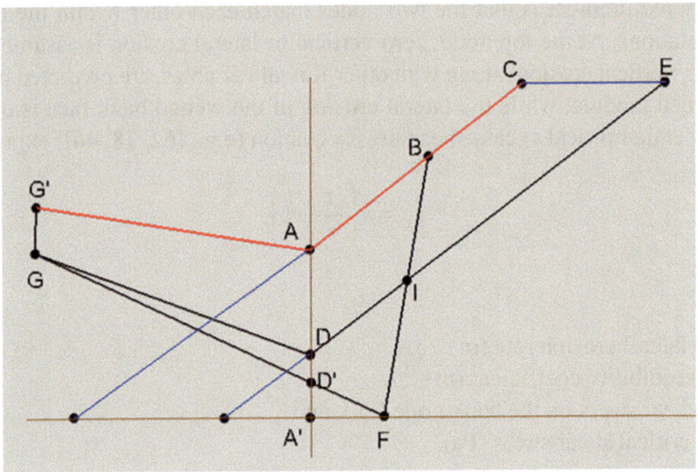

Fig. 9.1 Diagram of bank retreat computation with the uniform retreat module

that the total area (volume) of retreat equals the area computed by the basal erosion. Tests showed that the results are insensitive to the bank angle used and therefore, initial bank angles computed automatically by the model may also be used.

The bank retreat rate of the uniform retreat module may be derived to be expressed in an analytical form with the above mass conservation requirement, along with the linear shear stress distribution assumption. Its derivation is described next. Consider a bank shown in Fig. 9.1. The initial bank is G′ABC (G′AB is wetted and underwater while BC is dry and above water), and ABC is the straight bank face at the angle of repose. A is the bank toe node, C is the bank top, B is the intersect point between the bank and water surface, and G′ is the nearest mesh node in the stream adjacent to A. After a time period, erosion is assumed to occur: A is eroded vertically to A′ and G′ is to G; they are computed by the 2D model. By adding the lateral erosion of the toe node, A′ would move to its final position F. The new bank face after basal erosion would be GFBC due to the linear assumption of the shear stress along AB. Now, the new bank angle of FB exceeds the angle of repose; therefore, the bank material above B would "fail" to fill the toe area and a final bank face, with the angle of repose, would form. That is, GDIE would be the final bank face. The intercept point D between the new bank and the vertical line of AA′ may be above or below D′. In Fig. 9.1 and the derivation below, D is assumed to be above D′. Under such a scenario, D is assumed to be the final toe location. Mass (or area) conservation requires the following to be true (IBCE refers to the area contained within the polygon, etc.):

$$\text{IBCE} = \text{GDIFD}' \tag{9.2}$$

With some derivation, the bank retreat distance r_B (i.e., distance between C and E) may be computed by the following equation:

$$r_B = 0.5\frac{(h_0+\omega_V)(\omega_L+\dfrac{\omega_V}{\tan\alpha})+|\mathrm{GD}|\,\omega_V}{(H_0+\omega_V)+0.5|\mathrm{GD}|\tan\alpha}, \tag{9.3}$$

where

r_B is the bank retreat distance (m),
h_0 is the initial water depth at toe (m),
α is the initial bank height (m),
α is the bank angle or angle of repose (−),
ω_V is the vertical toe erosion distance predicted by the 2D model (m),
ω_L is the lateral toe erosion distance computed by equation (m), and
$|GD|$ is the horizontal distance between G and D (m).

The bank retreat distance r_B under the other scenario of D below D′ may be similarly derived; the final form is the same as Eq. (9.3) by setting $|GD| = 0$. Further, under the scenario of deposition at bank toe A, the bank would intrude towards the stream. The bank intrusion distance has also been derived similarly, and the final rate expression is also similar to Eq. (9.3) and they are not repeated. It is also noted that Eq. (9.3) may be viewed as a generalization of the Chen and Duan method [11].

4.4 Mechanistic Failure Module

Basal erosion may erode a bank to a point that the bank is too steep to remain stable and a gravitationally induced mass failure occurs. With the mechanistic failure module, geotechnical mass failure is explicitly computed through process-based models. The method follows the approach of Langendoen and Simon [46], but with some important differences. Details have been presented by Lai et al. [10, 84, 86]; and they are only briefly described next.

A bank is represented by an arbitrary number of bank nodes (independent of 2D mesh nodes along the bank), and it is assumed to consist of between one and an unlimited number of soil layers. Each bank layer may be assigned its own geotechnical properties. Basal erosion is carried out first and the computed lateral erosion is applied to the bank profile. After basal erosion, the potential for mass failure is evaluated by first finding the base of the failure plane on the bank face and the angle of the failure plane. The failure plan is found through a search algorithm so that the block above the plane is the most unstable. Block stability is measured using the factor of safety, the ratio of resisting to driving forces, through a force equilibrium analysis. Mass failure occurs if the factor of safety is less than one; otherwise, the bank is stable and no mass failure occurs.

The failure block is divided into many vertical slices. The force equilibrium analysis is first carried out on each individual slice; the force balance of the entire failure is then obtained by summing all forces on the slices—an approach developed by Langendoen and Simon [46]. In SRH-2D, it is assumed that that the groundwater table within the bank is horizontal as a constant elevation, pore-water pressures are distributed hydrostatically above and below the phreatic surface, and the bank is subject to planar or cantilever shear failures. In the force equilibrium analysis, the following forces are explicitly accounted for: (1) effective cohesion, describing the electrochemical force acting between charged clay minerals; (2) the weight of the soil block, a component of which acts to drive failure and a component of which acts to resist failure through friction; (3) the force produced by matric suction (negative pore-water pressure) on the unsaturated part of the failure plane; (4) the force caused by positive pore-water pressures on the saturated part of the failure plane; (5) the hydrostatic confining force provided by the water in the channel and acting on the bank surface; and (6) (when appropriate) inter-slice forces that act both normal to and parallel with the boundaries between vertical slices.

The factor of safety computation follows the Mohr–Coulomb shear strength criterion [115] which takes into account forces #1 to #4 acting on the shear plane at the base of slice *j*:

$$S_j = \frac{L_j}{F_s}\left[c_j' + (\sigma - \mu_a)_j \tan\varphi_j' + (\mu_a - \mu_w)_j \tan\varphi_j^b\right], \tag{9.4}$$

where

S_j is the driving force, a component of the weight of the soil block (kPa m),
L_j is the length of the slice base (m),
F_s is the factor of safety, defined as the ratio between the resisting and driving forces (–),
c' is the effective cohesion (kPa),
σ is the normal stress on the shear plane at the base of the slice (kPa),
μ_a is the pore-air pressure (kPa),
ϕ' is the effective angle of internal friction (°),
μ_w is the pore-water pressure (kPa), and
ϕ^b is the angle describing the increase in shear strength due to an increase in matric suction (°).

In Eq. (9.4), $(\mu_a - \mu_w)$ is the matric suction (kPa). For most analyses, the pore-air pressure can be set to zero. The value of ϕ^b varies with moisture content, but generally takes a value between 10° and 20°, with a maximum value of ϕ' under saturated conditions [116, 71].

The hydrostatic confining force #5 is taken into consideration by assuming that the surface water within the failure block is a material with no shear strength [117, 46]. Hence, the slip surface is extended vertically through the water and a horizontal hydrostatic force may be applied on the vertical portion of the slip surface. The inter-slice forces #6 that act both normal to and parallel with the boundaries between vertical slices are considered following the approach of Langendoen and Simon [46]. The details of the vertical slice approach are not described here and interested readers may refer to the Langendoen and Simon [46] or Lai et al. [10, 84, 86]. The extra input parameters with the mass failure modeling include (a) groundwater level, (b) elevations of bank layering, and (c) bank geotechnical parameters listed in Eq. (9.4).

4.5 Basal Cleanout Module

Large chunks of soil blocks are often deposited at the bank toe following a mass failure of cohesive banks. These blocks temporarily protect the bank from direct fluvial erosion, but over time are subject to subaerial weathering (when exposed) and gradual winnowing and eventual removal (when submerged). Many models ignored the bank toe protection features of the failed blocks, while others proposed a number of ways the impact be incorporated. For example, Pizzuto [118] and Nagata et al. [79] assumed that after failure, the fraction of failed material >0.062 mm settles at the angle of repose in an area of variable width, computed to ensure the conservation of mass. Darby and Delbono [119] assumed that the failed material settles at an angle of 35°, approximating the mean angle of repose, but the lateral extent of the deposit was limited to a one-bank height-wide region at the toe of the bank. This caused issues regarding mass conservation (*cf.* [78]), and thus Darby et al. [78] changed their formulation to permit the angle of the deposit to vary in order to ensure the conservation of mass. However, they also assumed that the failed material > 10 mm in size immediately became bed material, with the grain-size characteristics of the bed, and thus did not conserve mass within each grain-size class. While these assumptions may be valid for a few isolated cases, it is unsafe to impose them upon all failures under all flow conditions. SRH-2D therefore incorporated the basal cleanout process by placing failed materials into an invisible "tank" that has no topographic impact to the flow in the stream but is made available for the preferential basal erosion by size fractions following a mass failure [72]. That is, the basal erosion process must erode the sediment in the tank according to each size class first before the erosion of material in that size class from the wetted bank face is permitted. The tank approach explicitly accounts for the protection afforded by failed bank materials, does not make assumptions regarding the topographic form of failed blocks, and conserves the mass correctly. The use of "invisible" tank removes the need complexity needed to modify the local bank geometry. However, the method ignores the impact of blocks on near-bank flow conditions which may lead to an underestimation of the erosion force. There are other processes that are ignored in the current implementation in SRH-2D such as weathering and the establishment and proliferation of vegetation. These processes may change the erodibility of the failed block. It is suggested that the ignored processes and other uncertainty be taken into consideration in the selection of the erodibility coefficient for basal erosion. The calibration of the erodibility coefficient is recommended for most practical applications.

4.6 Coupling Procedure

Strategy and procedures are needed on how to couple the bank modules with the 2D mobile-bed model. This is probably one of the most important steps in model development as stability and ease of use of the geofluvial model depend on them, particularly for time-accurate continuous dynamic modeling. A general yet simple procedure is developed and used by SRH-2D.

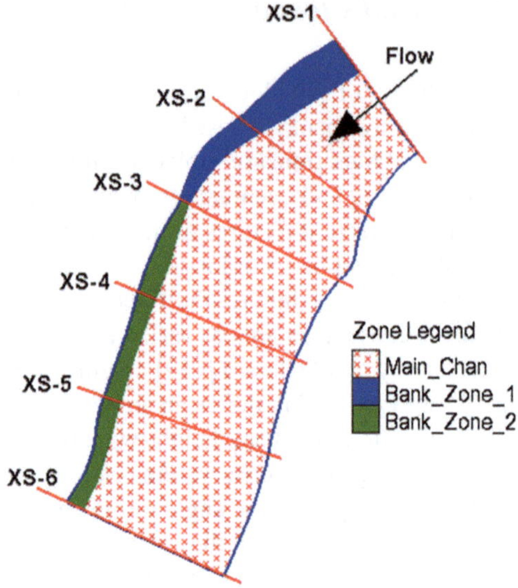

Fig. 9.2 Illustration of a channel reach for geofluvial modeling: right bank is subject to bank retreat

The SRH-2D model input and setup procedure for a coupled geofluvial modeling is made very similar to that for a nonbank mobile-bed modeling. An initial 2D mesh is generated along with regular boundary conditions specified. With geofluvial modeling, an extra model setup is to add one or multiple bank segments to the 2D mesh so that each bank segment is represented by a block of mesh cells (named bank zone). Each bank segment is a bank zone on the 2D model and each may contain an arbitrary number of banks that require bank retreat modeling. As an illustration, Fig. 9.2 shows a channel reach subject to geofluvial modeling whose right bank is to be simulated for bank retreat. Two bank segments are used in the example and each segment is represented with a bank zone consisting of 2D meh cells. Bank zone 1 contains three banks, while zone 2 has four banks. Each bank zone is defined by two edge lines. With the moving mesh approach to be described next, one edge is along the bank toe and the other along the bank top. A one-to-one toe-to-top correspondence is assumed. With the fixed mesh approach, the two represent simply the beginning and ending edges of the bank zone. But the zone has to be wide enough so that both toe and top nodes are contained within the zone all the time.

The above representation of banks on a 2D mesh is used only by the 2D mobile-bed module for computing appropriate flow hydraulics near the bank. The representation of each bank adopts a different method in which an arbitrary number of nodes may be used to describe the bank profile. The dual representations of a bank are used to improve the accuracy of bank retreat simulation. Large uncertainty may be resulted if only a 2D mesh is used for bank representation since only a few 2D mesh points may be feasible to use for steep banks. Another benefit is that bank layering and geotechnical properties of each bank layer may be specified for each bank accurately, independent of the 2D mesh. Therefore, the dual representation strategy makes SRH-2D easy to use with geofluvial modeling, but with improved accuracy.

A key remaining issue is how to update the feedback between the morphological changes predicted by the bank module and the 2D mobile-bed model during a continuous dynamic simulation. Two approaches are adopted by SRH-2D: the moving mesh and the fixed mesh approaches. The moving mesh approach adopts the arbitrary Lagrangian–Eulerian (ALE) formulation of Lai and Przekwas [120]. With this approach, the longitudinal mesh lines are aligned with the bank toe and top initially. The bank line "alignment" is enforced continuously throughout bank retreat by moving the 2D mesh along with the bank. The moving mesh approach "captures" the bank lines, leading to a better bank representation and more accurate retreat computation. The main drawback is that the 2D mesh has to be remeshed over the simulation period whenever a bank is moving, and the remeshing can only be done automatically by the model. The moving mesh approach has the potential to distort the mesh too much to cause model instability. Therefore, moving mesh is the choice only if no significant bank retreat is expected, e.g., less than one channel width. The fixed mesh approach does not move the mesh in planform in response to erosion, deposition, or bank retreat. With this approach, bank toe and top are found through a "fitting" (interpolation) procedure using nearby mesh nodes. A fixed mesh removes the mesh distortion issue, but the bank toe and top are only approximately represented by the 2D mesh. The fitting may lead to a high uncertainty that is proportional to the mesh density. Often, a much refined 2D mesh is required by the fixed mesh approach to capture bank retreat reasonably, increasing the computational cost significantly. The fixed mesh method is recommended only if bank retreat is anticipated to be large (e.g., more than one channel width); and a refined mesh should be used to reduce model uncertainty.

The fixed mesh approach requires tedious bookkeeping by the model, but is straightforward otherwise. No further discussion is necessary. The moving mesh approach is developed following the ALE formulation of Lai and Przekwas [120]. With this formulation, a mesh may be moved in an arbitrary manner. The governing equations are slightly modified from the fixed mesh formulation and they may be expressed in integral form for an arbitrarily moving mesh cell as:

$$\frac{d}{dt}\int_A h dA + \int_S h(\vec{V} - \vec{V}_g) \bullet d\vec{s} = 0 \tag{9.5a}$$

$$\frac{d}{dt}\int_A h\vec{V} dA + \int_S h\vec{V}(\vec{V} - \vec{V}_g) \bullet d\vec{s} = \int_S h\vec{\vec{\sigma}} \bullet d\vec{s} + \int_A \vec{S}_V dA \tag{9.5b}$$

$$\frac{d}{dt}\int_A h\phi dA + \int_S h\phi(\vec{V} - \vec{V}_g) \bullet d\vec{s} = \int_S h\vec{q} \bullet d\vec{s} + \int_A S_s dA, \tag{9.5c}$$

where

t is the time (s),
h is the water depth (m),
A is the area (m^2),

$\vec{V}$ is the fluid flow velocity vector (m/s),
$\vec{V}_g$ is the velocity vector of the moving mesh (m/s),
$\vec{s}$ is the edge length of a cell (m),
ϕ is the a scalar variable that is being transport (ϕ),
$\vec{\vec{\sigma}}$ is the stress tensor (m^2/s^2),
$\vec{q}$ is the scalar flux vector (ϕm/s),
$\vec{S}_V$ is the source/sink term of the momentum equation (m^2/s^2), and
S_S is the source/sink term of the scalar equation (ϕ m/s).

In the above Eq. (9.5a) is mass conservation, (9.5b) is momentum conservation, and (9.5c) is for transport of a scalar (e.g., sediment concentration for each size class). Integration over A denotes an arbitrary moving mesh cell, and over S is the side of the cell with the vector representing the unit normal. The grid velocity is computed using a geometric constraint, called the space conservation, written as:

$$\frac{d}{dt}\int_A dA = \int_S \vec{V}_g \bullet d\vec{s}. \tag{9.6}$$

The procedure developed by Lai and Przekwas [120] is used to compute the grid velocity accurately using Eq. (9.6). Once the grid velocity is computed, the discretization and solution algorithms are the same as the nonmoving-mesh cases. With the ALE method, the main flow and sediment variables represented by the mesh cells are automatically computed in a time-accurate manner; there is no need for additional interpolations except for derived variables such as bed topography.

4.7 Solution Flow and Data Exchange

The timescale of the bank retreat process is much longer than that of in-stream fluvial processes and thus the time step of the bank module is generally much larger than the 2D mobile-bed model. In a typical simulation, the 2D mobile-bed modeling is carried out first assuming fixed banks; flow hydraulics and vertical bed changes are predicted. The mobile-bed modeling proceeds in its own time until it reaches the bank time step to activate the bank module. The model then carries out the following steps: (a) time average the toe shear stress, near-bank water elevation, toe vertical erosion, etc. over the duration of the bank time step and transfer these to each bank; (b) distribute the shear stress along the wetted bank; (c) perform combined basal erosion and mass failure computation using the analytical rate expression with the uniform retreat module, and skip the rest of the steps, (d) through (f); (d) perform basal erosion by computing the lateral erosion volume (actually area for the bank cross section) of the wetted bank; (e) apply the lateral erosion volume to the cleanout tank first if not empty, and then apply to the wetted bank face (if any) and deform the bank accordingly; and (f) check the geotechnical stability of the bank, update the new bank geometry if mass failure occurs, and add the failed blocks to the cleanout tank. After all banks are simulated for bank retreat, the toe and top retreat distances are transferred back to the mobile-bed model. The 2D mesh is moved and deformed to capture the new bank location with the moving mesh approach, or the new bank location is found (fitted) using the nearby mesh nodes of the fixed mesh. In addition, the sediment volume in size classes removed during the basal erosion step is added to the stream for transport by the mobile-bed model. The coupling between the bank and the mobile-bed modules is done in a decoupled manner although the mobile-bed module itself is a coupled model [121].

5 Model Validation

The geofluvial model SRH-2D has been tested and validated extensively [10, 84–87]. Two cases are reported in this section to validate and verify the uniform retreat and mechanistic failure modules. They are the laboratory meander channel tested by Nagata et al. [79] and a bend on Goodwin Creek, Mississippi.

5.1 *Uniform Retreat Module: Flume Case Study*

The uniform retreat module, both the moving mesh and the fixed mesh versions, is validated and verified using the laboratory meander cases of Nagata et al. [79]. The experiments were carried out in a tilting flume having a length of 10 m, width of 1 m, and depth of 0.2 m. The initial meander channel was in the form of the sine-generated curve having wavelength of 2 m and maximum angle of 30° from the longitudinal direction (Fig. 9.3). The channel cross section was trapezoidal and

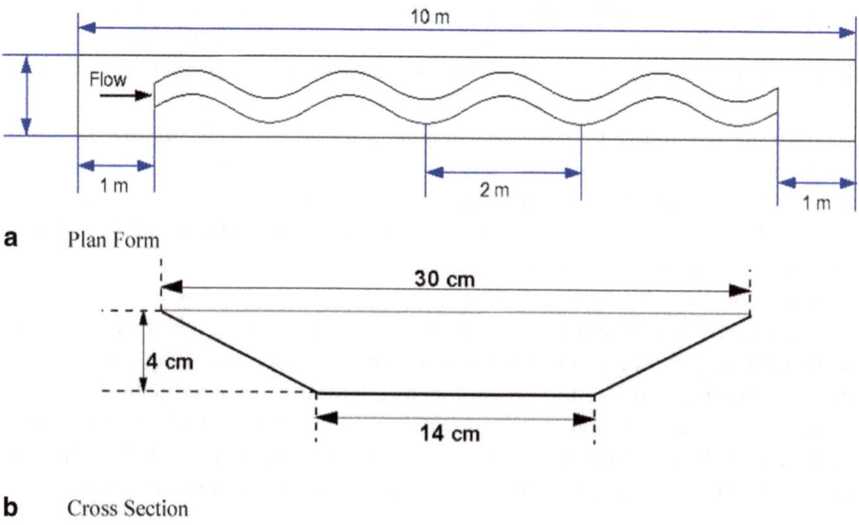

Fig. 9.3 Flume configuration and initial meander channel form of the Nagata et al. case [79]

uniform along the channel initially; its dimensions are marked in Fig. 9.3. Four channel wavelengths were used in the experiment, but data were measured only in the second from the upstream. The numerical model simulates only the first three wavelengths since the results in the fourth do not impact the second.

Two simulation runs are carried out. Run 1 has a flow discharge of 1980 cm^3/s, initial bed slope of 1/300, and initial water depth of 3 cm; Run 3 has a flow discharge of 1000 cm^3/s, initial bed slope of 1/100, and initial water depth of 1.42 cm. Both the bed and bank consisted of fairly uniform sand with a mean diameter of 1.42 mm ($\sqrt{d_{84}/d_{16}}=1.28$). Sediment was fed at the upstream end of the channel continuously in the experiments, and the channel evolution was initiated and the meander form was measured.

Both the moving and fixed mesh approaches are used for simulation. The moving mesh scenario has an initial mesh of 2684 cells, consisting of 123 longitudinal points and 23 lateral points (Fig. 9.4). This initial mesh is deforming while banks are retreating, but the total number of mesh cells and mesh topology remain unchanged. Two fixed meshes are used: Run 1 uses a mesh of 5612 cells and Run 3 has a mesh of 11,424 cells (see Fig. 9.4). The fixed mesh approach needs more mesh cells due to accuracy concern as discussed before. Also, the solution domain needs to be larger than the moving mesh scenario to take into account the potential bank retreat.

Model inputs include the following. The bed and bank are set to have uniform sand of a diameter of 1.42 mm. The upstream discharge is imposed as the boundary condition and the sediment feed rate is estimated using the Engelund and Hansen [122] equation so that there is no net erosion and deposition at the upstream section. The only downstream boundary condition is the water elevation that is specified based on the measured data. The same Engelund and Hansen equation is used to compute the entrainment rate in the nonequilibrium sediment transport partial differential equation. The bedload adaptation length uses the Philip–Sutherland formula [123] which was found suitable for sandy beds; and the active layer thickness is ten times the sediment diameter. A uniform Manning's roughness coefficient of $n=0.0168$ is used; it is estimated using the grain shear stress expression $n=d^{1/6}/20$ (d is the mean sediment diameter).

Both the left and right banks are simulated for retreat with the following bank properties. The critical shear stress is computed, by assuming the Shields number of 0.027, to be 0.62 Pa. The bank erodibility coefficient is calibrated and a value of $k=2.0\times10^{-4}\,\mathrm{ms}^{-1}$ is found and used for all model runs. The initial bank slope is assumed to be maintained during bank retreat.

Model results with Run 1 are shown in Figs. 9.5 and 9.6. Figure 9.5 compares the initial mesh at $t=0$ and the final mesh at $t=125$ min, along with contours of bed elevation changes. Figure 9.6 compares the predicted bank retreating process, using both the moving and fixed mesh approaches, with the measured data. The results show that bank erosion starts at the downstream half of the outer bend and extends into the inner bend, while bar deposition occurs on the downstream half of the inner bend. The three sets of bank retreat results, two simulated and one measured, agree

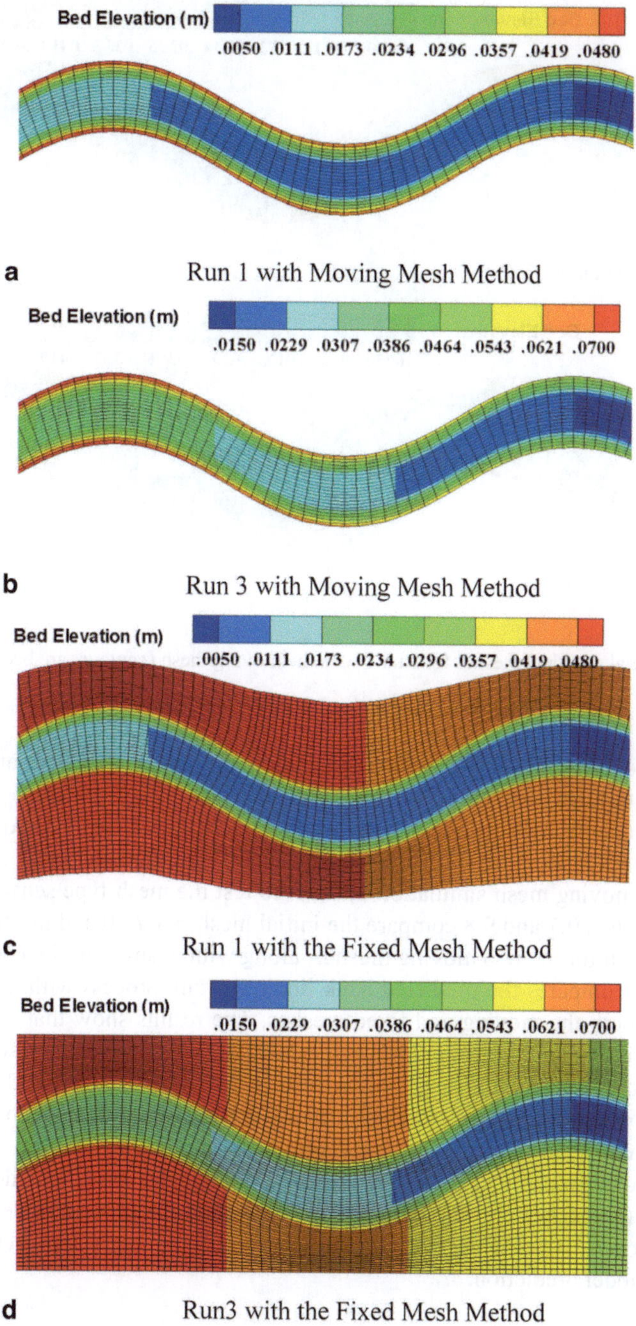

Fig. 9.4 Initial meshes used for Run 1 and Run 3 using both the moving and fixed mesh approaches (contour represents the initial bed elevation)

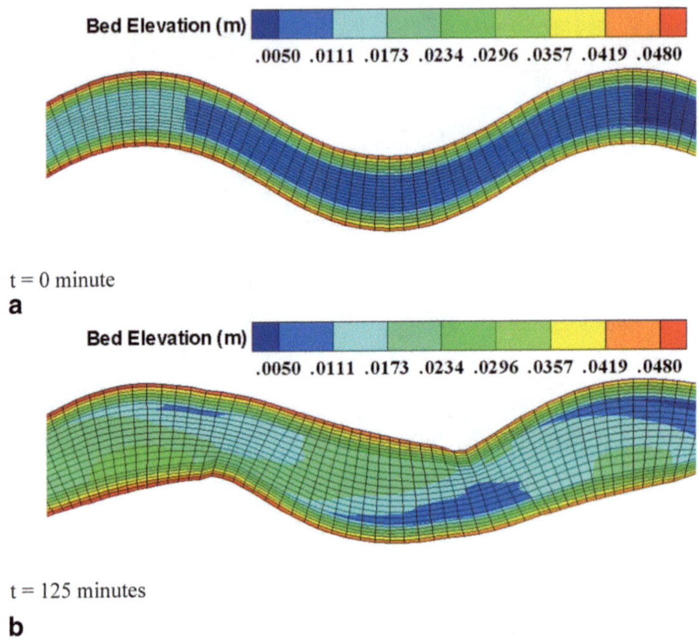

Fig. 9.5 Initial and final meshes for Run 1 with the moving mesh (*contours* are bed elevation)

with each other reasonably. There is some underprediction of the retreat amount by the model though.

Model results with Run 3 are shown in Figs. 9.7, 9.8 and 9.9. An additional mesh, the mixed mesh having 10,944 quadrilateral and triangular mesh cells, is added for the moving mesh simulation in order to test the mesh-type sensitivity of the model. Figures 9.7 and 9.8 compare the initial meshes at $t=0$ and the final meshes at $t=110$ min for the two moving meshes, along with changes of the bed elevation. Figure 9.9 compares the predicted bank line retreating process with the measured data with both the moving and fixed meshes. The results show that the predicted bank retreat of all numerical models agrees with the measured data reasonably well. Different meshes produce very similar results, which indicates that the errors introduced by the mesh type and mesh density are relatively small with the meshes adopted. Also, both the moving and fixed mesh approaches predict similar bank retreat results despite their vastly different methodologies used. This demonstrates that both approaches are implemented correctly, and the procedure developed for each approach works well. Both the moving and fixed mesh approaches may be used for model prediction.

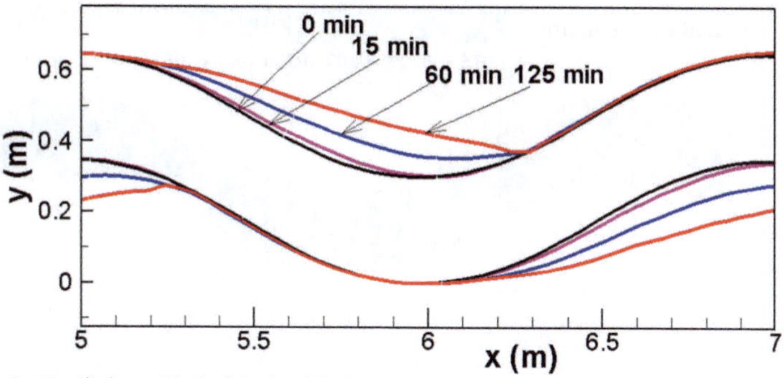

a Prediction with the Moving Mesh

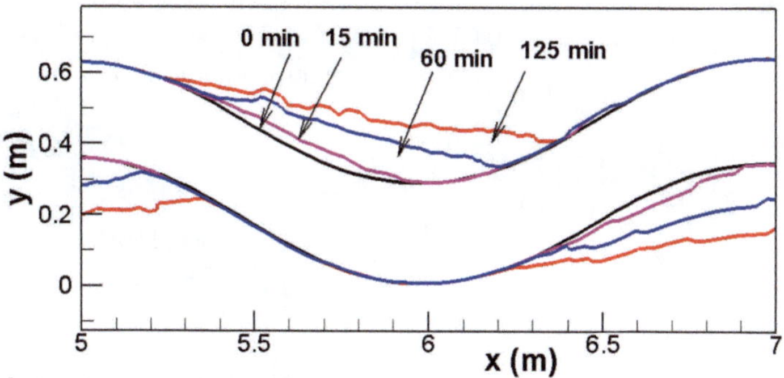

b Prediction with the Fixed Mesh

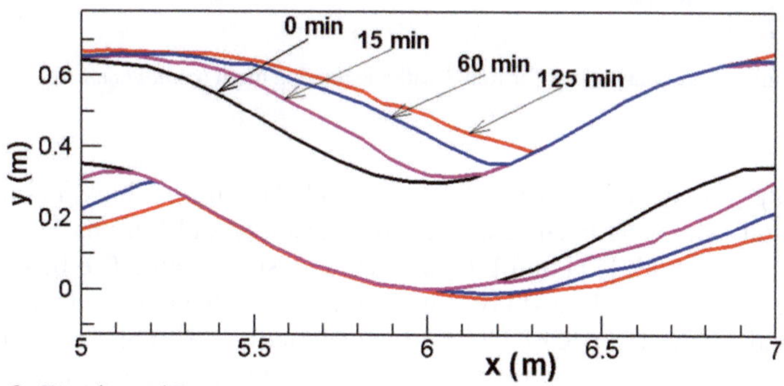

c Experimental Data

Fig. 9.6 Comparison of predicted and measured bank retreat for Run 1

5.2 Mechanistic Failure Module: Goodwin Creek Bend

The mechanistic failure module is validated and verified by simulating the geofluvial processes of a bend on Goodwin Creek, Mississippi, for the period between March 1996 and August 2001. This site is selected because of the wealth of data available from a long-term stream bank failure monitoring study carried out since 1996. Bank retreat data, as well as other hydrological and geotech-

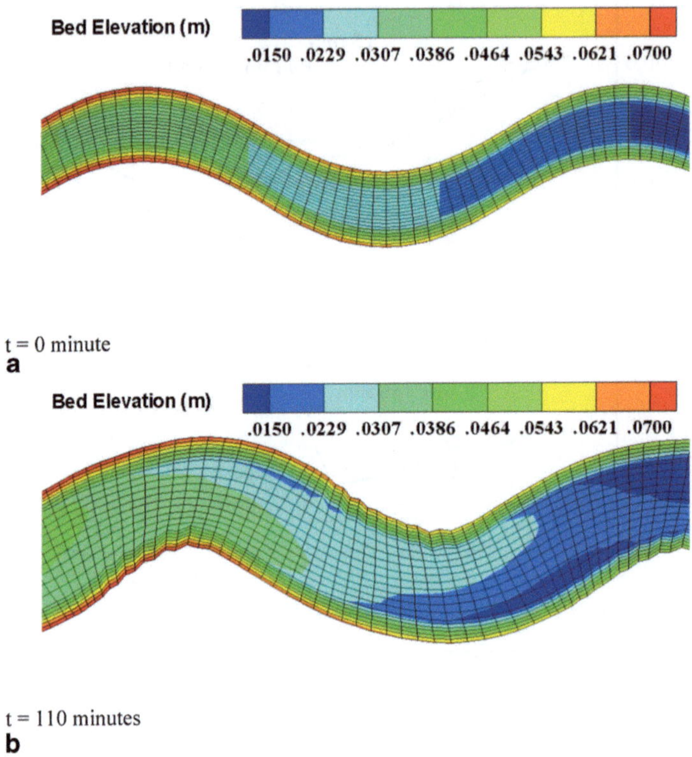

Fig. 9.7 Initial and final meshes for Run 3 with the moving mesh (*contours* are bed elevation)

nical data, are available over a 6-year period. The morphology and dynamics of the study site have been described by Grissinger and Murphy [124], and of late by Simon et al. [71] and Langendoen and Simon [46]. The bend selected by the SRH-2D modeling study was also numerically simulated using the 1D geofluvial model (CONCEPTS) by Langendoen and Simon [46]. The SRH-2D modeling study, therefore, may also shed light on whether the use of a 2D mobile-bed module improves the simulation of bank retreat in comparison to a 1D module. A wide range of model tests and verifications have been carried out and results were documented by Lai et al. [10]. Only the final verification results are presented next.

5.2.1 Model Input and Setup

An initial 2D mesh is generated to cover the solution domain of the Goodwin Creek bend as shown in Fig. 9.10a. The initial channel bathymetry is based on the survey data made in March, 1996, and the bed elevation is shown in Fig. 9.10b. A block of 2D mesh, the red box of Fig. 9.10a along the right bank, is designated as the bank

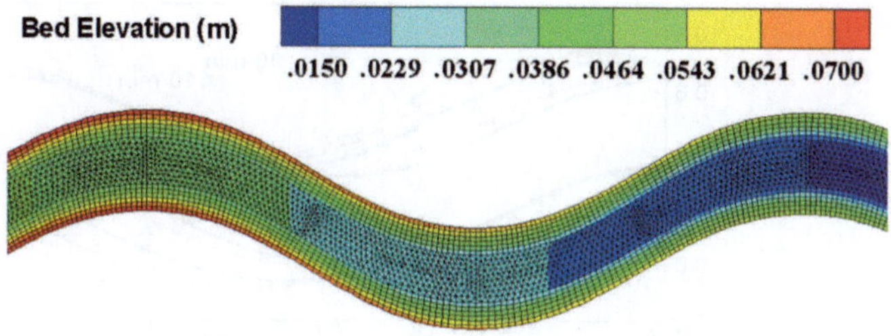

t = 0 minute
a

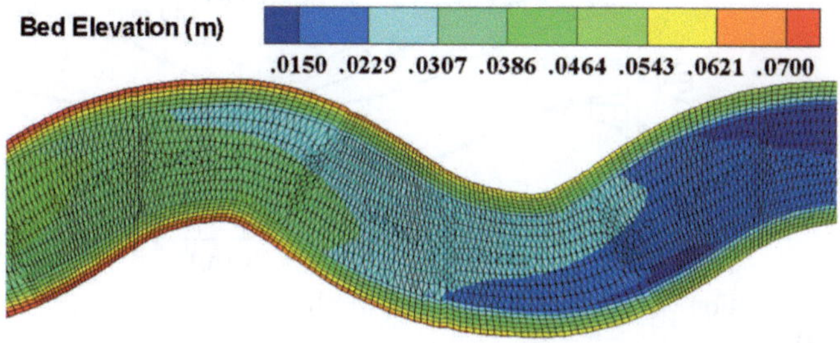

t = 110 minutes
b

Fig. 9.8 Initial and final meshes for Run 3 with the mixed, moving mesh (*contours* are bed elevation)

zone; and 11 banks (lateral lines within the bank zone) are selected for bank retreat modeling. These 11 banks correspond to the 11 cross sections surveyed over the years; this way, a direct comparison between the model results and the measured data may be made. The bank zone is defined by two mesh lines (edges): one corresponds to the bank toe and the other is along the bank top. In addition to the representation of these banks on the 2D mesh, each bank profile is also represented separately using representative measured nodes, much more than the six nodes of the 2D mesh, to improve the accuracy of the bank retreat computation. The only coupling requirement between the 2D mesh and the bank profile is that the toe and top nodes match each other. Note that the number of longitudinal 2D mesh points in the bank zone is not related to the number of banks, and is usually much larger than the number of banks.

The inputs for the 2D mobile-bed module include the following. At the upstream boundary (XS-1 in Fig. 9.10), the recorded time series discharge (Fig. 9.11a) and the sediment supply rates computed using the Wilcock and Crowe equation are applied [125]. At the downstream (XS-11 in Fig. 9.10), the stage-discharge rating curve, developed by Langendoen and Simon [46] from coincident stage and discharge records measured in 2001, is enforced (Fig. 9.11b). The measured grain-size

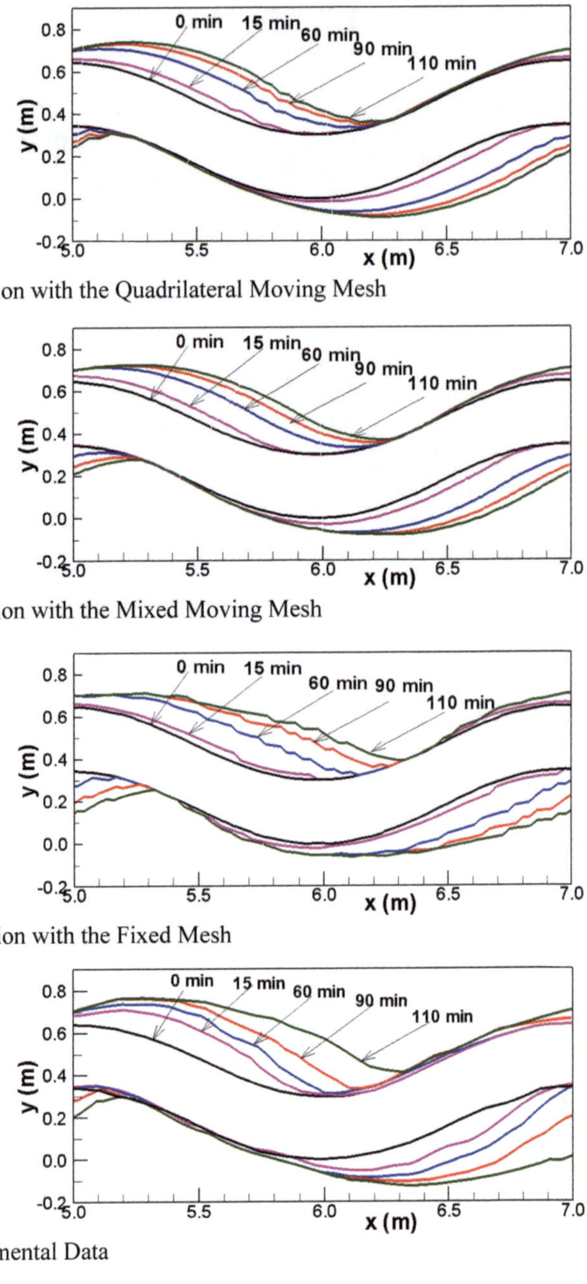

Fig. 9.9 Comparison of predicted and measured bank retreat for Run 3

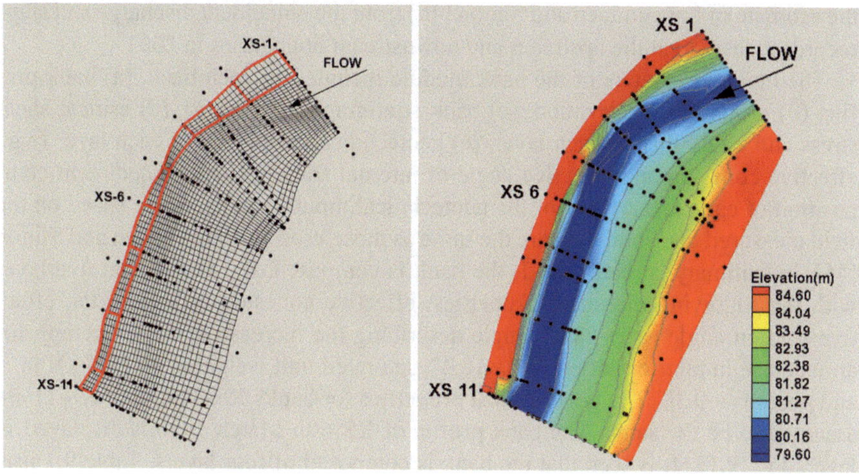

a Initial Mesh **b** Initial Bathymetry

Fig. 9.10 The initial 2D mesh and bathymetry of modeling at the Goodwin Creek bend; the red box on the left figure is the bank zone and 11 lateral lines represent the banks for retreat modeling. **a** Initial mess, **b** Initial bathymetry

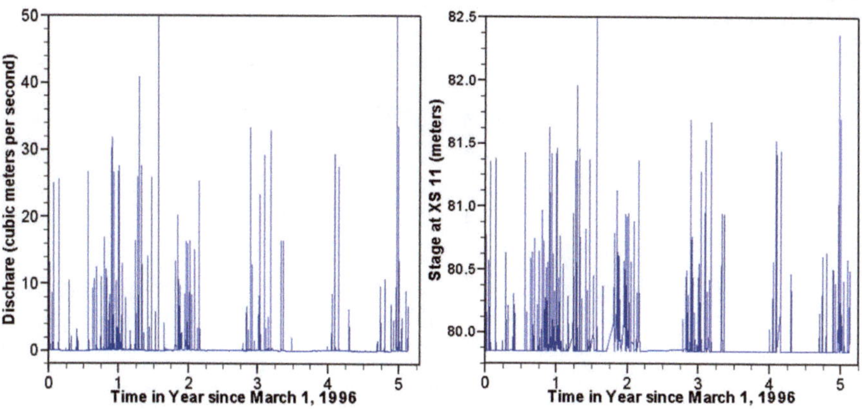

a Discharge **b** Stage at XS 11

Fig. 9.11 Recorded flow discharge through the bend and the stage at XS-11 during the simulation period

composition is segregated into nine size classes and the measured data are applied as the initial streambed gradation. The bed material at the site is bimodal with peaks at 0.5 and 22.6 mm, with a median grain size of 6.7 mm and a gradation coefficient of 8.2 [46]. The transport of each size class is governed by its own nonequilibrium equation that used the Wilcock and Crowe equation [125] as the erosional rate term. In addition, a uniform Manning's roughness coefficient of 0.032 is used based on the estimate of Langendoen and Simon [46] from the coincident discharge and stage records measured at the upstream and downstream boundaries in 2001.

The input parameters of the bank module include, for each bank, (a) bank profile, (b) groundwater elevation, (c) bank stratigraphy (layering), (d) critical shear stress and erodibility of each layer, (e) geotechnical properties of each layer (e.g., effective cohesion and effective angle of internal friction), and (f) sediment composition of each layer. Most of the geotechnical input parameters are based on the field measured data, and they are the same as those used by Langendoen and Simon [46]. In summary, at bank XS-1, the bank is composed of a single cohesive layer with the following measured properties: effective cohesion (c') = 4.5 kPa, effective friction angle (ϕ') = 28.6°, angle describing the increase in shear strength for an increase in matric suction (ϕ^b) = 10.4°, saturated unit weight (γ_s) = 19.4 kN m^{-3}, and porosity = 0.38. The geotechnical properties for banks XS-2 through XS-11 are assumed to be the same. The bank profile of XS-6 in March 1996 is displayed in layers Fig. 9.12. It is seen that the bank is composed of four layers. Table 9.1 lists the geotechnical properties measured for each bank layer, while Table 9.2 lists the sediment compositions of each layer, segregated into nine size classes.

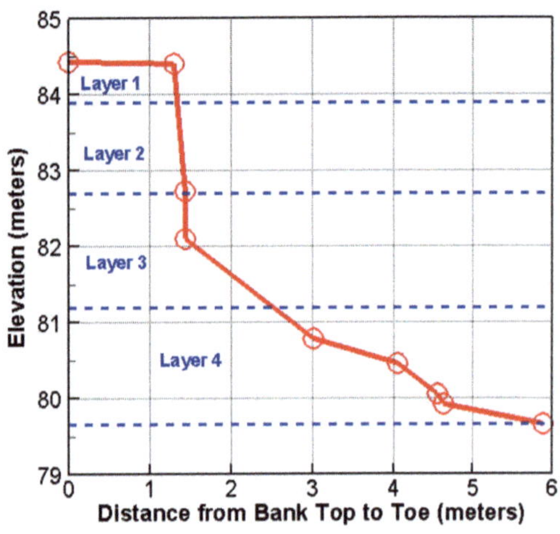

Fig. 9.12 Bank profile and its layering (*stratigraphy*) at XS-6

Table 9.1 Bank stratigraphy and geotechnical properties for XS-2 to XS-11 at the Goodwin Creek bend

Layer	Depth below surface (m)	Porosity	Saturated unit weight (kNm^{-3})	Friction angle (°)	Angle ϕ^b (°)	Cohesion (kPa)
1	0–0.5	0.489	16.9	33.1	17.0	1.41
2	0.5–1.7	0.489	19.3	28.1	10.2	2.70
3	1.7–3.2	0.380	19.9	27.0	17.0	6.30
4	>3.2	0.320	21.0	35.0	17.0	1.0

Table 9.2 Size range of each sediment size class and the corresponding volumetric fractions (%) on the initial channel bed and within each bank layer

Size range (mm)	<0.01	0.01–0.0625	0.0625–0.25	0.25–1.0	1–2	2–8	8–16	16–32	32–128
Initial bed	0.17	0.48	1.3	27.6	6.1	17.56	16.54	21.0	9.25
Layer 1	11.77	75.46	9.95	1.41	1.4	0.01	0	0	0
Layer 2	11.77	75.64	9.95	1.41	1.4	0.01	0	0	0
Layer 3	11.64	35.54	25.7	13	12.07	2.05	0	0	0
Layer 4	0	0	20	16	14	30	15	5	0

Other bank module inputs are as followings. The critical shear stress of 5.35 Pa is used with all banks. This value represents the median of 16 nonvertical and vertical jet tests [126, 127] conducted on failed bank materials at the study site (16th and 84th percentiles 0.17 and 24.6 Pa, respectively). This is different from the approach taken by the 1D modeling of Langendoen and Simon [46] who took the critical shear stress as a calibration parameter and it was allowed to change at different banks. Herein, the erodibility coefficient is selected as a calibration parameter that was kept the same for all eleven banks. The fact that only a single erodibility value is used for all banks makes SRH-2D more attractive since modeling would become unpractical if the erodibility of each bank needs to be calibrated. The final calibrated erodibility value was 1.2×10^{-7} ms^{-1}, which may be compared with the median erodibility of 1.47×10^{-6} ms^{-1} based on a widely varying range of measured data at the Goodwin Creek. The slight difference in values may reflect the fact that the model erodibility may also take other factors into consideration as discussed before. Finally, the groundwater elevation was constant at 82.3 m throughout the simulation. This approximates the top of a less permeable soil layer containing manganese nodules [71, 128].

The simulation is carried out for a period from March 1996 to February 2001 using the moving mesh approach. The time step of the 2D mobile-bed module is 5 s, while the time step of the bank module is the time needed for a given amount of water flowing through the bend. A volume of 4000 m^3 is used. Sensitivity analysis indicated that the model results were insensitive to further reduction of these time steps.

5.2.2 Results and Discussion

The predicted bank retreat from March 1996 to February 2001 is compared with the measured data in Fig. 9.13 on a planar view; the initial and final 2D meshes are plotted in Fig. 9.14. Detailed comparisons between the predicted and measured bank profiles are made in Fig. 9.15 in which individual bank retreat processes are shown at different times.

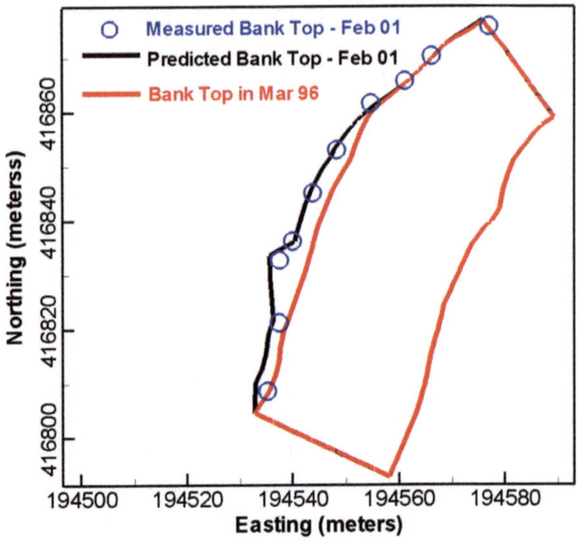

Fig. 9.13 Comparison of predicted and measured bank retreat distance from March 1996 to February 2001

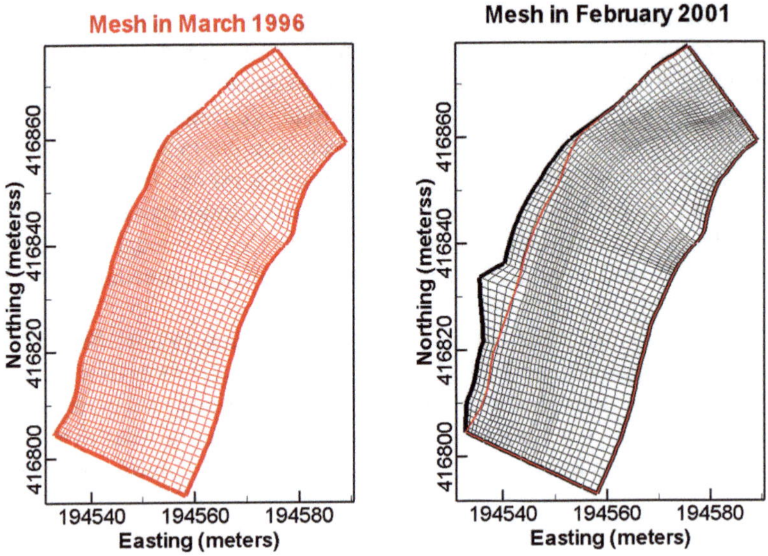

Fig. 9.14 Initial (March 1996) and final (February 2001) 2D meshes

Overall agreements between the model prediction and the measured data are satisfactory, considering that the physical processes of bank retreat at the site are very complex. The site is characterized by tall, steep, multilayer, cohesive banks, and the bank retreat is subject to influences of many processes including mass failure due to fluvial, seepage, and tension crack processes. The site is also frequently

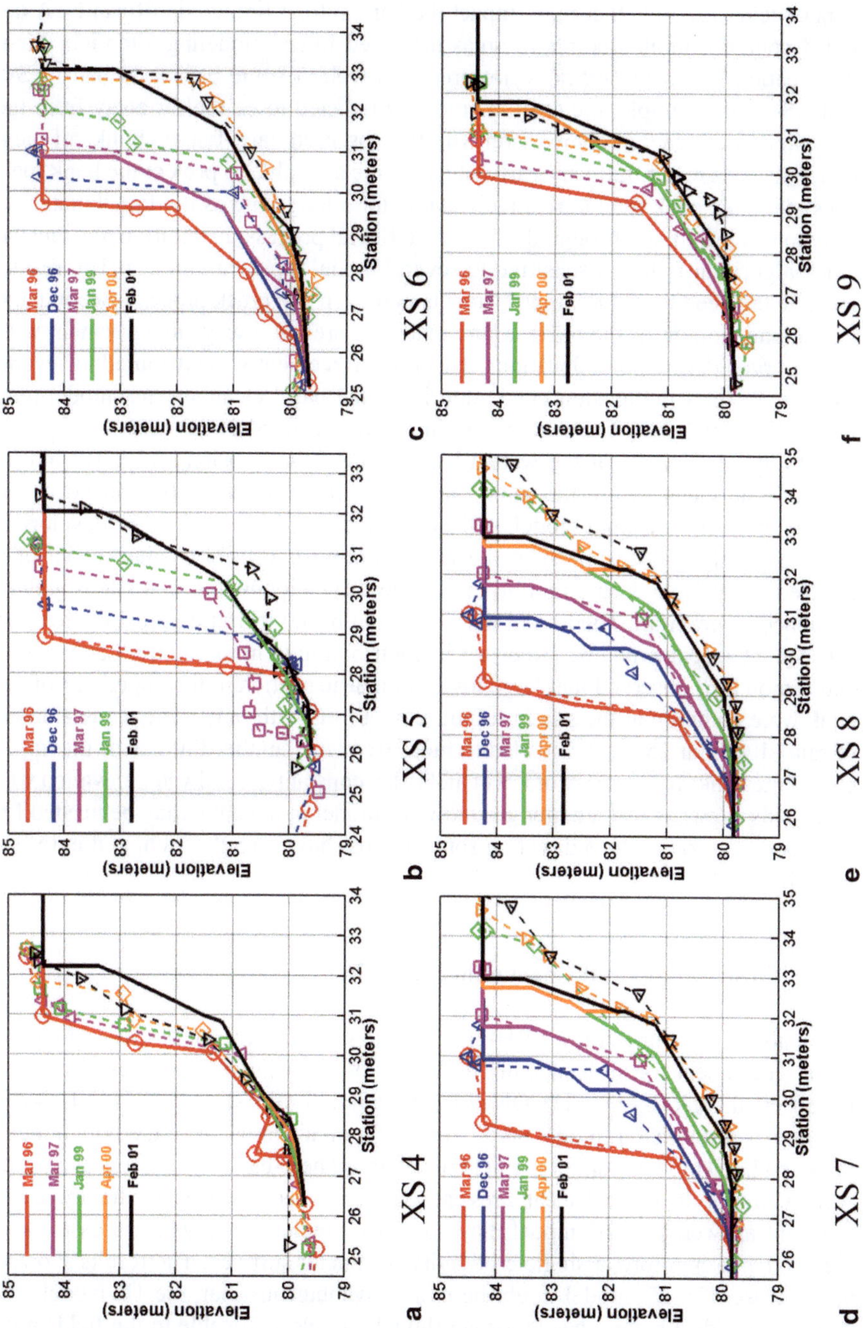

Fig. 9.15 Comparison of predicted (*solid lines*) and measured (*dash lines with symbols*) bank retreat at XS-4 through XS-9 (*the same color corresponds to the same time*) **a** XS-4, **b** XS-5, **c** XS-6, **d** XS-7, **e** XS-8, **f** XS-9

experiencing rainfalls that may impact the bank retreat timing significantly. A detailed comparison also points to areas that need future modeling attention. First, the location of tension crack is not predicted well at some banks. At bank XS-5 and XS-9, for example, the tension crack is predicted to be too far away from the bank in earlier years in comparison with the observed data. Large chunks of bank materials, as a result, fail very early, and become available to protect the bank toes. This made the bank remain relatively stable for a long time as the failed materials have to be cleared out through the hydraulic fluvial process. In addition, the tension crack algorithm is found sensitive to several model input parameters. The results indicate the need for future improvements in the tension crack process. Second, the timing and amount of the mass failure materials are not always in agreement with the observed data at some locations. This poor agreement was attributed to several sources. Besides the tension crack algorithm, the basal cleanout tank model may be partly responsible. The tank model implemented in SRH-2D assumes that the erodibility coefficients of the wetted bank during basal erosion and failed blocks during basal cleanout are the same—an assumption that could be relaxed in the future. Or, a new cleanout model, still simple and practical, needs to be developed. More probably, the mismatch of the failure timing is caused by the impact of a perched water table during rainfall events [46]. The Goodwin Creek site is greatly influenced by rainfall. The loss of matric suction from infiltrating precipitation and subsequent seepage may be important in contributing to mass failure. Langendoen and Simon [46] observed that large losses of matric suction in the upper part of the bank were common at the site in response to storms with only a moderate amount of rainfall (about 25 mm). This led to more frequent, smaller failures of the upper part of the bank. In the study of Lai et al. [10], a constant groundwater elevation was used. This suggests that a better groundwater model in the bank may be needed for the Goodwin Creek site, a direction some authors have already attempted in recent years [e.g., 129].

5.2.3 Observations and Concluding Remarks

In the above modeling study, SRH-2D used the same critical shear stress and erodibility coefficient for all banks, but different critical shear stress values were used for different banks by the 1D CONCEPT modeling for the same case [46]. However, similar, if not better, model predictions were obtained between the two models. In general, 1D models cannot predict the enhanced near-bank shear stress at a bend. A reduction of the critical shear stress at some cross sections was an attempt to get the right answer. The 2D model may correctly predict the fluvial processes near the bend and therefore, a single erodibility parameter sufficed. The results showed that the use of a 2D model might be more advantageous than the 1D model. 2D modeling holds a better chance for geofluvial models applicable to the field cases as the determination of the calibration parameters can be reach based, instead of individual bank based.

Sensitivity studies have been carried out by Lai et al. [10] in an effort to isolate the most important input parameters and to provide guidance for model input and setup. Geofluvial modeling involves many input parameters; some are measurable but with high uncertainty and others need to be estimated and/or calibrated. Therefore, Lai et al. [10] developed a recommended calibration procedure for practical geofluvial modeling in the field based on sensitivity study. The modeling procedure may be briefly summarized as follows: (a) Simulate and calibrate the nonbank case first with the same 2D mesh to be used by the with-bank model; (b) select the reach-based erodibility coefficient as the primary calibration parameter while others are obtained from in situ measurements, or using empirical relations, or based on past experience; (c) erodibility is calibrated to match the observed bank retreat distance over a time period with known retreat data; and (d) the calibrated model is finally used for project application simulation study. The critical shear stress can be taken as a secondary input parameter. It may be estimated using the Shields diagram for noncohesive banks, and measured or estimated values for cohesive banks. Model results are less sensitive to the critical shear stress with cohesive banks. If geotechnical properties of a cohesive bank are unavailable, the uniform retreat module may be used as it eliminates the need of geotechnical properties for mass failure analysis. For many engineering applications, the uniform retreat module may be sufficient.

Bank erosion modeling at the Goodwin Creek is very challenging since banks are tall and steep, impacted by both positive and negative pore-water pressures and seepage, and consist of multilayer cohesive materials. In addition, the structure of the massive silt and meander belt alluvium units promotes the development of large vertical tension cracks [124] that seem to dominate the shape of the bank profile. Despite these complexities, SRH-2D has been able to reproduce the overall bank retreat process, as well as the formation of the new bank profiles with tension cracks. Overall, predicted results agree reasonably with the measured data over the 5-year period simulated. It is shown that the model is suitable for bank retreat prediction over an annual or multiyear hydrograph by using an average constant groundwater elevation. For the correct prediction of the mass failure timing over a precipitation event, model improvement is needed with water table variation incorporated.

6 Model Application: Chosui River in Taiwan

Chosui River watershed is located in central Taiwan; it is the longest river in Taiwan, 186.6 km in length, covering an area of 3157 km^2. A description of the river hydrology, as well as geomorphic features, is provided by Lai et al. [10]. The selected study reach is near the river mouth; it is relatively flat with an average slope of 1:900.

6.1 Model Calibration

A calibration study is carried out for the river reach under study during the time period of July 2004 to August 2007. The solution domain is shown in Fig. 9.16. The upstream boundary is located about 2.9 km upstream of the Ziqiang bridge, and the downstream boundary is about 1.8 km downstream of the Xibin bridge. The solution domain covers about 16.7 km of river channel with an average width of 1.9 km. The upstream boundary is located within a bend and therefore the model results near the upstream boundary may possess high uncertainty. The mesh generated for the calibration study consists of 16,698 mixed quadrilateral and triangular mesh cells. The model simulation starts from July 2004 with the measured bed elevation in April 2004 used as the initial topography (Fig. 9.17).

The flow roughness and bed sediment gradation are needed for 2D mobile-bed modeling. The flow roughness is represented by the Manning's coefficient (n) and a constant value of $n=0.027$ is used based on an estimate of the field data. Initial bed gradation is from the survey data made in December 2010. Five data points within the study area are available; results are plotted in Fig. 9.18a. It is seen that the grain-size spatial variation is small; the medium diameter is about 0.35 ~ 0.43 mm within the reach. A comparison with the previous year measured data showed that the 2010 gradation data are representative of all previous year bed conditions. Bed and bank sediments are divided into a total of five sediment size classes: (a) 0.0625–0.125 mm; (b) 0.125–0.25 mm; (c) 0.25–0.5 mm; (d) 0.5–2.0 mm; and (e) 2.0–32 mm. The Engelund–Hansen [122] equation is used as the erosional rate potential

Fig. 9.16 Solution domain selected for the Chosui river modeling (aerial photo is in August 2007)

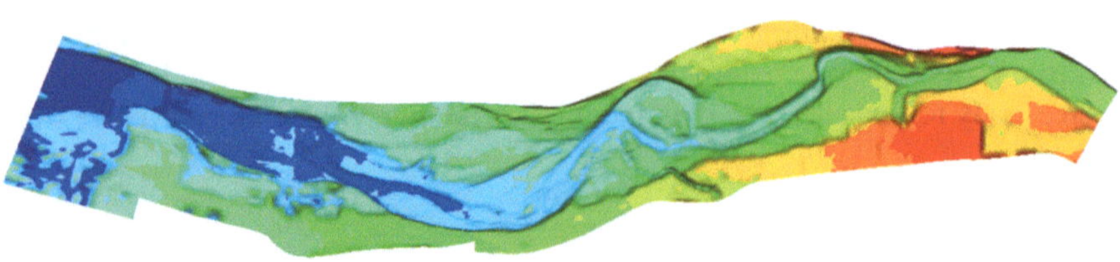

Fig. 9.17 Bed elevation surveyed in April 2004 and used as the initial topography of the numerical model

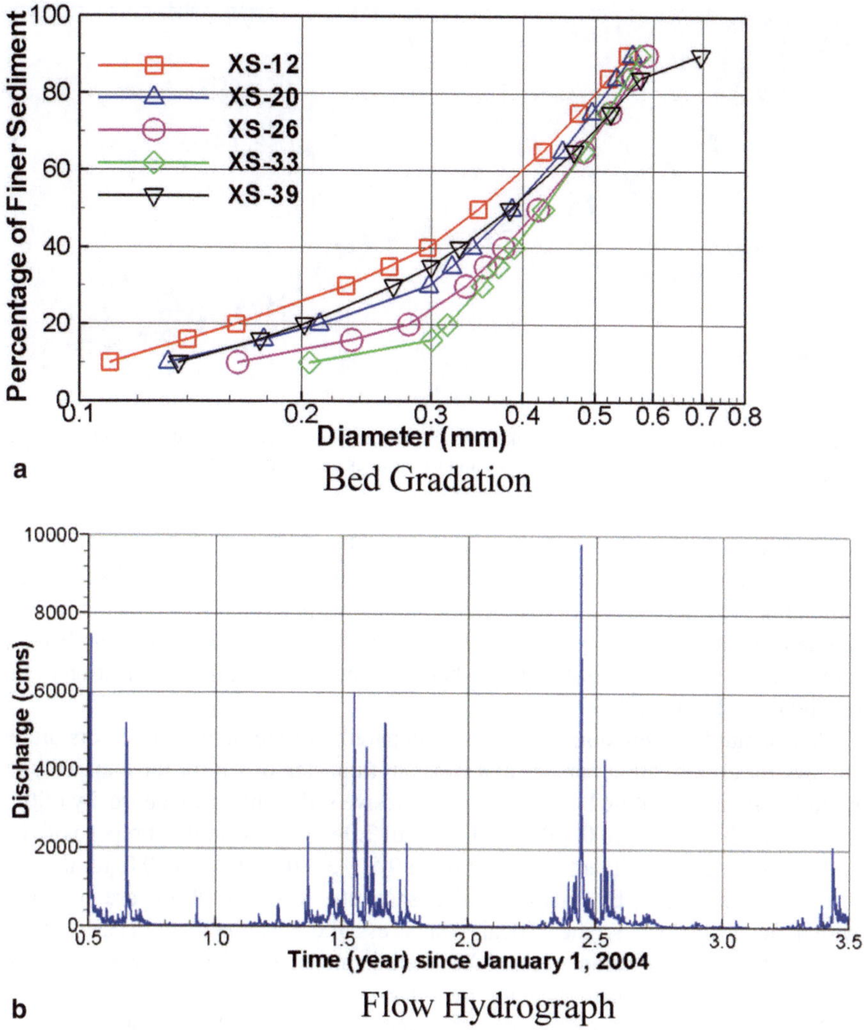

Fig. 9.18 Bed gradation measured in 2010 at five locations and flow hydrograph from July 2004 to August 2007 through the study reach. **a** Bed gradation, **b** flow hydrograph

in the sediment transport equation as it is suitable for sandy rivers. The hourly recorded discharge at the Ziqiang bridge (see Fig. 9.18b) is applied as the upstream boundary condition. In addition, the sediment supply rate at the upstream boundary is estimated by using 50 % of the capacity computed by the Engelund–Hansen equation. No measured sediment flux data are available and the 50 % was calibrated. The stage at the downstream boundary is based on the rating curve developed using the measured data downstream of the Xibin bridge.

Bank retreat modeling is carried out for a bank segment on the right upstream of the Xibin bridge. The bank zone, formed by the bank toe and top bank lines, is

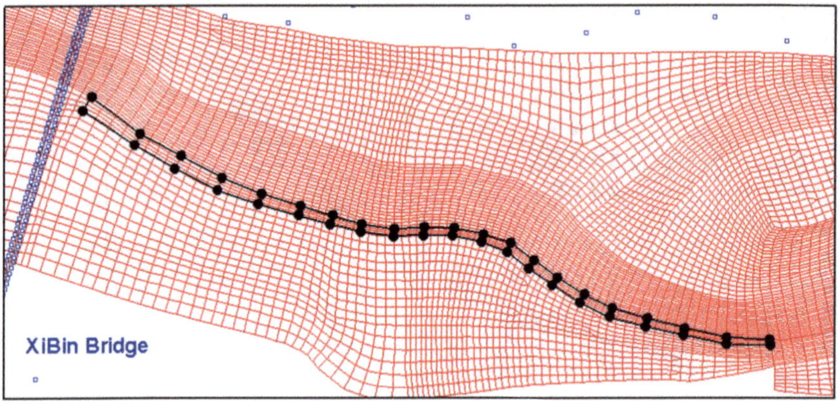

Fig. 9.19 A zoom-in view of the bank zone (*black polygon*) used for retreat modeling; *upper black dots* represent bank toes of all banks and lower ones are the top nodes

represented by a mesh block shown by the black box in Fig. 9.19. Twenty-one bank profiles, shown as black circular symbols in Fig. 9.19, are chosen for bank retreat simulation. During the bank erosion simulation, the materials eroded from the bank are deposited to the nearby main channel while the mesh surrounding the bank is moved and deformed according to the bank retreat (see a detailed description of the procedure in Lai et al. [130]).

A separate bank erosion input file is prepared and the input parameters are as follows for all retreating banks: (a) a constant time step of 1 h for the bank erosion module; (b) all retreating banks have noncohesive sediments with a porosity of 0.3, the critical shear stress of 0.5 Pa, erodibility of 3.0×10^{-5}, the angle of repose of 30°, and the volumetric sediment composition of 0.5, 0.35, 0.15, 0.0, and 0.0, for the five size classes, respectively. Only the critical shear stress and erodibility are important with noncohesive banks. The critical shear stress of 0.5 Pa is estimated based on the 0.045 critical Shields parameter and 0.65 mm diameter. The erodibility parameter is calibrated with the measured bank retreat data.

The predicted change of bed elevation during the period of July 2004 to August 2007 is compared with the measured data in Fig. 9.20. Model results are less satisfactory upstream of Ziqiang bridge. This is expected since the upstream boundary is located within a bend and results near the boundary are subject to high uncertainty due to the boundary conditions applied. Immediately downstream of the Ziqiang bridge, the model predicts the erosion and deposition pattern well, including the correct prediction of the "straightening" trend. Both the first left and right sand bars downstream of Ziqiang are eroded while the opposite main channel experiences deposition. The model predicts the potential erosion in the "Left-Upstream zone" (see Fig. 9.20b for the labels). The predicted amount of erosion though is much less than the measured value due probably to the neglect of bank erosion. It is possible that the area is also impacted by the tributary flow nearby which is not incorporated by the model due to unknown data. Erosion in the "Right-Middle" zone is not pre-

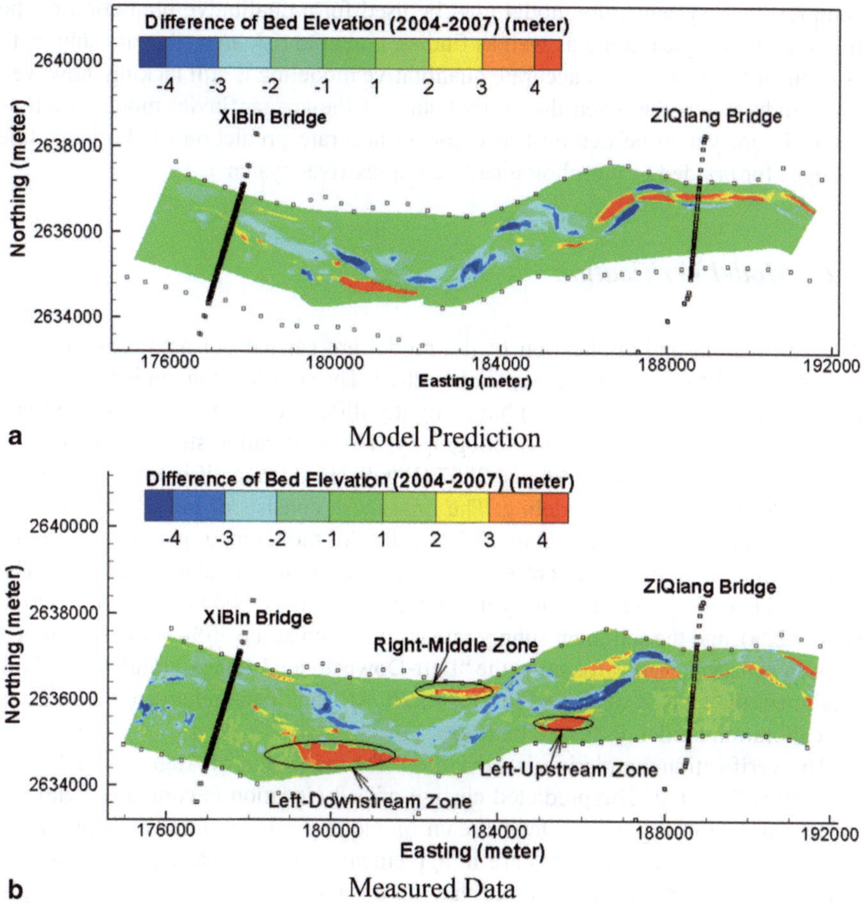

Fig. 9.20 Predicted and measured net erosion (*positive*) and deposition (*negative*) depth from July 2004 to August 2007 with the calibration model. **a** Model prediction, **b** measured data

dicted at all by the model. Attempts have been made to simulate this zone with the bank erosion module added. However, such efforts are not successful since there is not enough water flowing towards the area. The bank erosion experienced in this zone is probably due to undocumented events. Possible causes include: (a) local peculiar topographic features that caused more water directed towards the area; (b) impact of the tributary flow on the opposite side of the bank; and (c) bank failure due to nonfluvial processes. Bank erosion in the "Left-Downstream" zone is explicitly simulated with the bank module. The model results show that the model predicts the bank retreat reasonably in this area although the amount of bank retreat is underpredicted. Uncertainties in the measured topography and bank properties, along with the morphological complexity of the area, contribute to the model uncertainty.

The calibration modeling demonstrates that SRH-2D is capable of predicting the overall erosion and deposition patterns and bank erosion process downstream. For complex river systems, the model may be used for a qualitative prediction of the future morphological trend as well as finding potential risk areas that are subject to erosion and deposition. An accurate quantitative modeling is still lacking, however. This study shows that even the current state-of-the-art geofluvial models such as SRH-2D are yet to be demonstrated for an accurate prediction of the bank line changes for braided or morphologically complex river systems.

6.2 Model Verification

The verification and application of the model are carried out next, covering the time period of August 2007 to September 2010. The solution domain is the same as the calibration model (Fig. 9.16) but a slightly different 2D mesh is generated due to the use of different initial bed topography. The verification study topography is based on the survey data in August 2007 (Fig. 9.21). Also, a different bank zone is needed for bank erosion modeling. The final mesh consists of mixed quadrilateral and triangular cells with a total of 19,395 cells. All model input parameters are the same as those used by the calibration model except for the boundary conditions. The upstream boundary uses the hourly discharge from August 2007 to September 2010 (Fig. 9.22a), and the sediment supply rate is computed as the 50 % of the Engelund–Hansen equation. Bank erosion in the "Left-Downstream" zone is simulated and the mesh near the bank is displayed in Fig. 9.22b. All bank parameters are the same as the calibration model discussed before.

The verification simulation is carried out for the period of August 31 2007, to September 22 2010. The predicted change of bed elevation is compared with the measured data in Fig. 9.23. Downstream of Ziqiang bridge, the overall predicted erosion and deposition patterns are in agreement with the measured data qualitatively. The measured data suggested that there is continued erosion along the right bank during the verification period; this, however, is missed by the numerical model—a phenomenon similar to the calibration modeling. Bank erosion in the "Left-Downstream" zone is predicted by the model, but the amount of bank retreat is way underpredicted.

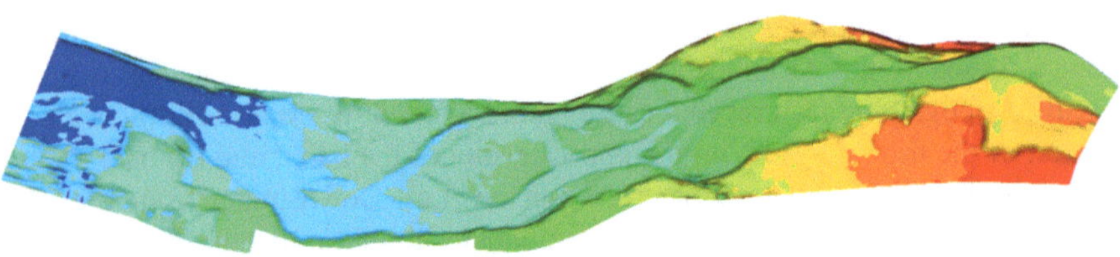

Fig. 9.21 Bed elevation surveyed in August 2007; it is used as the initial topography of the verification model

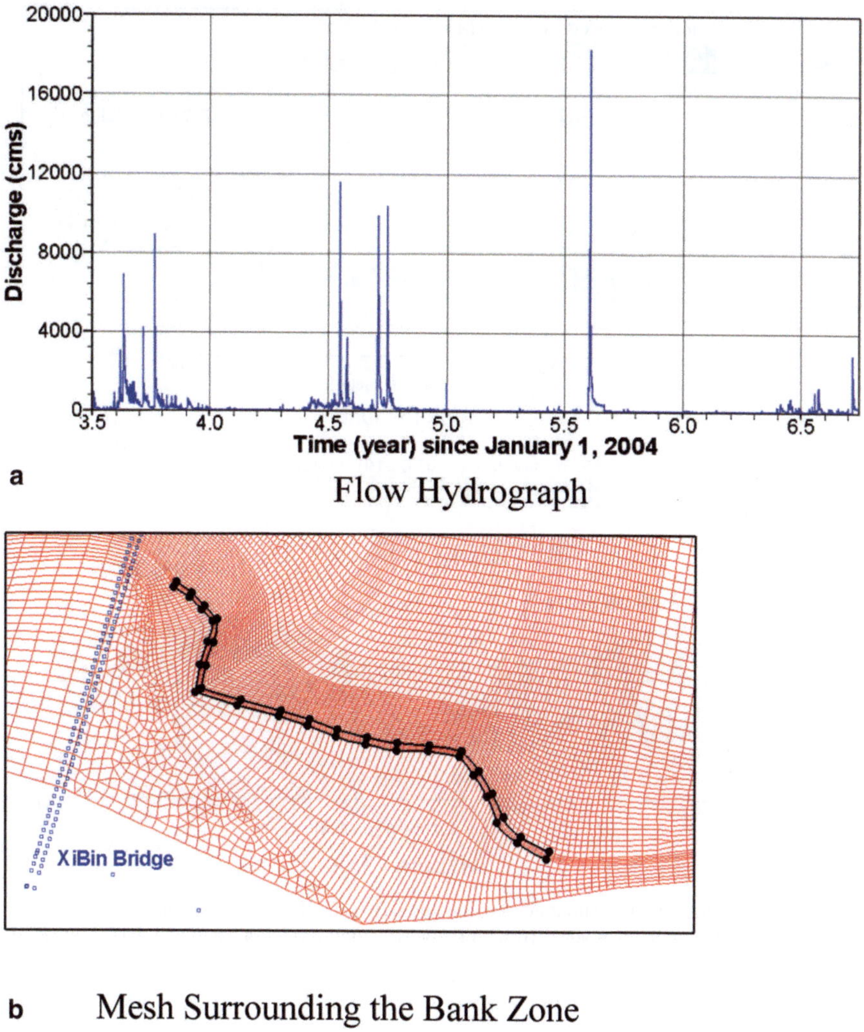

Fig. 9.22 Hydrograph between August 2007 and September 2010 through the study reach and a zoom-in view of the mesh near the bank. **a** Flow hydrograph, **b** mesh surrounding the bank zone

Verification modeling of the Chosui River reach during August 2007 and September 2010 confirms all the major findings of the calibration study. Overall, the geofluvial model is capable of predicting the bed elevation change and bank retreat. Agreement with the measured data, however, is satisfactory only in a quantitative manner.

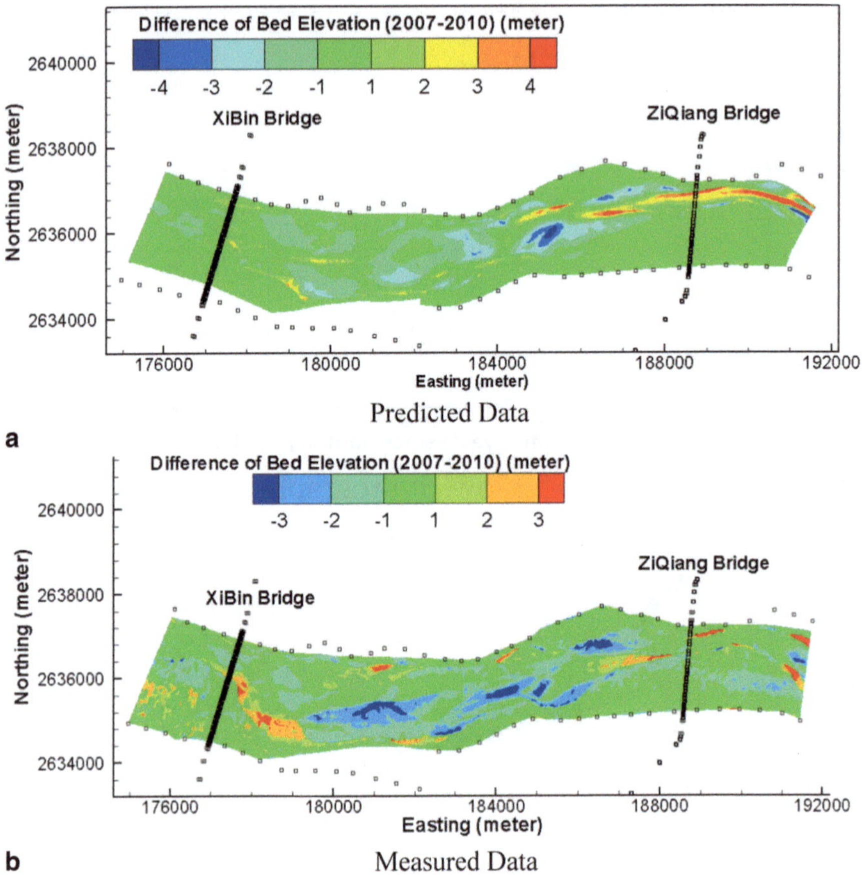

Fig. 9.23 Predicted and measured net erosion (*positive*) and deposition (*negative*) depth (*meter*) from August 2007 to September 2010 with the verification model. **a** Predicted data, **b** measured data

6.3 Summary

The geofluvial model SRH-2D has been applied to compute the river morphological changes of a Chosui River reach. The reach is from the Ziqiang bridge to the Xibin bridge encompassing about 16.7 km in length. The study is used as a test of the model for its ability to simulate complex river systems with rapid lateral shift. The modeling is conducted in two steps: a calibration study over the period of July 2004 to August 2007, and a verification study over the period from August 2007 to September 2010. The study demonstrates that the geofluvial model may be successfully developed and applied to practical rivers for morphological and bank erosion predictions. It is shown that the proposed coupling procedure is flexible and robust; it makes the model setup process general and easy. The study also shows that the coupled model is stable; a minimized set of calibration parameters are needed for practical applications.

Both the calibration and verification studies show that the coupled numerical model is capable of predicting the bed elevation change and bank retreat at the Chosui River reach. Agreement with the measured data is good quantitatively in some areas and only qualitatively in the others. The uncertainty in bed gradation spatial distribution and lack of such data are partially responsible to the less satisfactory model predictions. The study shows that even the current state-of-the-art geofluvial models are not yet able to make accurate quantitative predictions of bank line changes for braided or morphologically dynamic river systems such as the Chosui River. For such complex systems, accurate measured data of bed gradation distribution are important. Without such data, the model may be used for a qualitative prediction of the future morphological trend and for finding potential erosion and deposition trouble spots. The geofluvial model that couples both bank erosion and mobile-bed models is more ideally suited for relative comparison among different engineering plans for river maintenance, flood protection, restoration, and other river projects.

7 Model Application: Trinity River in California

Trinity River is located in northern California. A 42-mile reach of the river has been subject to extensive restoration and rehabilitation effort since a decade ago. This application study is to use the geofluvial model SRH-2D to assist the design of the Upper Junction City (UJC) channel rehabilitation project by predicting potential channel morphological changes by the project implementations. The effort has been reported in details by Lai [130] recently.

7.1 Modeling Scenarios and Model Inputs

A total of three topographic data sets were used for the study. Two sets were surveyed under the existing condition: one in April 2009 (the "pre-erosion" condition) and another in August 2011 (the "post-erosion" condition, PC). A third topographic data set was created representing the project changes to the site according to the design (named the "design construction" (DC) condition).

The modeling was carried out in two steps. First, a calibration was carried out under the existing condition between 2009 pre-erosion and 2011 post-erosion; the scenario was named the 2009 "pre-erosion baseline" (PB) case. Second, the calibrated model was applied to assess the impact of the proposed channel rehabilitation project on channel morphology. Two scenarios were simulated: the 2011 "PC" and the 2012 "DC" condition. The PC runs utilized the 2011 post-erosion terrain as the initial condition while the DC runs started from the "DC" terrain of the river rehabilitation project. Since the DC terrain was close to the 2011 PC and all model parameters were the same, a comparison of the PC and DC modeling results provided the necessary data to assess the impact of the DC on channel morphology.

The solution domain covered about 4000 ft in channel length and an average 700 ft in width. Three meshes were generated corresponding to the three modeling scenarios (PB, PC, and DC). The PB and DC meshes, along with the corresponding terrains, are shown in Figs. 9.24 and 9.25, respectively.

The Manning's coefficient was estimated from the previous studies on the Trinity River: 0.035 in the main channel and bare floodplain and 0.085 for vegetated zones. The bed sediment gradation was based on the survey data. Sediments in the main channel and the bare floodplain had a medium diameter (d_{50}) of about 29 mm, while sediments in the vegetated zones allowed only deposition.

Time-accurate unsteady simulations were carried out for all scenarios using the recorded hydrograph from April 29, 2009, to September 3, 2011 (USGS gaging station 11526250), shown in Fig. 9.26, with three spring runoffs. In addition to the discharge, the upstream sediment supply rates were based on the rating curves developed from the 2006–2007 sediment flux data. The downstream water surface elevation (stage) was based on a stage-discharge rating curve generated using the HEC-RAS model computation.

A total of nine sediment size classes were used to represent bed materials ranging from 0.25 to 181 mm. Each size class is governed by the nonequilibrium transport equation with the sediment erosion rate computed using the Gaeuman et al. equation [131]. Other model inputs included: (a) time step = 5 s; (b) active layer thickness = 0.15 ft (five times d_{50} and 1.5 times d_{90}); and (c) bed-load adaptation length = 80 m.

The final set of input parameters were related to the bank properties. Ten bank locations were surveyed and data were documented by Cardno ENTRIX [132]. In the numerical model, only the left bank segment containing four survey banks (UJC-C to UJC-F) was selected for bank retreat modeling, since survey data analysis from the 2009 to 2011 terrain data showed that there was limited bank movement at other locations. Sixteen banks (cross sections) were used to simulate the bank segment retreat (see Fig. 9.27). An analysis of the survey data showed that the bank segment consisted of essentially uniform and noncohesive materials. Therefore, the uniform retreat module was used in which the key input parameters were critical shear stress and erodibility. Both the critical shear stress and the erodibility coefficient were based on the survey data reported by Cardno ENTRIX [132] and their values may be found in the report by Lai et al. [130]. The bank module time step is 6 h, while the mobile-bed model used the time step of 5 s.

7.2 *Calibration Results*

The PB scenario was developed for calibration study under the existing condition. The measured net erosion and deposition depth between the 2009 pre-erosion and 2011 post-erosion terrains is displayed in Fig. 9.28. The survey data suggested that

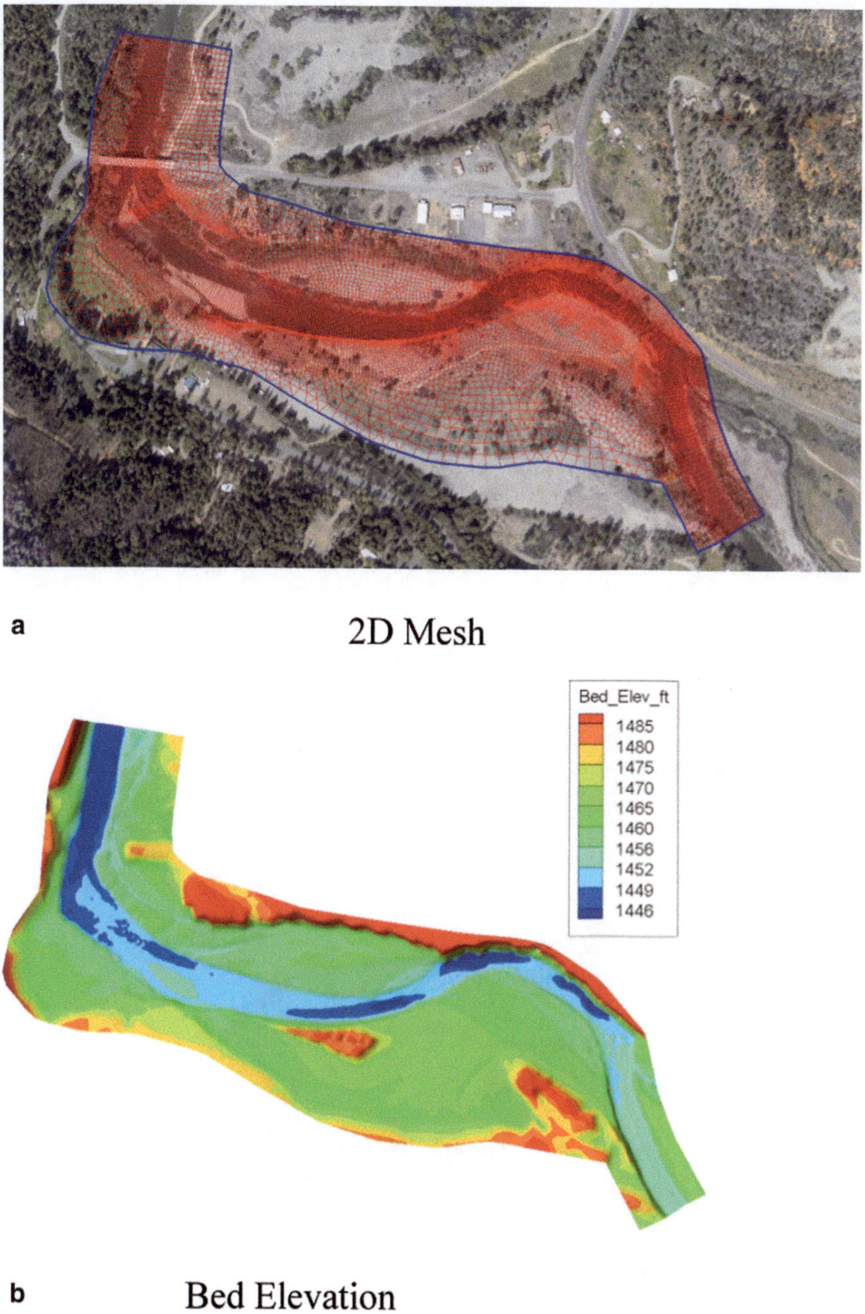

Fig. 9.24 2D mesh for the PB (*pre-erosion baseline*) scenario with the 2009 "pre-erosion" terrain. **a** 2D Mesh, **b** bed elevation

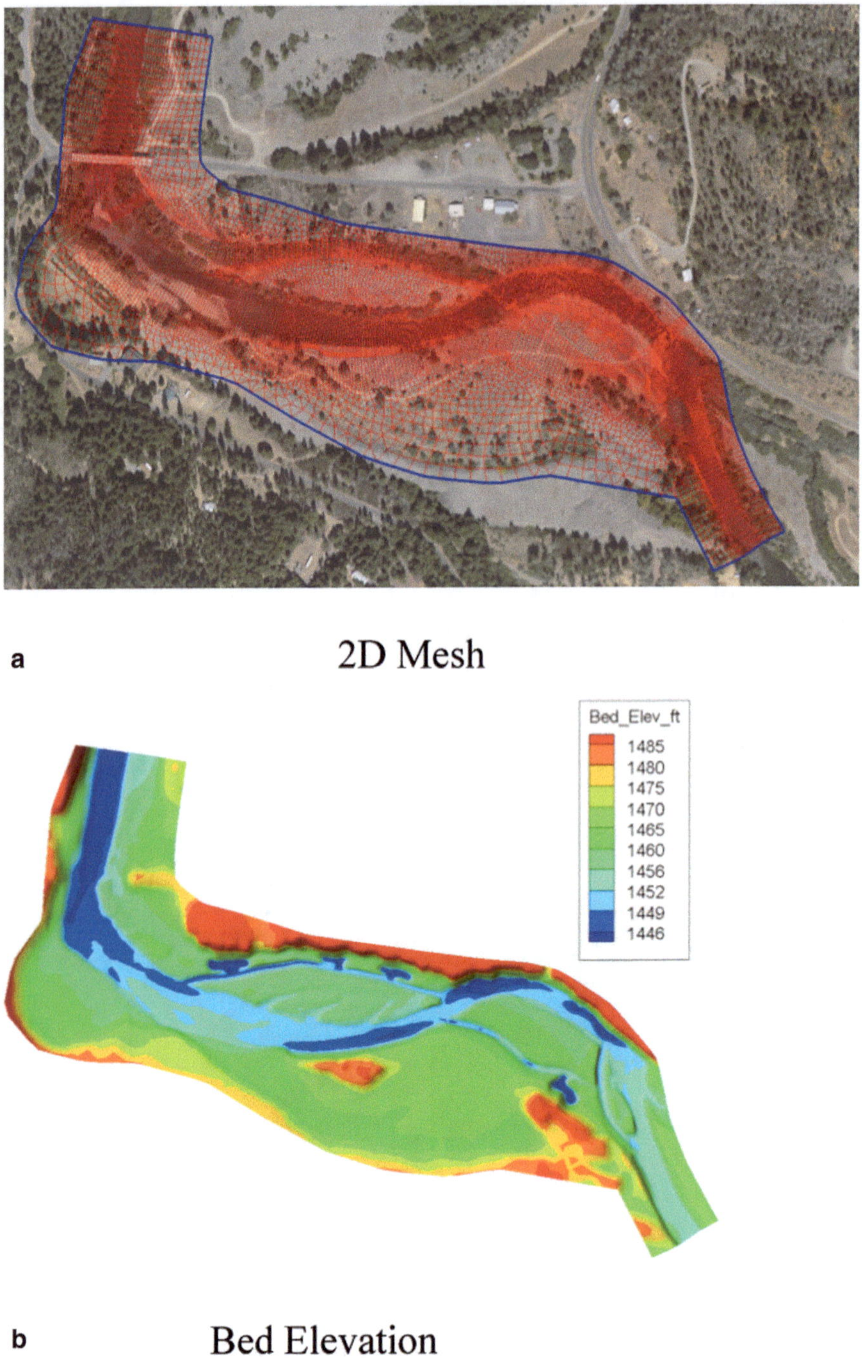

Fig. 9.25 2D mesh for the DC (*design construction*) scenario with the 2012 "design construction" terrain. **a** 2D Mesh, **b** bed elevation

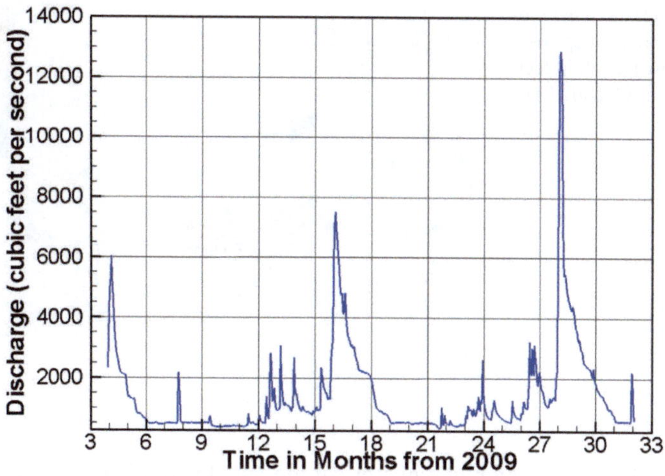

Fig. 9.26 Daily discharge from April 29, 2009, to September 3, 2011, at the UJC site. UJC Upper Junction City

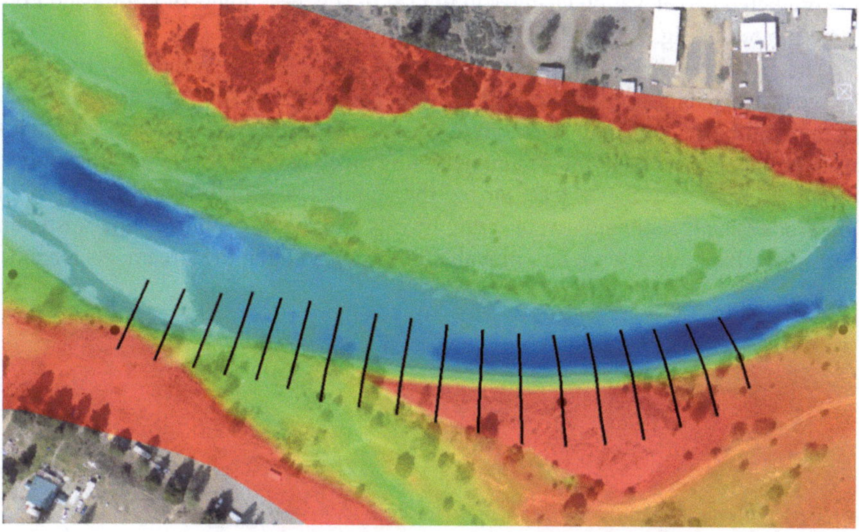

Fig. 9.27 Sixteen banks (*black lines*) simulated for bank retreat

the three pools, marked as "Pool 1," "Pool 2," and "Pool 3," were subject to deposition while a section of the left bank, marked as "Bank Erosion," experienced significant bank erosion. Two areas of comparisons were made between the model predictions and the survey data: (a) ability of the model to simulate the riffle-pool system and (b) the ability of the model to simulate bank retreat. It has been known (e.g., [133]) that the sediment transport and morphologic modeling of riffle-pool

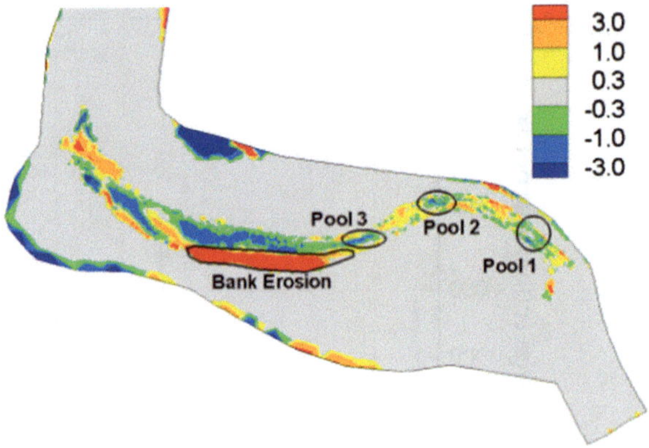

Fig. 9.28 Measured net erosion (*positive*) and deposition (*negative*) depth in feet between the 2009 and 2011 terrains

systems is challenging with any numerical models. In addition, few existing morphological models are available to simulate lateral bank erosions.

The predicted net erosion and deposition are compared with the measured data in Fig. 9.29; zoom-in views of the same plots are displayed in Fig. 9.30. In addition, the predicted bed elevation changes in time at the deepest points of Pool 1 and Pool 2 are plotted in Fig. 9.31.

The model results showed that bank erosion on the left was reasonably predicted by the model using simply the estimated bank parameters despite that no adjustment of the erodibility or critical shear stress was attempted. The eroded bank sediments were mostly deposited in the stream near the eroded bank and were not transported to downstream very far, at least during the simulation period. More than 50% of the eroded bank sediments were large gravels and small cobbles in the upstream part of the bank; they were less movable once deposited in the stream. As a result,

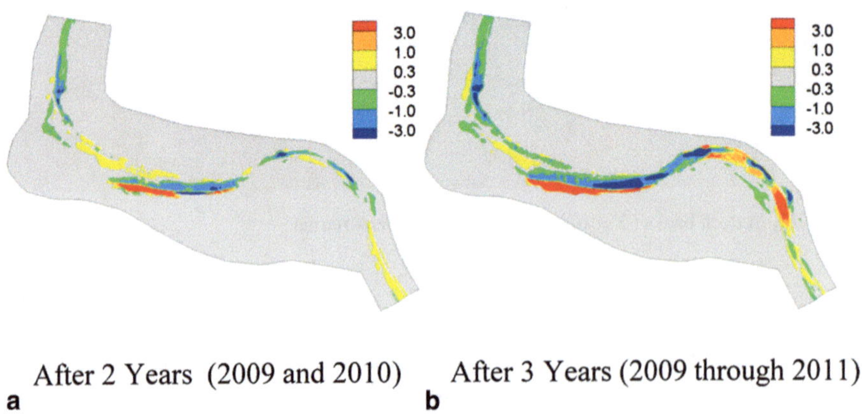

Fig. 9.29 Predicted net erosion (*positive*) and deposition (*negative*) depth in feet with the PB calibration scenario. **a** After 2 years (2009 and 2010), **b** after 3 years (2009 through 2011). *PB* pre-erosion baseline

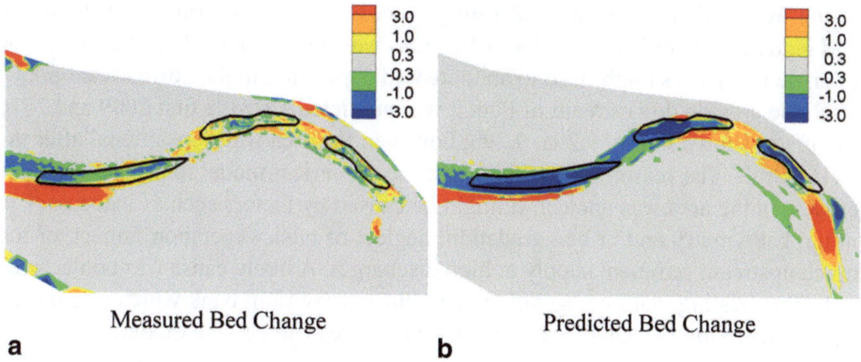

Fig. 9.30 Zoom-in views of the measured and predicted pool-filling after 3-year runoffs with the PB calibration scenario. **a** Measured bed change, **b** predicted bed change. *PB* pre-erosion baseline

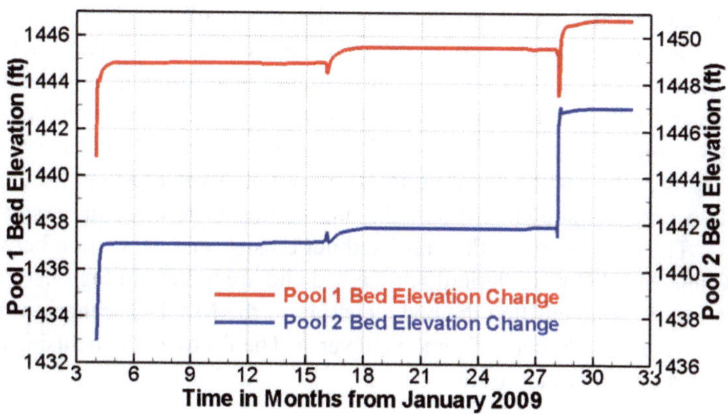

Fig. 9.31 Predicted bed elevation variations in time at the deepest points of Pool 1 and Pool 2 with the PB calibration scenario. *PB* pre-erosion baseline

the impact of the eroded bank sediments on the downstream half of the reach was relatively small.

Upstream of the bank erosion zone, the model predicted erosion and deposition patterns in a qualitative sense. Two main discrepancies were identified between the model prediction and measured data. First, the model predicted much more deposition than the measured data in the three pools. Second, the riffle erosion downstream of Pool 2 was not predicted by the model. The predicted bed elevation changes at the deepest points of the first two pools are plotted in Fig. 9.31. A total of 5 and 8 ft of deposition were predicted in Pool 1 and 2, respectively; the corresponding measured depositions were approximately 2.2 and 3.2 ft. This overprediction of the pool filling process has been a consistent problem with any depth-averaged numerical model since such models do not take the horizontal vortices into consideration. 3D models may be needed to improve the predictions. Other factors might contribute to the overprediction of the pool filling process. A high uncertainty in the initial bed gradation specification of the riffle areas was one of them. The overprediction of erosion at riffles might lead to an increased deposition in the downstream pool.

Riffle erosion downstream of Pool 2 was predicted after the first 2009 and 2010 runoffs as shown in Fig. 9.29a, 9.29b; however, the area was depositional after the 2011 runoff. The reason for the failure of the numerical model to predict the riffle erosion in the area was unclear. It might be caused by factors such as inaccuracy in initial bathymetry and/or bed gradation, neglect of bank vegetation impact, or too much upstream sediment supply at high discharges. A likely cause was conjectured to be the existing mature vegetation along the nearby right bank which was underwater only at high discharges in 2011 but not represented by the model.

7.3 *Morphological Assessment of the Proposed Project*

The calibrated model was applied to assess the impact of the rehabilitation project on the stream morphology; two scenarios were created as discussed before. The two, named MA-PC and MA-DC, used the same model inputs as the calibration model except for the initial terrain. Therefore, the differences between the two were due to the project design only. MA-PC is the run with the 2011 PC scenario while MA-DC run is the 2012 DC scenario. The predicted net erosion and deposition of the two runs are shown in Fig. 9.32 and zoom-in views of the results are in Fig. 9.33. The differences of the predicted net erosion and deposition between the MA-DC and MA-PC were digitally processed and are plotted in Fig. 9.34. Positive depth in Fig. 9.34 means that the bed elevation of the MA-DC scenario was lower than that of the MA-PC scenario and vice versa. The following conclusions may be drawn based on the model results:

- Deposition was predicted in both the left- and right-side channels as seen in Fig. 9.32a. The model results after 2009 and 2010 runoffs are further plotted in Fig. 9.35. A comparison of the predicted deposition in the two side channels between Figs. 9.35 and 9.32 showed that the predicted deposition occurred mainly during the 2011 runoff. The peak discharge is 6040 and 7520 cfs, respectively, in 2009 and 2010, while the peak is 12,900 cfs in 2011. Therefore, the predicted side channel deposition was mainly due to flows when flow was higher than 12,000 cfs.
- The primary side channel deposition that might be of concern for the project was the entrance zone of the downstream right-side channel. An increased deposition risk was expected with increased flows higher than 12,000 cfs.
- The predicted side channel deposition may not be a concern for other locations of the two side channels. For the upstream left-side channel, the entrance to the side channel was predicted to have erosion (Fig. 9.33) if the initial bed materials in the entrance were the same as the main channel ($d_{50}=29$ mm). This entrance erosion may not be a problem since the actual construction used much coarser sediments. Slight deposition was predicted downstream of the entrance zone in

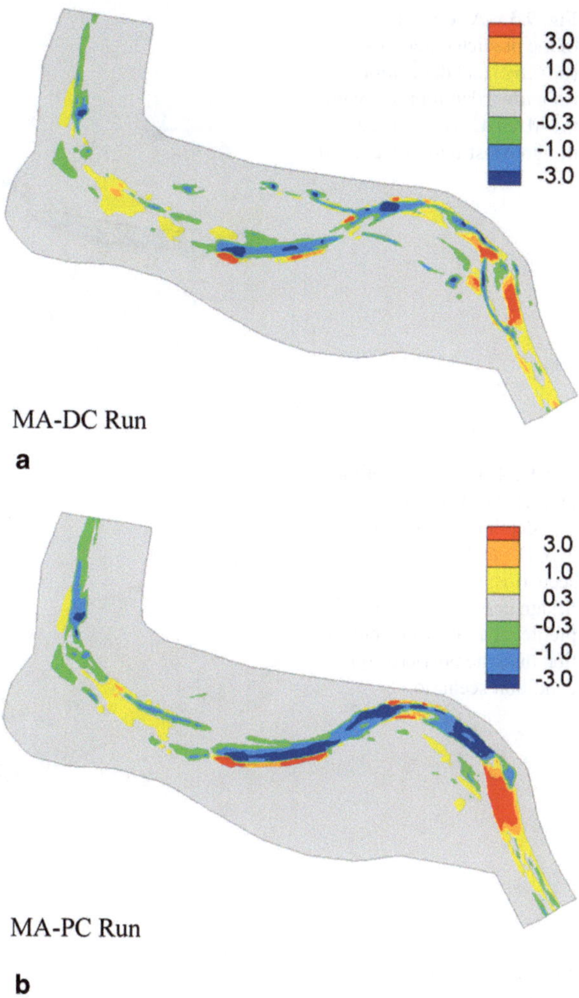

Fig. 9.32 Predicted net erosion (*positive*) and deposition (*negative*) depth in feet with the 2012 design construction and the 2011 post-erosion condition scenarios. **a** MA-DC run, **b** MA-PC run

the left-side channel (including the pool). However, only fine sediments, less than 10 mm, were deposited based on the predicted medium sediment size distribution in Figs. 9.36 and 9.37. For the downstream right-side channel, deposition is limited to the entrance and three side channel pools. Again, only fine sediments were predicted to deposit and they were not considered to be a concern.

- A major potential impact of the DC, based on the model results, is that the main channel downstream of the design island might experience less deposition in some areas and more erosion in others (see Fig. 9.34). However, the left bank zone was not predicted to experience a higher rate of lateral erosion than the 2011 PC due to the DC (compare results in Fig. 9.32). On the contrary, the DC was predicted to lead to slightly less bank erosion. Less deposition in the stream near the bank erosion zone was probably due to less bank erosion

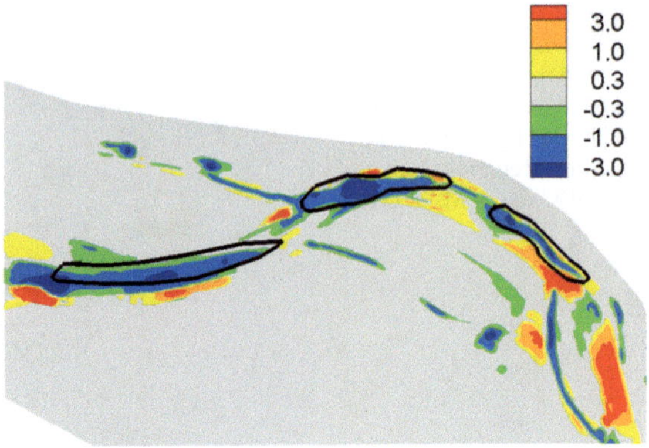

Fig. 9.33 A zoom-in view of the predicted net erosion (*positive*) and deposition (*negative*) depth in feet with the MA-DC scenario (2012 design construction scenario)

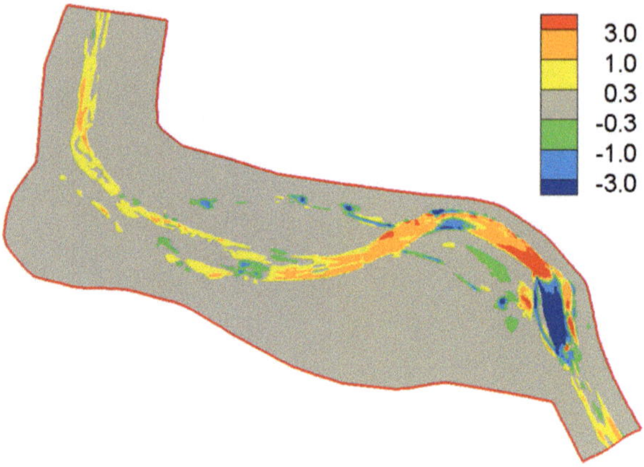

Fig. 9.34 Difference of the predicted erosion and deposition depth in feet between MA-DC and MA-PC scenarios; positive if the design construction scenario predicted a lower bed elevation than the post-erosion condition scenario

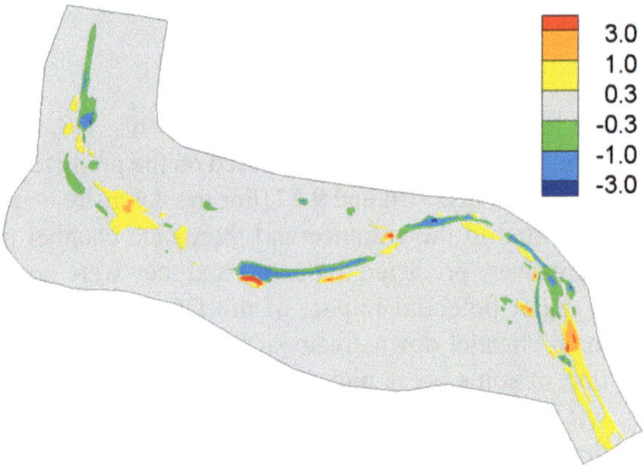

Fig. 9.35 Predicted net erosion (*positive*) and deposition (*negative*) depth in feet with the MA-DC (design construction) scenario after 2009 and 2010 runoffs

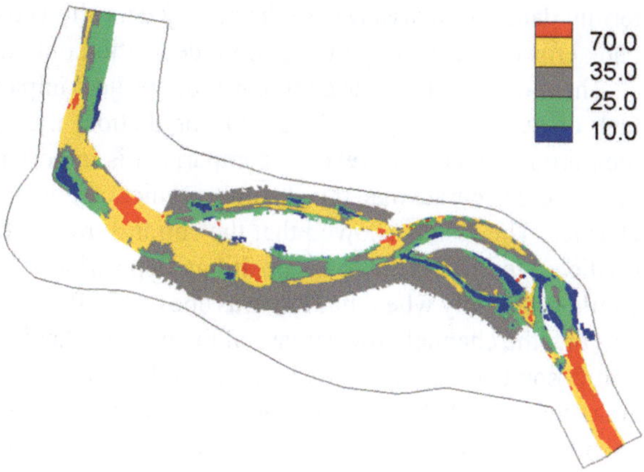

Fig. 9.36 Predicted medium sediment diameter on the streambed in August 2011

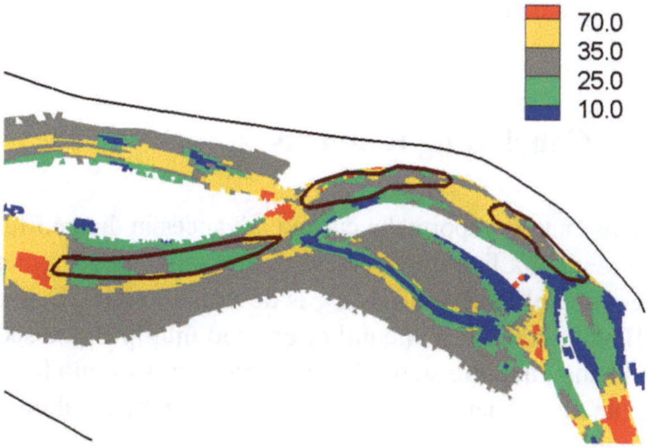

Fig. 9.37 A zoom-in view of the predicted medium sediment diameter on the streambed in August 2011

predicted with the DC scenario. Note that the modeling did not consider other bank sections.

- The model predicted some deposition in the left-split channel of the design island and erosion on the right-split channel. In view that the 2011 post-erosion scenario was predicted to be erosional in the same area, the right-split channel erosion may not be a concern. The left-split channel deposition might be an important risk to consider since the deposited sediment sizes were not small. Based on the sediment size prediction in Fig. 9.36, the deposited sediments in the left-split channel had d_{50} around 15–17 mm. It is possible that the neglect of lower flows (<4000 cfs) contributed to the predicted deposition in the left-split channel.

7.4 Summary

The calibration study of the geofluvial model developed for the UJC site was started from the 2009 "PB" condition and model results were compared with the 2011 terrain data. Comparison with the survey data showed that the model predicted the bank erosion reasonably but the pool deposition only qualitatively.

The model was then used to carry out project impact assessment. Even with the lack of quantitative pool deposition prediction, the model was still adequate for the purpose since only relative comparison is important. Relative comparisons are often more accurate than the absolute prediction with regard to channel morphology changes. The results showed that the entrance region of the right-side channel was predicted to have a risk of deposition. Deposition in this entrance zone, however, would occur only when the flow was above 12,000 cfs. A potential impact of the DC was that the channel downstream of the design island might experience less deposition in some areas and more erosion in others, resulting in deeper channels in most areas than the 2011 post-erosion terrain condition. However, the simulated bank erosion zone was not predicted to experience more erosion under the DC condition; a slight decrease in the bank erosion rate was predicted instead. The left-split channel was predicted to experience deposition with the DC; the deposited sediment size was predicted to be around d_{50} of 15–17 mm.

8 Concluding Remarks

This chapter reports the current advances in the geofluvial modeling. In addition to providing a literature review, the research, development, and application of a new geofluvial model, SRH-2D, is described. SRH-2D is based on 2D depth-averaged flow hydraulics, nonequilibrium and multiple-size sediment transport in the main channel, and the state-of-the-art bank erosion modules. The model was developed with the primary objective of providing a practical tool for engineering applications in alluvial rivers. A version of SRH-2D, with the flow and mobile-bed modeling capability, has already been publicly available at the following website: http://www.usbr.gov/pmts/sediment/model/srh2d/index.html. The SRH-2D model has been widely used for project applications.

A literature review focused on both the empirical models and the process-based models. The review finds that empirical models are relatively simple to use and will continue to be useful for project planning purpose. However, users need to be aware of their applicability ranges. They are useful mostly for the long-term estimate of an equilibrium channel. They are inappropriate for short- and medium-term prediction of geomorphic responses of a stream to disturbances; they are particularly inappropriate for unstable fluvial channels. The process-based bank erosion models hold a better promise to predict unsteady morphological processes and bank line changes. They are more accurate to use to estimate the local retreat of stream banks over a short term and medium term. Frequently, process models are incorporated into analytical or numerical models. Among all process-based models, some are uncoupled to any numerical models in the stream and others are coupled to less general mobile-bed models. The uncoupled models are simple to use and may provide a quick assessment for projects. However, they often need to be calibrated with more parameters with a limited predictive capability. The review shows that a coupled mobile-bed and bank erosion model holds the best promise for a more general assessment and possesses the predicative capability. The coupled model, however, is rarely available, less mature, and more difficult to apply. Their practical use is yet to be demonstrated. Previous studies suggested that good and accurate flow and sediment transport data are necessary to calibrate and verify the process-based bank erosion models.

A new geofluvial model, SRH-2D, is described which incorporates the state-of-the-art methods and algorithms. Two bank erosion modules are described, tested, and verified: the uniform retreat and the mechanistic failure modules. The uniform retreat module treats the entire bank as one unit, whether the bank has a single or multiple layers of materials. An analytical retreat rate was derived incorporating the combined effect of basal erosion and mass failure. The rate equation is general in that both retreat and encroachment are taken into consideration. The mechanistic failure module simulates the basal erosion and mass failure separately. It is applicable to single- or multilayer banks with cohesive or noncohesive layer materials, with both planar (slab) and cantilever failures. The mechanistic failure model, however, requires more input parameters related to the geotechnical properties of the bank. For the coupled bank erosion and mobile-bed modeling to work, a coupling procedure is needed between the bank modules and the mobile-bed module; such a procedure is also described and discussed. Both the moving and the fixed mesh approaches are developed which are easy to set up and provide the needed variable exchange. The procedure has the following features: (a) A 2D mesh is used to represent both bank zones and main channel while a separate bank profile is tracked with its geometry; (b) an arbitrary number of lateral mesh lines in the bank zones may be selected as retreating bank modeling; and (c) a separate input file may be used to run the bank module that specifies for each bank the geometry, materials, and properties to compute basal erosion and geotechnical failure.

The geofluvial model SRH-2D has been tested, validated, verified with both laboratory and field cases. The model has also been applied to practical restoration projects to demonstrate its usage and its effectiveness in assisting project designs.

Acknowledgments The bank erosion module development was sponsored by the Science and Technology Program of Reclamation and the Water Resources Agency of Taiwan. Much of the code within the mechanistic failure module was the result of a collaborative effort with Andrew Simon and Rob Thomas when they were at the USDA-ARS National Sedimentation Laboratory, Oxford, MS, and inspired by the work of Eddy Langendoen at the same laboratory. Technical discussion and peer reviews provided by Blair Greimann and Paul Makar at the US Bureau of Reclamation during the course of the study are greatly appreciated.

Glossary

Alluvial streams Refer to streams in which the bed and banks are made up of mobile sediment and/or soil. As a result, an alluvial stream is subject to continuous changes in planar forms and vertical shapes. They can assume a number of forms, e.g., meandering, braiding, wandering, and straight, based on the properties of their banks, the flows they experience, the local riparian ecology, and the amount, size, and type of sediment that they carry.

Basal cleanout It is the process of removing the mass-failure-produced bank materials that may provide protections of bank toe.

Basal erosion The term refers to direct removal of bank materials laterally by flowing water. In addition to shear force, basal erosion is also dependent on bank properties, bank protective materials, vegetation/debris, and soil characteristics.

Cantilever failure It refers to the collapse of an overhanging bank block into a stream, often due to significant undercutting. The failure tends to occur on banks with composite layers of fine/coarse and/or cohesive/noncohesive materials.

Dry granular flow It refers to a process in which individual sediment grains roll, slide, and bounce down the bank. It typically occurs on noncohesive banks near the angle of repose.

Geofluvial models A geofluvial model refers to one that is capable of coupling in-channel flow and sediment-routing models with bank erosion and mass wasting algorithms.

Geomorphic processes They may encompass a wide range of processes related to landforms and stream forms and the processes that shape them. In this report, they specifically refer to the processes that shape the streams.

Hydraulic fluvial processes Refer to the processes that are responsible for bank retreat due to flowing water.

Mass failure A type of bank failure resulting in a block of the bank sliding or toppling into the stream in a single event.

Mobile-bed models Refer to numerical models that may be used to predict the sediment transport in streams, along with the associated streambed changes in shape.

Piping failure It refers to the collapse of part of a bank due to seepage flows causing selective removal of bank layers. The failure is usually caused by preferential groundwater flows along inter-bedded saturated bank layers; the bank has lenses of noncohesive materials sandwiched between layers of finer cohesive materials.

Planar failure It refers to the sliding and forward toppling of a deep-seated bank mass into the stream. Planar failure occurs often on steep, fine-grained cohesive banks.

Rotational failure It refers to a deep-seated movement of bank material both downward and outward along a curved slip surface. After the failure, the upper surface of the slipped block is typically tilted inward toward the bank.

Shallow slide It refers to a process in which a layer of material moves along a plane parallel to the bank surface. This failure often occurs on banks where soils have low and varied cohesion and the bank is moderately steep.

Wet earthflow It refers to a process where the soil of a bank flows as a highly viscous liquid. The flowing material is extremely weak and easily removed by hydraulic fluvial process, even at lower flows.

References

1. Environment Agency. (1999). *Waterway bank protection: A guide to erosion assessment and management* (p. 235). Bristol: R and D Publication 11, Environment Agency.
2. Florsheim, J. L., Mount, J. F., & Chin, A. (2008). Bank erosion as a desirable attribute of rivers. *BioScience, 58*(6), 519–529.
3. van der Mark, C. F., van der Sligte, R. A. M., Becker, A., Mosselman, E., & Verheij, H. J. (2012). *A method for systematic assessment of the morphodynamic response to removal of bank protection. River Flow 2012*. International Conference on Fluvial Hydraulics, 5–7 Sep, San Jose, Costa Rica, Sep.
4. Lai, Y. G. (2007). *Erosion analysis upstream of the San Cacia diversion dam on the Rio Grande*. Technical Report, Technical Service Center, Bureau of Reclamation, Denver CO.
5. Wilkin, D. C., & Hebel, S. J. (1982). Erosion, redeposition, and delivery of sediment to Midwestern streams. *Water Resources Research, 18*(4), 1278–1282.
6. Simon, A., Rinaldi, M., & Hadish, G. (1996). *Channel evolution in the loess area of the Midwestern United States*. Proceedings, Sixth Federal Interagency Sedimentation Conference, Las Vegas, III-86–93.
7. Kronvang, B., Grant, R., & Laubel, A. L. (1997). Sediment and phosphorus export from a lowland catchment: Quantification of sources. *Water Air and Soil Pollution, 99*(1-4), 465–476.
8. Howard, A., Raine, S. R., & Titmarsh, G. (1998). *The contribution of stream bank erosion to sediment loads in Gowrie Creek, Toowoomba*. ASSI National Soils Conference, Brisbane, Vol. 2004, pp. 491–493.
9. Merritt, W. S., Letcher, R. A., & Jakeman, A. J. (2003). A review of erosion and sediment transport models. *Environmental Modeling and Software, 18*, 761–799.
10. Lai, Y. G. (2013). *A Coupled Stream Bank Erosion and Two-Dimensional Mobile-Bed Model*. Technical Report No. SRH-2013-07, Technical Service Center, Bureau of Reclamation, Denver, CO.
11. Chen, D., & Duan, J. G. (2006). Modeling width adjustment in meandering channels. *Journal of Hydrology, 321*, 59–76.
12. Dunne, T., & Leopold, L. B. (1978). *Water in environmental planning* (p. 818). San Francisco: W.H. Freeman.
13. Montgomery, D. R., & Gran, K. B. (2001). Downstream variations in the width of bedrock channels. *Water Resources Research, 37*(6), 1841–1846.
14. Leopold, L. B., & Maddock, T. (1953). *The hydraulic geometry of stream channels and some physiographic implications*. US Geological Survey Professional Paper 252, US Geological Survey, p. 57.
15. Hey, R. D., & Thorne, C. R. (1986). Stable channels with mobile gravel beds. *Journal of Hydraulic Engineering ASCE, 112*, 671–689.
16. Schumm, S. A. (1968). *River adjustment to altered hydrologic regime-Murumbidgee river and paleochannels, Australia*. U.S. Geological Survey Professional Paper 598.
17. Ferguson, R. I. (1973). Channel pattern and sediment type. *Area, 5*, 38–41.
18. Prosser, I., Rustomji, P., Young, B., Moran, C., & Hughes, A. (2001). *Constructing river basin sediment budgets for the national land and water resources audit*. CSIRO Land and Water Technical Report Number 15/01, CSIRO Land and Water, Canberra.
19. Dickinson, W. T., Rudra, R. P., & Wall, G. J. (1989). Nomographs and software for field and bank erosion. *Journal of Soil and Water Conservation, 44*, 596–600.
20. Hooke, J. M. (1980). Magnitude and distribution of rates of river bank erosion. *Earth Surface Processes and Landforms, 5*, 143–157.
21. Nanson, G. C., & Hickin, E. J. (1983). Channel migration and incision on the Beatton River. *Journal of Hydraulic Engineering ASCE, 109*, 227–327.
22. Task Committee, A. S. C. E. (1998). River width adjustment. I: Process and mechanism. *Journal of Hydraulic Engineering ASCE, 124*(9), 881–902.
23. Rosgen, D. L. (1996). *Applied river morphology*. Pagosa Springs: Wildland Hydrology Inc.

24. Rosgen, D. L. (2001). *A practical method of computing streambank erosion rate*. Proceedings of the Seventh Federal Interagency Sedimentation Conference, 25–29 March 2001, Reno. Nevada, 2, 9–15.
25. Larsen, E. W. (1995). *Mechanics and modeling of river meander migration*. PhD Dissertation. Berkeley, CA, University of California.
26. Simon, A., Doyle, M., Kondolf, M., Shields, F. D. Jr., Rhoads, B., & McPhillips, M. (2007). Critical evaluation of how the Rosgen classification and associated "Natural Channel Design" methods fail to integrate and quantify fluvial processes and channel response. *Journal of the American Water Resources Association, 43*(5), 1117–1131.
27. Yang, C. T. (1976). Minimum unit stream power and fluvial hydraulics. *Journal of Hydraulic Engineering ASCE, 102,* 769–784.
28. Chang, H. H. (1979). Minimum stream power and river channel patterns. *Journal of Hydrology, 41,* 301–327.
29. Kirkby, M. J. (1977). Maximum sediment efficiency as a criterion for alluvial channels. In K. J. Gregory (Ed.), *River channel changes* (pp. 429–442). Chichester: Wiley.
30. White, W. R., Bettess, R., & Paris, E. (1982). Analytical approach to river regime. *Journal of Hydraulic Engineering ASCE, 108,* 1179–1193.
31. Millar, R. G., & Quick, M. C. (1993). Effect of bank stability on geometry of gravel rivers. *Journal of Hydraulic Engineering ASCE, 119*(12), 1343–1363.
32. Millar, R. G., & Quick, M. C. (1998). Stable width and depth of gravel bed rivers with cohesive banks. *Journal of Hydraulic Engineering ASCE, 124*(10), 1005–1013.
33. Williams, G. P. (1978). *Hydraulic geometry of river cross sections-theory of minimum variance*. Professional Paper 1029. US Geological Survey. pp. 47.
34. Huang, H. Q., & Nanson, G. C. (2000). Hydraulic geometry and maximum flow efficiency as products of the principle of least action. *Earth Surface Processes and Landforms, 25,* 1–16.
35. Eaton, B. C., Church, M., & Millar, R. G. (2004). Rational regime model of alluvial channel morphology and response. *Earth Surface Processes and Landforms, 29,* 511–529.
36. Glover, R. E., & Florey, Q. L. (1951). *Stable channel profiles* (Vol. 235). Denver: US Bureau of Reclamation.
37. Lane, E. W. (1955). Design of stable channels. *Transmission of the ASCE, 120,* 1234–1260.
38. Parker, G. (1978). Self-formed straight rivers with equilibrium banks and mobile bed. Part 2. The gravel river. *Journal of Fluid Mechanics, 89*(1), 127–146.
39. Ikeda, S., Parker, G., & Kimura, Y. (1988). Stable width and depth of straight gravel rivers with heterogeneous bed materials. *Water Resources Research, 24*(5), 713–722.
40. Ikeda, S., & Izumi, N. (1990). Width and depth of self-formed straight gravel rivers with bank vegetation. *Water Resources Research, 26*(10), 2353–2364.
41. Kovacs, A., & Parker, G. (1994). A new vectorial bedload formulation and its application to the time evolution of straight river channels. *Journal of Fluid Mechanics, 267,* 153–183.
42. Vigilar, G. G., & Diplas, P. (1997). Stable channels with mobile bed: Formulation and numerical solution. *Journal of Hydraulic Engineering ASCE, 123*(3), 189–199.
43. Sun, T., Meakin, P., JØssang, T., & Schwarz, K. (1996). A simulation model for meandering rivers. *Water Resources Research, 32,* 2937–2954.
44. ASCE Task Committee on Hydraulics, Bank Mechanics and Modeling of River Width Adjustment (1998). River width adjustment, II: Modeling. *Journal of Hydraulic Engineering ASCE, 124*(9), 903–917.
45. Rinaldi, M., & Darby, S. E. (2008). Modelling river-bank-erosion processes and mass failure mechanisms: Progress towards fully coupled simulations. In H. Habersack, H. Piégay, & M. Rinaldi (Eds.), *Gravel-bed rivers 6: From process understanding to river restoration; developments in earth surface processes 11* (pp. 213–239). Amsterdam: Elsevier.
46. Langendoen, E. J., & Simon, A. (2008). Modeling the evolution of incised streams. II: Streambank erosion. *Journal of Hydraulic Engineering ASCE, 134*(7), 905–915.
47. Motta, D., Abad, J. D., Langendoen, E. J., & Garcia, M. H. (2012). A simplified 2D model for meander migration with physically-based bank evolution. *Geomorphology, 163–164,* 10–25.

48. Hasegawa, K. (1977). *Computer simulation of the gradual migration of meandering channels*. Proceedings of the Hokkaido Branch. Japan Society of Civil Engineering, pp. 197–202 (in Japanese).
49. Ikeda, S., Parker, G., & Sawai, K. (1981). Bend theory of river meanders. Part I. Linear development. *Journal of Fluid Mechanics, 112,* 363–377.
50. Parker, G. (1983). Theory of meander bend deformation. In C. M. Elliott (Ed.), *River Meandering Proceedings of ASCE River '83 Conference*, ASCE, pp. 722–731.
51. Blondeaux, P., & Seminara, G. (1985). A unified bar-bend theory of river meanders. *Journal of Fluid Mechanics, 157,* 449–470.
52. Johannesson, H., & Parker, G. (1989). Velocity redistribution in meandering rivers. *Journal of Hydraulic Engineering ASCE, 115*(8), 1019–1039.
53. Odgaard, A. J. (1989). River meander model I development. *Journal of Hydraulic Engineering ASCE, 115*(11), 1433–1450.
54. Sun, T., Meakin, P., & JØssang, T. (2001a). Meander migration and the lateral tilting of floodplains. *Water Resources Research, 37*(5), 1485–1502.
55. Sun, T., Meakin, P., & JØssang, T. (2001b). A computer model for meandering rivers with multiple bed load sediment sizes I theory. *Water Resources Research, 37*(8), 2227–2241.
56. Zolezzi, G., & Seminara, G. (2001). Downstream and upstream influence in river meandering. Part 1. General theory and application to overdeepening. *Journal of Fluid Mechanics, 438,* 183–211.
57. Lancaster, S. T., & Bras, R. L. (2002). A simple model of river meandering and its comparison to natural channels. *Hydrological Processes, 16*(1), 1–26.
58. Rüther, N., & Olsen, N. R. B. (2007). Modelling free-forming meander evolution in a laboratory channel using three-dimensional computational fluid dynamics. *Geomorphology, 89*(3-4), 308–319.
59. Crosato, A. (1989). *Meander migration prediction. Excerpta, GNI, 4. Libreria Progetto* (Vol. 4, pp. 169–198). Padova: GNI, Gruppo Nazionale Idraulica.
60. Mosselman, E. (1998). *Mathematical modeling of morphological processes in rivers with erodible cohesive banks. Communications on Hydraulic and Geotechnical Engineering, 92.* Delft: Delft University of Technology.
61. Spruyt, A., van der Sligte, R. A. M., van der Mark, C. F., Sieben, A., & Mosselman, E. (2012). Development of a rapid assessment tool for river bank erosion. River Flow 2012, International Conference on Fluvial Hydraulics, Sept. 5-7, 2012, San Jose, Costa Rica.
62. Osman, M. A., & Thorne, C. R. (1988). Riverbank stability analysis: I. Theory. *Journal of Hydraulic Engineering ASCE, 114*(2), 134–150.
63. Thorne, C. R., & Abt, S. R. (1993). Analysis of riverbank stability due to toe scour and lateral erosion. *Earth surface Processes and Landforms, 18,* 835–843.
64. Darby, S. E., & Thorne, C. R. (1996a). Numerical simulation of widening and bed deformation of straight sand-bed rivers, I: Model development. *Journal of Hydraulic Engineering ASCE, 122,* 184–193.
65. Darby, S. E., & Thorne, C. R. (1996b). Stability analysis for steep, eroding, cohesive riverbanks. *Journal of Hydraulic Engineering ASCE, 122,* 443–454.
66. Darby, S. E., Gessler, D., & Thorne, C. R. (2000). Computer program for stability analysis of steep, cohesive riverbanks. *Earth Surface Process and Landforms, 25*(2), 175–190.
67. Simon, A., Curini, A., Darby, S. E., & Langendoen, E. J. (1999). Streambank mechanics and the role of bank and near-bank processes in incised channels. In S. E. Darby & A. Simon (Eds.), *Incised river channels* (pp. 123–152). Chichester: Wiley.
68. Rinaldi, M., & Casagli, N. (1999). Stability of streambanks formed in partially saturated soils and effects of negative pore water pressures: The Sieve River (Italy). *Geomorphology, 26*(4), 253–277.
69. Dapporto, S., Rinaldi, M., Casagli, N., & Vannocci, P. (2003). Mechanisms of riverbank failure along the Arno River, Central Italy. *Earth Surface Process and Landforms, 28,* 1303–1323.

70. Pollen, N., & Simon, A. (2005). Estimating the mechanical effects of riparian vegetation on stream bank stability using a fiber bundle model. *Water Resources Research, 41*(7), W0702510. doi:10.1029/2004WR003801.
71. Simon, A., Curini, A., Darby, S. E., & Langendoen, E. J. (2000). Bank and near-bank processes in an incised channel. *Geomorphology, 35*(3-4), 193–217.
72. Langendoen, E. J. (2000). *CONCEPTS—Conservational channel evolution and pollutant transport system*. Research Report 16, USDA Agricultural Research Service National Sedimentation Laboratory, Oxford, MS.
73. Langendoen, E. J., Wells, R. R., Thomas, R. E., Simon, A., & Bingner, R. L. (2009). Modeling the evolution of incised streams. III: Model application. *Journal of Hydraulic Engineering ASCE, 135*(6), 476–486.
74. Darby, S. E., Rinaldi, M., & Dapporto, S. (2007). Coupled simulations of fluvial erosion and mass wasting for cohesive river banks. *Journal of Geophysical Research, 112*(F03022). doi:10.1029/2006JF000722.
75. Rinaldi, M., & Darby, S. E. (2008). Modelling river-bank-erosion processes and mass failure mechanisms: Progress towards fully coupled simulations. In H. Habersack, H. Piégay, & M. Rinaldi (Eds.), *Gravel-bed Rivers 6: From process understanding to river restoration; developments in earth surface processes 11* (pp. 213–239). Amsterdam: Elsevier.
76. Rinaldi, M., Mengoni, B., Luppi, L., Darby, S. E., & Mosselman, E. (2008). Numerical simulation of hydrodynamics and bank erosion in a river bend. *Water Resources Research, 44*(W09428). doi:10.1029/2008WR007008.
77. Luppi, L., Rinaldi, M., Teruggi, L. B., Darby, S. E., & Nardi, L. (2009). Monitoring and numerical modelling of riverbank erosion processes: A case study along the Cecina River (Central Italy). *Earth Surface Processes and Landforms, 34,* 530–546. doi:10.1002/esp.1754.
78. Darby, S. E., Alabyan, A. M., & van de Wiel, M. J. (2002). Numerical simulation of bank erosion and channel migration in meandering rivers. *Water Resources Research, 38,* 1163–1183.
79. Nagata, N., Hosoda, T., & Muramoto, Y. (2000). Numerical analysis of river channel processes with bank erosion. *Journal of Hydraulic Engineering ASCE, 126*(4), 243–252.
80. Duan, G., Wang, S. S. Y., & Jia, Y. (2001). The applications of the enhanced CCHE2D model to study the alluvial channel migration processes. *Journal of Hydraulic Research, IAHR, 39,* 469–480.
81. Simon, A., Pollen-Bankhead, N., & Thomas, R. E. (2011). Development and application of a deterministic bank stability and toe erosion model for stream restoration. In *Stream restoration in dynamic fluvial systems: Scientific approaches, analyses, and tools*. American Geophysical Union, Geophysical Monograph Series 194 (pp. 453–474).
82. Darby, S. E., Spyropoulos, M., Bressloff, & Rinaldi, M. (2004). *Fluvial bank erosion in meanders: A CFD modelling approach.* In Proceedings 5th International Symposium on Ecohydraulics, Madrid, Spain 12–17 Sept 2004, International Association of Hydraulic Engineering and Research, pp. 268–273.
83. Olsen, N. R. B. (2003). Three-dimensional CFD modeling of self-forming meandering channel. *Journal of Hydraulic Engineering ASCE, 129*(5), 366–372.
84. Lai, Y. G., Thomas, R., Ozeren, Y., Simon, A., Greimann, B. P., & Wu, K. (2014). Modeling of multi-layer cohesive bank erosion with a coupled bank stability and mobile-bed model. Under review by *Geomorphology.*
85. Lai, Y. G., & Wu, K. (2013). *Modeling of vertical and lateral erosion on the Chosui River, Taiwan*. ASCE World Environmental and Water Resources Congress, Cincinnati, Ohio, 19–23 May.
86. Lai, Y. G., Thomas, R., Ozeren, Y., Simon, A., Greimann, B. P., & Wu, K. (2012). *Coupling a two-dimensional model with a deterministic bank stability model*. Albuquerque, New Mexico: ASCE World Environmental and Water Resources Congress, May 20–24.
87. Lai, Y. G., Greimann, B. P., & Wu, K. (2012). *Modeling bank erosion in fluvial channels*. Costa Rica: 2012 River Flow, Sept 3–8.
88. Yang, C. T. (2014). Sediment transport, river morphology, and river engineering. In L. K. Wang & C. T. Yang (Eds.), *Modern water resources engineering: Handbook of environmental engineering* (Vol. 15, pp. 339–372, Chapter 6). NYC: Humana Press.

89. Watson, A. J., & Basher, L. R. (2006). *Stream bank erosion: A review of processes of bank failure, measurement and assessment techniques, and modeling approaches*. Landcare ICM Report No. 2005-2006/01, Landcare Research, Lincoln, Private Bag 6, Nelson, New Zealand.
90. Thorne, C. R., & Tovey, N. K. (1981). Stability of composite river banks. *Earth Surface Processes and Landforms, 6,* 469–484.
91. Lawler, D. M., Thorne, C. R., & Hooke, J. M. (1997). Bank erosion and instability. In C. R. Thorne, R. D. Hey, & M. D. Newson (Eds.), *Applied fluvial geomorphology for river engineering and management* (pp. 137–172). Chichester: Wiley.
92. Fox, G. A., & Wilson, G. V. (2010). The role of subsurface flow in hillslope and streambank erosion: A review of status and research needs. *Soil Science Society of America Journal, 74*(3), 717–733.
93. Thorne, C. R. (1982). Processes and mechanisms of river bank erosion. In R. D. Hey, J. C. Bathurst & C. R. Thorne (Eds.), *Gravel-bed rivers* (pp. 227–271). Chichester: Wiley.
94. Thorne, C. R., & Osman, A. M. (1988). Riverbank stability analysis, II: Applications. *Journal of Hydraulic Engineering ASCE, 114*(2), 151–172.
95. Carson, M. A., & Kirkby, M. J. (1972). *Hillslope form and process*. Cambridge: Cambridge University Press.
96. Thorne, C. R. (1978). *Processes of bank erosion in river channels*. Ph.D. thesis, University of East Anglia, UK.
97. Lai, Y. G. (2014). *SRH-2D manual: Mobile-bed modeling*. Denver: Technical Service Center, Bureau of Reclamation.
98. Lai, Y. G., & Randle, T. J. (2007). *Bed evolution and bank erosion analysis of the Palo Verde Dam on the Lower Colorado River*. Technical Report, Technical Service Center, Bureau of Reclamation, Denver, CO.
99. Greimann, B. P., Lai, Y. G., & Huang, J. (2008). Two-dimensional total sediment load model equations. *Journal of Hydraulic Engineering ASCE, 134*(8), 1142–1146.
100. Lai, Y. G., & Greimann, B. P. (2008). *Modeling of erosion and deposition at meandering channels*. World Environmental & Water Resources Congress, 12–16 May 2008, Honolulu, Hawaii.
101. Lai, Y. G., & Greimann, B. P. (2010). Predicting contraction scour with a two-dimensional depth-averaged model. *Journal of Hydraulic Research, IAHR, 48*(3), 383–387.
102. Lai, Y. G. (2011). *Prediction of channel morphology upstream of Elephant Butte Reservoir on the Middle Rio Grande*. Technical Report No. SRH-2011-04. Technical Service Center, Bureau of Reclamation, Denver, CO.
103. Lai, Y. G., Greimann, B. P., & Wu, K. (2011). Soft bedrock erosion modeling with a two-dimensional depth-averaged model. *Journal of Hydraulic Engineering ASCE, 137*(8), 804–814.
104. Lai, Y. G. (2008). *SRH-2D theory and user's manual version 2.0*. Denver: Technical Service Center, Bureau of Reclamation.
105. Lai, Y. G. (2010). Two-dimensional depth-averaged flow modeling with an unstructured hybrid mesh. *Journal of Hydraulic Engineering ASCEI, 136*(1), 12–23.
106. Lai, Y. G., Weber, L. J., & Patel, V. C. (2003). Non-hydrostatic three-dimensional method for hydraulic flow simulation—Part I: Formulation and verification. *Journal of Hydraulic Engineering ASCE, 129*(3), 196–205.
107. Hasegawa, K. (1981). Bank-erosion discharge based on a nonequilibrium theory. *Proceeding of the JSCE, Tokyo, 316,* 37–50 (in Japanese).
108. Partheniades, E. (1965). Erosion and deposition of cohesive soils. *Journal of the Hydraulics Division ASCE, 91*(1), 105–139.
109. Ariathurai, R., & Arulanandan, K. (1978). Erosion rates of cohesive soils. *Journal of Hydraulic Engineering ASCE, 104*(2), 279–283.
110. Guo, J., & Julien, P. Y. (2005). Shear stress in smooth rectangular open-channel flows. *Journal of Hydraulic Engineering ASCE, 131*(1), 30–37.
111. Kean, J. W., & Smith, J. D. (2006a). Form drag in rivers due to small-scale natural topographic features: 1. Regular sequences. *Journal of Geophysical Research, 111*(F04009). doi:10.1029/2006JF000467.

112. Kean, J. W., & Smith, J. D. (2006b). Form drag in rivers due to small-scale natural topographic features: 2. Irregular sequences. *Journal of Geophysical Research, 111*(F04010). doi:10.1029/2006JF000490.
113. Khodashenas, S. R., El Kadi Abderrezzak, K., & Paquier, A. (2008). Boundary shear stress in open channel flow: A comparison among six methods. *Journal of Hydraulic Research, 46*(5), 598–609.
114. Kean, J. W., Kuhnle, R. A., Smith, J. D., Alonso, C. V., & Langendoen, E. J. (2009). Test of a method to calculate near-bank velocity and boundary shear stress. *Journal of Hydraulic Engineering ASCE, 135*(7), 588–601.
115. Fredlund, D. G., Morgenstern, N. R., & Widger, R. A. (1978). The shear strength of unsaturated soils. *Canadian Geotechnical Journal, 15,* 313–321.
116. Fredlund, D. G., & Rahardjo, H. (1993). *Soil mechanics of unsaturated soils*. New York: Wiley.
117. Huang, Y. H. (1983). *Stability analysis of earth slopes*. New York: Van Nostrand Reinhold Company.
118. Pizzuto, J. E. (1990). Numerical simulation of gravel river widening. *Water Resources Research, 26,* 1971–1980.
119. Darby, S. E., & Delbono, I. (2002). A model of equilibrium bed topography for meander bends with erodible banks. *Earth Surface Process and Landforms, 27*(10), 1057–1085.
120. Lai, Y. G., & Przekwas, A. J. (1994). A finite-volume method for fluid flow simulations with moving boundaries. *Computational Fluid Dynamics, 2,* 19–40.
121. Kassem, A. A., & Chaudhry, M. H. (1998). Comparison of coupled and semicoupled numerical models for alluvial channels. *Journal of Hydraulic Engineering ASCE, 124*(8), 794–802.
122. Engelund, F., & Hansen, E. (1972). *A monograph on sediment transport in alluvial streams*. Copenhagen: Teknish Forlag, Technical Press.
123. Phillips, B. C., & Sutherland, A. J. (1989). Spatial lag effects in bedload sediment transport. *Journal of Hydraulic Research, 27*(1), 115–133.
124. Grissinger, E. H., & Murphey, J. B. (1983). Present channel stability and late quaternary valley deposits in northern Mississippi. In J. D. Collingson & J. Lewin (Eds.), *Modern and Ancient Fluvial Systems* (pp. 241–250). Oxford: Blackwell Scientific.
125. Wilcock, P. R., & Crowe, J. C. (2003). Surface-based transport model for mixed-size sediment. *Journal of Hydraulic Engineering ASCE, 129*(2), 120–128.
126. Hanson, G. J. (1990). Surface erodibility of earthen channels at high stress, Part II—developing an in situ testing device. *Transactions of the American Society of Agricultural Engineers, 33*(1), 132–137.
127. Hanson, G. J., & Simon, A. (2001). Erodibility of cohesive streambeds in the loess area of the midwestern USA. *Hydrological Processes, 15*(1), 23–38.
128. Cancienne, R. M., Fox, G., & Simon, A. (2008). Influence of seepage undercutting on the stability of root-reinforced streambanks. *Earth Surface Processes and Landforms, 33*(11), 1769–1786.
129. Langendoen, E. J. (2010). Assessing post-dam removal sediment dynamics using the CONCEPTS computer model. In Proceedings of the 2nd Joint Federal Interagency Conference, Las Vegas, NV, June 27–July 1, 2010, 12 p.
130. Lai, Y. G. (2013). *Coupled 2D Morpho-Dynamic and bank erosion modeling at the upper junction city channel rehabilitation project site, Trinity River, CA*. Technical Report No. SRH-2013-09, Technical Service Center, Bureau of Reclamation, Denver, CO.
131. Gaeuman, D., Andrews, E. D., Krause, A., & Smith, W. (2009). Predicting fractional bed load transport rates: Application of the Wilcock-Crowe equations to a regulated gravel bed river. *Water Resources Research, 45,* W06409. doi:10.1029/2008WR007320.
132. Cardno ENTRIX. (2012). *Bank-stability analysis of the Upper Junction City Reach Trinity River, CA*. Project Report by Cardno Entrix, Project 30088130, Dec 2012.
133. Logan, B., Nelson, J., McDonald, R., & Wright, S. (2010). *Mechanics and modeling of flow, sediment transport and morphologic change in riverine lateral separation zones*. 2nd Joint Federal Interagency Conference, Las Vegas, NV, June 27–July 1.

Chapter 10
Environmental Water Engineering Glossary

Mu-Hao Sung Wang and Lawrence K. Wang

Contents

Abstract Technical and legal terms commonly used by environmental water engineers are introduced. This chapter covers mainly the glossary terms used in the following two books:

1. Advances in Water Resources Engineering and
2. Modern Water Resources Engineering,

The above two books form a miniseries in the field of water resources engineering.

Keywords Water resources engineering · Environmental engineering · Glossary · Hydraulics · Hydrology · Watershed · Dredging · Salinity · Sensitivity · Land application · Receiving waters · Modeling · Water pollution control

M.-H. S. Wang (✉) · L. K. Wang
Rutgers University, New Brunswick, NJ, USA
e-mail: lenox.institute@gmail.com

Lenox Institute of Water Technology, Newtonville, NY, USA

L. K. Wang
e-mail: lawrencekwang@gmail.com

C. T. Yang, L. K. Wang (eds.), *Advances in Water Resources Engineering*,
Handbook of Environmental Engineering, Volume 14, DOI 10.1007/978-3-319-11023-3_10

1 Environmental Water Engineering Glossary [1–10]

Abiotic Nonliving, pertaining to physicochemical factors only.

Ablation The process by which ice and snow waste away owing to melting and evaporation.

Absorption The entrance of water into the soil or rocks by all natural processes. It includes the infiltration of precipitation or snowmelt, gravity flow of streams into the valley alluvium (see bank storage) into sinkholes or other large openings, and the movement of atmospheric moisture.

Abutment Connecting an element or structure used for support.

Acequia Acequias are gravity-driven waterways, similar in concept to a flume. Most are simple ditches with dirt banks, but they can be lined with concrete. They were important forms of irrigation in the development of agriculture in the American Southwest. The proliferation of cotton, pecans, and green chili as major agricultural staples owe their progress to the acequia system.

Acid A substance that has a pH of less than 7, which is neutral. Specifically, an acid has more free hydrogen ions (H^+) than hydroxyl ions (OH^-).

Acid Neutralizing Capacity (ANC) The equivalent capacity of a solution to neutralize strong acids.

Acid Rain or Acid Precipitation Precipitation having a pH lower than the pH range commonly found in natural waters, caused by absorption from the atmosphere of sulfur dioxide gas and nitrogen oxides gas, which then form sulfuric acid and nitric acid, respectively, in solution.

Action Level The level of toxic substances (such as lead or copper) which, if exceeded, triggers treatment or other requirements that a water system must follow.

Activated Sludge (1) A flocculent assemblage of microorganisms, nonliving organic matter, and inorganic materials. (2) Sludge floc produced in raw or settled wastewater by the growth of bacteria and other organisms in the presence of dissolved oxygen and accumulated in sufficient concentration by returning floc previously formed.

Activated Sludge Process A biological wastewater treatment process in which a mixture of wastewater and activated sludge is agitated and aerated. The activated sludge is subsequently separated from the mixed liquor (aerated wastewater) by sedimentation clarification (secondary sedimentation clarifier) or flotation clarification (secondary dissolved air flotation, DAF, clarifier). The separated activated sludge (biosolids) is partially wasted and partially returned to the aeration basin as needed. The clarified effluent is the activated sludge process effluent.

Acute Health Effect An immediate (i.e., within hours or days) effect that may result from exposure to certain drinking-water contaminants (e.g., pathogens).

Adsorption (1) A physicochemical, passive, and reversible process that attracts and adsorbs molecules of a gas, liquid, or dissolved substance to a surface of adsorbent. (2) The interaction of an analyte with the surface of a matrix.

Advisory A nonregulatory document that communicates risk information to those who may have to make risk management decisions. For example, a fish consumption advisory may recommend that people limit or avoid eating certain species of fish caught from certain lakes, rivers, or coastal waters. In some cases, advisories may include recommendations for specific groups (such as infants, children, the elderly, or women who are pregnant or may become pregnant).

Aeration A gas transfer process that allows for the absorption of gas (such as oxygen) by liquid (such as water or wastewater).

Aerobic Living, acting, or occurring in the presence of oxygen.

Aggradation The process by which streambeds, floodplains, and the bottoms of other water bodies are raised in elevation by the deposition of material eroded and transported from other areas. It is the opposite of degradation.

Aggregation Soil particles bound to each other by moist clay, organic matter, organic compounds, and fungal structures. Well-aggregated soils are more stable and less susceptible to erosion.

Agricultural and Animal Waste Waste generated by the production and harvest of crops or trees or the rearing of animals. Animal waste is a subset of agricultural waste and includes waste (e.g., feed waste, bedding and litter, and feedlot and paddock runoff) from livestock, dairy, and other animal-related agricultural and farming practices.

Agricultural Land (1) Land suitable for agricultural production, both crops and livestock. (2) Land on which a food, feed, or fiber crop is grown. This includes range land or land used as pasture.

Agronomic Rate The whole sludge application rate designed to (a) provide the amount of nitrogen needed by a crop or vegetation grown on the land and (b) minimize the amount of nitrogen in the sewage sludge that passes below the root zone of the crop or vegetation grown on the land to the groundwater.

Air Dissolving Tube (ADT) A cylindrical pressure vessel (pressurization tank) into which both gas (such as air) and water (or waste stream) are introduced for dissolving the gas under high pressure. Subsequent release of the pressurized water (or waste stream) under normal low-atmospheric pressure will generate extremely fine gas bubbles.

Air Pollutant Any substance in air that could, in high-enough concentration, harm humans, animals, vegetation, or material. Air pollutants can include almost any natural or artificial composition of matter capable of being airborne—solid particles, liquid droplets, gases, or a combination thereof. Air pollutants are often grouped in categories for ease in classification; some of the categories are sulfur compounds, volatile organic compounds, particulate matter, nitrogen compounds, and radioactive compounds.

Air Quality Index (AQI) An index for reporting daily air quality that characterizes air pollution levels and associated health effects that might be of concern. The US Environmental Protection Agency (USEPA) calculates the AQI for five criteria pollutants. AQI values range from 0 to 500; the higher the AQI value, the greater the level of air pollution and the greater the health concern. AQI values below 100 are generally thought of as satisfactory. When AQI values are above 100, air quality is considered to be unhealthy—at first for certain sensitive groups of people, then for everyone as AQI values get higher.

Air Quality System (AQS) The USEPA's electronic repository of ambient air monitoring data collected by USEPA, state, local, and tribal air pollution control agencies from thousands of monitoring stations. The AQS contains monitoring data, descriptive information about monitoring stations, and data quality assurance and quality control information.

Air Toxics Air pollutants that cause or may cause cancer or other serious health effects, such as reproductive effects or birth defects, or adverse environmental and ecological effects. Examples of toxic air pollutants include benzene (found in gasoline), perchloroethylene (emitted from some dry-cleaning facilities), and methylene chloride (used as a solvent by a number of industries). Air toxics are also known as hazardous air pollutants.

Algae Any of a group of chlorophyll-bearing aquatic plants with no true leaves.

Algal Bloom (1) A sudden, excessive growth of algae in a water body. (2) Rapid and flourishing growth of algae.

Algorithm Step-by-step procedure for calculations.

Alkaline Sometimes water or soils contain an amount of alkali (strongly basic) substances sufficient to raise the pH value above 7.0 and be harmful to the growth of crops.

Alkalinity The capacity of water for neutralizing an acid solution.

Allochthonous Pertaining to turbidity-causing substances that originate from outside and are carried into a reservoir or other surface water. Such substances include humus, organic detritus, chlloidal substances, silts, aquatic plants, and small animals.

Alluvial (1) It is of alluvium. (2) Pertains to alluvium deposited by a stream or flowing water (usually sand).

Alluvial Aquifer Aquifer formed by sedimentation of gravel, sand, silt, or clay materials deposited in river channels or on floodplains.

Alluvial River A river that has formed its channel by the process of aggradation. The sediment that it carries is similar to that in its bed and banks.

Alluvial Stream A stream whose channel boundary is composed of alluvium, and which generally changes its cross section and bed form due to the interaction of the flow and mobile boundary adjustment.

Alluvial Streams Refer to streams in which the bed and banks are made up of mobile sediment and/or soil. As a result, an alluvial stream is subject to continuous changes in planar forms and vertical shapes. They can assume a number of forms, e.g., meandering, braiding, wandering and straight, based on the properties of their banks, the flows they experience, the local riparian ecology, and the amount, size, and type of sediment that they carry.

Alluvium Deposits of clay, silt, sand, gravel, or other particulate material that has been deposited by a stream or other body of running water in a streambed, on a floodplain, on a delta, or at the base of a mountain.

Alluvium Sediments deposited by running water.

Ambient It is surrounding on all sides.

Ambient Monitoring Monitoring within natural systems (e.g., lakes, rivers, estuaries, wetlands) to determine existing conditions.

Ammonia Nitrogen A common way to report ammonia concentration (expressed as nitrogen).

Ammonification A process of formation of ammonia nitrogen from reduced organic nitrogen compounds.

Anabranch A diverging branch of a river which reenters the main stream.

Anaerobic (1) Capable of living or acting in the absence of oxygen. (2) Without oxygen; water and sediment environments without oxygen produce, for example, chemical conditions that precipitate and permanently store many metals from water and that release dissolved phosphorus to the water.

Anchor Ice Ice in the bed of a stream or upon a submerged body or structure.

Anion Negatively charged ion that is attracted to the anode in electrolysis.

Annual Flood The highest peak discharge in a water year.

Annual Flood Series A list of annual floods.

Annual Pollutant Loading Rate (APLR) The maximum amount of a pollutant that can be applied to a unit area of land during a 365-day period. This term describes pollutant limits for sewage sludge that is given away or sold in a bag or other container for application to the land.

Annual Whole Sludge Application Rate The maximum amount of sewage sludge on a dry weight basis that can be applied to a land application site during a 365-day (1-year) period.

Anoxic Pertaining to conditions of oxygen deficiency.

Antecedent Precipitation Index An index of moisture stored within a drainage basin before a storm.

Anthropogenic Originating from humans; not naturally occurring.

Aphotic Below the level of light penetration in water

Appropriation Doctrine The system for allocating water to private individuals used in most Western states. The doctrine of prior appropriation was in common use throughout the arid west as early settlers and miners began to develop the land. The prior appropriation doctrine is based on the concept of "First in Time, First in Right." The first person to take a quantity of water and put it to beneficial use has a higher priority of right than a subsequent user. Under drought conditions, higher-priority users are satisfied before junior users receive water. Appropriative rights can be lost through nonuse; they can also be sold or transferred apart from the land. This system contrasts with Riparian Water Rights.

Aquaculture (1) Farming of plants and animals that live in water, such as fish, shellfish, and algae; (2) a process for removing pollutants from water through the use of aquatic plants (such as, water hyacinths) in pond contaminants. The contaminants are either synthesized by, or bioaccumulated in, the aquatic plants, which ultimately are harvested for disposal.

Aquaculture, Living Machine System A man-made wastewater treatment system which adapts and enhances the ecological processes in a series of tidal wetland cells or basins. Each cell or basin is filled with special gravel that promotes the development of micro-ecosystems. A computer controls the fill and drain cycles, alternating anoxic (without oxygen) and aerobic (with oxygen) conditions. As wastewater moves through the system, the cells are alternately flooded and drained to create multiple tidal cycles each day, much like one finds in nature, resulting in high-quality reusable water.

Aquaculture, Natural and Constructed Wetland Systems The aquatic wastewater treatment systems involve the production of algae and higher plants (both submerged and emergent), invertebrates, and fish for wastewater treatment and water conservation. A wastewater treatment by natural and constructed wetland systems is generally accomplished by sprinkling or flood irrigating the wastewater into the wetland area or by passing the wastewater through a system of shallow ponds, channels, basins, or other constructed areas where the emergent aquatic vegetation has been planted or naturally occurs and is actively growing. The treated wastewater is totally reused in a natural environment achieving almost 100 % water conservation. The vegetation produced as a result of the system's operation may or may not be removed and can be utilized for various purposes: (1) composted for use as source of fertilizer/soil conditioner, (2) dried or otherwise processed for use as animal feed supplements, or digested to produce methane.

Aquaculture, Water Hyacinth System Wastewater treatment by aquaculture water hyacinth system is accomplished by passing the wastewater through a hyacinth-covered basin where the plants remove nutrients, biochemical oxygen demand (BOD)/chemical oxygen demand (COD)/total organic carbon (TOC), suspended solids, heavy metals, etc. Batch treatment and flow-through systems, using single and multiple cell units, are all possible. The treated wastewater is reused in a natural environment, or recharged to the underground becoming new groundwater. Hyacinths harvested from these systems can be used as a fertilizer/soil conditioner after composting, an animal feed, and a source of methane when anaerobically digested.

Aquatic Associated with water.

Aqueduct A pipe, conduit, or channel designed to transport water from a remote source, usually by gravity.

Aquifer (1) A natural underground geologic formation, often of sand or gravel, that is water bearing. A geological formation or structure that stores and/or transmits water, such as to wells and springs. Use of the term is usually restricted to those water-bearing formations capable of yielding water in sufficient quantity to constitute a usable supply for people's uses. (2) Underground water-saturated unconsolidated formation from which groundwater can be extracted from a well.

Aquifer (Confined) Soil or rock below the land surface that is saturated with water. There are layers of impermeable material both above and below it, and it is under pressure so that when the aquifer is penetrated by a well, the water will rise above the top of the aquifer.

Aquifer (Unconfined) An aquifer whose upper water surface (water table) is at atmospheric pressure, and thus is able to rise and fall.

Aquifer Pumping and Recharge Groundwater management practice in which, after being pumped, the aquifer is recharged with surface water at prescribed locations.

Aquifer Storage and Recovery Groundwater management practice in which each operating well is provided with a pump, which is able to extract water during periods of water need for irrigation, and inject water when surface water is available for storage.

Area of Cropland An area of cropland that has been subdivided into several strips and is not a single field. Rather, each strip represents an individual field unit.

Area Source A source of air pollution that is released over an area that cannot be classified as a point source. Area sources can include vehicles and other small engines, small businesses, and household activities, or biogenic sources such as a forest that releases hydrocarbons.

Area-Capacity Curve A graph showing the relation between the surface area of the water in a reservoir and the corresponding volume.

Arid Pertaining to climatic conditions or a soil that lacks humidity.

Arid Climate A climate characterized by less than 10 in. (25.4 cm) of annual rainfall.

Artesian Water Groundwater that is under pressure when tapped by a well and is able to rise above the level at which it is first encountered. It may or may not flow out at ground level. The pressure in such an aquifer commonly is called artesian pressure, and the formation containing artesian water is an artesian aquifer or confined aquifer (see flowing well).

Artificial Recharge A process where water is put back into groundwater storage from surface-water supplies such as irrigation, or induced infiltration from streams or wells.

Assimilation Transformation of absorbed nutrients into living matter.

Autochthonous Material derived from within a habitat, such as through plant growth

Average Discharge In the annual series of the Geological Survey's reports on surface-water supply—the arithmetic average of all complete water years of record whether or not they are consecutive. Average discharge is not published for less than 5 years of record. The term "average" is generally reserved for average of record and "mean" is used for averages of shorter periods, namely, daily mean discharge.

Backwater Water backed up or retarded in its course as compared with its normal or natural condition of flow. In stream gaging, a rise in stage produced by a temporary obstruction such as ice or weeds, or by the flooding of the stream below. The difference between the observed stage and that indicated by the stage–discharge relation is reported as backwater.

Bagged Sewage Sludge Sewage sludge that is sold or given away in a bag or other container (i.e., either an open or closed receptacle containing 1 t or less of sewage sludge).

Bank The margins of a channel. Banks are called right or left as viewed facing in the direction of the flow.

Bank Migration Lateral shifting of the banks of a stream course.

Bank Storage The water absorbed into the banks of a stream channel, when the stages rise above the water table in the bank formations, then returns to the channel as effluent seepage when the stages fall below the water table.

Bank-Full Stage A stage at which a stream first overflows its natural banks.

Basal Cleanout It is the process of removing the bank materials that may provide protections of bank toe.

Basal Erosion The term refers to direct removal of bank materials laterally by flowing water. In addition to shear force, basal erosion is also dependent on bank properties, bank protective materials, vegetation/debris, and soil characteristics.

Base A substance that has a pH of more than 7, which is neutral. A base has less free hydrogen ions (H^+) than hydroxyl ions (OH^-).

Base Discharge (For Peak Discharge) In the Geological Survey's annual reports on surface-water supply, the discharge above which peak discharge data are published. The base discharge at each station is selected so that an average of about three peaks a year will be presented (see also partial-duration flood series).

Base flow A sustained flow of a stream in the absence of direct runoff. It includes natural and human-induced streamflows. A natural base flow is sustained largely by groundwater discharges (also see base runoff).

Base Runoff A sustained or fair-weather runoff. In most streams, base runoff is composed largely of groundwater effluent. The term base flow is often used in the same sense as base runoff. However, the distinction is the same as that between streamflow and runoff. When the concept in the terms base flow and base runoff is that of the natural flow in a stream, base runoff is the logical term.

Baseline A reference condition against which changes or trends are judged—usually a set of conditions that exist at a particular point in time.

Basic Hydrologic Data Include inventories of features of land and water that vary only from place to place (topographic and geologic maps are examples), and records of processes that vary with both place and time. (Records of precipitation, streamflow, groundwater, and quality-of-water analyses are examples.)

Basic Hydrologic Information It is a broader term that includes surveys of the water resources of particular areas and a study of their physical and related economic processes, interrelations, and mechanisms.

Basic-Stage Flood Series See partial duration flood series.

Bed Forms Wave-like irregularities found on the bottom (bed) of a stream that are related to flow characteristics. They are given names such as "dunes," "ripples," and "anti-dunes." They are related to the transport of sediment, and they interact with the flow because they change the roughness of the streambed. An analog to streambed forms is desert sand dunes.

Bed Load Material moving on or near the stream bed by rolling, sliding, and sometimes making brief excursions into the flow a few diameters above the bed, i.e., jumping. The term "saltation" is sometimes used in place of "jumping." Bed load is bed material that moves in continuous contact with the bed; contrast with suspended load.

Bed Material The sediment mixture of which the bed is composed. In alluvial streams, bed-material particles are liable to be moved at any moment or during some future flow condition. Bed material may include grain sizes that travel as both bed load and suspended load.

Bedrock The solid rock beneath the soil and superficial rock. A general term for solid rock that lies beneath the soil, loose sediments, or other unconsolidated material.

Benchmark A concentration or other accepted measure against which environmental conditions are compared.

Benefit Maximization The process of increasing benefits to the greatest extent possible within constraints such as limitation on financial resources.

Benefits A good, service, or attribute of a good or service that promotes or enhances the well-being of an individual, an organization, or a natural system.

Benthic (1) Pertaining to the bottom of a water body; (2) pertaining to the bottom or bottom environment of a water body.

Benthos Those organisms that live on the bottom of a body of water.

Best Available Technology (1) A method that has been determined to be the most effective, practical means of preventing or reducing pollution from nonpoint and point sources. (2) The water treatment(s) that the government (such as the USEPA) certifies to be the most effective for removing a contaminant.

Bioaccumulation The uptake of contaminants from all sources including direct sorption to living biological bodies such as fish or microorganisms

Bioaccumulative Compound A compound that tends to accumulate in tissues and build up in food webs. Some bioaccumulative compounds can potentially have adverse effects on ecosystems or human health.

Bioavailability The existence of a chemical in a form that it can be readily integrated into.

Bioavailable The state of a toxicant such that there is increased physicochemical access to the toxicant by an organism. The less the bioavailability of a toxicant, the less its toxic effect on an organism.

Biochemical Oxygen Demand (BOD) The amount of oxygen required to decompose a given amount of organic matter.

Biodegradable Subject to degradation (breakdown) of complex organic substances into simple organic and inorganic substances by biological action.

Biodegradation The biological process of breaking down complex organic substances into simple organic and inorganic substances by a biogenic source.

Biogenic Source An air emission source created by some sort of biological activity. Examples include emissions resulting from microbial activity in soils and emissions from trees and other vegetation. Emissions from biogenic sources are a subset of emissions from natural sources (see natural source).

Biological Balance The interrelationships among organisms, including the structure of food webs and the ability of ecological systems to maintain themselves over time. Balance is a dynamic characteristic, rather than a fixed state.

Biological Diversity The variety and variability among living organisms and the ecological complexes in which they occur. Though it most often refers to the numbers of species, the term can apply to levels of organization ranging from genes to ecosystems.

Biological Oxidation A process by which living organisms in the presence of oxygen convert organic matter into a more stable or a mineral form.

Biomagnification (1) The step-by-step concentration of chemicals in successive levels of a food chain: (2) A process by which a compound increases in concentration in an organism's tissue as the food level increases. Organisms at the top of the food chain have a higher concentration of the compound.

Biomarker A molecular or cellular indicator (or "marker") of an event or condition (exposure, effect, susceptibility) in a biological system or sample. It is the product of an interaction between a contaminant and some target molecule or cell.

Biomarker of Effect A measure of disease progression, representing a measurable alteration at the molecular, cellular, or some other structural level in the body that can be recognized as a potential or an established adverse health effect. Such a biomarker can indicate a biological response or health effect related to a chemical or other stressor; however, it is not always possible to link a biomarker with exposure to a single substance.

Biomarker of Exposure The level of a contaminant or its metabolite collected from the body or from substances produced or excreted within biological systems. In humans, this measurement can reflect the amount of the contaminant that is stored in the body, and is sometimes referred to as the body burden. It indicates the level of exposure.

Biomarker of Susceptibility A measurement of individual factors that can affect response to environmental agents. Examples include enzymes whose presence or absence may reflect a particular genetic condition.

Biomass The total weight of matter incorporated into (living and/or dead) organisms.

Biomonitoring The measurement of human tissues or excreta from biological systems for direct or indirect evidence of exposure to chemical, biological, or radiological substances.

Biosolids (1) Biological mass (biomass) collected from bioreactors, such as activated sludge, trickling filters, rotating biological reactors, lagoons, sequencing batch bioreactors, fluidized beds, septic tanks, etc. (2) Biosolids are solids, semisolids, or liquid materials, resulting from biological treatment of domestic sewage that has been sufficiently processed to permit these materials to be safely land applied. The term of biosolids was introduced by the wastewater treatment industry in the early 1990s and has been recently adopted by the USEPA to distinguish high quality, treated sewage sludge from raw sewage sludge and from sewage sludge containing large amounts of pollutants.

Biota The fauna and flora of a habitat or region.

Biotic Environment The biological component of an ecosystem, including plants and animals.

Biotransformation The permanent changing of a substance from one chemical identity to the bottom, thus usually supporting rooted aquatic plants.

Boundary Conditions Definitions or statements of conditions or phenomena at spatial or temporal boundaries of a model. Water levels, flows, sediment concentrations, etc., that are specified at the boundaries of the area being modeled. A specified tailwater elevation and incoming upstream discharge are typical boundary conditions.

Braided Channel A stream that is characterized by random interconnected channels divided by islands or bars. Bars that divide the stream into separate channels at low flow are often submerged at high flow.

Braiding of River Channels Successive division and rejoining (of river flow) with accompanying islands is the important characteristic denoted by the synonymous terms, braided or anastomosing stream. A braided stream is composed of anabranches.

Bulk Modulus Material resistance to uniform compression.

Bulk Sewage Sludge Sewage sludge that is not sold or given away in a bag or other container for application to the land.

Cantilever Failure It refers to the collapse of an overhanging bank block into a stream, often due to significant undercutting. The failure tends to occur on banks with composite layers of fine/coarse and/or cohesive/noncohesive materials.

Capillary Action The means by which liquid moves through the porous spaces in a solid, such as soil, plant roots, and the capillary blood vessels in our bodies due to the forces of adhesion, cohesion, and surface tension. Capillary action is essential in carrying substances and nutrients from one place to another in plants and animals.

Carbonaceous Containing carbon and derived from organic substances such as coal, coconut shells, and organic waste.

Catastrophic Failure Sudden and total failure of some system from which recovery is impossible.

Catchment Area See drainage basin.

Cation Positively charged ion that is attracted to the cathode in electrolysis.

Ceiling Concentration Limits (CCL) The ceiling concentration limits are the maximum concentrations of the nine trace elements allowed in biosolids to be land applied. Sewage sludge exceeding the ceiling concentration limit for even one of the regulated pollutants is not classified as biosolids and, hence, cannot be land applied.

Cementation Hardening and welding of sediments by the precipitation of mineral matter in the pore spaces. Affects porosity and permeability.

CFD Short for computational fluid dynamics, it is a discipline that uses numerical methods and algorithms to solve fluid mechanics problems with computers.

CGNS A standard for the storage and retrieval of digital data produced in computational fluid dynamics (CFD) applications. It stands for CFD general notation system chain or food web.

Channel (1) A conduit formed by the flow of water and debris. The time and volume characteristics of water or debris can be altered by man, climate change, or by alterations in protective vegetal cover on the land of the watershed. The stream channel adjusts to the new set of conditions. (2) A natural or artificial waterway that periodically or continuously contains moving water.

Channel (Watercourse) An open conduit either naturally or artificially created which periodically or continuously contains moving water, or which forms a connecting link between two bodies of water. River, creek, run, branch, anabranch, and tributary are some of the terms used to describe natural channels. Natural channels may be single or braided (see braiding of river channels). Canal and floodway are some of the terms used to describe artificial channels.

Channel Storage The volume of water at a given time in the channel or over the flood plain of the streams in a drainage basin or river reach. Channel storage is great during the progress of a flood event.

Channelization The practice of straightening a waterway to remove meanders and make water flow faster. Sometimes concrete is used to line the sides and bottom of the channel.

Chloride It is one of the major inorganic anions, or negative ions, in saltwater and freshwater. It originates from the dissociation of salts, such as sodium chloride or calcium chloride, in water. Chlorides are binary compounds of chlorine. 1 g/L of chloride concentration = 1 g/L of chlorinity = 1.80655 ppt (or g/L) of salinity in seawater.

Chlorinity Chlorinity is defined in relation to salinity as follows: Salinity = 1.80655 (Chlorinity). Although chlorinity is not equivalent to chloride concentration, the factor for translating a chloride determination in seawater to include bromide, for example, is only 1.0045 based on the molecular weights and the relative amounts of the two ions. Therefore, for practical purposes, chloride (in mg/g of solution, or g/L of solution) is nearly equal to chlorinity in seawater. Or 1 g/L of chloride concentration = 1 g/L of chlorinity in seawater. For the wastewater, a knowledge of the ions responsible for the solution's electrical conductivity is necessary to correct for the ions impact on oxygen solubility and use of the tabular value or the equation is inappropriate unless the relative composition of the wastewater is similar to seawater.

Chlorophyll The green, photosynthetic pigments of plants.

Chronic Health Effect The possible result of exposure over many years to a drinking-water contaminant at levels above its maximum contaminant level (MCL).

Clamshell Dredge A mechanical equipment that excavates with a clamshell bucket suspended by cables from a forward-extending boom that can swing about the bow of the dredge. Spuds, extending to the bottom, are used for keeping the dredge stable in the proper working position. The boom and bucket are swung around to empty the bucket into a scow or a small barge positioned alongside the dredge. The clamshell dredge is designed for maintenance dredging in river channels where soft or cohesive underwater materials are to be removed periodically. This type of dredge is exceptionally useful for deep digging and for dredging in close quarters alongside structures.

Clarity A measure of the amount of particles suspended in water; determined by using a disk or turbidity test.

Class I Sludge Management Facility Publicly owned treatment works (POTWs), required to have an approved pretreatment program under 40 CFR 403.8(a), including any POTW located in a state that has elected to assume local pretreatment program responsibilities under 40 CFR 403.10(e). In addition, the regional administrator or, in the case of approved state programs, the regional administrator in conjunction with the state director, has the discretion to designate any treatment works treating domestic sewage (TWTDS) as a class I sludge management facility.

Clean Water Act (CWA) The US law, codified generally as 33 USC 1251–1387, that establishes a regulatory and enforcement program administered by the USEPA to control pollutant discharges into the US waters.

Cleanup Action taken to deal with a release (or threat of release) of a hazardous substance that could affect humans and/or the environment. This term is sometimes used interchangeably with the terms "remedial action," "removal action," "response action," and "corrective action."

Climate The sum total of the meteorological elements that characterize the average and extreme condition of the atmosphere over a long period of time at any one place or region of the earth's surface. The collective state of the atmosphere at a given place or over a given area within a specified period of time.

Climate Change A term sometimes used to refer to all forms of climatic inconsistency; because the earth's climate is never static, the term is more properly used to imply a significant change from one climatic condition to another. In some cases, "climate change" has been used synonymously with "global warming." Scientists, however, tend to use "climate change" in the wider sense to also include natural changes in climate.

Climatic Year A continuous 12-month period during which a complete annual cycle occurs, arbitrarily selected for the presentation of data relative to hydrologic or meteorologic phenomena. The climatic year is usually designated by the calendar year during which most of the 12 months occur (see water year).

Cloudburst A torrential downpour of rain, which by its spottiness and relatively high intensity suggests the bursting and discharge of a whole cloud at once.

Coastal Waters Waters at the interface between terrestrial environments and the open ocean. Many unique habitats lie in coastal waters—for example, estuaries, coastal wetlands, seagrass meadows, coral reefs, mangrove and kelp forests, and upwelling areas.

Coliform A group of related bacteria whose presence in drinking water may indicate contamination by disease-causing, pathogenic microorganisms.

Colloid (1) A dispersion of particles larger than small molecules and that do not settle. (2) A small, discrete solid particle in water that is not dissolved, but is suspended and will not settle for very long periods of time due to molecular bombardment.

Colloidal Matter See colloid.

Combined Sewers and Combined Sewer Overflow (CSO) Pipes that carry both storm water and household sewage, to sewage treatment plants. During a big storm, they may overflow and dump untreated sewage into streams, lakes, and coastal waters. These overflows are called combined sewer overflows or CSOs.

Commercial Water Use Water used for motels, hotels, restaurants, office buildings, other commercial facilities, and institutions. Water for commercial uses comes from both public-supplied sources, such as a county water department, and self-supplied sources, such as local wells.

Community In ecology, an assemblage of populations of different species within a specified location in space and time. Sometimes, a particular subgrouping may be specified, such as the fish community in a lake or the soil arthropod community in a forest.

Community Water System A water system which supplies drinking water to 25 or more of the same people year-round in their residences.

Compliance The act of meeting all state and federal drinking-water regulations.

Computational Cell Each individual point, or volume, of the lattice (computational grid) transforming the continuous real-world domains into its discrete counterpart, suitable for numerical evaluation and implementation on digital computers.

Concentration Time See time of concentration.

Concordant Flows Flows at different points in a river system that have the same recurrence interval, or the same frequency of occurrence. It is most often applied to flood flows.

Condensation The process by which water vapor changes from the vapor state into the liquid or solid state. Waterdrops on the outside of a cold glass of water are condensed water. It is the reverse of evaporation.

Condition of Ecology The state of a resource, generally reflecting a combination of physical, chemical, and biological characteristics such as temperature, water clarity, chemical composition, or the status of biological communities. The condition of fresh surface waters, groundwater, wetlands, coastal waters, recreational waters, and consumable fish and shellfish. (Also see ecological condition.)

Conductivity Measure of the ease at which an electric charge or heat can pass through a material.

Confined Aquifer Aquifer delimited above and below by low permeable or impermeable formations, such as aquitards or aquicludes.

Conjunctive Use Combined management of surface water and groundwater resources that optimizes common benefits by trading off water demand against water supply, while complying with a series of constraints of technical, physical, environmental, and economical nature.

Connectivity Degree to which movement is facilitated or impeded.

Conservation Storage Storage of water for later release for useful purposes such as municipal water supply, power, or irrigation in contrast with storage capacity used for flood control.

Conservative Chemical A chemical that essentially maintains its concentration in passing through a unit treatment process. For instance, coagulation can remove turbidity and organic carbon, but it does not reduce the chloride ion concentration; thus, chloride ion is called a conservative chemical in this circumstance.

Conservative Constituent See conservative chemical.

Constructed Wetland or Created Wetland A wetland at a site where it did not formerly occur. The constructed/created wetlands are designed to meet a variety of human benefits including, but not limited to, the treatment of water pollution discharges (e.g., municipal wastewater, storm water, etc.), and the mitigation of wetland losses permitted under Section 404 of the Clean Water Act.

Construction and Demolition Debris Waste materials generated during the construction, renovation, and demolition of buildings, roads, and bridges. Construction and demolition debris often contain bulky, heavy materials such as concrete, wood (from buildings), asphalt (from roads and roofing shingles), gypsum (from drywall), metals, bricks, glass, plastics, building components (doors, windows, plumbing fixtures), and trees, stumps, earth, and rock from clearing sites.

Consumer (1) An organism that consumes another; (2) the user.

Consumptive Use (1) The quantity of water absorbed by the crop and transpired or used directly in the building of plant tissue together with that evaporated from the cropped area. (2) The quantity of water transpired and evaporated from a cropped area or the normal loss of water from the soil by evaporation and plant transpiration. (3) The quantity of water discharged to the atmosphere or incorporated in the products of the process in connection with vegetative growth, food processing, or an industrial process. (4) The part of water withdrawn that is evaporated, transpired by plants, incorporated into products or crops, consumed by humans or livestock, or otherwise removed from the immediate water environment. Also referred to as water consumed.

Consumptive Use, Net (1) The consumptive use decreased by the estimated contribution by rainfall toward the production of irrigated crops; (2) the net consumptive use is sometimes called crop irrigation requirement.

Consumptive Waste The water that returns to the atmosphere without benefiting man.

Contaminant (1) Anything found in the environment (including microorganisms, minerals, chemicals, radionuclides, etc.) which may be harmful to human health; (2) any physical, chemical, biological, or radiological substance or matter that has an adverse effect on air, water, or soil.

Contaminated Land Land that has been polluted with hazardous materials and requires cleanup or remediation. Contaminated lands include sites contaminated as a result of improper handling or disposal of toxic and hazardous wastes, sites where improper handling or accidents release toxic or hazardous materials that are not wastes, and sites where toxics may have been deposited by wind or flooding.

Contents The volume of water in a reservoir. Unless otherwise indicated, reservoir content is computed on the basis of a level pool and does not include bank storage.

Contouring Farming practice of tilling sloped farmland along the lines of constant elevation.

Control A natural constriction of the channel, a long reach of the channel, a stretch of rapids, or an artificial structure downstream from a gaging station that determines the stage–discharge relation at the gage. A control may be complete or partial. A complete control exists where the stage–discharge relation at a gaging station is entirely independent of fluctuations in stage downstream from the control. A partial control exists where downstream fluctuations have some effect upon the stage–discharge relation at a gaging station. A control, either partial or complete, may also be shifting. Most natural controls are shifting to a degree, but a shifting control exists where the stage discharge relation experiences frequent changes owing to impermanent bed or banks.

Conventional Wastewater Treatment System A wastewater treatment system which includes preliminary treatment (screening, grit chamber, pH adjustment, oil and grease removal), primary clarification (primary sedimentation), bioreactor (aeration basin, trickling filter, rotating biological reactor, fluidized bed, etc.), and secondary clarification (secondary sedimentation).

Conveyance A measure of the flow capacity of a channel section. Flow is directly proportional to conveyance for a steady uniform flow.

Conveyance Loss The water that is lost in transit from a pipe, canal, or ditch by leakage or evaporation. Generally, the water is not available for further use; however, leakage from an irrigation ditch, for example, may percolate to a groundwater source and be available for further use.

Convolution Integral The integral of the product of two functions, with one function reversed and shifted with respect to another. The integral is often used to model conditions of continuous, as opposed to discrete, superposition of effect.

Copepods A large subclass of usually minute, mostly free-swimming aquatic crustaceans.

Correlation The process of establishing a relation between a variable and one or more related variables. Correlation is simple if there is only one independent variable; multiple, if there is more than one independent variable. For gaging station records, the usual variables are the short-term gaging-station record and one or more long-term gaging-station records.

Correlative Estimate A discharge determined by correlation. A correlative estimate represents a likely value of the discharge for any particular period—commonly a month—according to a specified method of analysis.

Cost Minimization The process of reducing costs to the lowest possible amount, given constraints such as requirements that a specified level of benefits or other resources be attained or provided.

Counterions The ion that accompanies an ionic species in order to maintain electric neutrality.

Criteria Pollutants A group of six widespread and common air pollutants that the USEPA regulates on the basis of standards set to protect public health or the environment (see National Ambient Air Quality Standards). The six criteria pollutants are carbon monoxide, lead, nitrogen dioxide, ozone, particulate matter, and sulfur dioxide.

Crop Group Individual farm fields that are managed in the same manner, with the similar yield goals, are called a crop group.

Crop Management Crop management involves crop group identification, crop nitrogen deficit determination, crop nitrogen fertilizer rate calculation, and crop yield optimization.

Crop Nitrogen Deficit (CND) Crop nitrogen deficit (CND) is equal to anticipated crop nitrogen fertilizer rate (CNFR) minus all past plant-available nitrogen (PAN) sources (PAN-past) and current planned nonbiosolids PAN sources (PAN-plan), in the unit of lb N/acre. Previous biosolids carryover nitrogen is included in this calculation.

Crop Nitrogen Fertilizer Rate (CNFR) CNFR is a rate (lb N/acre) = (Yield; UNFR), where UNFR is the unit nitrogen fertilizer rate (lb N/unit crop yield), and Yield is the crop harvested, or crop yield (bu/acre or ton/acre).

Crop Year The basic time management unit is often called the crop year or planting season. The crop year is defined as the year in which a crop receiving the biosolids/manure treatment is harvested. For example, fall applications of biosolids/manure in 2000 intended to provide nutrients for a crop to be harvested in 2001 are earmarked for crop year 2001. Likewise, biosolids/manure applied immediately prior to planting winter wheat in October 2000 should be identified as fertilizer intended for crop year 2001 because the wheat will be harvested in the summer of 2001.

Crop Yield It is the crop harvested in the unit of bu/acre or ton/acre.

Cross Section The shape of the channel in which a stream flows on a line perpendicular to the flow or banks.

Cross-Sectional Area The wetted area of a cross section perpendicular to the direction of flow.

Crustacean A large class of arthropods that bear a horny shell.

Cryology The science of ice and snow.

Cryptosporidium A microorganism commonly found in lakes and rivers which is highly resistant to disinfection, and has caused several large outbreaks of gastrointestinal illness, with symptoms that include diarrhea, nausea, and/or stomach cramps.

Cumulative Pollutant Loading Rate (CPLR) CPLR is equal to the total amount of pollutant that can be applied to a site in its lifetime by all bulk biosolids applications meeting ceiling concentration limits (CCL). It is the maximum amount of an inorganic pollutant that can be applied to an area of land. This term applies to bulk sewage sludge that is land applied.

Current Meter An instrument for measuring the speed of flowing water. The Geological Survey uses a rotating cup meter.

CWA § 101 The objective of the Clean Water Act (CWA) is to restore and maintain the chemical, physical, and biological integrity of the nation's waters.

CWA § 303d This section of the CWA requires the states to identify waters that do not or are not expected to meet applicable water quality standards with technology-based controls alone. The waters impacted by thermal discharges are also to be identified. After the identification and priority ranking of water quality-limited waters are completed, the states are to develop total maximum daily loads (TMDLs) at a level necessary to achieve the applicable state water quality standards.

CWA § 314 This section of the CWA establishes the Clean Lakes Program, which supports activities from initial identification of potential water quality problems through post-restoration monitoring. Cooperative grants provide funding for these activities.

CWA § 319 This section of the CWA requires the states to develop nonpoint source (NPS) control programs. The USEPA awards grants to implement approved programs that include, as appropriate, nonregulatory and regulatory programs for enforcement, technical assistance, financial assistance, education, training, technology transfer, and demonstration projects.

CWA § 320 This section of the CWA establishes that National Estuary Program (NEP), a demonstration program designed to show how estuaries and their living resources can be protected through comprehensive, action-oriented management. Participation in the NEP is limited to estuaries determined by the USEPA administrator to be of "national significance" after nomination by the governors of the states in which the estuaries are located.

CWA § 402 This section of the CWA establishes the National Pollutant Discharge Elimination System (NPDES), which provides for the issuance of point source permits to discharge any pollutant or a combination of pollutants, after an opportunity for public hearing.

CWA § 404 The discharges of dredged or fill material into wetlands is regulated under this section of the CWA. Permits may be issued after a notice and an opportunity for public hearings.

Cycle A regularly recurring succession of events such as the cycle of the seasons. The use of cycle to describe a group of wet years followed or preceded by a group of dry years is to be avoided.

Cyclic Well Operation A groundwater management practice in which each well is operated at fixed rate for a prescribed period of each year and turned off afterwards.

Dam A barrier that impounds water.

Dead Storage The volume in a reservoir below the lowest controllable level.

Decision Variables The independent variables of an optimization problem that a decision maker can control and choose in order to best achieve management goals.

Decomposers The bacteria and fungi that break down organic detritus. Decomposers (bacteria and fungi).

Degradation The process by which streambeds, floodplains, and the bottoms of other water bodies are lowered in elevation by erosion of material. It is the opposite of aggradation.

Deleted NPL Site A site that has been deleted from the Superfund National Priorities List (NPL) because its cleanup goals have been met, and there is no further need for federal action (see Superfund and National Priorities List).

Denitrification A biochemical process of conversion of nitrite nitrogen and nitrate nitrogen to molecular nitrogen, nitrogen dioxide, or a mixture of these two gases, under reducing conditions in the absence of free dissolved oxygen.

Density A physical property of all matter and is given as the mass per unit volume.

Deoxygenation It is a process for depletion of the dissolved oxygen in a liquid either under natural conditions associated with the biochemical oxidation of organic matter present or by the addition of chemical reducing agents.

Dependable Yield, n Years The minimum supply of a given water development that is available on demand, with the understanding that lower yields will occur once in *n* years, on the average.

Depletion The progressive withdrawal of water from surface or groundwater reservoirs at a rate greater than that of replenishment (see recession curve and streamflow depletion).

Deposit Material left in a new position by a transporting agent such as earthquake, gravity, human activity, ice, water current, or wind.

Deposition The mechanical or chemical processes through which sediments accumulate in a (temporary) resting place.

Depression Storage The volume of water contained in natural depressions in the land surface, such as puddles.

Depth of Flow The vertical distance from the bed of a stream to the water surface.

Depuration (1) The excretion of contaminant by an organism. (2) The action or process of freeing impurities.

Desalination The removal of salts from saline water to provide freshwater. This method is becoming a more popular way of providing freshwater to populations.

Designated Use A simple narrative description of water quality expectations or water quality goals. A designated use is a legally recognized description of a desired use of the water body, such as (1) support of communities of aquatic life, (2) body contact recreation, (3) fish consumption, and (4) public drinking water supply. These are uses that the state or an authorized tribe wants the water body to be healthy enough to fully support. The US Clean Water Act (CWA) requires that waterbodies attain or maintain the water quality needed to support designated uses.

Desorption The process by which chemicals are detached and released from solid surfaces; the opposite of adsorption.

Detachment Process of tearing loose soil particles.

Detention Time The average length of time a water molecule or a suspended particle remains in a tank or chamber. Mathematically, it is the volume of water (L^3) in the tank divided by the flow rate (L^3/T) through the tank.

Detritus Dead organic matter.

Diatom Any of class of minute algae with cases of silica.

Dielectric Constant The ratio of permittivity of a substance to the permittivity of free space.

Digital Terrain Model Also called a digital elevation model (DEM), it is a three-dimensional digital representation of a terrain's surface.

Dipper Dredge A mechanical equipment that has a power-operated dipper stick which, by sliding through the center plane of a boom, allows the operator to control the movement of the dipper bucket in any direction. Spuds are used for maintaining a stable working position. The boom and dipper bucket are swung around to empty the bucket into a scow that is waiting alongside the dredge. The dipper dredge is a heavy-duty excavator designed for breaking up hard compacted clay and ledge rock with its bucket. New work dredging is usually done with this type of dredge.

Direct Runoff The runoff entering stream channels promptly after rainfall or snow-melt. Superposed on base runoff, it forms the bulk of the hydrograph of a flood.

Discharge In its simplest, concept discharge means outflow; therefore, the use of this term is not restricted as to a course or location, and it can be applied to describe the flow of water from a pipe or from a drainage basin. If the discharge occurs in some course or channel, it is correct to speak of the discharge of a canal or of a river. It is also correct to speak of the discharge of a canal or stream into a lake, a stream, or an ocean (see also streamflow and runoff). The data in the reports of the Geological Survey on surface water represent the total fluids measured. Thus, the terms discharge, streamflow, and runoff represent water with the solids dissolved in it and the sediment mixed with it. Of these terms, discharge is the most comprehensive. The discharge of drainage basins is distinguished as follows: (1) yield = total water runout or crop, includes runoff plus underflow; (2) runoff = that part of water yield that appears in streams; and (3) stream flow = the actual flow in streams, whether or not subject to regulation, or underflow. Each of these terms can be reported in total volumes (such as acre–feet) or time rates (such as cubic feet per second or acre–feet per year). The differentiation between runoff as a volume and streamflow as a rate is not accepted.

Discharge of Waste The volume of a waste fluid or solid passing a cross section of a stream per unit time.

Discharge Rating Curve See stage–discharge relation.

Disinfectant A chemical (commonly chlorine, chloramine, or ozone) or physical process (e.g., ultraviolet light) that kills microorganisms such as bacteria, viruses, and protozoa.

Dissolved Air Flotation (DAF) A process in which air is dissolved into water under high pressure and is subsequently released into the bottom of a treatment unit to float solid particles. Upon release, the lower pressure in the treatment unit results in the formation of fine air bubbles that collect particles as they rise to the surface. The floated particles are then skimmed for subsequent processing. After the particles are collected and removed, the remaining process water is purified.

Dissolved Air Flotation Clarifier (DAF Clarifier) A flotation clarifier (DAF clarifier) which is used for separation of particles from water by the flotation force of gas bubbles.

Dissolved Gas Flotation (DGF) A process which is similar to DAF, except that different kinds of gases (instead of air) can be used for generation of gas bubbles.

Dissolved Gases The sum of gaseous components, such as oxygen, nitrogen, carbon dioxide, methane, hydrogen sulfide, etc., that are dissolved in water.

Dissolved Oxygen (DO) The concentration of oxygen in water, which is often expressed in units of parts per million (ppm) or milligrams per liter (mg/L).

Dissolved Solids The constituents in water that can pass through a 0.45-μm pore-diameter filter.

Distressed Watershed It is a watershed which has aquatic life and health that is impaired by nutrients (nitrogen and phosphorus) from agricultural land uses, such as land application. Threats to public health, drinking water supplies, recreation, and public safety are also taken into consideration if a watershed is designated as a distressed watershed.

Distribution Graph (Distribution Hydrograph) A unit hydrograph of direct runoff modified to show the proportions of the volume of runoff that occurs during successive equal units of time.

Distribution System A network of pipes leading from a treatment plant to customers' plumbing systems.

Diurnal Pertaining to daily occurrence.

Diversion The taking of water from a stream or other body of water into a canal, pipe, or other conduit.

Domain A set of input or argument values for which a function is defined.

Domestic Septage Either a liquid or solid material removed from a septic tank, cesspool, portable toilet, type III marine sanitation device, or similar treatment works that receives only domestic sewage. This does not include septage resulting from treatment of wastewater with a commercial or industrial component.

Domestic Water Use Water used for household purposes, such as drinking, food preparation, bathing, washing clothes, dishes, dogs, flushing toilets, and watering lawns and gardens. About 85 % of domestic water is delivered to homes by a public-supply facility, such as a county water department. About 15 % of the nation's population supply their own water, mainly from wells.

Double-Mass Curve A plot on an arithmetic cross-section paper of the cumulated values of one variable against the cumulated values of another or against the computed values of the same variable for a concurrent period of time.

Drainage Area The drainage area of a stream at a specified location is that area, measured in a horizontal plane, which is enclosed by a drainage divide.

Drainage Basin (1) A part of the surface of the earth that is occupied by a drainage system, which consists of a surface stream or a body of impounded surface water together with all tributary surface streams and bodies of impounded surface water; (2) the land area where precipitation runs off into streams, rivers, lakes, and reservoirs; (3) a land feature that can be identified by tracing a line along the highest elevations between two areas on a map, often a ridge. Large drainage basins, like the area that drains into the Mississippi River contain thousands of smaller drainage basins. Also called a "watershed."

Drainage Density The length of all channels above those of a specified stream order per unit of drainage area.

Drainage Divide The rim of a drainage basin (see watershed).

Drawdown A lowering of the groundwater surface caused by pumping.

Dredge A mechanical equipment that is used for dredging, such as clamshell dredge, dipper dredge, pipeline dredge, and hopper dredge.

Dredge Containment A constructed structure, such as a diked disposal area, which is built on coastal marshes, or into appropriate shore promontories or offshore islands, for confinement of dredged material.

Dredged Material The material that is excavated or dredged from surface waters, such as stream/river, lake, reservoir, or seawater.

Dredging (1) A mechanical method for removing unwanted aquatic plants and/or mud or the sediment from the bottom of river/stream, lake, reservoir, or coastal area. (2) The dredged material.

Drinking Water Quality Refers to whether contaminants are present in water that people drink, including water from the tap, private wells, hauled water, untreated surface water sources, and bottled water, at levels that could affect human health.

Drinking-Water Standards Regulations that the government, such as the USEPA, sets to control the level of contaminants in the nation's drinking water. Enforceable standards include maximum contaminant levels (MCLs) and treatment techniques (TTs; see separate entries for each). Drinking-water standards apply to all public water systems (PWS; see public water system).

Drip Irrigation A common irrigation method where pipes or tubes filled with water slowly drip onto crops. Drip irrigation is a low-pressure method of irrigation and less water is lost to evaporation than high-pressure spray irrigation.

Drop Structure A natural or man-placed structure that disrupts the continuous surface flow pattern in a river or stream by producing a pooling of water behind the structure and a rapid drop in the surface gradient for water flowing over the structure; used to improve habitat conditions for aquatic life and to increase the air (especially oxygen) content of water.

Drought A period of deficient precipitation or runoff extending over an indefinite number of days, but with no set standard by which to determine the amount of deficiency needed to constitute a drought. Thus, there is no universally accepted quantitative definition of drought; generally, each investigator establishes his own definition. When in an area that is ordinarily classed as humid, natural vegetation becomes desiccated or defoliates unseasonably and crops fail to mature owing to lack of precipitation, or when precipitation is insufficient to meet the needs of established human activities, drought conditions may be said to prevail. Although water for irrigation or other uses in arid areas is always limited, special shortages in such areas are also regarded as droughts. Unsatisfactory distribution of precipitation throughout the year may be as effective a factor in causing a drought as a shortage in the total amount. Temperature and wind may also play an important part, especially in relation to the damage done.

Dry Granular Flow It refers to a process in which individual sediment grains roll, slide, and bounce down the bank. It typically occurs on noncohesive banks near the angle of repose.

Dunes Bed forms with triangular profile that advance downstream due to net deposition of particles on the steep downstream slope. Dunes move downstream at velocities that are small, relative to the stream flow velocity.

Duration Curve See flow-duration curve for one type.

Dust Bowl Period of severe dust storms (1930s) that greatly damaged the ecology and agriculture of the US and Canadian prairies.

Dynamic Equilibrium A state of relative balance between processes having opposite effects.

Dynamic Viscosity It is the property of a fluid whereby it tends to resist relative motion within itself. It is the shear stress, i.e., the tangential force on unit area, between two infinite horizontal planes at unit distance apart, one of which is fixed while the other moves with unit velocity. In other words, it is the shear stress divided by the velocity gradient, i.e., $(N/m^2)/(m/s/m) = N\ s/m^2$.

Earthen Structure Refers to a host of artificial structures constructed using earthen (soil, rock, etc.) materials.

Ecological Condition A term referring to the state of the physical, chemical, and biological characteristics of the environment, and the processes and interactions that connect them.

Ecological Connectivity A term referring to the connected system of open space throughout an ecosystem and adjacent ecosystems. Includes the presence of ecotones, the transitional regions between ecosystems.

Ecological Processes The metabolic functions of ecosystems—energy flow; elemental cycling; and the production, consumption, and decomposition of organic matter.

Ecological System A hierarchically nested area that includes all living organisms (people, plants, animals, and microorganisms), their physical surroundings (such as soil, water, and air), and the natural cycles that sustain them.

Ecology (1) The study of the relationships between the environment and the living organisms and beings present. (2) The study of the interrelationships of organisms with and within their environment.

Economic Lifetime Period over which an asset is expected to be useable, with normal repairs and maintenance.

Ecoregion (1) An ecological region that has broad similarities with respect to soil, relief, and dominant vegetation. (2) An area within which the ecosystems—and the type, quality, and quantity of environmental resources—are generally similar. An ecoregion can serve as a spatial framework for the research, assessment, management, and monitoring of ecosystems and ecosystem components. Several different classification schemes have been developed, at various resolutions.

Ecosystem (1) The interacting system of a particular biological community and its nonliving environmental surroundings, or a class of such systems (e.g., forests or wetlands). (2) A biotic community and its (living and nonliving) environment considered together.

Ecotoxicology The study of the quantitative effects of chemicals on ecological system, particularly in terms of definite harmful actions and degrees of safety.

Eddy Viscosity A method to model the transfer of the momentum caused by turbulent eddies that is mathematically similar to the momentum transfer due to molecular diffusion, and that consists in replacing the fluid viscosity v by an effective turbulent viscosity, v_t, also called the eddy viscosity.

Effective Precipitation (Rainfall) (1) That part of the precipitation that produces runoff; (2) a weighted average of current and antecedent precipitation that is "effective" in correlating with runoff; (3) as described by the US Bureau of Reclamation, that part of the precipitation falling on an irrigated area that is effective in meeting the consumptive use requirements.

Effluent The water that flows from a sewage treatment plant after it has been treated.

Electrode An electrical conductor used to make contact with a nonmetallic part of a circuit.

Embankment An artificial bank raised above the immediately surrounding land to redirect or prevent flooding by a river, lake, or sea.

Emergent Plant An aquatic plant, usually rooted, which has portions above water for part of the year, but does not tolerate prolonged inundation.

Emission Factor The relationship between the amount of pollution produced by a particular source and the amount of raw material processed. For example, an emission factor for a blast furnace making iron might be pounds of particulates emitted per ton of raw materials processed.

Emission Inventory A listing, by source and pollutant, of the amount of air pollutants discharged into the atmosphere. Emission inventories can be based on emissions estimates, emissions measurements, or both.

End State Any one of a number of ecosystem characteristics observed at a point in time. The term is commonly used to represent the results of ecological processes.

Endpoint A biological or ecological characteristic that is the basis for evaluation or measurement.

Energy Cycling The movement, or flow, and storage of energy among production and use components of ecological and physiological systems.

Energy Dissipation The transformation of mechanical energy into heat energy. In fluids, this is accomplished by viscous shear. The rate of energy dissipation in flowing fluids varies with the scale and degree of the turbulence. Baffles, the hydraulic jump, and other damping methods may be used to dissipate energy.

Enhancement An activity increasing one or more natural or artificial wetland functions. For example, the removal of a point source discharge impacting a wetland.

Environment The sum total of all the external conditions that act on an organism.

Ephemeral Gully Erosion A topographically driven erosion caused by runoff concentration within a few natural waterways or swales. Typically, these features are larger than rills and may be erased/removed by tillage operations.

Ephemeral Waters Water bodies (e.g., streams or wetlands) that contain water for brief periods, usually in direct response to a precipitation event. Ephemeral waters generally flow for a shorter time period than intermittent waters, although in some cases the terms are used interchangeably (see intermittent waters).

Epilimnion The well-mixed surficial layer of a lake; above the hypolimnion (see thermal stratification).

Epiphytes Plants that grow on other plants, but are not parasitic.

Equilibrium A steady state in a dynamic system, with the outflow balancing the inflow.

Erodibility Susceptibility of a soil to be detached and transported. Defined as the amount of soil loss per unit exogenic force of rainfall and overland flow.

Erosion (1) The process in which a material is worn away by a stream of liquid (water) or air, often due to the presence of abrasive particles in the stream. (2) The process by which soil and rock are removed by the action of wind and water. (3) The wearing away of the land surface or stream boundaries by detachment and movement of soil and rock fragments through the action of moving water or other geological agents.

Erosion Control A mechanical practice of preventing or controlling wind or water erosion in agriculture, land development, and construction.

Erosivity The power of a storm or surface flow to erode the soil, usually determined from storm characteristics such as rainfall intensity and energy or flow volume and flow gradient.

Estuary A place where freshwater and saltwater mix, such as a bay, salt marsh, or where a river enters an ocean.

Eutrophic Aquatic systems with high nutrient input and high plant growth.

Eutrophication Enrichment of an aquatic ecosystem with nutrients (nitrogen, phosphorus) that accelerate biological productivity (growth of algae and weeds) and an undesirable accumulation of algal biomass.

Evaporation (1) The process by which water is changed from the liquid or the solid state into the vapor state; (2) the process of liquid water becoming water vapor, including vaporization from water surfaces, land surfaces, and snow fields, but not from leaf surfaces (see transpiration). In hydrology, evaporation is the vaporization that takes place at a temperature below the boiling point.

Evaporation Opportunity (Relative Evaporation) The ratio of the rate of evaporation from a land or water surface in contact with the atmosphere, to the evaporativity under existing atmospheric conditions. It is the ratio of the actual to potential rate of evaporation, generally stated as a percentage. The opportunity for a given rate of evaporation to continue is determined by the available moisture supply.

Evaporation Pan An open tank used to contain water for measuring the amount of evaporation. The US Weather Bureau class A pan is 4 (1.22 m) in diameter and 10 in. (25.4 cm) deep, set up on a timber grillage so that the top rim is about 16 in. from the ground. The water level in the pan during the course of observation is maintained between 2 and 3 in. (5.08 and 7.62 cm) below the rim.

Evaporation, Total The sum of water lost from a given land area during any specific time by transpiration from vegetation and building of plant tissue; by evaporation from water surfaces, moist soil, and snow; and by interception. It has been variously termed "evaporation," "evaporation from land areas," "evapotranspiration," "total loss," "water losses," and "fly off."

Evaporativity (Potential Rate of Evaporation) The rate of evaporation under the existing atmospheric conditions from a surface of water that is chemically pure and has the temperature of the atmosphere.

Evapotranspiration (1) The water withdrawn from a land area by evaporation from water surfaces and moist soil and plant transpiration; (2) the sum of evaporation and transpiration; (3) the combined conversion of water to water vapor and loss resulting from both evaporation and transpiration.

Evapotranspiration, Potential See potential evapotranspiration.

Exceptional Quality Sewage Sludge Sewage sludge that meets the most stringent limits for the three sludge quality parameters. In gaging sewage sludge quality, USEPA determined that three main parameters of concern should be considered: (1) pollutant levels, (2) the relative presence or absence of pathogenic organisms, such as salmonella and *Escherichia coli* bacteria, enteric viruses, or viable helminth ova, and (3) the degree of attractiveness of the sewage sludge to vectors, such as flies, rats, and mosquitoes, that could potentially come in contact with pathogenic organisms and spread disease. Given these three variables, there can be a number of possible sewage sludge qualities. The term exceptional quality (EQ), which does not appear in the Part 503 regulation, is used to describe sewage sludge that meets the highest quality for all three of these sewage sludge quality parameters.

Excessive Rainfall See rainfall, excessive.

Exemption The state or USEPA permission for a water system not to meet a certain drinking-water standard. An exemption allows a system additional time to obtain financial assistance or make improvements in order to come into compliance with the standard. The system must prove that: (1) there are compelling reasons (including economic factors) why it cannot meet an maximum contaminant level (MCL) or treatment technique (TT); (2) it was in operation on the effective date of the requirement, and (3) the exemption will not create an unreasonable risk to public health. The state must set a schedule under which the water system will comply with the standard for which it received an exemption.

Exogenic External forcing.

Exposure For humans, the amount of a chemical, physical, or biological contaminant at the outer boundary of the body available for exchange or intake via inhalation, ingestion, or skin or eye contact.

Extent The amount and distribution of a resource, which may be measured in terms of spatial area, volume, depth, or flow (e.g., for water resources). Report on the Environment (ROE) questions address the extent of fresh surface waters, groundwater, wetlands, and coastal waters.

Extraction and Mining Waste Soil and rock generated during the process of gaining access to the ore or mineral body, as well as water that infiltrates the mine during the extraction process. This category also includes certain wastes associated with the beneficiation of ores and minerals, including wastes from the following activities: crushing, grinding, washing, dissolution, crystallization, filtration, sorting, sizing, drying, sintering, pelletizing, briquetting, calcining to remove water and/or carbon dioxide, roasting in preparation for leaching (except where the roasting/leaching sequence produces a final or intermediate product that does not undergo further beneficiation or processing), gravity concentration, magnetic separation, electrostatic separation, floatation, ion exchange, solvent extraction, electrowinning, precipitation, amalgamation, heap, dump, vat, tank, and in situ leaching.

Farm Field The farm field is the basic management unit used for all farm nutrient management, and defined as "the fundamental unit used for cropping agricultural products."

Fauna The animals of a habitat or region.

Feed Crop Crops produced primarily for consumption by animals. These include, but are not limited to, corn and grass. For a crop to be considered a feed crop, it has to be produced for consumption by animals (e.g., grass grown to prevent erosion or to stabilize an area is not considered a feed crop).

Fiber Crop Crops, such as flax and cotton, that were included in Part 503 because products from these crops (e.g., cotton seed oil) may be consumed by humans.

Field Capacity See field-moisture capacity.

Field-Moisture Capacity The quantity of water which can be permanently retained in the soil in opposition to the downward pull of gravity.

Field-Moisture Deficiency The quantity of water, which would be required to restore the soil moisture to field-moisture capacity.

Final National Priorities List (NPL) Site A site that has been formally added to the Superfund NPL (see Superfund and National Priorities List).

Finished Water The water that has been treated and is ready to be delivered to customers.

Finite Aquifer A constant-width aquifer, delimited by two parallel boundaries.

Firn (Firn Snow) Old snow on the top of glaciers, granular and compact, but not yet converted into ice. It is a transitional stage between snow and ice.

Firn Line The highest level to which the fresh snow on a glacier's surface retreats during the melting season. The line separating the accumulation area from the ablation area.

Flood (1) An overflow or inundation that comes from a river or other body of water, and causes or threatens damage; (2) any relatively high streamflow overtopping the natural or artificial banks in any reach of a stream; (3) a relatively high flow as measured by either gage height or discharge quantity; (4) an overflow of water onto lands that are used or usable by man and not normally covered by water. Floods have two essential characteristics: The inundation of land is temporary; and the land is adjacent to and inundated by an overflow from a river, stream, lake, or ocean. (5) The inundation of a normally dry area caused by high flow, or overflow of water in an established watercourse, such as a river, stream, or drainage ditch; or ponding of water at or near the point where the rain fell.

Flood Control All methods used to reduce or prevent the detrimental effects of floodwaters.

Flood Control Storage Storage of water in reservoirs to abate flood damage.

Flood Crest See flood peak.

Flood Event See flood wave.

Flood Peak The highest value of the stage or discharge attained by a flood; thus, peak stage or peak discharge. Flood crest has nearly the same meaning, but since it connotes the top of the flood wave, it is properly used only in referring to stage—thus, crest stage, but not crest discharge.

Flood Plain (1) A strip of relatively smooth land bordering a stream, built of sediment carried by the stream and dropped in the slack water beyond the influence of the swiftest current. It is called a living flood plain if it is overflowed in times of high water; but a fossil flood plain if it is beyond the reach of the highest flood. (2) The lowland that borders a river, usually dry but subject to flooding. (3) That land, outside of a stream channel, described by the perimeter of the maximum probable flood. (4) A strip of relatively flat and normally dry land alongside a stream, river, or lake that is covered by water during a flood. (5) That part of a river valley that is covered in periods of high (flood) water.

Flood Plane The position occupied by the water surface of a stream during a particular flood. Also, loosely, the elevation of the water surface at various points along the stream during a particular flood.

Flood Profile A graph of elevation of the water surface of a river in flood, plotted as ordinate, against distance, measured in the downstream direction, plotted as abscissa. A flood profile may be drawn to show the elevation at a given time, crests during a particular flood, or to show the stages of concordant flows.

Flood Routing The process of determining progressively the timing and shape of a flood wave at successive points along a river.

Flood Stage (1) The elevation at which overflow of the natural banks of a stream or body of water begins in the reach or area in which the elevation is measured; (2) the gage height of the lowest bank of the reach in which the gage is situated. The term "lowest bank" is, however, not to be taken to mean an unusually low place or break in the natural bank through which the water inundates an unimportant and small area. The stage at which overflow of the natural banks of a stream begins to cause damage in the reach in which the elevation is measured (see also bankfull stage).

Flood Wave A distinct rise in stage culminating in a crest and followed by recession to lower stages.

Flood Zone The land bordering a stream which is subject to floods of about equal frequency; for example, a strip of the flood plain subject to flooding more often than once but not as frequently as twice in a century.

Flood, 100-Year A 100-year flood does not refer to a flood that occurs once every 100 years, but to a flood level with a 1 % chance of being equaled or exceeded in any given year.

Flood, Maximum Probable The largest flood for which there is any reasonable expectancy in this climatic era.

Flood-Control Storage Storage of water in reservoirs to abate flood damage (see retarding reservoir).

Flood-Frequency Curve (1) A graph showing the number of times per year on an average, plotted as abscissa, that floods of magnitude, indicated by the ordinate, are equaled or exceeded; (2) a similar graph but with recurrence intervals of floods plotted as abscissa.

Floodplain A normally dry land adjacent to a body of water which is susceptible to periodic inundation by floodwaters.

Floods Above a Base See partial-duration flood series.

Floodway (1) The channel of a river or stream and the parts of the floodplain adjoining the channel that are reasonably required to efficiently carry and discharge the floodwater or flood flow of a river or stream; (2) a part of the floodplain otherwise leveed, reserved for emergency diversion of water during floods. A part of the floodplain which, to facilitate the passage of floodwater, is kept clear of encumbrances. The channel of a river or stream and those parts of the floodplains adjoining the channel, which are reasonably required to carry and discharge the floodwater or flood flow of any river or stream.

Flora Plants of a habitat or region

Flotation A process method for removing coagulated particles by introducing gas bubbles to attach and float particles to the surface. The collected particles are skimmed and discharged to solids processing. After the particles are collected and removed, the remaining processed water is purified.

Flotation Cell A tank, clarifier, or reactor which is used for releasing gas bubbles, floating the coagulated particles to the water surface, skimming off the float (floated scum) from the water surface, and discharging the flotation-treated effluent.

Flotation/Filtration Cell A tank or reactor which is a combination of both flotation clarifier and filterer, and is used for releasing gas bubbles, floating the coagulated particles to the water surface, skimming off the float (floated scum) from the water surface, filtering the flotation effluent, and discharging the filtered effluent.

Flow-Duration Curve A cumulative frequency curve that shows the percentage of time that specified discharges are equaled or exceeded.

Flowing Well/Spring A well or spring that taps groundwater under pressure so that water rises without pumping. If the water rises above the surface, it is known as a flowing well.

Flume An open conduit of masonry, wood, or metal constructed on a grade and sometimes elevated. It is sometimes called an aqueduct.

Fluvial (1) In this chapter, pertaining to streams; (2) of or pertaining to one or more rivers; produced by the action of a river or stream; growing, living, or existing in or near a river or stream.

Fluvial Deposit The sediment deposited by the action of rivers. It is also called an alluvial deposit.

Fluvial Erosion Erosion caused by the action of rivers.

Fluvial Index River index.

Fluvial Process The process of one or more rivers causing erosion, sedimentation, etc.

Fluvial Sediment Particles derived from rocks or biological materials that are transported by, suspended in, or deposited by streams.

Food Chain Animals linked by linear predator–prey relationships with plants or detritus.

Food Crop The crops consumed by humans. These include, but are not limited to, fruits, grains, vegetables, and tobacco.

Food Web This is similar to food chain, but implies cross connections.

Forage Fish The fish eaten by other fish.

Forest Influences The effects resulting from the presence of forest or brush upon climate, soil water, runoff, streamflow, floods, erosion, and soil productivity.

Forest Land A tract of land thick with trees and underbrush.

Formation The process of orderly grouping.

Fossil Fuel Combustion Waste The waste from the combustion of oil, natural gas, or petroleum coke; the combustion of coal at electric utilities and independent power-producing facilities, nonutilities, and facilities with fluidized bed combustion technology; or the combustion of mixtures of coal and other fuels (i.e., coburning of coal with other fuels) where coal is at least 50 % of the total fuel.

Frazil (Frazil Ice) A French–Canadian term for fine spicular ice, derived from the French for cinders which this variety of ice most resembles. When formed in saltwater, it is known as lolly ice. It is composed of fine particles which, when first formed, are colloidal and not seen in the water in which they are floating.

Frequency Number of occurrences of a repeating event per unit time.

Fresh Water See freshwater.

Freshwater Water that contains less than 1000 mg/L of dissolved solids; generally, more than 500 mg/L of dissolved solids is undesirable for drinking and many industrial uses.

Froude Number (F_r) (1) It is a numerical quantity used in open-channel flow studies or in cases in which the free surface plays an essential role in influencing motion. It is given by the equation of $Fr = V^2/(gL)$, where V is the characteristic velocity, g is the gravitational constant, and L is the characteristic linear dimension. (2) A dimensionless quantity defined as the ratio of a characteristic velocity to a gravitational wave velocity.

Functions of Wetland The role wetlands serve which are of value to society or the environment.

Gage Height (1) The water-surface elevation referred to some arbitrary gage datum. Gage height is often used interchangeably with the more general term stage although gage height is more appropriate when used with a reading on a gage. (2) The height of the water surface above the gage datum (zero point). Gage height is often used interchangeably with the more general term, stage, although gage height is more appropriate when used with a gage reading.

Gaging Station (1) A particular site on a stream, canal, lake, or reservoir where systematic observations of gage height or discharge are obtained (see also stream-gaging station); (2) a site on a stream, lake, reservoir, or other body of water where observations and hydrologic data are obtained. The US Geological Survey measures stream discharge at gaging stations.

Gametes See gametogenesis.

Gametogenesis The formation of male and female sex cells (sperm or ova), which are also referred to as gametes.

Gas Adsorption An adsorption process that attracts and adsorbs a gas to a liquid (such as water), or a solid (such as granular activated carbon).

Geofluvial Models A geofluvial model refers to one that is capable of coupling in-channel flow and sediment-routing models with bank erosion and mass wasting algorithms.

Geographic Information System (GIS) A tool that links spatial features commonly seen on maps with information from various sources ranging from demographics to pollutant sources.

Geomorphic Processes They may encompass a wide range of processes related to landforms and stream forms and the processes that shape them. In this report, they specifically refer to the processes that shape the streams.

Geomorphology The geologic study of the evolution and configuration of landforms.

Geophone A device that converts ground movement into voltage.

Geophysics The study of earth materials using quantitative physical methods.

Geotechnical The study of the behavior of earth materials.

Geyser A geothermal feature of the earth where there is an opening in the surface that contains superheated water that periodically erupts in a shower of water and steam.

Giardia Lamblia A microorganism frequently found in rivers and lakes, which, if not treated properly, may cause diarrhea, fatigue, and cramps after ingestion.

Giardiasis A disease that results from an infection by the protozoan parasite *Giardia intestinalis*, caused by drinking water that is either not filtered or not chlorinated. The disorder is more prevalent in children than in adults and is characterized by abdominal discomfort, nausea, and alternating constipation and diarrhea.

Glacier (1) A huge mass of ice, formed on land by the compaction and recrystallization of snow, that moves very slowly downslope or outward due to its own weight; (2) bodies of land ice that consist of recrystallized snow accumulated on the surface of the ground, and that move slowly downslope.

Global Climate Change See climate change.

Glover Solution Analytical model developed by Glover and Balmer (1954) to estimate the stream depletion rate due to pumping from a point source in a semi-infinite, homogeneous, and connected alluvial aquifer.

Godunov Scheme A conservative finite-volume numerical scheme used in the solution of partial differential equations, which solves exact or approximate Riemann problems at intercell boundaries.

Grain Size See particle size.

Graphical User Interface A type of computer user interface that allows the user to interact with a computer program using pointing hardware devices, graphical icons, and other visual indicators.

Greenhouse Gas (GHG) Any gas that absorbs infrared radiation in the atmosphere. Greenhouse gases include water vapor, carbon dioxide (CO_2), methane (CH_4), nitrous oxide (N_2O), halogenated fluorocarbons (HCFCs), ozone (O_3), perfluorinated carbons (PFCs), and hydrofluorocarbons (HFCs).

Gray Water Wastewater from clothes washing machines, showers, bathtubs, hand washing, lavatories and sinks.

Groundwater (1) The water that flows or seeps downward and saturates soil or rock, supplying springs and wells. The upper surface of the saturate zone is called the water table; (2) the water stored underground in rock crevices and in the pores of geologic materials that make up the earth's crust; (3) the supply of freshwater that is found under the earth's surface in underground rock formations or soil; (4) the water in the ground that is in the zone of saturation, from which wells, springs, and groundwater runoff are supplied; (5) the water that systems pump and treat from aquifers (natural reservoirs below the earth's surface).

Groundwater, Confined Groundwater under pressure significantly greater than atmospheric, with its upper limit the bottom of a bed with hydraulic conductivity distinctly lower than that of the material in which the confined water occurs.

Groundwater, Unconfined The water in an aquifer that has a water table that is exposed to the atmosphere.

Groundwater Outflow That part of the discharge from a drainage basin that occurs through the groundwater. The term "underflow" is often used to describe the groundwater outflow that takes place in valley alluvium (instead of the surface channel) and thus is not measured at a gaging station.

Groundwater Recharge The inflow of water to a groundwater reservoir from the surface. The infiltration of precipitation and its movement to the water table is one form of natural recharge. Also, the volume of water added by this process.

Groundwater Runoff That part of the runoff which has passed into the ground, has become groundwater, and has been discharged into a stream channel as spring or seepage water (see also base runoff and direct runoff).

Gully Erosion Slope incisions by flowing water that erodes soil to form channels deeper than 30 cm. Typically, these features require extensive and expensive treatment methods to abate.

Guttation The loss of water in liquid form from the uninjured leaf or stem of the plant, principally through water stomata.

Habitat (1) The environment in which a population of plants or animals occurs. (2) The environment occupied by individuals of a particular species, population, or community.

Hardness A water-quality indication of the concentration of alkaline salts in water, mainly calcium and magnesium. If the water you use is "hard" then more soap, detergent, or shampoo is necessary to raise a lather.

Hazardous Air Pollutants See air toxics.

Hazardous Waste Waste with properties that make it dangerous or potentially harmful to human health or the environment. The universe of hazardous wastes is large and diverse. Hazardous wastes can be liquids, solids, contained gases, or sludge. They can be the byproducts of manufacturing processes or simply discarded commercial products, like cleaning fluids or pesticides. Hazardous waste is regulated under the Resource Conservation and Recovery Act (RCRA) Subtitle C (see RCRA hazardous waste for the regulatory definition). The states can identify additional wastes as hazardous beyond those identified by the USEPA.

Head The difference between the pool height and tailwater height. Usually expressed in feet of head, or in lbs./sq. in. (meter of head or in kg/sq meter).

Head Drawdown A state variable consisting of the reduction of hydraulic head in an aquifer due to groundwater pumping.

Head Loss The decrease in total head caused by friction, entrance, and exit losses.

Head Cut A step change in bed topography. The primary position of soil detachment within rills and gullies.

Headwater(s) (1) The source and upper reaches of a stream; also the upper reaches of a reservoir. (2) The water upstream from a structure or point on a stream. (3) The small streams that come together to form a river. Also, may be thought of as any and all parts of a river basin except the mainstream river and main tributaries.

Health Advisory An EPA document that provides guidance and information on contaminants that can affect human health and that may occur in drinking water, but which EPA does not currently regulate in drinking water.

Health-Based Standards Standards based on contaminant concentrations in environmental media or exposure doses that are likely to be without an appreciable risk of adverse health effects in humans. (Some health-based standards allow for consideration of technological and cost limitations.)

Heat Budget, Annual (of a Lake) The amount of heat necessary to raise the water from the minimum temperature of winter to the maximum temperature of summer.

Heavy Metals Trace elements are found in low concentrations in the environment, such as water, soil, or biosolids. They are commonly referred to as either "heavy metals" or "trace elements" (e.g., copper, molybdenum, and zinc) which are nutrients needed for plant or animal growth in low concentrations, but all of these elements can be toxic to humans, animals, or plants at high concentrations. Possible hazards associated with a buildup of trace elements in the soil include their potential to cause phytotoxicity (i.e., injury to plants) or to increase the concentration of potentially hazardous substances in the food chain. Federal and state regulations have established standards for the following nine trace elements: arsenic (As), cadmium (Cd), copper (Cu), lead (Pb), mercury (Hg), molybdenum (Mo), nickel (Ni), selenium (Se), and zinc (Zn).

Hopper Dredge A self-propelled ship that is equipped with dray heads for excavating the sediments, pumps for sucking up and transporting the dredging, and hoppers (movable containers) for settling and storing the dredging. When operating, the underwater sediments are excavated and eventually settled in the hoppers. Excess water with suspended fines overflows at the tops of the hoppers. Dredging continues until a sufficient load of sediments has been accumulated in the hoppers. The dredge itself will then transport its load to a disposal site where the hoppers are emptied by opening the discharge doors or by pumping the dredging out through a discharge pipe. Hopper dredges are efficient in excavating a thin layer of loose sediment covering extensive areas.

Humic Pertaining to the partial decomposition of leaves and other plant material

Hydraulic (1) Referring to water or other fluids in motion. (2) Study of the mechanical properties of liquids.

Hydraulic Conductivity A physical property representing the ability of the porous medium to conduct water.

Hydraulic Fluvial Processes Refer to the processes that are responsible for bank retreat due to flowing water.

Hydraulic Fracturing A process of aquifer development in which fluid is injected at pressures that exceed the tensile stress of the aquifer, causing cracks to develop and propagate in the formation. These cracks serve as conduits for liquid flow to a production well. This process can be used in petroleum (nature gas) recovery. It can also be used for increasing water production in rock-like aquifers, or for contaminant recovery.

Hydraulic Head A state variable consisting of the mechanical energy per unit weight of water at any point in the aquifer and at any time.

Hydraulic Loading The amount (volume) of water applied to a given water or wastewater treatment process, usually expressed as the volume per unit time, or volume per unit time per unit surface area.

Hydraulics A science that studies water or other fluids in motion.

Hydrodynamics The study of the movement of water

Hydroelectric Power Water Use The use of water in the generation of electricity at plants where the turbine generators are driven by falling water.

Hydrograph (1) A graph showing stage, flow, velocity, or other property of water with respect to time. (2) A graph showing the water level (stage), discharge, or other property of a river volume with respect to time.

Hydrologic Budget An accounting of the inflow to, outflow from, and storage in, a hydrologic unit, such as a drainage basin, aquifer, soil zone, lake, reservoir, or an irrigation project.

Hydrologic Cycle (1) The cyclic transfer of water vapor from the earth's surface via evapotranspiration into the atmosphere, from the atmosphere via precipitation back to earth, and through runoff into streams, rivers, and lakes, and ultimately into the oceans; (2) a convenient term to denote the circulation of water from the sea, through the atmosphere, to the land; and thence, with many delays, back to the sea by overland and subterranean routes, and in part by way of the atmosphere; also the many short circuits of the water that is returned to the atmosphere without reaching the sea.

Hydrologic Equation (1) The equation balancing the hydrologic budget. (2) The water inventory equation (Inflow = Outflow + Change in Storage) which expresses the basic principle that during a given time interval the total inflow to an area must be equal to the total outflow plus the net change in storage.

Hydrologic Model A conceptual or physically based procedure for numerically simulating a process or processes which occur in a watershed.

Hydrologic System A system dealing with the properties, distribution, and circulation of water.

Hydrology (1) The science encompassing the behavior of water as it occurs in the atmosphere, on the surface of the ground, and underground. (2) The science that relates to the water of the earth. (3) The science treating of the waters of the earth, their occurrence, distribution, and movements. (4) The science dealing with the properties, distribution, and circulation of water both on the surface and under the earth. In practice, the study of the water of the oceans and the atmosphere is considered part of the sciences of oceanography and meteorology. (5) The study of flow of water over the earth's surface. (6) The applied science concerned with the waters of the earth, their occurrences, distribution, and circulation through the unending hydrologic cycle of precipitation, consequent runoff, infiltration, and storage; eventual evaporation; and so forth. It is concerned with the physical and chemical reaction of water with the rest of the earth, and its relation to the life of the earth.

Hydrous Aluminum Phyllosilicates Clay minerals.

Hyetograph Graphical representation of rainfall intensity against time.

Hypolimnion The lower layer of a stratified water body, below the well-mixed zone (see thermal stratification).

Hypoxia The occurrence of low dissolved oxygen concentrations in water. Hypoxia is generally defined with respect to saturation; because saturation levels vary with temperature and salinity, the concentration that defines hypoxia may vary seasonally and geographically. In practice, scientists often use a threshold of two parts per million, the generally accepted minimum required for most marine life to survive and reproduce.

Impaired Water Body A water body that does not meet the criteria that support its designated use.

Impermeable Layer A layer of solid material, such as rock or clay, which does not allow water to pass through.

Impervious Surface (1) A hard surface area that either prevents or retards the entry of water into the soil mantle or causes water to run off the surface in greater quantities or at an increased rate of flow. (2) A paved or other hard surface that does not allow water to penetrate. Common impervious surfaces include rooftops, walkways, patios, driveways, parking lots, storage areas, concrete or asphalt paving, and gravel roads in contact.

Inactive Storage Capacity The portion of capacity below which the reservoir is not normally drawn, and which is provided for sedimentation, recreation, fish and wildlife, aesthetic reasons, or for the creation of a minimum controlled operational or power head in compliance with operating agreements or restrictions.

Index A single number, derived from two or more environmental variables, that is intended to simplify complex information. For example, the Index of Biological Integrity combines several metrics of benthic community condition into a single index score.

Index Period In the USEPA's aquatic resource monitoring, a term used to describe the portion of the year when data are collected. The index period is often selected based on ecological considerations.

Indicator A numerical value derived from the actual measurements of a stressor, state or ambient condition, exposure, or human health or ecological condition over a specified geographic domain, whose trends over time represent or draw attention to the underlying trends in the condition of the environment.

Indicator Organism An indicator organism (e.g. fecal coliform) is a nonpathogenic organism whose presence implies the presence of pathogenic organisms. Indicator organisms are selected to be conservative estimates of the potential for pathogenicity.

Individual Field Unit An area of cropland that has been subdivided into several strips is not a single field. Rather, each strip represents an individual field unit.

Industrial Nonhazardous Waste The waste generated from processes associated with the production of goods and products, such as electric power generation and manufacturing of materials such as pulp and paper, iron and steel, glass, and concrete. This waste usually is not classified as municipal solid waste by the federal government, but some states may classify it as such if it enters the municipal solid waste stream.

Industrial Source A term used in this report to describe air emission sources of industrial origin. The report breaks industrial sources down into contributions from selected industries, as appropriate.

Industrial Water Use Water used for industrial purposes in such industries as steel, chemical, paper, and petroleum refining. Nationally, water for industrial uses comes mainly (80 %) from self-supplied sources, such as a local wells or withdrawal points in a river, but some water comes from public-supplied sources, such as the county/city water department.

Infiltration (1) The flow of water from the land surface into the subsurface. (2) The flow of a fluid into a substance through pores or small openings. It connotes flow into a substance in contradistinction to the word percolation, which connotes flow through a porous substance.

Infiltration Capacity The maximum rate at which the soil, when in a given condition, can absorb falling rain or melting snow.

Infiltration Index An average rate of infiltration, in inches per hour, equal to the average rate of rainfall such that the volume of rainfall at greater rates is equal to the total direct runoff.

Infinite Aquifer An ideal aquifer characterized by an infinite lateral extension.

Influent Anything flowing into a water body;

Injection Well Refers to a well constructed for the purpose of injecting treated wastewater directly into the ground. Wastewater is generally forced (pumped) into the well for dispersal or storage into a designated aquifer. Injection wells are generally drilled into aquifers that do not deliver drinking water, unused aquifers, or below freshwater levels.

Innovative Flotation Wastewater Treatment System A wastewater treatment system which includes preliminary treatment (screening, grit chamber, pH adjustment, oil and grease removal), primary flotation clarification (primary DAF), bioreactor (aeration basin, trickling filter, rotating biological reactor, fluidized bed, etc.), and secondary flotation clarification (secondary DAF).

Inorganic Pertaining to matter that is neither living nor organic.

Inorganic Contaminants Mineral-based compounds such as metals, nitrates, and asbestos. These contaminants are naturally occurring in some water, but can also get into water through farming, chemical manufacturing, and other human activities. The USEPA has set legal limits on 15 inorganic contaminants.

Instability The tendency to behave in an unpredictable, changeable, or erratic manner.

Interbasin Water Transfer The physical transfer of water from one watershed to another.

Interception The process and the amount of rain or snow stored on leaves and branches and eventually evaporated back to the air. Interception is equal to the precipitation on the vegetation minus the stem flow and through fall.

Interferometric Techniques in which waves are superimposed in order to extract information about the waves.

Intermittent Waters Water bodies (e.g., streams or wetlands) that contain water for part of each year, due to precipitation events and some groundwater contributions. Intermittent streams and wetlands typically contain water for weeks or months, while "ephemeral" streams and wetlands contain water for briefer periods, but in some cases these terms are used interchangeably (see ephemeral waters).

Interrill Erosion The process of soil detachment by raindrops and transport in thin sheet flow.

Intrinsic Soil Properties Field observable soil attributes like texture, structure, organization, color, features, and consistence.

Invasive Species A nonindigenous plant or animal species that can harm the environment, human health, or the economy.

Invertebrate Animals lacking a backbone.

Ionic Conductance The movement of an ion from one site to another.

Irrigated Area The gross farm area upon which water is artificially applied for the production of crops, with no reduction for access roads, canals, or farm buildings.

Irrigation (1) The controlled application of water for agricultural purposes through man-made systems to supply water requirements not satisfied by rainfall. Here is a quick look at some types of irrigation systems; (2) The controlled application of water to arable lands to supply water requirements not satisfied by rainfall..

Irrigation Efficiency The percentage of water applied that can be accounted for in soil-moisture increase.

Irrigation Requirement The quantity of water, exclusive of precipitation, that is required for crop production. It includes surface evaporation and other economically unavoidable wastes.

Irrigation Water Use The water application on lands to assist in the growing of crops and pastures or to maintain vegetative growth in recreational lands, such as parks and golf courses.

Irrigation, Supplemental See supplemental irrigation.

Iso-Erodent Lines Lines on a map that join points having the same value for erosivity.

Isohyet See isohyetal line.

Isohyetal Line (Isohyet) A line drawn on a map or chart joining points that receive the same amount of precipitation.

Kinematic Viscosity It is the dynamic viscosity of a fluid divided by its density, i.e., W/m^2.

Kinetic Processes Description of the dynamic rate and mode of change in the transformation

Kinetic Reaction A physical, chemical, or biological transformation/reaction that describes the dynamic rate and mode of change.

k-e Model A turbulence model based on solving two differential transport equations, one for the turbulent kinetic energy *k* and the other for the rate of turbulent dissipation *e*.

Lag Variously defined as time from the beginning (or center of mass) of rainfall to the peak (or center of mass) of runoff.

Land Application Land application is defined as the spreading, spraying, injection, or incorporation of liquid or semi-liquid organic substances, such as sewage sludge, biosolids, livestock manure, compost, septage, legumes, and other types of liquid organic waste, onto or below the surface of the land to take advantage of the soil-enhancing qualities of the organic substances. These organic substances are land applied to improve the structure of the soil. It is also applied as a fertilizer to supply nutrients to crops and other vegetation grown in the soil. The liquid or semi-liquid organic substances are commonly applied to agricultural land (including pasture and range land), forests, reclamation sites, public contact sites (e.g., parks, turf farms, highway median strips, golf courses), lawns, and home gardens (see spray irrigation for land application of wastewater).

Land Application Site An area of land on which sewage sludge is applied to condition the soil or to fertilize crops or vegetation grown in the soil.

Land Treatment Unit A site where physical, chemical, and biological processes occurring in the topsoil layers (e.g., naturally occurring soil microbes and sunlight) are used to treat and contain waste. Hazardous waste is applied directly to the soil surface or incorporated into the upper layers of the soil, where its constituents are degraded, transformed, or immobilized. Liner systems or leachate collection and removal systems are not required for land treatment units. The closure consists primarily of placing a vegetative cover over the unit and certifying that hazardous constituent levels in the treatment zone do not exceed background levels.

Landfill A disposal site for solid wastes spread in layers, compacted to the smallest practical volume, and covered by material (e.g., soil). Landfills are designed to isolate waste from the surrounding environment (e.g., groundwater, rain, air). Landfills are subject to requirements that include installing and maintaining a final cover, operating leachate collection and removal systems, maintaining and monitoring the leak detection system, groundwater monitoring, preventing storm water run-on and -off, and installing and protecting surveyed benchmarks.

Landscape Comprises visible features of an area of land, including physical elements (i.e., landforms), living elements (i.e., land cover), human elements (i.e., land use), and transitory elements (i.e., climate).

Leaching The process by which soluble materials in the soil, such as salts, nutrients, pesticide chemicals, or contaminants, are washed into a lower layer of soil or are dissolved and carried away by water.

Lentic Waters Ponds or lakes (standing water).

Levee (1) A natural or man-made earthen barrier along the edge of a stream, lake, or river. Land alongside rivers can be protected from flooding by levees. (2) Elongated naturally occurring ridge or artificially constructed wall that regulates water levels.

Limiting Factor An environmental factor that limits the growth of an organism; the factor.

Limnetic Community The area of open water in a fresh water lake, providing the habitat for fish, phytoplankton and zooplankton.

Limnetic Zone (1) The open water zone of a lake or pond from the surface to the depth of light penetration. (2) A body of fresh water beyond the outer border of the littoral zone and to a depth that light penetrates.

Limnology (1) That branch of hydrology pertaining to the study of lakes. (2) The study of inland waters.

Lipids The structural components of the cell that are fatty or waxy.

Littoral Zone (1) A portion of a fresh water body extending from the shoreline lakeward to the limit of occupancy of rooted plants. (2) A strip of land along a shoreline between the high- and low-water levels.

Livestock Operation A facility that raises animals such as cows, sheep, or hogs. Fecal coliform bacteria are present in livestock waste.

Livestock Water Use The water used for livestock watering, feed lots, dairy operations, fish farming, and other on-farm needs.

Living Machine System A patented, man-made aquaculture wetland waste treatment system which adapts and enhances the ecological processes in a series of tidal wetland cells or basins. Each cell or basin is filled with special gravel that promotes the development of micro-ecosystems. A computer controls the fill and drain cycles, alternating anoxic (without oxygen) and aerobic (with oxygen) conditions. As wastewater moves through the system, the cells are alternately flooded and drained to create multiple tidal cycles each day, much like one finds in nature, resulting in high-quality reusable water.

Local Scour An erosion caused by an abrupt change in the flow direction or velocity. Examples include erosion around bridge piers, downstream of stilling basins, at the end of dikes or vanes, and near snags.

Long-Period Variations Secular when a cycle or a change in trend is completed within a century, climatic when the period of change runs through centuries or a few millennia, and geologic when the period runs into geological time.

Lotic Waters Flowing waters, as in streams and rivers.

Low-Flow Frequency Curve A graph showing the magnitude and frequency of minimum flows for a period of given length. Frequency is usually expressed as the average interval, in years, between recurrences of an annual minimum flow equal to or less than that shown by the magnitude scale.

Lysimeter A structure containing a mass of soil, and designed to permit the measurement of water draining through the soil.

Macrofauna Animals visible to the naked eye.

Macroinvertebrate An organism that lacks a backbone and is large enough to be seen with the naked eye.

Macrophytes Large (non-microscopic), usually rooted, aquatic plants.

Manure Any wastes discharged from livestock.

Marginal Costs The incremental cost of increasing output of a good or service by a small amount.

Mass Balance An equation that accounts for the flux of mass going into a defined area.

Mass Curve A graph of the cumulative values of a hydrologic quantity (such as precipitation or runoff), generally as ordinate, plotted against time or date as abscissa (see double-mass curve, and residual-mass curve).

Mass Failure A type of bank failure resulting in a block of the bank sliding or toppling into the stream in a single event.

Mass Transfer Coefficient (1) A constant of proportionality that is specific to an individual compound and is used in a mass transfer expression to determine equilibrium conditions between two phases. (2) Mass transfer coefficients are determined experimentally; the units will depend on the nature of the mathematical expression and the phase transfer.

Maximum Contaminant Level (MCL) The highest level of a contaminant that the USEPA allows in drinking water. MCLs ensure that drinking water does not pose either a short-term or long-term health risk. The USEPA sets MCLs at levels that are economically and technologically feasible. MCLs are enforceable standards. Some states set MCLs which are more strict than the USEPAs.

Maximum Contaminant Level Goal (MCLG) The level of a contaminant at which there would be no risk to human health. This goal is not always economically or technologically feasible, and the goal is not legally enforceable.

Maximum Probable Flood See flood, maximum probable.

Maximum Spillway Discharge Spillway discharge (ft^3/s or m^3/s) when the reservoir is at maximum designed water surface elevation.

Mean Velocity The discharge divided by the wetted area of a cross section.

Meander The winding of a stream channel.

Meander Amplitude The distance between points of maximum curvature of successive meanders of opposite phase in a direction normal to the general course of the meander belt, measured between centerlines of channels.

Meander Belt The area between the lines drawn tangential to the extreme limits of fully developed meanders.

Meander Breadth The distance between the lines used to define the meander belt.

Meander Length The distance in the general course of the meanders between the corresponding points of successive meanders of the same phase. Twice the distance between successive points of inflection of the meander wave.

Meandering A planform (alluvial) process that generates a series of bends of alternate curvature connected by straight reaches.

Meandering Stream (1) One that follows its natural course creating winding curves. (2) An alluvial stream characterized in planform by a sequence of alternating bends. The bends are usually a result of alluvial processes rather than the nature of the terrain.

Medical Waste Any solid waste generated in the diagnosis, treatment, or immunization of human beings or animals, in research pertaining thereto, or in the production or testing of biologicals, excluding hazardous waste identified or listed under 40 CFR Part 261, or any household waste as defined in 40 CFR Subsection 261.4(b)(1).

Meromictic Lake A lake in which some water remains partly or wholly unmixed with the main water mass at circulation periods is said to be meromictic. The process leading to a meromictic state is termed meromixis. The perennially stagnant deep layer of a meromictic lake is called the monimolimnion. The part of a meromictic lake in which free circulation can occur is called the mixolimnion. The boundary between the monimolimnion and the mixolimnion is called the chemocline.

Mesotrophic The term describes reservoirs and lakes that contain moderate quantities of nutrients and are moderately productive in terms of aquatic animal and plant life.

Metal Mining Sector Metal mining facilities that fall within the Standard Industrial Classification Code 10 and must report to the Toxics Release Inventory (TRI) in accordance with Section 313 of the Emergency Planning and Community Right to Know Act.

Microorganisms The tiny living organisms that can be seen only with the aid of a microscope. Some microorganisms can cause acute health problems when consumed in drinking water. Also known as microbes.

Migration Movement of an organism from one location to another

Mineralization Most nitrogen exists in biosolids/manure as organic N, principally contained in proteins, nucleic acids, amines, and other cellular material. These complex molecules must be broken apart through biological degradation for nitrogen to become available to crops. The conversion of organic N to inorganic-N forms is called mineralization.

Mining Water Use The water use during quarrying rocks and extracting minerals from the land.

Mixed Liquor The liquid consisting of wastewater, microorganisms, nutrients, oxygen, suspended solids, etc., in the aeration basin of activated sludge process.

Mobile Source A term used to describe a wide variety of vehicles, engines, and equipment that generate air pollution and that move, or can be moved, from place to place. "On-road" sources are vehicles used on roads to transport passengers or freight. "Non-road" sources include vehicles, engines, and equipment used for construction, agriculture, transportation, recreation, and many other purposes.

Mobile-Bed Models Refer to numerical models that may be used to predict the sediment transport in streams, along with the associated streambed changes in shape.

Model (1) An idealized representation of a system; and (2) schematic description of a system that accounts for known or inferred properties and may be used to study system characteristics.

Moisture The water diffused in the atmosphere or the ground.

Moisture Equivalent The ratio of (1) the weight of water which the soil, after saturation, will retain against a centrifugal force 1000 times the force of gravity, to (2) the weight of the soil when dry. The ratio is stated as a percentage.

Molecular Diffusion A physical–chemical process whereby mobile compounds (dissolved or suspended in another compound) move from areas of high concentration to areas of low concentration.

Molecular Diffusion Coefficient of Oxygen (a) A coefficient (D_L) for diffusion of oxygen into water in a natural aquatic system; (b) the molecular diffusion coefficient (D_L) of oxygen at 20 °C = 1.76×10^{-4} m^2/day. (c) the molecular diffusion coefficient (D_L) of oxygen at any temperature = $(1.76 \times 10^{-4}$ m^2/day$) \times (1.037^{T-20})$.

Monitoring Testing that water systems must perform to detect and measure contaminants. A water system that does not follow the USEPA's monitoring methodology or schedule is in violation, and may be subject to legal action.

Morphology A scientific study of form and structure.

Mortality Death.

Mortality Rate A rate expressing the proportion of a population who die of a specific disease or of all causes during a stated period of time, usually a year.

Mudflow A well-mixed mass of water and alluvium which, because of its high viscosity and low fluidity as compared with water, moves at a much slower rate, usually piling up and spreading over the fan like a sheet of wet mortar or concrete.

Multipurpose Reservoir A reservoir constructed and equipped to provide storage and release of water for two or more purposes such as flood control, power development, navigation, irrigation, recreation, pollution abatement, domestic water supply, etc.

Multi-Reservoir System More than one reservoir connected with some hydraulic relationship or some operation objective in common, which becomes a system. Multi-reservoir system operation can obtain more benefits than surplus of the ones from individual operation of all member reservoir.

Municipal Solid Waste Waste from homes, institutions, and commercial sources consisting of everyday items such as product packaging, grass clippings, furniture, clothing, bottles and cans, food scraps, newspapers, appliances, consumer electronics, and batteries. (Excluded from this category are municipal wastewater treatment sludge, industrial process wastes, automobile bodies, combustion ash, and construction and demolition debris.)

Municipal Use of Water The various uses of water in developed urban areas, including domestic use, industrial use, street sprinkling, fire protection, etc.

Municipal Water System A water system that has at least five service connections or which regularly serves 25 individuals for 60 days; also called a public water system (PWS).

MUSCL Short for Modified Upwind Scheme for Conservation Laws, it is a method to describe (reconstruct) the states of the variables in a computational cell based on the cell-averaged states (and their gradients) obtained in the previous time step.

Narrative Criteria Nonnumeric descriptions of desirable or undesirable water quality conditions.

National Ambient Air Quality Standards (NAAQS) Standards established by the USEPA that apply to outdoor air throughout the country. The Clean Air Act established two types of national air quality standards. Primary standards set limits to protect public health, including the health of "sensitive" populations such as asthmatics, children, and the elderly. Secondary standards set limits to protect public welfare, including protection against decreased visibility and damage to animals, crops, vegetation, and buildings. The USEPA has set NAAQS for the six criteria pollutants.

National Indicator A Report on the Environment (ROE) indicator for which nationally consistent data are available, and which helps to answer an ROE question at a national scale. Some National Indicators also present data broken down by USEPA region (see ROE indicator).

National Pollutant Discharge Elimination System (NPDES) Permit The regulatory agency document issued by either a federal or state agency that is designed to control all discharges of pollutants from point sources into US waterways. The NPDES permits regulate discharges into navigable waters from all point sources of pollution, including industries, municipal water and wastewater treatment plants, power plants, sanitary landfills, large agricultural feed lots, and return irrigation flows.

National Priorities List (NPL) The USEPA's list of the most serious uncontrolled or abandoned hazardous waste sites identified for possible long-term remedial action under Superfund (see superfund).

National Water Quality Inventory A report that the USEPA prepares every 2 years summarizing information from states about the quality of the nation's waters.

Natural Source A term used in this report to describe any air emissions source of natural origin. Examples include volcanoes, wild fires, wind-blown dust, and releases due to biological processes (see biogenic source).

Navier–Stokes Equations Partial differential equations arising from Newton's second law (conservation of momentum) that describe the motion of fluids.

Nephelometric Turbidity Unit (NTU) A unit of measure for the turbidity of water. Essentially, a measure of the cloudiness of water as measured by a nephelometer. Turbidity is based on the amount of light that is reflected off particles in the water.

National Geodetic Vertical Datum (NGVD) NGVD. (1) As corrected in 1929, a vertical control measure used as a reference for establishing varying elevations. (2) Elevation datum plane previously used by the Federal Emergency Management Agency (FEMA) for the determination of flood elevations. FEMA currently uses the North American Vertical Datum Plane.

NGVD of 1929 A geodetic datum derived from a general adjustment of the first-order level nets of the USA and Canada. It was formerly called "Sea Level Datum of 1929" or "mean sea level" in the USGS series of reports. Although the datum was derived from the average sea level over a period of many years at 26 tide stations along the Atlantic, Gulf of Mexico, and Pacific Coasts, it does not necessarily represent local mean sea level at any particular place. NPS pollution is the contamination that occurs when rainwater, snowmelt, or irrigation washes off plowed fields, city streets, or suburban backyards. As this runoff moves across the land surface, it picks up soil particles and pollutants, such as nutrients and pesticides.

Nitrate Nitrogen A common way to report nitrate concentration (expressed as nitrogen).

Nitrification A process of formation of nitrate nitrogen from reduced inorganic nitrogen compounds, such as ammonia nitrogen. Nitrification in the natural environment is carried out primarily by autotrophic bacteria.

Nitrite Nitrogen A common way to report nitrite concentration (expressed as nitrogen).

Nitrogen Nutrient that is essential to plants and animals.

Nomograph A two-dimensional diagram designed to allow the approximate graphical computation of a function.

Nonindigenous Species A species that has been introduced by human action, either intentionally or by accident, into an area outside its natural geographical range; also called an alien, exotic, introduced, or nonnative species. Certain nonindigenous species are considered “invasive” (see invasive species).

Nonpoint Source (NPS) Diffuse pollution source; a source without a single point of origin or not introduced into a receiving stream from a specific outlet. The pollutants are generally carried off the land by rainfall or snowmelt moving over and through the ground and carrying natural and human-made contaminants into lakes, rivers, streams, wetlands, estuaries, other coastal waters, and groundwater. Common NPSs are agriculture, forestry, urban areas, mining, construction, dams, channels, land disposal, saltwater intrusion, and city streets.

NPS Pollution (1) Pollution discharged over a wide land area, not from one specific location. These are forms of diffuse pollution caused by sediments, nutrients, organic and toxic substances originating from land-use activities, which are carried to lakes and streams by surface. (2) NPS pollution is the contamination that occurs when rainwater, snowmelt, or irrigation washes off plowed fields, city streets, or suburban backyards. As this runoff moves across the land surface, it picks up soil particles and pollutants, such as nutrients and pesticides.

Nonproduction-Related Waste The waste that is not production related; for example, waste associated with catastrophic events and cleanup actions. Toxic chemicals in nonproduction-related waste must be reported to the TRI (see toxics release inventory).

Nonpublic Water System A water system that does not provide water for human consumption through at least 15 service connections, or regularly serve at least 25 individuals, for at least 60 days per year.

Nontidal Stream/River A stream/river in which water level and flow direction will not fluctuate and will not be affected by the action of lunar and solar forces upon the rotating earth.

Nontransient Noncommunity Water System A type of public water system (PWS) that supplies water to 25 or more of the same people at least 6 months per year in places other than their residences. Some examples are schools, factories, office buildings, and hospitals that have their own water systems (see public water system).

Normal A central value (such as arithmetic average or median) of annual quantities for a 30-year period ending with an even 10-year period, thus 1921–1950; 1931–1960, and so forth. This definition accords with that recommended by the Subcommittee on Hydrology of the Federal Inter-Agency Committee on Water Resources.

Normal Year A year during which the precipitation or stream flow approximates the average for a long period of record.

Numeric Criteria Numeric descriptions of desirable or undesirable water quality conditions.

Nutrients (1) Nutrients are elements required for plant growth that provide biosolids with most of their economic value. These include nitrogen (N), phosphorus (P), potassium (K), calcium (Ca), magnesium (Mg), sodium (Na), sulfur (S), boron (B), copper (Cu), iron (Fe), manganese (Mn), molybdenum (Mo), and zinc (Zn). (2) Any substance assimilated by living things that promotes growth. (3) Substances necessary for the growth of all living things, such as nitrogen, carbon, potassium, and phosphorus. Too many nutrients in water bodies can contribute to algal blooms. The term is generally applied to nitrogen and phosphorus but is also applied to other essential and trace elements. (4) Chemical elements essential to life.

Objective Function The mathematical function of decision variables and state variables, whose minimization or maximization constitutes the objective of the optimization problem.

Oil and Gas Production Waste Gas and oil drilling muds, oil production brines, and other waste associated with exploration for, or development and production of, crude oil or natural gas.

Omnivorous Feeding on a variety of organisms and organic detritus on the bottom.

Onsite Treatment The water or waste treatment on the project site (see treatment).

Open Channel Flow Flow of water or a liquid with its surface exposed to the atmosphere. The conduit may be an open channel or a closed conduit flowing partly full.

Open Water Disposal An approved waste (or dredged material) dumping site in an open water, such as lake or coastal area, or ocean.

Optimization The selection of a best element, with regard to some specific criterion, expressed as an objective, from a finite or infinite set of feasible alternatives.

Optimization Constraints Set of equations and inequalites defining the set of feasible alternatives within which the solution to the optimization problem is sought.

Organic Chemical Compounds containing carbon.

Organic Contaminants Carbon-based chemicals, such as solvents and pesticides, which can get into water through runoff from cropland or discharge from factories. The USEPA has set legal limits on 56 organic contaminants.

Organic Matter (1) Plant and animal residues, or substances made by living organisms. All are based upon carbon compounds. (2) Matter composed of organic compounds that has come from the remains of once-living organisms such as plants and animals and their waste products in the environment.

Osmosis The movement of water molecules through a thin membrane. The osmosis process occurs in our bodies and is also one method of desalinating saline water.

Outfall The place where a sewer, drain, or stream discharges; the outlet or structure through which reclaimed water or treated effluent is finally discharged to a receiving water body.

Outflow Channel A natural stream channel which transports reservoir releases.

Outlet An opening through which water can be freely discharged from a reservoir.

Overland Flow The flow of rainwater or snowmelt over the land surface toward stream channels. After it enters a stream, it becomes runoff.

Overtopping To extend or rise over or beyond the top of a dam or levee.

Overturn (1) The complete circulation or mixing of the upper and lower waters of a lake. (2) The phenomenon of vertical circulation that occurs in large bodies of water. It is also known as turnover.

Oxidation A reaction between molecules, ordinarily involves gain of oxygen.

Oxygen Demand The need for molecular oxygen to meet the needs of biological and chemical processes in water. Even though very little oxygen will dissolve in water, it is extremely important in biological and chemical processes.

Oxygen Depletion The exhaustion of oxygen by chemical or biological use.

Ozone Depletion The destruction of the stratospheric ozone layer, which shields the earth from ultraviolet radiation harmful to life. This destruction of ozone is caused by the breakdown of certain chlorine- and/or bromine-containing compounds (chlorofluorocarbons or halons). These compounds break down when they reach the stratosphere and then catalytically destroy ozone molecules.

Ozone-Depleting Substance Any compound that contributes to stratospheric ozone depletion (see ozone depletion).

Parameter A measurable, variable quantity as distinct from a statistic.

Partial-Duration Flood Series A list of all flood peaks that exceed a chosen base stage or discharge, regardless of the number of peaks occurring in a year (also called basic-stage flood series, or floods above a base).

Particle Size (1) The diameter, in millimeters, of suspended sediment or bed material. Particle-size classifications are: (1) clay = 0.00024–0.004 mm; (2) silt = 0.004–0.062 mm; (3) sand = 0.062–2.0 mm; and (4) gravel = 2.0–64.0 mm. (3) A linear dimension, usually designated as "diameter," used to characterize the size of a particle.

Particulates Small pieces of material (such as sand) floating in the water.

Pasture Land on which animals feed directly on feed crops such as legumes, grasses, or grain stubble.

Pathogen (1) A disease-causing organism; (2) a disease-producing agent; usually applied to a living organism. Generally, any viruses, bacteria, or fungi that cause disease.

Peak Flow The maximum instantaneous discharge of a stream or river at a given location. It usually occurs at or near the time of maximum stage.

Pelagic Zone Open water with no association with the bottom.

Per Capita Use The average amount of water used per person during a standard time period, generally per day.

Percolation (1) The movement of water through the openings in rock or soil; (2) the entrance of a portion of the stream flow into the channel materials to contribute to the groundwater replenishment; (3) the movement, under hydrostatic pressure, of water through the interstices of a rock or soil, except the movement through large openings such as caves.

Percolation, Deep In irrigation or farming practice, the amount of water that passes below the root zone of the crop or vegetation.

Periphyton Microscopic plants and animals that are firmly attached to solid surfaces of logs, rocks, etc. under water.

Permeability (1) The ability of a material to allow the passage of a liquid, such as water through rocks. Permeable materials, such as gravel and sand, allow water to move quickly through them, whereas unpermeable material, such as clay, do not allow water to flow freely. (2) Measure of the ability of a porous material (such as soil) to allow fluids to pass through it. (3) A measure of the resistance that is encountered when forming an electric field in a medium.

Pervious Surface A surface which allows water to soak into it.

pH (1) A measure of the degree of acidity or alkalinity of a substance. (2) A symbol for expressing the degree to which a solution is acidic or basic. It is based on a scale from 0 (very acid) to 14 (very basic). Water with a pH of 7 is neutral; lower pH levels indicate increasing acidity, while pH levels higher than 7 indicate increasingly basic solutions. The pH of biosolids is often raised with alkaline materials to reduce the pathogen content and attraction of disease-spreading organisms (vectors). High pH (greater than 11) kills virtually all pathogens and reduces the solubility, biological availability, and mobility of most metals. Lime also increases the gaseous loss (volatilization) of the ammonia form of nitrogen (ammonia-N), thus reducing the N-fertilizer value of biosolids.

Phosphorus A nutrient that is essential to plants and animals.

Photic Zone The upper layer of a body of water delineated by the depth to which enough sunlight can penetrate to permit photosynthesis.

Photosynthesis The conversion of light energy to chemical energy. At night, this process reverses—plants and algae suck oxygen out of the water.

Phreatic Surface Commonly called the water table. Location where the pressure head is zero.

Phytoplankton Small, mostly microscopic algae floating in the water column

Piping Soils being washed out of an earthen structure through an unfiltered exit.

Piping Failure It refers to the collapse of part of a bank due to seepage flows causing selective removal of bank layers. The failure is usually caused by preferential groundwater flows along interbedded saturated bank layers; the bank has lenses of noncohesive materials sandwiched between layers of finer cohesive materials.

Pipline Dredge A mechanical equipment that excavates with a revolving cutter surrounding the intake end of a suction pipe. The dredged materials are sucked up and transported by a pumping unit through a trailing pipeline to an appropriate disposal site. The dredge is generally equipped with two stern spuds and forward anchors to swing the hull around one of the stern spuds. Pipeline dredges are designed for excavating clayey, sandy, or silty bottoms.

Planform The shape and size of the channel and overbank features as viewed form directly above.

Planar Failure It refers to the sliding and forward toppling of a deep-seated bank mass into the stream. Planar failure occurs often on steep, fine-grained cohesive banks.

Plankton Small organisms floating in the water.

Plant Available Nitrogen (PAN) Only a portion of the total nitrogen present in biosolids/manure is available for plant uptake. This plant available nitrogen (PAN) is the actual amount of N in the biosolids/manure that is available to crops during a specified period.

Planting and Harvesting Periods The cycle of crop planting and harvesting periods, not the calendar year, dictates the timing of biosolids and manure land application activities. Winter wheat and perennial forage grasses are examples of crops that may be established and harvested in different calendar years.

Planting Season The basic time management unit is often called the crop year or planting season. The crop year is defined as the year in which a crop receiving the biosolids/manure treatment is harvested.

Point Bar Deposits of sediment that occur on the convex side or inside of channel bends. Their shape may vary with changing flow conditions, but they do not move significantly relative to the bends. However, the general magnitude and location of the bars vary with discharge and sediment load.

Point Source A stationary location or fixed facility from which pollutants are discharged; any single identifiable source of pollution, such as a pipe, ditch, ship, ore pit, or factory smokestack.

Point-Source Pollution The water pollution coming from a single point, such as a sewage-outflow pipe.

Pollutant (1) A contaminant in a concentration or amount that adversely alters the physical, chemical, or biological properties of the natural environment. (2) Any substance introduced into the environment that may adversely affect the usefulness of a resource or the health of humans, animals, or ecosystems. For most environmental media, this term is commonly understood to refer to substances introduced by human activities. In the case of air, the convention is to include substances emitted from natural sources as well (see air pollutant).

Pollutant Concentration Limits (PCL) PCL are the maximum concentrations of heavy metals for biosolids whose trace element pollutant additions do not require tracking (i.e., calculation of cumulative pollutant loading rate (CPLR). PCL are the most stringent pollutant limits included in the US Federal Regulation Part 503 for land application. Biosolids meeting pollutant concentration limits are subject to fewer requirements than biosolids meeting ceiling concentration limits.

Polychlorinated Biphenyls (PCBs) A group of synthetic, toxic industrial chemical compounds once used in making paint and electrical transformers, which are chemically inert and not biodegradable. PCBs were frequently found in industrial wastes, and subsequently found their way into surface and groundwaters. As a result of their persistence, they tend to accumulate in the environment. In terms of streams and rivers, PCBs are drawn to sediment, to which they attach and can remain virtually indefinitely. Although virtually banned in 1979 with the passage of the Toxic Substances Control Act, they continue to appear in the flesh of fish and other animals.

Pond A small, shallow lake.

Pondage Small-scale storage at a waterpower plant to equalize daily or weekly fluctuations in river flow or to permit irregular hourly use of the water for power generation to accord with fluctuations in load.

Pool (1) A deep reach of a stream; (2) in streams, a relatively deep area with low velocity; in ecological systems, the supply of an element or compound, such as exchangeable or weather-able cations or adsorbed sulfate, in a defined component of the ecosystem. The reach of a stream between two riffles. Natural streams often consist of a succession of pools and riffles.

Pool–Riffle Ratio The ratio of stream surface area covering pools to stream surface area covering riffles in a given segment of stream.

Population (1) In ecology, a group of interbreeding organisms occupying a particular space. In other contexts, including human health, this term generally refers to the number of humans living in a designated area. (2) A group of organisms of the same species.

Porosity (1) A measure of the water-bearing capacity of subsurface rock. With respect to water movement, it is not just the total magnitude of porosity that is important, but the size of the voids and the extent to which they are interconnected, as the pores in a formation may be open, or interconnected, or closed and isolated. For example, clay may have a very high porosity with respect to potential water content, but it constitutes a poor medium as an aquifer because the pores are usually so small. (2) A measure of the empty pore spaces in a material. Defined as the ratio between the volume of voids (i.e., empty pore space) and the total volume.

Potable water The water of a quality suitable for drinking.

Potential The energy stored in a system of forcefully interacting physical entities.

Potential Evapotranspiration Water loss that will occur if at no time there is a deficiency of water in the soil for use of vegetation.

Potential Natural Water Loss The water loss during years when the annual precipitation greatly exceeds the average water loss. It represents the approximate upper limit to water loss under the type and density of vegetation native to a basin, actual conditions of moisture supply, and other basin characteristics, whereas potential evapotranspiration represents the hypothetical condition of no deficiency of water in the soil at any time for use of the type and density of vegetation that would develop.

Potential Rate of Evaporation See evaporativity.

Precipitation (1) Precipitation includes rain, snow, hail, sleet, dew, and frost. (2) As used in hydrology, precipitation is the discharge of water, in liquid or solid state, out of the atmosphere, generally upon a land or water surface. It is the common process by which atmospheric water becomes surface or subsurface water. The term "precipitation" is also commonly used to designate the quantity of water that is precipitated. Precipitation includes rainfall, snow, hail, and sleet, and is therefore a more general term than rainfall. (3) As used in hydrology, precipitation is the discharge of water, in a liquid or solid state, out of the atmosphere, generally onto a land or water surface. It is the common process by which atmospheric water becomes surface, or subsurface water. The term "precipitation" is also commonly used to designate the quantity of water that is precipitated. Precipitation includes rainfall, snow, hail, and sleet, and is therefore a more general term than rainfall.

Precursor In photochemistry, any compound antecedent to a pollutant. For example, volatile organic compounds (VOCs) and nitrogen oxides react in sunlight to form ozone or other photochemical oxidants. As such, VOCs and nitrogen oxides are precursors.

Predator An organism, usually an animal, that kills and consumes other organisms.

Preparer Either the person who generates sewage sludge during the treatment of domestic sewage in a treatment works or the person who derives a material from sewage sludge pressures.

Prey An organism killed and at least partially consumed by a predator.

Primacy State A state that has the responsibility and authority to administer the USEPA's drinking water regulations within its borders. The state must have rules at least as stringent as the USEPA's.

Primary DAF A dissolved air flotation (DAF) clarifier which is used for primary wastewater treatment before the wastewater reaches the secondary wastewater treatment bioreactors, such as aeration basins, trickling filters, rotating biological reactors, etc. The objective of the primary wastewater treatment is removal of suspended solids, oil, and grease from the raw wastewater.

Primary Pollutant Any pollutant that is emitted into the atmosphere directly from its source and which retains the same chemical form. An example of a primary pollutant is dust that blows into the air from a landfill.

Primary Wastewater Treatment The first stage of the wastewater-treatment process where mechanical methods, such as filters and scrapers, are used to remove pollutants. The solid material in sewage also settles out in this process.

Prior Appropriation Doctrine The system for allocating water to private individuals used in most Western states. The doctrine of prior appropriation was in common use throughout the arid West as early settlers and miners began to develop the land. The prior appropriation doctrine is based on the concept of "First in Time, First in Right." The first person to take a quantity of water and put it to beneficial use has a higher priority of right than a subsequent user. The rights can be lost through nonuse; they can also be sold or transferred apart from the land. Contrasts with riparian water rights.

Priority Chemicals A set of chemicals, found in the nation's products and wastes, that USEPA targets for voluntary reduction (or recovery and recycling if they cannot be eliminated or reduced at the source).

Producer An organism that can synthesize organic matter using inorganic materials

Production The amount of organic material produced by biological activity.

Production-Related Waste The sum of a facility's production-related onsite waste releases, onsite waste management (recycling, treatment, and combustion for energy recovery), and offsite transfers for disposal, treatment, recycling, or energy recovery. Toxic chemicals in production-related waste must be reported to the TRI (see toxics release inventory).

Productivity The rate of production of organic matter.

Productivity, Secondary The rate of production by consumers.

Productivity, Primary The rate of production by plants.

Public Contact Site Land with a high potential for contact by the public, including public parks, ball fields, cemeteries, nurseries, turf farms, and golf courses.

Public Notification An advisory that the USEPA requires a water system to distribute to affected consumers when the system has violated MCLs or other regulations. The notice advises consumers what precautions, if any, they should take to protect their health.

Public Supply Water withdrawn by public governments and agencies, such as a county water department, and by private companies that is then delivered to users. Public suppliers provide water for domestic, commercial, thermoelectric power, industrial, and public water users. Most people's household water is delivered by a public water supplier. The systems have at least 15 service connections (such as households, businesses, or schools) or regularly serve at least 25 individuals daily for at least 60 days out of the year.

Public Water System (PWS) A system that provides water for human consumption through at least 15 service connections, or regularly serves at least 25 individuals, for at least 60 days per year. PWSs are divided into three categories (see community water system, nontransient noncommunity water system, and transient noncommunity water system). Examples of PWSs include municipal water companies, homeowner associations, schools, businesses, campgrounds, and shopping malls. There are more than 170,000 PWSs providing water from wells, rivers and other sources to about 250 million Americans.

Public Water Use The water supplied from a public-water supply and used for such purposes as firefighting, street washing, and municipal parks and swimming pools.

Radioactive Waste Waste-containing substances that emit ionizing radiation. Radioactive waste is classified by regulation according to its source and/or content. The types of waste that are typically considered "radioactive waste" include high-level waste, low-level waste, mixed low-level waste, transuranic waste (i.e., elements heavier than uranium), and certain wastes from the extraction and processing of uranium or thorium ore. Spent nuclear fuel, which is produced as a result of the controlled nuclear fission process in nuclear reactors, is considered a nuclear material rather than radioactive waste.

Radionuclides Any man-made or natural element that emits radiation and that may cause cancer after many years of exposure through drinking water.

Rain Liquid precipitation.

Rainfall The quantity of water that falls as rain only. Not synonymous with precipitation.

Rainfall excess The volume of rainfall available for direct runoff. It is equal to the total rainfall minus interception, depression storage, and absorption.

Rainfall Intensity The measure of the amount of rain per unit area and unit time.

Rainfall Simulator A tool that produces controlled parameter (intensity, duration, drop size) rainstorms over a confined soil surface. Also known as a rainulator

Rainfall, Excessive Rainfall in which the rate of fall is greater than certain adopted limits, chosen with regard to the normal precipitation (excluding snow) of a given place or area. In the US Weather Bureau, it is defined, for states along the southern Atlantic coast and the Gulf coast, as rainfall in which the depth of precipitation is 0.90 in. at the end of 30 min and 1.50 in. at the end of an hour, and for the rest of the country as rainfall in which the depth of precipitation at the end of each of the same periods is 0.50 and 0.80 in., respectively.

Range Land Open land with indigenous vegetation.

Rating Curve A drawn curve showing the relation between gage height and discharge of a stream at a given gaging station.

Raw Water A water in its natural state, prior to any treatment for drinking.

RCRA Cleanup Baseline A priority subset of the universe of facilities that are subject to cleanup under the Resource Conservation and Recovery Act (RCRA) due to past or current treatment, storage, or disposal of hazardous wastes, and that have historical releases of contamination.

RCRA Hazardous Waste A national regulatory designation for certain wastes under the RCRA. Some wastes are given this designation because they are specifically listed on one of four RCRA hazardous waste lists (see http://www.epa.gov/epaoswer/osw/hazwaste.htm). Other wastes receive this designation because they exhibit at least one of four characteristics which are ignitability, corrosivity, reactivity, or toxicity.

Reach (1) The length of channel uniform with respect to discharge, depth, area, and slope. (2) The length of a channel for which a single gage affords a satisfactory measure of the stage and discharge. (3) The length of a river between two gaging stations. (4) More generally, any length of a river..

Reaeration (1) The physical–chemical reaction by which oxygen is absorbed back into water. This is dependent, among other things, upon turbulence intensity and the water depth. (2) An aeration process by which oxygen in air is absorbed back into natural water, such as stream water and lake water. (3) A natural process of oxygen exchange between the atmosphere and a natural water body in contact with the atmosphere. Typically, the net transfer of oxygen is from the atmosphere and into the water, since dissolved oxygen levels in most natural waters are below saturation. When photosynthesis produces supersaturated dissolved oxygen levels, however, the net transfer is back into the atmosphere. (4) Reaeration process is modeled as the product of reaeration coefficient multiplied by the difference between dissolved oxygen saturation and the actual dissolved oxygen concentration, that is: $F_c = K_2 (C_s - C) = K_L/H\,(C_s - C)$. Here, F_c = rate or flux of dissolved oxygen across the water body (M/L^3/T); C = dissolved oxygen concentration (M/L^3); C_s = saturation of dissolved oxygen concentration (M/L^3); K_2 = reaeration coefficient (1/T); H = water depth (L); and K_L = surface transfer coefficient (L/T).

Reaeration Rate Coefficient See reaeration coefficient.

Reaeration Coefficient A mass transfer coefficient (K_2) in reaeration process (see reaeration and mass transfer coefficient).

Reaeration Rate (1) The rate at which oxygen is absorbed back into water. This is dependent, among other things, upon turbulence intensity, temperature, and the water depth. (2) The reaeration rate is defined as the rate of dissolved oxygen across the water body $F_c = K_2\,(C_s - -C)$. Here, F_c = rate or flux of dissolved oxygen across the water body (M/L^3/T); C=dissolved oxygen concentration (M/L$^{3)}$; C_s=saturation of dissolved oxygen concentration (M/L^3); K_2=reaeration coefficient (1/T).

Receiving Water (1) A distinct water body that receives runoff, or wastewater discharges, such as streams, rivers, lakes, estuaries and oceans. (2) A river, lake, ocean, stream, or other body of water into which wastewater or treated effluent is discharged.

Recession Curve A hydrograph showing the decreasing rate of runoff following a period of rain or snowmelt. Since direct runoff and base runoff recede at different rates, separate curves, called direct runoff recession curves or base runoff recession curves, are generally drawn. The term "depletion curve" in the sense of base runoff recession is not recommended.

Recharge The water added to an aquifer. For instance, rainfall that seeps into the ground.

Reclaimed Wastewater Treated wastewater that can be used for beneficial purposes, such as irrigating certain plants.

Reclamation Site Drastically disturbed land, such as strip mines and construction sites that is reclaimed using sewage sludge.

Recurrence Interval (Return Period) The average interval of time within which the given flood will be equaled or exceeded once.

Recycled Water The water that is used more than one time before it passes back into the natural hydrologic system.

Regime "Regime theory" is a theory of the forming of channels in the material carried by the streams. As used in this sense, the word "regime" applies only to streams that make at least part of their boundaries from their transported load and part of their transported load from their boundaries, carrying out the process at different places and times in any one stream in a balanced or alternating manner that prevents unlimited growth or removal of boundaries. A stream, river, or canal of this type is called a "regime stream, river, or canal." A regime channel is said to be "in regime" when it has achieved average equilibrium; that is, the average values of the quantities that constitute regime do not show a definite trend over a considerable period—generally of the order of a decade. In unspecialized use "regime" and "regimen" are synonyms. Regimen of a stream—the system or order characteristic of a stream; in other words, its habits with respect to velocity and volume, form of and changes in channel, capacity to transport sediment, and amount of material supplied for transportation. The term is also applied to a stream which has reached an equilibrium between corrosion and deposition or, in other words, to a graded stream.

Regional Indicator An Report on the Environment (ROE) indicator that helps to answer an ROE question on a smaller-than-national geographic scale. A regional indicator may cover a topic for which nationally consistent data are unavailable, or it may present an issue that is of particular concern within a certain geographic area (see ROE indicator).

Regression Analysis (1) A statistical method in which an empirical relationship between an independent variable and a dependent variable can be determined such that the average tendency of the observed values and the average tendency of the predictive values given by the empirical equation may be identical. (2) A statistical method attempts to determine the best mathematical model, given the available data, to describe a dependent variable as a function of an independent variable or, in the case of multiple regression analysis, more than one independent variable.

Regression Coefficient (1) A derived coefficient in regression analysis that expresses the change in the dependent variable associated with a change in one or more independent variables. (2) It is referred to as the slope of the relationship between the variables, as the derivative in bivariate analysis, or as the partial derivative in multiple regression.

Regulation The artificial manipulation of the flow of a stream.

Remedial Action The actual construction or cleanup phase of a site cleanup.

Remote Sensing (1) The analysis and interpretation of images gathered through techniques that do not require direct contact with the subject. (2) A discipline that evolved from photogrammetry, remote sensing of the earth's resources uses aerial or space photographs, electronic scanners, and other devices to collect data about the earth's surface and subsurface.

Report on the Environment (ROE) A USEPA report which presents the best available indicators of information on national conditions and trends in air, water, land, human health, and ecological systems that address all questions USEPA considers mission critical to protecting our environment and human health.

Reregulating Reservoirs A reservoir for reducing diurnal fluctuations resulting from the operation of an upstream reservoir for power production.

Reservoir (1) A pond, lake, or basin, either natural or artificial, for the storage, regulation, and control of water. (2) A man-made facility for the storage, regulation, and controlled release of water. (3) A natural or artificial lake, storage pond, or impoundment from a dam which is used to store water. (4) An artificially impounded body of water.

Reservoir Inflow The amount of water entering a reservoir expressed in acre–feet per day or cubic feet per second.

Reservoir Operation Management of water release from reservoir to satisfy some specific objective.

Reservoir Regulation (or Operating) Procedure Operating procedures that govern reservoir storage and releases.

Reservoir Surface Area The area covered by a reservoir when filled to a specified level.

Reservoir Volume The volume of a reservoir when filled to a normal pool or water level.

Residual-Mass Curve A graph of the cumulative departures from a given reference such as the arithmetic average, generally as ordinate, plotted against time or date, as abscissa (see mass curve).

Resistance Opposition to the passage of electrical current.

Resistivity An intrinsic property of a material, measured as its resistance to current per unit length for a uniform cross section.

Respiration The biological oxidation of organic carbon with concomitant reduction of external oxidant and the production of energy. In aerobic respiration, O_2 is reduced to CO_2. Anaerobic respiration processes utilize NO_3^- (denitrification), SO_4^{2-} (sulfate reduction), or CO_2 (methanogenesis).

Restoration An activity returning a wetland from a disturbed or altered condition with lesser acreage or functions to a previous condition with greater wetland acreage or functions. For example, restoration might involve the plugging of a drainage ditch to restore the hydrology to an area that was a wetland before the installation of the drainage ditch.

Retarding reservoir An ungated reservoir for temporary storage of floodwater; sometimes called detention reservoir.

Retention time The average length of time a water molecule or a suspended particle remains in a tank or chamber. Mathematically, it is the volume of water (L^3) in the tank divided by the flow rate (L^3/T) through the tank.

Return Flow (1) That part of a diverted flow that is not consumptively used and returned to its original source or another body of water. (2) (Irrigation) drainage water from irrigated farmlands that reenters the water system to be used further downstream.

Return Flow (Irrigation) (1) Irrigation water that is applied to an area and which is not consumed in evaporation or transpiration and returns to a surface stream or aquifer. (2) That part of irrigation water that is not consumed by evapotranspiration and that returns to its source or another body of water. The term is also applied to the water that is discharged from industrial plants. Also called return water.

Reverse Osmosis (1) (Desalination) The process of removing salts from water using a membrane. With reverse osmosis, the product water passes through a fine membrane that the salts are unable to pass through, while the salt waste (brine) is removed and disposed. This process differs from electrodialysis, where the salts are extracted from the feedwater by using a membrane with an electrical current to separate the ions. The positive ions go through one membrane, while the negative ions flow through a different membrane, leaving the end product of freshwater. (2) (Water Quality) An advanced method of water or wastewater treatment that relies on a semi-permeable membrane to separate waters from pollutants. An external force is used to reverse the normal osmotic process resulting in the solvent moving from a solution of higher concentration to one of lower concentration.

Riffle (1) A rapid in a stream. (2) A shallow section in a stream where water is breaking over rocks or other partially submerged organic debris and producing surface agitation.

Rill Natural fluvial topographic feature. These channels are shallow and narrow, form in multiples, and parallel to each other.

Riparian (1) Areas next to or substantially influenced by water; (2) pertaining to the banks of a stream. These may include areas adjacent to rivers, lakes, or estuaries. These areas often include wetlands.

Riparian Water Rights The rights of an owner whose land abuts water. They differ from state to state and often depend on whether the water is a river, lake, or ocean. The doctrine of riparian rights is an old one, having its origins in English common law. Specifically, persons who own land adjacent to a stream have the right to make reasonable use of the stream. Riparian users of a stream share the stream flow among themselves, and the concept of priority of use (prior appropriation doctrine) is not applicable. Riparian rights cannot be sold or transferred for use on non-riparian land.

Risk A measure of the chance that damage to life, health, property, or the environment will occur.

Risk Assessment A methodology used to examine all possible risks involved with a particular product or organism. Risk assessment can be divided into four parts: identification of hazards, dose response (how much exposure causes particular problems such as cancer, convulsions, death, etc.), exposure assessment (determining how much exposure will be received by people during particular activities), and risk characterization (determining a probability that a risk will occur).

Risk Factor A characteristic (e.g., race, sex, age, obesity) or variable (e.g., smoking, occupational exposure level) associated with increased probability of an adverse effect.

River A natural stream of water of considerable volume, larger than a brook or creek.

River Morphology A study of the evolution and configuration of river.

River Training Enhancements made to the banks of a river to constrain the river to a desired course; the enhancements are usually in the form of realignment or regrading of the banks, and installation of bank protection structures such as rock revetments (riprap), dikes, weirs, or submerged vanes.

Riverine Pertaining to rivers.

ROE See "Report on the Environment."

ROE Indicator An indicator that meets the Report on the Environment (ROE) criteria and has been peer-reviewed (see "indicator").

Rotational Failure It refers to a deep-seated movement of bank material both downward and outward along a curved slip surface. After the failure, the upper surface of the slipped block is typically tilted inward toward the bank.

Rough Fish A non-sport fish, usually omnivorous in food habits.

Runge–Kutta Methods A family of implicit and explicit iterative methods used in temporal discretization for the approximation of solutions of ordinary differential equations.

Runoff (1) Excess water flow that occurs when the soil infiltration capacity is exceeded during a rainstorm event, meltwater, or other sources of flows over the land. (2) That part of the precipitation such as snowmelt, or irrigation water that appears in uncontrolled surface streams, drains, or sewers. It is the same as streamflow unaffected by artificial diversions, storage, or other works of man in or on the stream channels. Runoff may be classified as follows: (1) classification as to speed of appearance after rainfall or snow melting: direct runoff or base runoff; (2) classification as to source: surface runoff (see overland flow), storm seepage (storm inter), or groundwater runoff (see stream, gaining). It can collect pollutants from air or land and carry them to streams and other waterbodies. Also defined as the depth to which a drainage area would be covered if all of the runoff for a given period of time were uniformly distributed over it.

Runoff Plots Field plots of various size (standard Universal Soil Loss Equation; USLE plot size is 3.7 m wide and 18.3 m long) to monitor runoff volumes, soil loss, chemical transport, etc.

Runout See water yield.

Saline Water The water that contains significant amounts of dissolved solids. There are four types of water in terms of salinity: freshwater—less than 1000 parts per million (ppm); slightly saline water—from 1000 to 3000 ppm; moderately saline water—from 3000 to 10,000 ppm; and highly saline water—from 10,000 to 35,000 ppm.

Saline Water Intrusion The movement of saline groundwater into a formerly freshwater aquifer as a result of pumping in that aquifer usually near coastal areas where the source of saline water is the nearby ocean.

Salinity (1) Salinity is the total concentration of all ionic constituents present in the water. This is comprised mostly by chloride ions and sodium ions. Seawater may have other ions, such as potassium ions, magnesium ions, sulfate ions, etc.; (2) salinity is defined in relation to chlorinity as follows: salinity = 1.80655 (chlorinity); (3) salinity is traditionally defined as the total solids in water after all carbonates have been converted to oxides, all bromide and iodide have been replaced by chloride, and all organic matter has been oxidized. The new scale used to define salinity is based on the electrical conductivity of seawater relative to a specified solution of potassium chloride (KCl) and water (H_2O). The scale is dimensionless and the traditional dimensions of parts per thousand (i.e., mg/g of solution) no longer applies. The unit of salinity is ppt, or parts per thousand, or g/L.

Sample The water that is analyzed for the presence of the USEPA-regulated drinking-water contaminants. Depending on the regulation, the USEPA requires water systems and states to take samples from source water, from water leaving the treatment facility, or from the taps of selected consumers.

Sanitary Survey An on-site review of the water sources, facilities, equipment, operation, and maintenance of a PWSs for the purpose of evaluating the adequacy of the facilities for producing and distributing safe drinking water.

Saturation The ratio of the volume of a particular fluid to the total volume of the void space.

Scour (1) Concentrated erosive action by water. The enlargement of a flow section or creation of a depression by the removal of bed material through the action of moving water. (2) The action of a flowing water as it lifts and carries away the material on the sides or bottom of a waterway, conduit, or pipeline.

Scouring Velocity The minimum velocity necessary to dislodge stranded material from the boundary of a waterway, conduit, or pipeline by a fluid in motion.

Secchi Disk A black-and-white disk used to measure the clarity of water. The disk is lowered into the water until it cannot be seen and then the depth of the disk is measured. Septic system: A system that treats and disposes of household wastewater under the ground.

Secondary Currents (or Flow) The movement of water particles normal to the principal direction of flow.

Secondary DAF A dissolved air flotation (DAF) clarifier which is used for secondary wastewater clarification after the wastewater is discharged from the secondary wastewater treatment bioreactors, such as aeration basins, trickling filters, rotating biological reactors, etc. The objective of secondary wastewater clarification is separation of biosolids (microorganisms) from the bioreactor effluents.

Secondary Drinking Water Standards Nonenforceable federal guidelines regarding cosmetic effects (such as tooth or skin discoloration) or aesthetic effects (such as taste, odor, or color) of drinking water.

Secondary Pollutant Any pollutant that is formed by atmospheric reactions of precursor or primary emissions. An example of a secondary pollutant is ground-level ozone, which forms from chemical reactions involving airborne nitrogen oxides, airborne volatile organic compounds, and sunlight.

Secondary Wastewater Treatment Treatment (following primary wastewater treatment) involving the biological process of reducing suspended, colloidal, and dissolved organic matter in an effluent from primary treatment systems and which generally removes 80–95 % of the biochemical oxygen demand (BOD) and suspended matter. Secondary wastewater treatment may be accomplished by biological or chemical–physical methods. Activated sludge and trickling filters are two of the most common means of secondary treatment. It is accomplished by bringing together waste, bacteria, and oxygen in trickling filters or in the activated sludge process. This treatment removes floating and settleable solids and about 90 % of the oxygen-demanding substances and suspended solids. Disinfection is the final stage of secondary treatment.

Second-Foot Same as cfs, or cubic foot per second. This term is no longer used in published reports of the US Geological Survey.

Sediment (1) Fragmental material that originates from weathering of rocks and is transported by, suspended in, or deposited by water or air or is accumulated in beds by other natural agencies; (2) usually applied to material in suspension in water or recently deposited from suspension. In the plural, the word is applied to all kinds of deposits from the waters of streams, lakes, or seas. (3) Any mineral and/or organic matter deposited by water or air. (4) Naturally occurring soil and gravel material that is broken down by processes of weathering and erosion, and is subsequently transported by the action of wind, water, or ice, and/or by the force of gravity acting on the particle itself. In this chapter, sediment is the material (sand) transported by water.

Sediment Discharge (1) The mass or volume of sediment (usually mass) passing a stream cross section in a unit of time. The term may be qualified, for example, as suspended-sediment discharge, bed-load discharge, or total-sediment discharge. (2) The rate at which dry weight of sediment passes a section of a stream or is the quantity of sediment, as measured by dry weight, or by volume, that is discharged in a given time (see sediment load).

Sediment Load A general term that refers to material in suspension and/or in transport. It is not necessarily synonymous with either discharge or concentration. It may also refer to a particular type of load; e.g., total, suspended bed, or bed-material load.

Sediment Particle Solid fragments of mineral material in either a singular or an aggregate state.

Sediment Transport (Rate) See sediment discharge.

Sedimentary Rock Rock formed of sediment, and specifically: (1) sandstone and shale, formed of fragments of other rock transported from their sources and deposited in water; and (2) rocks formed by or from secretions of organisms, such as most limestones. Many sedimentary rocks show distinct layering, which is the result of different types of sediment being deposited in succession.

Sedimentation (1) The tendency for particles in suspension to settle out of the fluid in which they are entrained, and come to rest. (2) Refers to the gravitational settling of suspended particles that are heavier than water.

Sedimentation tanks or basins Wastewater tanks/basins in which floating scums are skimmed off and settled solids are removed for disposal.

Seepage (1) The slow movement of water through small cracks, pores, Interstices, etc., of a material into or out of a body of surface or subsurface water. (2) The loss of water by infiltration into the soil from a canal, ditches, laterals, watercourse, reservoir, storage facilities, or other body of water, or from a field. (3) A process of liquid leaking through a porous substance. (4) The interstitial movement of water that may take place through a dam, its foundation, or abutments.

Seiche The free oscillation of the bulk of water in a lake and the motion caused by it on the surface of the lake.

Seismic Refraction Seismic refraction investigates the subsurface by generating the arrival time and offset distance information to determine the path and velocity of the elastic disturbance in the ground.

Seismograph An instrument for measuring and recording the vibrations of earthquakes.

Self-Supplied Water The water withdrawn from a surface-water or groundwater source by a user rather than being obtained from a public supply. An example would be homeowners getting their water from their own well.

Semi-Infinite Aquifer Ideal aquifer characterized by an infinite lateral extension on one side and a rectilinear boundary on the other.

Sensitivity (1) In analytical testing, the lowest practical detection level; (2) in microbiological testing, the likelihood that the test result will be positive when the target organism is present; (3) in water resources engineering, the smallest changes of certain physical parameters that will affect hydraulic or hydrological model's solutions.

Sensitivity Analysis (1) A mathematical analysis of the sensitivity of the dependent variable in a mathematical expression as a function of variations in the value of any independent variables or coefficients associated with the independent variables. (2) A mathematical analysis which determines how much the value of Y is affected by changes in the values of a and b.

Septage Septage means the liquid and solid material pumped from a septic tank, cesspool, or similar domestic sewage treatment system, or a holding tank when the system is cleaned or maintained.

Septic Tank A tank used to detain domestic wastes to allow the settling of solids prior to distribution to a leach field for soil absorption. Septic tanks are used when a sewer line is not available to carry them to a treatment plant. A settling tank in which settled sludge is in immediate contact with sewage flowing through the tank, and wherein solids are decomposed by anaerobic bacterial action.

Settling Clarifier A sedimentation clarifier which is used for separation of particles from water by the settling force of particles.

Settling Pond (Water Quality) An open lagoon into which wastewater contaminated with solid pollutants is placed and allowed to stand. The solid pollutants suspended in the water sink to the bottom of the lagoon and the liquid is allowed to overflow out of the enclosure.

Sewage Sludge The solid, semi-solid, or liquid residue generated during the treatment of domestic sewage in a treatment works. Sewage sludge includes, but is not limited to, domestic septage, scum, and solids removed during the primary, secondary, or advanced wastewater treatment processes. The definition of sewage sludge also includes a material derived from sewage sludge (i.e., sewage sludge whose quality is changed either through further treatment or through mixing with other materials).

Sewage Treatment Plant A facility designed to receive the wastewater from domestic sources and to remove materials that damage water quality and threaten public health and safety when discharged into receiving streams or bodies of water. The substances removed are classified into four basic areas: (1) greases and fats; (2) solids from human waste and other sources; (3) dissolved pollutants from human waste and decomposition products; and (4) dangerous microorganisms. Most facilities employ a combination of mechanical removal steps and bacterial decomposition to achieve the desired results. Chlorine is often added to discharges from the plants to reduce the danger of spreading disease by the release of pathogenic bacteria.

Sewer A system of underground pipes that collect and deliver wastewater to treatment facilities or streams.

Shallow Slide It refers to a process in which a layer of material moves along a plane parallel to the bank surface. This failure often occurs on banks where soils have low and varied cohesion and the bank is moderately steep.

Shear Modulus The ratio of stress to strain that describes deformation that takes place when a force is applied parallel to one face of an object while the opposite face is held fixed.

Shear Stress (Boundary Shear Stress) Frictional force per unit or area exerted on a channel boundary by the flowing water. An important factor in the movement of bed material.

Shifting Control See control.

Siltation The deposition of silt-sized and clay-sized (smaller than sand-sized) particles.

Sinkhole A depression in the earth' surface caused by dissolving of underlying limestone, salt, or gypsum. Drainage is provided through underground channels that may be enlarged by the collapse of a cavern roof.

Site Characterization A location-specific or area-specific survey conducted to characterize physical, chemical, and/or biological attributes of an area; such surveys may be conducted at different times to provide information on how these attributes may change over time.

Skimming The diversion of water from a stream or conduit by a shallow overflow used to avoid diversion of sand, silt, or other debris carried as bottom load.

Snow A form of precipitation composed of ice crystals.

Snow Course A line or series of connecting lines along which snow samples are taken at regularly spaced points.

Snow Density The ratio between the volume of meltwater derived from a sample of snow and the initial volume of the sample. This is numerically equal to the specific gravity of the snow.

Snow, Quality of The ratio of heat of melting of snow, in calories per gram to the 80 cal/g for melting pure ice at 0 °C. Percentage by weight which is ice.

Snowline The general altitude to which the continuous snow cover of high mountains retreats in summer, chiefly controlled by the depth of the winter snowfall, and by the temperature of the summer.

Snowline, Temporary A line sometimes drawn on a weather map during the winter showing the southern limit of the snow cover.

Sodium Exchange Capacity Measure of the soil's ability to hold and release sodium.

Soil Conservation A set of management practices to prevent soil from being eroded.

Soil Erosion A natural process that occurs when soil is removed through the action of wind and/or water.

Soil Moisture (Soil Water) Water diffused in the soil, the upper part of the zone of aeration from which water is discharged by the transpiration of plants or by soil evaporation (see field-moisture capacity and field-moisture deficiency).

Soil Production The rate of bedrock weathering into soil as a function of soil thickness.

Soil Texture Refers to the size and size distribution of the particles that make up the soil.

Sole Source Aquifer An aquifer that supplies 50 % or more of the drinking water of an area.

Solubility The ability of a chemical (e.g., pollutant) to be dissolved into a solvent (e.g., water column).

Solute A substance that is dissolved in another substance, thus forming a solution.

Solution A mixture of a solvent and a solute. In some solutions, such as sugar water, the substances mix so thoroughly that the solute cannot be seen. But in other solutions, such as water mixed with dye, the solution is visibly changed.

Solvent A substance that dissolves other substances, thus forming a solution. Water dissolves more substances than any other, and is known as the "universal solvent."

Sounding A mechanism of probing the environment by sending out some kind of stimulus.

Source water Water in its natural state, prior to any treatment for drinking.

Spatial (1) Pertaining to an object, dataset, or activity that has a geographic component or is related to a location in a coordinate system. Spatial modeling and analysis are based on geographical relationships of data; (2) in biology or ecology, pertaining to the occurrence of or relationships among microorganisms, plants, or animals living in the same habitat, as in the spatial distribution of a particular species.

Spatial Variation Changes in water quality characteristics among various locations in a water body or among water bodies.

Specific Conductance A measure of the ability of water to conduct an electrical current as measured using a 1-cm cell and expressed in units of electrical conductance, i.e., Siemens per centimeter at 25 °C. Specific conductance can be used for approximating the total dissolved solids content of water by testing its capacity to carry an electrical current. In water quality, specific conductance is used in groundwater monitoring as an indication of the presence of ions of chemical substances that may have been released by a leaking landfill or other waste storage or disposal facility. A higher specific conductance in water drawn from downgradient wells when compared to upgradient wells indicates possible contamination from the facility.

Specific Elastic Storage Physical property of a porous medium, representing the volume of water released by per unit bulk volume and per unit decline in pressure head.

Specific Yield Physical property given by the unitless ratio of the volume of water added or removed directly from the saturated zone of the aquifer to the resulting change in the volume of aquifer below water.

Spillway (1) Structure used to provide controlled release of flows from a dam or levee into a downstream area; and (2) a structure over or through which excess or flood flows are discharged. If the flow is controlled by gates, it is a controlled spillway, if the elevation of the spillway crest is the only control, it is an uncontrolled spillway.

Spillway Crest The elevation of the highest point of a spillway.

Spoil Water The top portion of water is spoil water after the dredged material is settled.

Spray Irrigation (1) A method of land application by which wastewater is sprayed from nozzles onto land; (2) a common irrigation method where water is shot from high-pressure sprayers onto crops. Because water is shot high into the air onto crops, some water is lost to evaporation.

Spring A water body formed when the side of a hill, a valley bottom, or other excavation intersects a flowing body of groundwater at or below the local water table, below which the subsurface material is saturated with water.

Stable Resistant to change.

Stable Channel (1) A channel which is resistant to change. (2) A stream channel that does not change in a planform, cross section, or bed profile during a particular period of time (but may over longer periods of time).

Stage The height of a water surface above an established datum plane; also gage height.

Stage, Flood See flood stage.

Stage-Capacity Curve A graph showing the relation between the surface elevation of the water in a reservoir, usually plotted as ordinate, against the volume below that elevation, plotted as abscissa.

Stage-Discharge Curve (Rating Curve) A graph showing the relation between the gage height, usually plotted as ordinate, and the amount of water flowing in a channel, expressed as volume per unit of time, plotted as abscissa.

Stage-Discharge Relation The relation expressed by the stage-discharge curve.

Stakeholder An individual or an organization that has a stake in the outcome of the watershed plan.

State Variables The variables of an optimization problem that characterize the mathematical state of a dynamic system and depend upon the selection of the decision variables (stems or roots).

Stem Flow Rainfall or snowmelt led to the ground, down the trunks, or stems of plants.

Stochastic Pertaining to actions for which the results occur from probabilistic events.

Stochastic Approach An approach using a random variable dependent on a parameter, usually time.

Stochastic Process A process involving a random variable dependent on a parameter, such as time.

Storage (1) Water artificially impounded in surface or underground reservoirs, for future use. The term regulation refers to the action of this storage in modifying streamflow (see also conservation storage, total storage, dead storage, and usable storage). (2) The water naturally detained in a drainage basin, such as groundwater, channel storage, and depression storage. The term "drainage basin storage" or simply "basin storage" is sometimes used to refer collectively to the amount of water in natural storage in a drainage basin.

Storage Equation The equation for the conservation of mass.

Storage Ratio The net available storage divided by the mean flow for 1 year.

Storage, Bank See bank storage.

Storage, Conservation See conservation storage.

Storage, Dead See dead storage.

Storage, Depression See depression storage.

Storage, Total See total storage.

Storage, Usable See usable storage.

Storage-Required Frequency Curve A graph showing the frequency with which storage equal to or greater than selected amounts will be required to maintain selected rates of regulated flow.

Storm A disturbance of the ordinary average conditions of the atmosphere which, unless specifically qualified, may include any or all meteorological disturbances, such as wind, rain, snow, hail, or thunder.

Storm Seepage That part of precipitation which infiltrates the surface soil, and moves toward the streams as ephemeral, shallow, or perched groundwater above the main groundwater level. Storm seepage is usually part of the direct runoff.

Storm Sewer A sewer that carries only surface runoff, street wash, and snowmelt from the land. In a separate sewer system, storm sewers are completely separate from those that carry domestic and commercial wastewater (sanitary sewers) stream—a general term for a body of flowing water; natural water course containing water at least part of the year. In hydrology, it is generally applied to the water flowing in a natural channel as distinct from a canal.

Storm Flow See direct runoff.

Stratification The division of a water body into two or more depth zones due to temperature.

Stratosphere The layer of the atmosphere that starts about 6–9 miles above the earth's surface at midlatitudes and lies atop the troposphere. The stratosphere contains small amounts of gaseous ozone, which filters out about 99 % of the incoming ultraviolet radiation.

Stream A general term for a body of flowing water. In hydrology, the term is generally applied to the water flowing in a natural channel as distinct from a canal. More generally as in the term stream gaging, it is applied to the water flowing in any channel, natural or artificial. Streams in natural channels may be classified as follows in relation to time: (1) perennial stream is one which flows continuously; (2) intermittent or seasonal stream is one which flows only at certain times of the year when it receives water from springs or from some surface source such as melting snow in mountainous areas; and (3) ephemeral stream is one that flows only in direct response to precipitation, and whose channel is at all times above the water table. Streams in natural channels may be classified as follows in relation to space: (1) continuous stream is one that does not have interruptions in space; (2) interrupted stream is one which contains alternating reaches, that are either perennial, intermittent, or ephemeral. Streams in natural channels may also be classified as follows in relation to groundwater: (1) gaining stream is a stream or reach of a stream that receives water from the zone of saturation; (2) losing stream is a stream or reach of a stream that contributes water to the zone of saturation; (3) insulated stream is a stream or reach of a stream that neither contributes water to the zone of saturation nor receives water from it. It is separated from the zones of saturation by an impermeable bed; (4) perched stream is either a losing stream or an insulated stream that is separated from the underlying groundwater by a zone of aeration.

Stream Bank Erosion The removal of bank material primarily by hydraulic action.

Stream Depletion Rate An instantaneous flow rate with which a stream, idealized as a constant-head boundary, recharges a hydraulically connected aquifer in which groundwater pumping is occurring.

Stream Depletion Volume The cumulative volume with which a stream, idealized as a constant-head boundary, recharges a hydraulically connected aquifer in which groundwater pumping is occurring.

Stream Flow The water flowing in the stream channel. It is often used interchangeably with discharge.

Stream Gage A site along a stream where the stage (water level) is either read by the eye or measured with a recording equipment.

Stream Gaging The process and art of measuring the depths, areas, velocities, and rates of flow in natural or artificial channels.

Stream Meander The length of a stream channel from an upstream point to a downstream point divided by the straight-line distance between the same two points.

Stream Order A method of numbering streams as part of a drainage basin network. The smallest unbranched mapped tributary is called first order, the stream receiving the tributary is called second order, and so on. It is usually necessary to specify the scale of the map used. A first-order stream on a 1:62,500 map, may be a third-order stream on a 1:12,000 map. Tributaries which have no branches are designated as of the first order, streams which receive only first-order tributaries are of the second order, larger branches which receive only first-order and second-order tributaries are designated as third order, and so on, the main stream being always of the highest order.

Streambed The channel bottom of a stream, representing the lower boundary of the streamflow as well as the interface between the surface and subsurface flow.

Streamflow (1) The water discharge that occurs in a natural channel; (2) a more general term than runoff, streamflow may be applied to discharge whether or not it is affected by diversion or regulation. The discharge that occurs in a natural channel. Although the term discharge can be applied to the flow of a canal, the word streamflow uniquely describes the discharge in a surface stream course. The term "streamflow" is more general than runoff, as streamflow may be applied to discharge whether or not it is affected by diversion or regulation.

Streamflow Depletion The amount of water that flows into a valley, or onto a particular land area, minus the water that flows out of the valley or off from the particular land area.

Stream-Gaging Station A gaging station where a record of discharge of a stream is obtained. Within the Geological Survey, this term is used only for those gaging stations where a continuous record of discharge is obtained.

Stressor A physical, chemical, or biological entity that can induce adverse effects on ecosystems or human health.

Strip Cropping Growing crops in an arrangement of lines in order to reduce erosion.

Submeander Small meander contained with banks of main channel, associated with relatively low discharges.

Subsidence A dropping of the land surface as a result of groundwater being pumped. Cracks and fissures can appear in the land. Subsidence is virtually an irreversible process.

Substrate (1) A substance used by organisms for growth. (2) A substance acted upon by an enzyme. (3) The basic, abundant component of a medium.

Subsurface Pertaining to, formed, or occurring underneath the ground surface.

Subsurface Runoff See storm seepage.

Succession The replacement of one plant assemblage with another through time sufficient for photosynthesis.

Supercritical Flow A flow whose velocity is larger than the wave velocity, therefore with $Fr > 1$.

Superfund A program, operated under the legislative authority of the Comprehensive Environmental Response, Compensation, and Liability Act and the Superfund Amendments and Reauthorization Act, that funds and carries out the USEPA solid waste emergency and long-term removal and remedial activities. These activities include establishing the National Priorities List (NPL), investigating sites for inclusion on the list, determining their priority, and conducting and/or supervising cleanup and other remedial actions (see National Priorities List).

Superposition Principle This principle applies to linear systems and states that the response caused by two or more forcing terms at a generic location and time is equal to the sum of the responses associated with each forcing terms as this was acting individually.

Supplemental Irrigation Commonly, irrigation as carried on in humid areas. The term means that the irrigation water is supplementary to the natural rainfall rather than being the primary source of moisture as in the arid and semiarid West. Supplementary irrigation is used generally to prevent retardation of growth during periods of drought.

Supplemental Sources When irrigation water supplies are obtained from more than one source, the source furnishing the principal supply is commonly designated the primary source, and the sources furnishing the additional supplies, the supplemental sources.

Surcharge Capacity The volume of a reservoir between the maximum water surface elevation for which the dam is designed and the crest of an uncontrolled spillway, or the normal full-pool elevation of the reservoir with the crest gates in the normal closed position.

Surface Runoff That part of the runoff which travels over the soil surface to the nearest stream channel. It is also defined as that part of the runoff of a drainage basin that has not passed beneath the surface since precipitation. The terms groundwater runoff and surface runoff are classifications according to source. The terms base runoff and direct runoff are time classifications of runoff.

Surface Tension The attraction of molecules to each other on a liquid's surface. Thus, a barrier is created between the air and the liquid.

Surface Transfer Coefficient (a) A mass transfer coefficient (K_L) which governs the rate for transferring dissolved oxygen across the water surface, L/T; (b) a mass transfer coefficient which is defined as $K_L = (K_2)H$. Here, K_2 is reaeration coefficient (1/T); and H is the water depth (L; see mass transfer coefficient, reaeration coefficient, and reaeration).

Surface Water (1) The water on the surface of the earth such as in a stream, river, lake, or reservoir; (2) the water that systems pump and treat from sources open to the atmosphere, such as rivers, lakes, and reservoirs.

Survey A method of collecting quantitative and qualitative information.

Suspended Sediment Very-fine-soil particles that remain in suspension in water for a considerable period of time without contact with the bottom. Such material remains in suspension due to the upward components of turbulence and currents and/or by suspension.

Suspended Solids Solids that are not in true solution and that can be removed by filtration. Such suspended solids usually contribute directly to turbidity. As defined in waste management, these are small particles of solid pollutants that resist separation by conventional methods.

Suspended-Sediment Concentration The ratio of the mass of dry sediment in a water–sediment mixture to the mass of the water–sediment mixture. Typically expressed in milligrams of dry sediment per liter of water–sediment mixture.

Suspended-Sediment Discharge The quantity of suspended sediment passing a point in a stream over a specified period of time. When expressed in tons per day, it is computed by multiplying water discharge (in cubic feet per second) by the suspended-sediment concentration (in milligrams per liter) and by the factor 0.0027.

Sustainable Management Concept of keeping a system running indefinitely without depleting resources, while maintaining economic viability and providing for the needs of present and future generations.

Terrace (1) A piece of slope plane that has been cut into a series of successively receding flat, horizontal surfaces which resemble steps, for the purpose of decreasing erosion and surface runoff. (2) A berm or discontinuous segments of a berm, in a valley at some height above the floodplain, representing a former abandoned floodplain of the stream.

Tertiary Wastewater Treatment The selected biological, physical, and chemical separation processes to remove organic and inorganic substances that resist conventional treatment practices; the additional treatment of effluent beyond that of primary and secondary treatment methods to obtain a very high quality of effluent. A complete wastewater treatment system typically involves a three-phase treatment: (1) first, in the primary wastewater treatment, which incorporates physical aspects, the untreated raw wastewater is passed through a series of screens to remove solid wastes; (2) second, in the secondary wastewater treatment, which typically involves biological and chemical processes, the screened primary effluent wastewater is then passed through a series of holding and aeration tanks and ponds for removal of soluble organic pollutants; and (3) third, in the tertiary wastewater treatment, which consists of flocculation basins, clarifiers, filters, and chlorine basins or ozone or ultraviolet radiation processes, the secondary effluent wastewater is finally polished and ready for discharge into a receiving water.

Theis Equation Analytical model developed by Charles Vernon Theis for aquifer drawdown associated with two-dimensional radial flow to a point source in an infinite, homogeneous, confined aquifer.

Thermal Pollution A reduction in water quality caused by increasing its temperature, often due to disposal of waste heat from industrial or power generation processes. Thermally polluted water can harm the environment because plants and animals can have a hard time adapting to it.

Thermal Stratification (of A Lake) Vertical temperature stratification that shows the following: The upper layer of the lake, known as the epilimnion, in which the water temperature is virtually uniform; a stratum next below, known as the thermocline, in which there is a marked drop in temperature per unit of depth; and the lowermost region or stratum, known as the hypolimnion, in which the temperature from its upper limit to the bottom is nearly uniform.

Thermocline See thermal stratification.

Thermoelectric Power Water Use The water used in the process of the generation of thermoelectric power. Power plants that burn coal and oil are examples of thermoelectric-power facilities.

Threatened Waterbody A waterbody that is meeting standards but exhibits a declining trend in water quality such that it will likely exceed standards.

Throughfall In a vegetated area, the precipitation that falls directly to the ground or the rainwater or snowmelt that drops from twigs or leaves.

Tidal Pertaining to periodic water level fluctuations due to the action of lunar (moon) and solar (sun) forces upon the rotating earth.

Tidal Current A water current brought about or caused by tidal forces.

Tidal Stream/River A stream/river which is affected by the tidal current and its water level and flow direction fluctuate due to the action of lunar and solar forces upon the rotating earth.

Time of Concentration The time required for water to flow from the farthest point on the watershed to the gaging station.

TMDL Process The approach normally used to develop a TMDL for a particular water body or watershed. This process consists of five activities, including selection of the pollutant to consider, estimation of the water body's assimilative capacity, estimation of the pollution from all sources to the water body, predictive analysis of pollution in the water body and determination of total allowable pollution load, and allocation of the allowable pollution among the different pollution sources in a manner that water quality standards are achieved to permit growth of plants.

Tolerance An organism's capacity to endure or adapt to unfavorable conditions.

Tomography The process of imaging by sections or sectioning, through the use of a penetrating wave.

Top Width The width of a stream section at the water surface; it varies with stage in most natural channels.

Tortuosity The degree to which a path is curved.

Total Nitrogen It is the summation of ammonium nitrogen (NH_4^+–N), nitrate nitrogen (NO_3^-–N), nitrite nitrogen (NO_2^-–N), and organic nitrogen (organic-N). Usually, nitrite nitrogen is in negligible amount. Crops directly utilize nitrogen in its inorganic forms, principally nitrate-N and ammonium-N.

Total Gross Reservoir Capacity The total amount of storage capacity available in a reservoir for all purposes from the streambed to the normal water or normal water or normal pool surface level. It does not include surcharge, but does include dead storage.

Total Kjeldahl Nitrogen (TKN) TKN is the summation of ammonium nitrogen (NH_4^+–N) and organic nitrogen (organic-N).

Total Maximum Daily Load (TMDL) (1) The amount, or load, of a specific pollutant that a waterbody can assimilate and still meet the water quality standard for its designated use; (2) an estimate of the pollutant concentrations resulting from the pollutant loadings from all sources to a water body. The TMDL is used to determine the allowable loads and provides the basis for establishing or modifying controls on pollutant sources. For impaired water bodies, the TMDL reduces the overall load by allocating the load among current pollutant loads (from point and NPSs), background or natural loads, a margin of safety, and sometimes an allocation for future growth.

Total Solids (TS) Total solids (TS) include suspended and dissolved solids and are usually expressed as the concentration present in biosolids. TS depend on the type of wastewater process and biosolids' treatment prior to land application. Typical solids contents of various biosolids are: liquid (2–12 %), dewatered (12–30 %), and dried or composted (50 %).

Total Storage The volume of a reservoir below the maximum controllable level including dead storage.

Total Variation Diminishing (TVD) It is a property of certain discretization schemes used to solve hyperbolic partial differential equations that do not increase the total.

Toxic Chemical A chemical that can produce injury if inhaled, swallowed, or absorbed through the skin.

Toxicology The study of the quantitative effects of chemicals on biologic tissue, particularly in terms of definite harmful actions and degrees of safety.

Toxics Release Inventory (TRI) A database containing detailed information on nearly 650 chemicals and chemical categories that over 23,000 industrial and other facilities manage through disposal or other releases, recycling, combustion for energy recovery, or treatment.

Toxics Release Inventory (TRI) Chemicals The chemicals and chemical categories that appear on the current TRI toxic chemical list. As of December 2007, the TRI toxic chemical list contains 581 individually listed chemicals and 30 chemical categories (including three delimited categories containing 58 chemicals). The list of TRI chemicals is available at http://www.epa.gov/tri/chemical/index.htm.

Toxics Release Inventory (TRI) Facilities The facilities that are required by Section 313 of the Emergency Planning and Community Right to Know Act to report to the TRI. In the 2005 reporting year, approximately 23,500 facilities reported to the TRI.

Trace Elements Trace elements are found in low concentrations in biosolids. The trace elements of interest in biosolids are those commonly referred to as "heavy metals."

Transient Noncommunity Water System A water system which provides water in a place such as a gas station or a campground where people do not remain for long periods of time. These systems do not have to test or treat their water for contaminants which pose long-term health risks because fewer than 25 people drink the water over a long period. They still must test their water for microbes and several chemicals.

Transmissibility (Groundwater) The capacity of a rock to transmit water under pressure. The coefficient of transmissibility is the rate of flow of water, at the prevailing water temperature, in gallons per day, through a vertical strip of the aquifer one foot wide, extending the full saturated height of the aquifer under a hydraulic gradient of 100 %. A hydraulic gradient of 100 % means a 1-ft drop in head in 1 ft of flow distance.

Transpiration (1) The quantity of water absorbed and transpired and used directly in the building of plant tissue, in a specified time; (2) process by which water that is absorbed by plants, usually through the roots, is evaporated into the atmosphere from the plant surface, such as leaf pores (see evapotranspiration). It does not include soil evaporation. The process by which water vapor escapes from the living plant, principally the leaves, and enters the atmosphere. As considered practically, transpiration also includes guttation.

Treatment Any process that changes the physical, chemical, or biological character of a waste to make it less of an environmental threat. Treatment can neutralize the waste, recover energy or material resources from it, render it less hazardous, or make it safer to transport, store, or dispose of.

Treatment Technique (TT) A required process intended to reduce the level of a contaminant in drinking water.

Treatment Works A federally owned, publicly owned, or privately owned device or system used to treat (including recycle or reclaim) either domestic sewage or a combination of domestic sewage and industrial waste of a liquid nature.

Treatment Works Treating Domestic Sewage A publicly owned treatment works (POTW) or other sewage sludge or wastewater treatment system or device, regardless of ownership used in the storage, treatment, recycling, and reclamation of municipal or domestic sewage, including land dedicated for the disposal of sewage sludge.

Trend A statistical term referring to the direction or rate of increase or decrease in magnitude of the individual members of a time series of data when random fluctuations of individual members are disregarded.

Tributary A smaller river or stream that flows into a larger river or stream. Usually, a number of smaller tributaries merge to form a river.

Trophic Level All organisms that secure their food at a common step in the food chain.

Trophic State The state of nutrition (e.g., amount of nutrients) in a body of water.

Troposphere The layer of the atmosphere closest to the earth's surface. The troposphere extends from the surface up to about 6–9 miles.

Turbidity (1) The cloudy appearance of water caused by the presence of tiny particles. High levels of turbidity may interfere with proper water treatment and monitoring. (2) The amount of solid particles that are suspended in water and that cause light rays shining through the water to scatter. Thus, turbidity makes the water cloudy or even opaque in extreme cases. Turbidity is measured in NTU. (3) A measure of the degree of clarity of a solution. For cloudy water, turbidity would be high; for clear water, turbidity would be low.

Two-Stage DAF Two dissolved air flotation (DAF) clarifiers are connected in series for consecutive water or wastewater treatment using different chemicals and/or different operational conditions.

Unconfined Aquifer An aquifer whose upper boundary is the water table or phreatic surface.

Underflow The downstream flow of water through the permeable deposits that underlie a stream and that are more or less limited by rocks of low permeability.

Underground Injection or Well Injection The technology of placing fluids underground in porous formations of rocks, through wells or other conveyance systems. The fluids may be water, wastewater, or water mixed with chemicals. Regulations for disposing of waste this way vary depending on the type of waste. RCRA hazardous waste is placed in highly regulated (class 1) wells.

Unit Hydrograph The hydrograph of direct runoff from a storm uniformly distributed over the drainage basin during a specified unit of time; the hydrograph is reduced in the vertical scale to correspond to a volume of runoff of 1 in. from the drainage basin.

Unit Nitrogen Fertilizer Rate (UNFR) UNFR is a rate in lb-N per unit crop yield, where the unit can be either bushel or ton. (Note: 1 bu (US bushel) = 1.2444 ft^3; 1 British bushel = 1.2843 ft^3; 1 t (British ton) = 2000 lb; 1 T (metric ton) = 1000 kg)

Unit Plot A standard plot condition to determine soil erodibility. Conditions for the plot are LS factor (length-slope and steepness factor) = 1 (slope = 9 % and Length = 72.6 ft = 22.13 m), plot is fallow, tillage is up and down slope, and no conservation practices are applied (CP = 1).

Unsaturated Zone The zone immediately below the land surface where the pores contain both water and air, but are not totally saturated with water. These zones differ from an aquifer, where the pores are saturated with water.

Upland Any area that does not qualify as wetland because the associated hydrologic regime is not sufficiently wet to elicit development of vegetation, soils, and/or hydrologic characteristics associated with wetlands, or is defined as open waters.

Uptake The rate of reversible or irreversible buildup of a compound or element in an organism through inhalation, ingestion, absorption, or a combination of the above, with subsequent assimilation, utilization, clearance, or a combination of the above.

Urbanization The concentration of development in relatively small areas (cities and suburbs). The US Census Bureau defines "urban" as referring to areas with more than 1.5 people per acre.

Usable Storage The volume normally available for release from a reservoir below the stage of the maximum controllable level.

Validation/Verification Check of the behavior of a model against a set of prototype conditions that was not used for calibration.

Variance A state or the USEPA permission not to meet a certain drinking water standard. The water system must prove that: (1) it cannot meet an maximum contaminant level (MCL), even while using the best available treatment method, because of the characteristics of the raw water, and (2) the variance will not create an unreasonable risk to public health. The state or the USEPA must review, and allow public comment on, a variance every three years. The states can also grant variances to water systems that serve small populations and which prove that they are unable to afford the required treatment, an alternative water source, or otherwise comply with the standard.

Vector Attraction Characteristics (e.g., odor) that attract birds, insects, and other animals that are capable of transmitting infectious agents.

Vectors Vectors include rodents, birds, and insects that can transport pathogens away from the land application site.

Violation A failure to meet any state or federal drinking-water regulation.

Volatile Solids (VS) VS provide an estimate of the readily decomposable organic matter in biosolids and are usually expressed as a percentage of total solids. VS are an important determinant of potential odor problems at land application sites.

Volatilization (1) Ammonium-N in biosolids/manure can be significant, making up even half the initial plant available nitrogen (PAN) of biosolids/manure. The ammonium-N of biosolids/manure can vary widely depending on treatment and storage. Since ammonium-N is prone to volatilization (as ammonia gas, NH_3), the application method affects PAN. For instance, surface-applied biosolids are expected to lose half of their ammonium-N. Conversely, direct subsurface injection or soil incorporation of biosolids within 24 h minimizes volatilization losses. The conversion of ammonium-N to ammonia gas form (NH_3) is called volatilization.

Vulnerability Assessment An evaluation of drinking water source quality and its vulnerability to contamination by pathogens and toxic chemicals.

Wadeable Stream A stream, creek, or small river that is shallow enough to be sampled using methods that involve wading into the water. Wadeable streams typically include waters classified as first through fourth order in the Strahler stream order classification system.

Wastewater (1) The water that has been used in homes, industries, and businesses that is not for reuse unless it is treated. (2) The water derived from a municipal or industrial waste treatment plant.

Wastewater Renovation Treatment of wastewater for its reuse.

Wastewater-Treatment Return Flow The water returned to the environment by wastewater-treatment facilities.

Water Balance See hydrologic budget.

Water Conservation The promotion of the efficient use of water through the economically or socially beneficial lessening of water withdrawals, water use, or wastewater reduction. Conservation can forestall future water supply capacity needs and can be implemented on water supply as well as on water demand. It can consist of both temporary and permanent measures for improvement of both water quantity and water quality.

Water Content of Snow See water equivalent of snow.

Water Crop See water yield.

Water Cycle The circuit of water movement from the oceans to the atmosphere and to the earth and return to the atmosphere through various stages or processes such as precipitation, interception, runoff, infiltration, percolation, storage, evaporation, and transportation.

Water Equivalent of Snow Amount of water that would be obtained if the snow should be completely melted. Water content may be merely the amount of liquid water in the snow at the time of observation.

Water Loss The difference between the average precipitation over a drainage basin and the water yield from the basin for a given period. The basic concept is that water loss is equal to evapotranspiration, that is, water that returns to the atmosphere and thus is no longer available for use. However, the term is also applied to differences between measured inflow and outflow even where part of the difference may be seepage.

Water Pollution The addition into water of harmful or objectionable materials and substances in large-enough quantities to adversely affect the water's usefulness.

Water Pollution Control The removal of harmful or objectionable materials and substances from water in large-enough quantities to restore or maintain the water's usefulness.

Water Quality A term used to describe the chemical, physical, and biological characteristics of water, usually in respect to its suitability for a particular purpose.

Water Quality Impairment The description of a diminished strength or value based upon designated water use. Pollutants and sources are considered.

Water Quality Standards (WQS) (1) The combination of a designated use and the maximum concentration of a pollutant which will protect that use for any given body of water. For example, in a trout stream, the concentration of iron should not exceed 1 mg/L. (2) Standards that set the goals, pollution limits, and protection requirements for each water body. These standards are composed of designated (beneficial) uses, numeric and narrative criteria, and anti-degradation policies and procedures.

Water Recreation Diving, fishing, swimming, surfing, etc.

Water Requirement The quantity of water, regardless of its source, required by a crop in a given period of time, for its normal growth under field conditions. It includes surface evaporation and other economically unavoidable wastes.

Water Retarding Structure This is designed to hold back water to prevent downstream flooding.

Water Supply The provision of water by public utilities, commercial organizations, community endeavors, or by individuals, usually via a system of pumps and pipes.

Water Table (1) The top of the water surface in the saturated part of an aquifer; (2) the upper surface of a zone of saturation; (3) the boundary between the saturated and unsaturated zones. Generally, the level to which water will rise in a well (except artesian wells)

Water Treatment Plant A facility designed to receive and treat the raw surface water, raw groundwater, or rainwater for production of drinking water meeting the government's drinking-water standards, or for production of industrial water meeting the specific industrial water quality standards.

Water Use Water that is used for a specific purpose, such as for domestic use, irrigation, or industrial processing. Water use pertains to human's interaction with and influence on the hydrologic cycle, and includes elements, such as water withdrawal from surface- and groundwater sources, water delivery to homes and businesses, consumptive use of water, water released from wastewater-treatment plants, water returned to the environment, and instream uses, such as using water to produce hydroelectric power.

Water Well An underground structure constructed by digging, driving, boring, or drilling to access groundwater. The water is typically lifted to the surface through a pump.

Water Year (1) In Geological Survey reports dealing with surface-water supply, the 12-month period, October 1st through September 30th. The water year is designated by the calendar year in which it ends and which includes 9 of the 12 months. Thus, the year ended September 30, 1959, is called the "1959 water year." (2) The time period from October 1st through September 30th.

Water Yield (Water Crop or Runout) The runoff from the drainage basin, including groundwater outflow that appears in the stream plus groundwater outflow that bypasses the gaging station and leaves the basin underground. Water yield is the precipitation minus the evapotranspiration.

Watershed (1) A watershed is the area of land where all of the water that is under it or drains off of it goes into the same place at a lower elevation; (2) the land area from which water drains into a stream, river, or reservoir; (3) the land area that drains to a common waterway, such as a stream, lake, estuary, wetland, or ultimately the ocean; (4) the land area that drains water to a particular stream, river, or lake. It is a land feature that can be identified by tracing a line along the highest elevations between two areas on a map, often a ridge. Large watersheds, like the Mississippi River basin contain thousands of smaller watersheds; (5) the divide separating one drainage basin from another and in the past has been generally used to convey this meaning. However, over the years, use of the term to signify drainage basin or catchment area has come to predominate, although drainage basin is preferred. Drainage divide, or just divide, is used to denote the boundary between one drainage area and another.

Watershed Approach A flexible framework for managing water resource quality and quantity within specified drainage area, or watershed. This approach includes stakeholder involvement and management actions supported by sound science and appropriate technology.

Watershed Plan A document that provides assessment and management information for a geographically defined watershed, including the analyses, actions, participants, and resources related to the development and implementation of the plan.

Watershed Protection Approach (WPA) The USEPA's comprehensive approach to managing water resource areas, such as river basins, watersheds, and aquifers. WPA contains four major features: targeting priority problems, stakeholder involvement, integrated solutions, and measuring success.

Watertable The surface of an unconfined aquifer at which water pressure equals the atmospheric pressure.

Watt-hour (Wh) An electrical energy unit of measure equal to one watt of power supplied to, or taken from, an electrical circuit steadily for 1 h.

Wave Number Spatial frequency of a wave.

Well (Water) An artificial excavation put down by any method for the purposes of withdrawing water from the underground aquifers. A bored, drilled, or driven shaft, or a dug hole whose depth is greater than the largest surface dimension and whose purpose is to reach underground water supplies or oil, or to store or bury fluids below the ground.

Wellhead Protection Area The area surrounding a drinking water well or well field which is protected to prevent contamination of the well(s).

Wet Earth Flow It refers to a process where the soil of a bank flows as a highly viscous liquid. The flowing material is extremely weak and easily removed by hydraulic fluvial process, even at lower flows.

Wetland (1) An area that is inundated or saturated by surface or groundwater at a frequency and duration sufficient to support, and that under normal circumstances does support, a prevalence of vegetation typically adapted for life in saturated soil conditions. Wetlands generally include swamps, marshes, bogs, and similar areas; (2) an area where water covers the soil or is present either at or near the surface of the soil all year (or at least for periods of time during the year); (3) those areas that are inundated or saturated by surface or groundwater at a frequency and duration sufficient to support, and that under normal circumstances do support, a prevalence of vegetation typically adapted for life in saturated soil conditions. Wetland generally includes swamps, marshes, bogs, and similar areas.

Width/Depth Ratio The width to depth ratio describes a dimension of bank-full channel width to bank-full mean depth. Bank-full discharge is defined as the momentary maximum peak flow which occurs several days a year and is related to the concept of channel-forming flow.

Width/Meander Length Ratio The ratio of the average width of a stream or river over a reach divided by the average length over successive cycles of left and right bends of the stream or river.

Wildlife Refuge An area designated for the protection of wild animals, within which hunting and fishing are either prohibited or strictly controlled.

Withdrawal Water removed from a ground- or surface-water source for use.

Withdrawal Use of Water The water removed from the ground or diverted from a stream or lake for use.

Xeriscaping A method of landscaping that uses plants that are well adapted to the local area and are drought resistant. Xeriscaping is becoming more popular as a way of saving water at home.

Yield (1) It is the crop harvested in the unit of bu/acre or ton/acre; (2) mass per unit time per unit area.

Zone of Aeration The zone above the water table. Water in the zone of aeration does not flow into a well.

Zone of Saturation The zone in which the functional permeable rocks are saturated with water under hydrostatic pressure. The water in the zone of saturation will flow into a well, and is called groundwater.

Zooplankton Small aquatic animals, floating, usually with limited swimming capability.

References

1. The Institute of Ecology. (1974). *An ecological glossary for engineers and resource managers* (p. 50). Winchester: TIE Publication.
2. USEPA.(2004). *Modeling environmental fate and ecological effects in aquatic ecosystems.* Washington, DC: US Environmental Protection Agency.
3. Langbein, W. B., & Iseri, K. T. (1995). *Science in your watershed.* Washington, DC: US Government Printing Office.
4. Allaby, M. (1977). *A dictionary of the environment.* New York: Van Nostrand Reinhold.
5. Holum, J. R. (1977). *Topics and terms in environmental problems.* New York: Wiley.
6. Pankratz, T. (1991). *Dictionary of water and wastewater treatment trademarks and brand names.* Chelsea: Lewis Publishers.
7. Symons, J. M., Bradley, L. C., & Cleveland, T. C. (2000). *The drinking water dictionary.* Denver: American Water Works Association.
8. Wang, M. H. S., & Wang, L. K. (2014). Glossary and conversion factors for water resources engineers. In L. K. Wang & C. T. Yang (Eds.), *Modern water resources engineering* (pp. 759–851). New York: Humana Press-Springer.
9. APHA, AWWA, ASCE, & WPCF. (1969). *Glossary—water and wastewater control engineering.* Denver: American Water Works Association.
10. Wang, L. K. (1974). *Environmental Engineering Glossary* (P. 420). Buffalo NY: Calspan Corporation.

MIX
Papier aus verantwortungsvollen Quellen
Paper from responsible sources
FSC® C105338

If you have any concerns about our products,
you can contact us on
ProductSafety@springernature.com

In case Publisher is established outside the EU,
the EU authorized representative is:
Springer Nature Customer Service Center GmbH
Europaplatz 3, 69115 Heidelberg, Germany

Printed by Libri Plureos GmbH
in Hamburg, Germany